27

Springer Series in Chemical Physics

Edited by Robert Gomer

Springer Series in Chemical Physics

Editors: V. I. Goldanskii R. Gomer F. P. Schäfer J. P. Toennies

1 **Atomic Spectra and Radiative Transitions** By I. I. Sobelmann
2 **Surface Crystallography by LEED** Theory, Computation and Structural Results. By M. A. Van Hove, S. Y. Tong
3 **Advances in Laser Chemistry** Editor: A. H. Zewail
4 **Picosecond Phenomena** Editors: C. V. Shank, E. P. Ippen, S. L. Shapiro
5 **Laser Spectroscopy** Basic Concepts and Instrumentation By W. Demtröder 2nd Printing
6 **Laser-Induced Processes in Molecules** Physics and Chemistry Editors: K. L. Kompa, S. D. Smith
7 **Excitation of Atoms and Broadening of Spectral Lines** By I. I. Sobelman, L. A. Vainshtein, E. A. Yukov
8 **Spin Exchange** Principles and Applications in Chemistry and Biology By Yu. N. Molin, K. M. Salikhov, K. I. Zamaraev
9 **Secondary Ions Mass Spectrometry SIMS II** Editors: A. Benninghoven, C. A. Evans, Jr., R. A. Powell, R. Shimizu, H. A. Storms
10 **Lasers and Chemical Change** By A. Ben-Shaul, Y. Haas, K. L. Kompa, R. D. Levine
11 **Liquid Crystals of One- and Two-Dimensional Order** Editors: W. Helfrich, G. Heppke
12 **Gasdynamic Laser** By S. A. Losev
13 **Atomic Many-Body Theory** By I. Lindgren, J. Morrison
14 **Picosecond Phenomena II** Editors: R. M. Hochstrasser, W. Kaiser, C. V. Shank
15 **Vibrational Spectroscopy of Adsorbates** Editor: R. F. Willis
16 **Spectroscopy of Molecular Excitions** By V. L. Broude, E. I. Rashba, E. F. Sheka
17 **Inelastic Particle-Surface Collisions** Editors: E. Taglauer, W. Heiland
18 **Modelling of Chemical Reaction Systems** Editors: K. H. Ebert, P. Deuflhard, W. Jäger
19 **Secondary Ion Mass Spectrometry SIMS III** Editors: A. Benninghoven, J. Giber, J. László, M. Riedel, H. W. Werner
20 **Chemistry and Physics of Solid Surfaces IV** Editors: R. Vanselow, R. Howe
21 **Dynamics of Gas-Surface Interaction** Editors: G. Benedek, U. Valbusa
22 **Nonlinear Laser Chemistry** Multiple-Photon Excitation By V. S. Letokhov
23 **Picosecond Phenomena III** Editors: K. B. Eisenthal, R. M. Hochstrasser, W. Kaiser, A. Laubereau
24 **Desorption Induced by Electronic Transitions DIET I** Editors: N. H. Tolk, M. M. Traum, J. C. Tully, T. E. Madey
25 **Ion Formation from Organic Solids** Editor: A. Benninghoven
26 **Semiclassical Theories of Molecular Scattering** By B. C. Eu, M. L. Sink
27 **EXAFS and Near Edge Structure** Editors: A. Bianconi, L. Incoccia, S. Stipcich
28 **Atom in the Strong Light Field** By N. B. Delone, V. P. Krainov
29 **Gas Flow in Nozzles** By U. Pirumov, G. Roslyakov
30 **Theory of Slow Atomic Reactions** By E. E. Nikitin, S. Ya. Umanskii
31 **Reference Data on Atomic and Molecular Physics** By A. A. Radzig, B. M. Smirnov
32 **Adsorption Processes on Semiconductor and Dielectric Surfaces I** By V. F. Kiselev, O. V. Krylov

EXAFS and Near Edge Structure

Proceedings of the International Conference
Frascati, Italy, September 13–17, 1982

Editors:
A. Bianconi L. Incoccia S. Stipcich

With 316 Figures

Springer-Verlag
Berlin Heidelberg New York Tokyo 1983

Professor Antonio Bianconi

Dipartimento di Fisica, Università di Roma, La Sapienza, P. Aldo Moro 2,
I-00185 Roma, Italy

Dr. Lucia Incoccia

PULS-CNR, Laboratori Nazionali di Frascati, P.O. Box 13,
I-00044 Frascati, Italy

Dr. Stanislao Stipcich

INFN, Laboratori Nazionali di Frascati, P.O. Box 13,
I-00044 Frascati, Italy

Series Editors

Professor Vitalii I. Goldanskii

Institute of Chemical Physics
Academy of Sciences
Vorobyevskoye Chaussee 2-b
Moscow V-334, USSR

Professor Robert Gomer

The James Franck Institute
The University of Chicago
5640 Ellis Avenue
Chicago, IL 60637, USA

Professor Dr. Fritz Peter Schäfer

Max-Planck-Institut für
Biophysikalische Chemie
D-3400 Göttingen-Nikolausberg
Fed. Rep. of Germany

Professor Dr. J. Peter Toennies

Max-Planck-Institut für Strömungsforschung
Böttingerstraße 6-8
D-3400 Göttingen
Fed. Rep. of Germany

ISBN 978-3-642-50100-5 ISBN 978-3-642-50098-5 (eBook)

DOI 10.1007/978-3-642-50098-5

Library of Congress Cataloging in Publication Data. International Conference on EXAFS and Near Edge Structure (1st : 1982 : Frascati, Italy). EXAFS and near edge structure. (Springer series in chemical physics ; 27) Includes index. 1. Extended X-ray absorption fine structure–Congresses. 2. X-ray absorption near edge structure–Congresses. I. Bianconi, A. (Antonio), 1944- . II. Incoccia. L. (Lucia), 1950- . III. Stipcich, Stanislao. IV. Title. V. Title: E.X.A.F.S. and near edge structure. VI. Series. QC482.S6I56 1982 537.5'352 83-4761

Softcover reprint of the hardcover 1st edition 1983

Preface

The field of X-ray spectroscopy using synchrotron radiation is growing so rapidly and expanding into such different research areas that it is now difficult to keep up with the literature. EXAFS and XANES are becoming interdisciplinary methods used in solid-state physics, biology, and chemistry, and are making impressive contributions to these branches of science. The present book gives a panorama of the research activity in this field. It contains the papers presented at the International Conference on EXAFS and Near Edge Structure held in Frascati, Italy, September 13-17, 1982. This was the first international conference devoted to EXAFS spectroscopy (Extended X-ray Absorption Fine Structure) and its applications. The other topic of the conference was the new XANES (X-ray Absorption Near Edge Structure), which in terms of experimental and theoretical developments finally appears to have left its infancy.

The applications of EXAFS concern the determination of local structures in complex systems; we have therefore divided the subject matter into different parts on various types of materials: amorphous metals, glasses, solutions, biological systems, catalysts, and special crystals such as mixed valence systems and ionic conductors. EXAFS provides unique information for each kind of system, but the analysis of EXAFS data also poses special problems in each case. General problems of EXAFS data analysis are discussed, as well as developments in instrumentation for X-ray absorption using synchrotron radiation and laboratory EXAFS. The book also covers related techniques for local structure studies such as X-ray anomalous scattering, a new method using tunable synchrotron radiation and electron loss spectroscopies.

The popularity which these new methods have attained was demonstrated by the large number of scientists who attended the conference from all over the world and the wide variety of systems studied. Plans for a larger conference in Stanford in the summer of 1984 are an indication that the interest continues to grow.

The conference was planned by the Program Committee: A. Balzarotti (Italy), A. Bianconi (Italy), P. Eisenberger (USA), C.D. Garner (UK), S.S. Hasnain (UK), S. Mobilio (Italy), J.B. Pendry (UK), P. Rabe (W. Germany), D. Raoux (France), and E. Stern (USA), and the Advisory Committee: F. Bassani (Italy), B.W. Batterman (USA), A. Bienenstock (USA), J. Bordas (W. Germany), R. Hänsel (W. Germany), S. Heald (USA), S. Hosoya (Japan), P. Lagarde (France), and A. Meisel (DDR), and arranged by the members of the Organizing Committee: A. Bianconi, E. Burattini, L. Incoccia, S. Mobilio, C. Natoli, A. Reale, A. Savoia and S. Stipcich.

Our particular thanks go to Mrs. L. Invidia and Mrs. M. Macri for bearing the heavy burden of running the Conference Secretariat.

We gratefully acknowledge the sponsorship of the University of Rome "La Sapienza," the Consiglio Nazionale delle Ricerche, the Gruppo Nazionale di Struttura della Materia, and the Laboratori Nazionali di Frascati dell'INFN. Their financial support and the availability of their facilities were essential to the realization of the conference and the publication of this volume.

Frascati, January 1983 *A. Bianconi, L. Incoccia, and S. Stipcich*

Contents

Part III XANES

Part IV Special Crystalline Systems

Part V Liquids and Disordered Systems

Part VI Catalysts

Part VII Biological Systems

Part VIII Related Techniques – Anomalous Scattering

Part IX Related Techniques – Electron Energy Loss

Part X Instrumentation

Introduction
Historical Perspective of EXAFS and Near Edge Structure Spectroscopy

A. Bianconi

Dipartimento di Fisica, Università di Roma, La Sapienza, P. Aldo Moro 2, I-00185 Roma, Italy

X-ray spectroscopy provided, in its gold era, the basic information on the electronic structure of atoms and therefore the experimental basis of the quantum theory of atoms. After 1930's the progress of this spectroscopy has been limited by the lack of intense continuum radiation sources.

In the years around 1974 few people, working with synchrotron radiation in high energy elementary particles laboratories around the world, realized that the high energy ($E > 1$ GeV) electron storage rings were ideal sources for X-ray absorption spectroscopy.

Today new synchrotron radiation facilities with dedicated multiGeV storage rings built only as X-ray sources are coming into operation: SRS at Daresbury (UK), NSLS at Brookhaven (USA) ALLADIN in Wisconsin (USA), BESSY at Berlin (West Germany), and the Photon Factory in Japan is under construction. These new facilities enlarge the horizon of the synchrotron radiation research carried out up to now with the first storage rings, today mostly or completely dedicated, ADONE, Frascati (Italy), SPEAR, Stanford (USA), DORIS, Hamburg (West Germany), DCI and ACO, Paris (France), and VEPP-3, Novosibirsk (USSR).

Since 1920, any observed structure near the inner shell absorption threshold has been known as Kossel structure (1), assigned to bound excited electronic states. The observed structures beyond about 20 eV were commonly referred to as Kronig structure (2,3,4) in honor of the first investigator to propose their interpretation.

For many years, because of poor experimental spectra and the lack of comparison between theory and experiment, a controversy between "long-range order theories" formulated in terms of Block wave for the excited photoelectron final state and "short-range order theories" in terms of photoelectron scattering by neighboring atoms for the interpretation of the Kronig structure was going on (5,6,7). The revival of the interest on the Kronig structure began with the work of Sayers, Stern and Lytle where they pointed out both that a single scattering short-range order theory was adequate to explain the structures beyond 50 eV from the threshold (renamed EXAFS) and the potential of EXAFS modulations for determination of local geometry (8).

After the first experiment with an EXAFS apparatus on a X-ray beam line at a synchrotron radiation facility (9) and the rapid theoretical progress (10-12) we have seen an impressive development of EXAFS due to the fact that it is a relatively simple and direct method to obtain information on local structures around a selected atom in complex or disordered systems (13-17).

The field is so rapidly expanding that this year we could organize at Frascati the first International EXAFS Conference on the more recent advances in this field, where nearly all results came from synchrotron radiation laboratories.

On the other hand, the progress on the understanding the Kossel structure was slower than for EXAFS, because of the complex physical phenomena near the photoionization threshold. For many years (6,18) only the energy region in a 10 eV energy range above threshold, defined as the absorption feature at the lowest energy, has been possible to explain on the basis of the local electronic structure. Final state effects such the many-body effects in metals and core excitons in soft X-ray absorption spectra have been extensively studied (18).

Recently a third energy region extending from 10 eV to 40 eV beyond the threshold has been identified as due to multiple scattering resonances of the excited photoelectron, within the neighboring atoms, its application for local structure determination of surfaces and biological systems has been pointed out and it has been named XANES (19,20). The rapid theoretical progress (21,22) and the experimental evidence that the information extracted from XANES is complementary to that given by EXAFS has attracted recently a large interest in XANES. The papers presented at the Frascati Conference and reported in this book show that now this spectroscopy is leaving its infancy and is becoming a quantitative method for local structure studies giving unique information on geometrical details and also on bond lengths and the atomic effective charge.

References

1. W. Kossel, Z. Phys. 1, 119 (1920).
2. R. de L. Kronig, Z. Phys. 70, 317 (1931).
3. R. de L. Kronig, Z. Phys. 75, 468 (1932).
4. D.R. Hartree, R. de L. Kronig, and H. Petersen, Physica 1, 895 (1934).
5. A.H. Compton and S.K. Allison, "X Ray in Theory and Experiment", (D. Van Nostrand Comp. Inc., New York, 1935).
6. L.G. Parratt, Rev. Mod. Phys. 31, 616 (1959).
7. L.V. Azaroff, Rev. Mod. Phys. 35, 1012 (1963).
8. D.E. Sayers, E.A. Stern, and F.W. Lyttle, Phys. Rev. Letters 27, 1204 (1971).
9. B.M. Kincaid and P. Eisenberger, Phys. Rev. Letters 34, 1361 (1975).
10. C.A. Ashley and S. Doniach, Phys. Rev. B11, 1279 (1975).
11. P.A. Lee and J.B. Pendry, Phys. Rev. B11, 2795 (1975).
12. B.K. Teo and P.A. Lee, J. Am. Chem. Soc. 101, 2815 (1979).
13. E.A. Stern, Contemp. Phys. 19, 289 (1978).
14. D. Raoux, J.F. Sadoc, P. Lagarde, A. Sadoc, and P.A. Fontaine, Proc. LAM4 J. Physique C8,41, 207 (1980).
15. "Synchrotron Radiation Research", ed. by H. Winick and S. Doniach, (Plenum Press, New York, 1980).
16. P. Rabe and R. Haensel, in Festkorperprobleme (Advances in Solid State Physics) J. Treush editor (Vieweg, Braunschweig 1980), vol. XX, p. 43.
17. "EXAFS for Inorganic Systems", ed. by C.D. Garner and S.S. Hasnain, Daresbury publication DL/SCI/R17 (1981).
18. F.C. Brown, in ref. 15, p. 61 (1980).
19. A. Bianconi, Applications of Surface Science, 6, 392 (1980).
20. A. Bianconi, in ref. 17, p. 13 (1981).
21. C.R. Natoli, F.W. Kutzler, D.K. Misemer, and S. Doniach, Phys. Rev. A22, 1104 (1980).
22. P.J. Durham, J.B. Pendry, and C.H. Hodges, Solid State Commun. 38, 159 (1981).

Part I

Theoretical Aspects of EXAFS and XANES

The Transition Region Between XANES and EXAFS

J.B. Pendry

Imperial College of Science and Technology, Exhibition Road,
London SW7 2BX, England

X-ray absorption experiments have in recent years assumed an important role as a technique of structure determination [1,2]. The active ingredient for structural sensitivity is an electron emitted from an inner shell: the scattering of this electron modulates the absorption cross section. Figure 1 shows a typical X-ray absorption cross section and makes the somewhat arbitrary distinction between X-ray absorption near edge structure (XANES) and the extended structure (EXAFS). The distinction between these two regimes is in strength of electron scattering: at low electron energies cross sections are high and strong modulations in the cross section result. At higher energies figure 1 clearly shows much weaker modulations indicating weaker scattering. In fact, the experiments are misleading guides to the overall strength of scattering because the intensity of modulation is primarily sensitive to the backscattering factor. The electron is also forward scattered, resulting in changes to the phase of modulations in the EXAFS spectrum, a particularly insidious effect because it interferes with determinations of atomic coordinates. For classical EXAFS theory to be valid [3,4], the ejected electron must be only weakly scattered. Hence, a theory of XANES [5,6,7,8,9,10] must address the problem of multiple scattering. In the transition region only forward scatter-

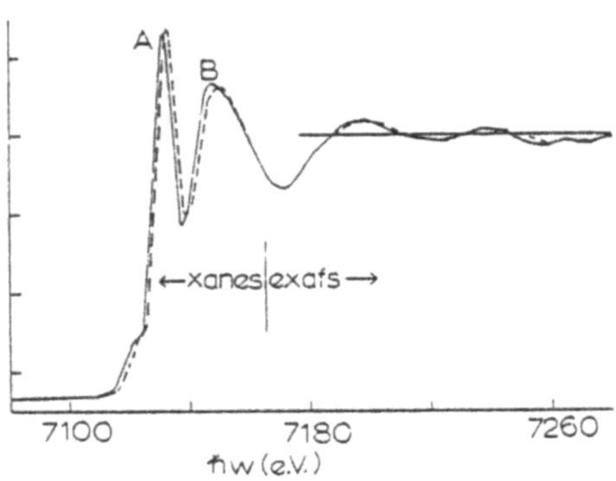

Fig. 1. The relative absorption coefficient for X-rays in the vicinity of the iron K edge shown for ——— $K_4Fe^{II}(CN)_6 \cdot 3H_2O$ and — — — $K_3Fe^{III}(CN)_6$.
The division of the spectrum into XANES and EXAFS is somewhat arbitrary

ing is strong, and we can draw on the experiences of LEED [12] to construct a perturbation theory in orders of back scattering - called renormalised forward scattering or RFS perturbation theory. For structures which are relatively open such that there is no 'shadowing' of atoms, the transition region terminates with the XANES region itself. However, forward scattering is always strong but is confined to an increasingly narrow cone of angles at higher energies. Thus scattering from two atoms in line of sight with the excited atoms will always suffer from multiple scattering. More commonly, effects are small beyond 100-200 e.V. above threshold.

The importance of understanding the more complex effects in the transition region is twofold. Firstly, we may be faced with a situation where the only data available lie in a transition region so to extract an accurate structure multiple scattering corrections must be applied. Secondly, we may wish to make use of the fact that there is an intrinsically different sort of information present in spectra taken in the multiple scattering regime which is particularly important for liquids and glasses.

To put the matter in simple terms : if we have a liquid compound of atoms A and B for which the pair function $g_2(R)$ has been found from a weak scattering diffraction experiment, this tells us the radius of the first coordinating shell, but not the arrangement of atoms in that shell, nor does it tell anything about ABA and BAB bond angles which involve at least the three-atom correlation function.

The result that only the pair distribution function can be found for liquids and glasses is a completely general one for weak scattering probes and follows for any situation in which only one scattering event is needed to describe the process. Thus X-ray diffraction, neutron scattering, EXAFS, high energy electron diffraction are all incapable of providing more than the pair distribution function and many potential subtleties of the disordered state are inaccessible to us. Suppose that our liquid was composed of well-defined molecular units, A_3B_3, which retained their structural integrity : it would be impossible to deduce the existence of these units from the diffraction data alone. Yet other properties of the liquid would be vitally affected by the molecular nature of the units, albeit in an indirect and complex fashion.

Thus the very property of weak scattering which makes an experiment so easily interpreted immediately removes the possibility of knowing the complete truth from that experiment. Radiation which scatters n times in a solid will provide information about up to the 2 nth order correlations. Unfortunately it is not possible to switch on scattering order by order, and a strong scattering probe involves several orders of scattering : infinite order in principle. It is the great virtue of X-ray absorption experiments that we can measure the extended structure for information about pair distributions, but when further information is desired, turn to the near edge or transition region to see the effects of more subtle correlations in the liquid [5]. Although interpretation of these experiments is not simple it is now possible, even for quite complex materials, and computer codes are readily available in well-documented form [6].

One mechanism whereby multiple scattering gives sensitivity to bond angles occurs when the emitting atom and two further atoms, let us say carbon and oxygen, are arranged linearly, so that oxygen assumed to be furthest from the emitter is shadowed by carbon. Except that the shadow analogy is the opposite of the truth. Atoms are attractive to electrons and tend to act as lenses, focussing the wave function of the far side of the atom and the oxygen scattering is intensified by the presence of carbon. This effect was noted by Petersson et al. [13] in XPS experiments. They measured the angle dependence of the electrons emitted from the K shell of a carbon atom, part of a carbon monoxide molecule adsorbed oxygen end up on a metal surface. The XPS signal was shown to be intensified in the direction of the CO bond - see Figure 2. We expect that XANES will be especially sensitive to bonds which lie along the line of sight from the central atom. The focussing effect is an especially powerful multiple scattering mechanism and may extend into the EXAFS region, interfering with the conventional single scattering interpretation of data, as Lee and Pendry were able to show.

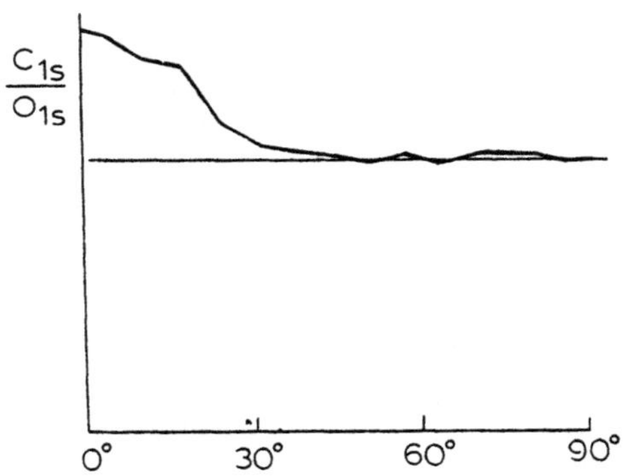

Fig. 2. Experimental ratio of the K shell XPS signals for carbon and oxygen from a carbon monoxide molecule adsorbed carbon-end down on a surface. The signal from carbon is enhanced by focussing through the oxygen atoms near normal emission

The finite path length of the excited electrons suggests an approach based on a finite cluster of atoms, not much larger than the mean free path. Thus any assumptions about long range order are eliminated. Dill and Dehmer [8,9] investigated the problem for small molecules and found that a multiple scattering theory based on a one-electron potential described the many features observed in spectra. The approach was extended to more complex systems such as $GeCl_4$ [10] and MoO_4 [11]. In these papers use was made of multiple scattering methods for molecular systems. However, the clusters we need to simulate a solid are large and some advances in technique are called for.

I shall first describe a formulation of the cluster technique based on LEED theory [6] which is a general solution to the multiple scattering problem, and then show how in the transition region where forward scattering dominates RFS perturbation theory can be used.

In electron diffraction from a surface, the surface is decomposed into a series of layers, each only one or two atoms thick. Scattering by the separate layers is found and the diffraction problem solved by putting together the layers to make a complete surface. By analogy the cluster is decomposed into shells of atoms and as a first step scattering by a single shell is calculated. We need to know how a wave radiating from the centre of the shell is transmitted or reflected by the shell. T^{00} and T^{0I} describe the amplitudes of transmission and reflection. These quantities are matrices because the wave can have s p d ... character in its angular behaviour. Another pair of matrices T^{II} and T^{I0} describing transmission and reflection by the shell of a wave incident for the outside complete the description of scattering (see figure 3 for an illustration of this point).

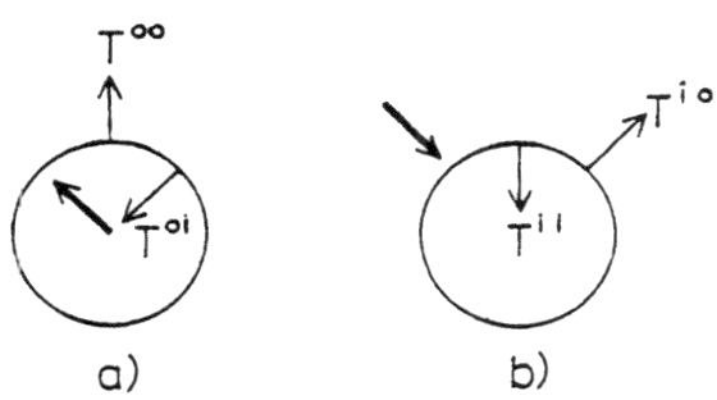

Fig. 3. Definitions of shell transmission and reflection matrices when a wave is incident from a) inside, and b) outside the shell

Using these quantities, we proceed to calculate how all the shells acting together reflect the electron trying to escape from the central atom of the cluster - i.e. how the surroundings interfere with the escape probability. The so-called "wallpaper algorithm" is used to solve the problem. Suppose that there are N shells, counting outwards from the centre. Start with the Nth shell alone. We already know how it reflects electrons trying to escape

$$\tau_N^{0I} = T_N^{0I} \quad . \tag{1}$$

Next introduce the (N-1)th shell. These two shells acting together have a combined reflection matrix

$$\tau_{N-1}^{0I} = T_{N-1}^{0I} + T_{N-1}^{00} (1-\tau_N^{0I} T_{N-1}^{I0})^{-1} \tau_N^{0I} T_{N-1}^{II}, \tag{2}$$

the second term accounting for processes in which the electron is trapped for a time between the Nth and (N-1)th shells. Equation (2) can be reapplied to calculate

$$\tau_{N-2}^{0I} = T_{N-2}^{0I} + T_{N-2}^{00} (1-\tau_{N-1}^{0I} T_{N-2}^{I0})^{-1} \tau_{N-1}^{0I} T_{N-2}^{II} , \tag{3}$$

which describes the reflection by three shells acting together. So we continue until we have the quantity we seek τ_1^{0I}. From this we can calculate corrections from the surroundings to a wavefield which in the absence of all the shells the central atom would be

$$\sum_{\ell m} a^{(o)}_{o\ell m}\, h^{(1)}_{\ell}\,(Kr)\, Y_{\ell m}(\theta\phi) \tag{4}$$

where K is the wave vector of the electron. I shall not discuss the atomic processes which give rise to absorption of X-rays by an isolated atom.

The effect of the cluster is to reflect back these waves, which can further scatter from the central atom; giving a total wavefield

$$\sum_{\ell m} a_{o\ell m}\, h^{(1)}_{\ell}\, Y_{\ell m} + b_{o\ell m}\, h^{(2)}_{\ell}\, Y_{\ell m} \quad , \tag{5}$$

where the vectors

$$a_o = a^{(0)}_o\, (1-\tau_1^{OI}\, t^{IO})^{-1} \tag{6}$$

$$b_o = a_o\, \tau^{OI} \tag{7}$$

and t is the central atom reflection coefficient. Hence the net outward flux of current is

$$\sum_{\ell m} K^{-1}(|a_{o\ell m}|^2 - |b_{o\ell m}|^2) = \sigma, \tag{8}$$

σ being the rate at which the atom absorbs X-rays.

The point of breaking up the cluster into shells is a computational one. If there are N shells with n atoms per shell, because the calculation involves inversion of matrices, each shell will take computer time proportional to n^3 to handle and the total time is $N\,n^3$. If the cluster had been handled as a single unit computer time would have been proportional to $N\,n^3$. Further savings in time can be had if the system has some symmetry about the central atom. Overall it is possible in this way to handle clusters of several hundred atoms. Computer codes are available based on this algorithm [6]. Despite the improvements in computational speed made by introducing a shell structure, the calculations are large ones, particularly in the transition region where more atomic phase shifts are needed to describe the strong peak in forward scattering.

The plan is to introduce a zeroth order wavefield that correctly takes account of all forward scattering events, i.e. ones in which the electron is scattered from one side of a shell to another described by T^{OO} or T^{II}. The corrections will be made in orders of back scattering described by T^{OI} and T^{IO} (assumed to be weak) but at all stages retaining all orders of forward scattering.

The wave emitted from the central atom is

$$\sum_{\ell m} a^{(0)}_{o\ell m}\, h^{(1)}_{\ell}\,(Kr)\, Y_{\ell m}(\theta\phi) \quad , \tag{9}$$

hence the zeroth order back scattering the wave amplitudes between the jth and (j+1)th shells is

$$\sum_{\ell m} a^{(0)}_{j\ell m}\, h^{(1)}_{\ell}\,(Kr)\, Y_{\ell m}(\theta\phi) \tag{10}$$

where

$$a^{(0)}_{j} = a^{(0)}_{j-1}\, T^{00}_{j} \quad . \tag{11}$$

If the cluster is terminated after N shells, we next proceed to calculate the first-order corrections from backscattering between the jth and (j + 1)th shells :

$$\sum_{\ell m} b^{(1)}_{j\ell m}\, h^{(2)}_{\ell}\,(Kr)\, Y_{\ell m}\,(\theta\phi) \tag{12}$$

starting at the outermost shell

$$b^{(1)}_{N} = 0 \tag{13}$$

$$b^{(1)}_{j} = a^{(0)}_{j}\, T^{0I}_{j+1} + b^{(1)}_{j+1}\, T^{II}_{j+1},\ j \neq N \tag{14}$$

and continuing until we find $b^{(1)}_{o}$. Second-order corrections start at the origin and work outwards

$$a^{(2)}_{o} = b^{(1)}_{o}\, t^{IO} \tag{15}$$

$$a^{(2)}_{j} = b^{(1)}_{j} T_{j}^{\,IO} + a^{(2)}_{j} T^{00}_{j}, \quad j \neq 0 \quad . \tag{16}$$

Further refinement of the system following the same pattern can be applied to as many orders as desired until we have accurate values for

$$a_{o} = \sum_{j\ \mathrm{even}} a^{(j)}_{o} \tag{17}$$

$$b_{o} = \sum_{j\ \mathrm{odd}} b^{(j)}_{o} \tag{18}$$

from which the absorption correction can be calculated according to equation (8).

The procedure is convergent provided that back scattering is small, and is extremely efficient. In contrast to the wallpaper algorithm which scales with cluster and shell size as $N\, n^3$, the RFS procedure scales as $N\, n^2$ for each order of back scattering. Usually three orders give accurate results in LEED calculations.

I expect the method to be even more convergent in the cluster case. In fact, the major item when RFS theory is used is the calculation of the shell scattering matrices.

Overall, both XANES and now the transition region have proven tractable to theoretical interpretation. Too much should not be made of the gains to be had from studying these ranges of the spectrum, but for the cases where a more detailed picture of the structure is required than provided by the pair correlation function, they should provide a unique and valuable probe.

References

1. D.E. Sayers, E.A. Stern, F.W. Lytle, Phys. Rev. Lett. 27, 1204 (1971)
2. P. Eisenberger and B.M. Kincaid, Chem. Phys. Lett. 36, 134 (1975)
3. C.A. Ashley and S. Doniach, Phys. Rev. B 11, 1279 (1975)
4. P.A. Lee and J.B. Pendry, Phys. Rev. B 11, 2795 (1975)
5. P.J. Durham, J.B. Pendry and C.H. Hodges, Sol. St. Comm. 38, 159 (1981)
6. P.J. Durham, J.B. Pendry and C.H. Hodges, Comp. Phys. Comm. 25, 193 (1982)
7. J.E. Muller, O. Jepsen, O.K. Andersen, J.W. Wilkins, Phys. Rev. Lett. 40, 720 (1975)
8. D. Dill and J.L. Dehmer, J. Chem. Phys. 61, 692 (1974)
9. J.L. Dehmer and D. Dill, J. Chem. Phys. 65, 5327 (1976)
10. C.R. Natoli, D.K. Misemer, S. Doniach, F.W. Kutzler, Phys. Rev. A22, 1104 (1980)
11. F.W. Kutzler, C.R. Natoli, D.K. Misemer, S. Doniach, K.O. Hodgson, J. Chem. Phys. 73, 3274 (1980)
12. J.B. Pendry, Low Energy Electron Diffraction (Academic : London) (1974)
13. L.G. Peterson, S. Kono, N.F.T. Hall, C.S. Fadley and J.B. Pendry, Phys. Rev. Lett. 42, 1545 (1979)

Bond Angle Determination by EXAFS: A New Dimension

Boon-Keng Teo
Bell Laboratories
Murray Hill, New Jersey 07974, U. S. A.

While single scattering EXAFS formalism can provide structural information about the local environment of the absorbing atoms in terms of radial distribution functions (distances), angular information is generally not available; except, perhaps, for polarization dependent measurements on single crystals. Furthermore, the very same advantageous characteristics of EXAFS (short-range, single-scattering) are also its serious limitations: distance determinations out to only *ca* 4Å. The situation, however, changes dramatically when atoms (including the X-ray absorbing atom and its neighbors) are arranged in a linear or nearly colinear fashion. In such cases, EXAFS contributions from neighboring atoms as far as 8Å can be observed. For these systems, both the amplitude and the phase of the EXAFS of a more distant neighbor are significantly affected by the intervening atom(s). In particular, the amplitude is greatly enhanced and is therefore commonly called "focusing" effect [1]. The short-range single-scattering theory of EXAFS fails in these situations and one must take into account multiple scattering processes involving the intervening atoms.

Recently, a new EXAFS formulation which takes into account the effect of multiple scattering has been developed [2]. Theoretical scattering amplitude and phase functions have also been calculated for various scattering angles [2,3]. Combination of the new multiple scattering formalism and the new theoretical functions facilitate the understanding of the "focusing" effect as well as the assessment of the relative importance of various multiple scattering pathways as the scattering angle varies. It also provides strong evidence that multiple scattering processes involving three atoms (including the absorber and two neighboring atoms) are important in determining the EXAFS of the distant shells, especially at large bond angles and low k values. Based on these, a new method for interatomic angle determinations by EXAFS has been devised which is applicable to systems with bond angles greater than *ca* 100°. The method requires no single crystal measurements and is applicable to wide varieties of materials (polycrystalline or amorphous solids, liquids and solutions, gases, surfaces, polymers, etc.) [4].

In this chapter, a brief description of the multiple scattering theory and bond angle determination will be given for various systems containing three (ABC), four (AB_1B_2C), or multiple ($AB_1 \cdots B_nC$) atoms at different geometries. We note that the problem of multiple scattering in EXAFS has also been considered by LEE and PENDRY [5], ASHLEY and DONIACH [6], BENI, LEE, and PLATZMAN [7], LEE and BENI [8], and more recently by BOLAND, CRANE, and BALDESCHWIELER [9].

Multiple Scattering in EXAFS

The single-electron single-scattering theory of EXAFS makes use of the fact that in most cases, multiple scattering is not important. This assumption is generally valid if one considers that multiple scattering processes can be accounted for by adding all scattering paths that originate and terminate at the central atom (absorber). Each of these processes then behaves like $\sin(2kr_{eff})$ where $2r_{eff}$ is the total scattering path length which is much larger than that of the backscattering(s) from the nearest neighbors. Thus, multiple scatterings will give rise to rapidly oscillatory waves in k space which tend to cancel out. The amplitude of these waves are also significantly attenuated by the large scattering path lengths, making it relatively unimportant in comparison with the direct backscattering.

On the other hand, multiple scattering in EXAFS can become important when atoms are arranged in an approximately colinear array. In such cases, the outgoing photoelectron is strongly *forward-scattered* by the intervening atom, resulting in a significant amplitude enhancement. In fact, both the amplitude and the phase are modified by the intervening atom(s). The effect, however, drops off very rapidly for bond angles smaller than *ca* 150°. For these systems, it is necessary to take into account multiple scatterings involving the intervening atom(s).

Scattering Amplitude and Phase

Let us first define a scattering angle β at atom B (and similarly γ at atom C) which is related to the bond angle A-B-C (α) (*cf.* Fig. 1) by $\beta = 180° - \alpha$. For any arbitrary angle β, the scattering of an electron by an atom B can be described by $f(\beta,k)$. If one approximates the spherical wave of the outgoing photoelectron by a plane wave, one obtains

$$f(\beta,k) = F(\beta,k)e^{i\theta(\beta,k)} \quad (1)$$

where $F(\beta,k)$ and $\theta(\beta,k)$ are the scattering amplitude and phase functions, respectively. For *backscattering* where $\beta = 180° = \pi$ radians, (1) becomes $F(\pi,k) \equiv F(k)$ and $\theta(\pi,k) = \theta(k)$ [10].

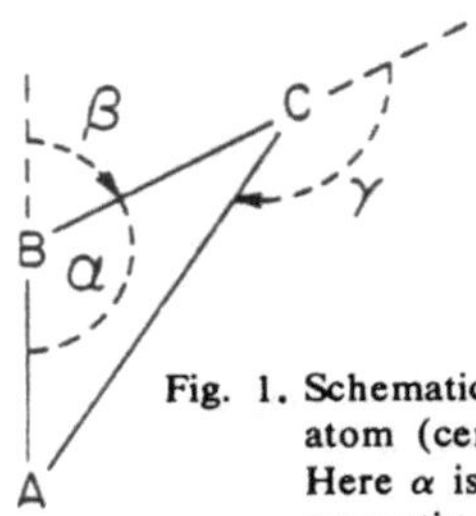

Fig. 1. Schematic representation of a three-atom *ABC* system where *A* is the X-ray absorbing atom (central atom), *B* is the nearest neighbor, and *C* is the next nearest neighbor. Here α is the A-B-C bond angle and β and γ are scattering angles at atoms B and C, respectively

For the formulation of the multiple scattering and the establishment of the bond angle determination method, these scattering amplitude and phase functions must be known accurately. In principle, these functions can be determined from studying the EXAFS of a large number of compounds with varying bond angles (one per each type of intervening atoms). In practice, however, it is extremely difficult to find model compounds with basically identical structural framework differing only in bond angles. Even if two or three such bond angles are known, it is still not enough to establish accurately $F(\beta,k)$ or $\theta(\beta,k)$ as a function β. It is therefore necessary to resort to theoretical calculations for these functions.

Using Lee and Beni's theory [8], we have calculated the $F(\beta,k)$ and $\theta(\beta,k)$ functions for the majority of elements in the Periodic Table [3]. In Fig. 2, we show the dependence of scattering amplitude $F(\beta,k)$ on β for a few representative k values for copper. For all k values, the amplitude attenuates rapidly from its maximum value at β=0° to $F \lesssim 1$ at $\beta \gtrsim 30°$. Hence, forward scattering is strongly "focussed" within a 30° cone. At low k values ($k \lesssim 9$ Å^{-1} for these elements), the amplitude exhibits both maxima and minima. For example, at k = 3.78 Å^{-1}, $F(\beta,k)$ functions reach their minima at $\beta \approx 36°$, 80°, and 141° for copper. The corresponding maxima occur at $\beta \approx 0°$, 52°, 108°, 180°. The maxima correspond to β values where multiple scattering effect is most important and vice versa. The number of these extrema increases with increasing atomic number Z, corresponding to the sampling of the increasing number of electronic shells in the scattering atom. The positions of these extrema (except the maxima at β=0° and 180°) also change somewhat (generally to lower β values) as k increases. At sufficiently large k values ($k \gtrsim 9$ Å^{-1} for these elements), these extrema vanish, resulting in monotonically decreasing amplitude functions. These characteristic features of $F(\beta,k)$ and $\theta(\beta,k)$, as functions of β, k, and Z, form the theoretical basis for bond angle determination by EXAFS.

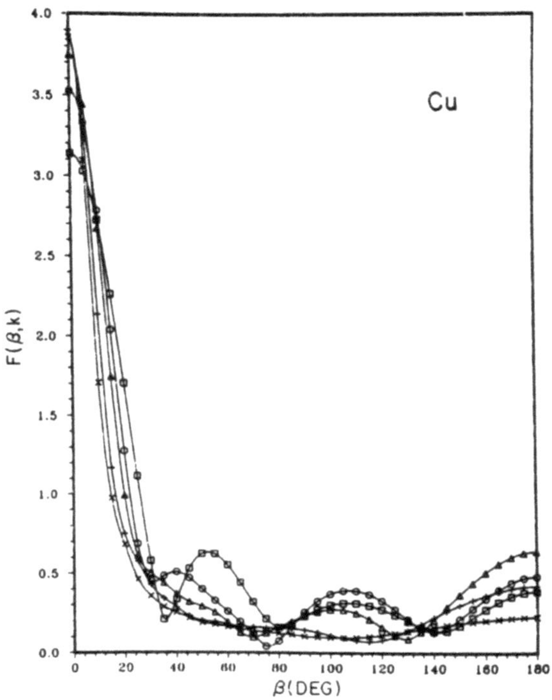

Fig. 2. Scattering amplitude $F(\beta,k)$ in Å *vs* scattering angle β in degree for copper at different k values where $k = 3.7795(\square)$, $5.6692(\bigcirc)$, $7.5589(\Delta)$, $11.3384(+)$, $15.1178(X)$ Å^{-1}

ABC Systems

Consider a three-atom array *A*-*B*-*C* (Fig. 1) where *A* is the central atom (absorber), *B* is the nearest neighbor (the intervening atom), and *C* is the next nearest neighbor. For such a system, the EXAFS of the absorber *A* comprises of two contributions, one from the *backscattering* of *B* and the other from the *backscattering* of *C* via the intermediary atom *B*. We shall designate these two contributions as *AB* and *AC*, respectively. These two contributions can generally be separated by Fourier Filtering. The former can be described quite adequately by the *backscattering* from the atom *B* with the single-electron single-scattering theory. The latter, if it is affected by multiple scattering involving the intervening atom *B*, must be treated with a generalized formulation which takes into account three pathways shown schematically in Fig. 3. The first pathway (I) is the direct backscattering from *A* to *C* and back (*viz.* $A \rightarrow C \rightarrow A$). The second pathway (II) is the multiple scattering via atom *B* around the triangle in either directions (*viz.* $A \rightarrow B \rightarrow C \rightarrow A$ or $A \rightarrow C \rightarrow B \rightarrow A$). This term should therefore be counted twice. The third pathway (III) is the multiple scattering again via the intervening atom *B*, involving the pathway $A \rightarrow B \rightarrow C \rightarrow B \rightarrow A$. The EXAFS for an unoriented sample (using unpolarized X-ray source), corresponding roughly to the *A*-*C* distance, is then the sum of these three terms:

$$\chi^{ABC}(k) = \frac{(-1)^l}{k} \operatorname{Im} e^{i2\delta_l'(k)} \left\{ \frac{f_C(\pi,k)}{r_{AC}^2} e^{-2r_{AC}/\lambda} e^{i2kr_{AC}} e^{-2\sigma_{AC}^2 k^2} \right.$$

$$+ \frac{2\cos(\gamma-\alpha) f_B(\beta,k) f_C(\gamma,k)}{r_{AB}\, r_{BC}\, r_{AC}} e^{-2\left[\frac{r_{AB}+r_{BC}+r_{AC}}{2}\right]/\lambda} e^{i2k\left[\frac{r_{AB}+r_{BC}+r_{AC}}{2}\right]} e^{-(\sigma_{AB}^2 + \sigma_{BC}^2 + \sigma_{AC}^2)k^2}$$

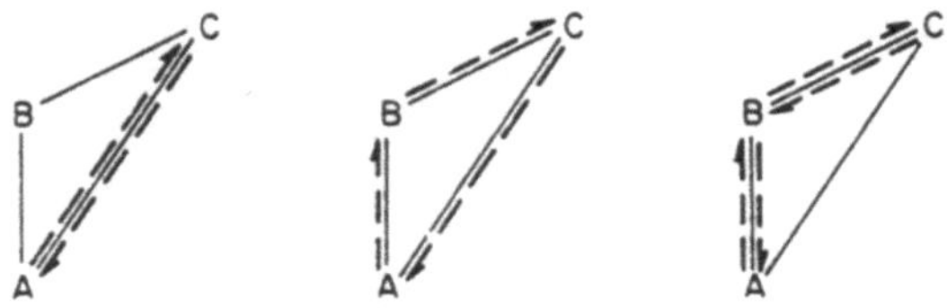

I II III

Fig. 3. Schematic representation of three scattering pathways for a three-atom *ABC* system. Each of these pathways originates and terminates at the absorbing atom *A*. Pathway I is the direct backscattering from *A* to *C* and back. Pathway II is the multiple scattering via B around the triangle in either direction (only one is shown) and pathway III is the multiple scattering via B in both outgoing and incoming trips

$$+\frac{f_B(\beta,k)^2 f_C(\pi,k)}{r_{AB}^2 r_{BC}^2}\, e^{-2(r_{AB}+r_{BC})/\lambda}\, e^{i2k(r_{AB}+r_{BC})}\, e^{-2(\sigma_{AB}^2+\sigma_{BC}^2)k^2}\Bigg\} \qquad (2)$$

Here we assume that the Debye-Waller factors for the three pathways can be approximated by $e^{-2\sigma_{AC}^2 k^2}$, $e^{-(\sigma_{AB}^2+\sigma_{BC}^2+\sigma_{AC}^2)k^2}$, and $e^{-2(\sigma_{AB}^2+\sigma_{BC}^2)k^2}$ for I, II and III, respectively. $\cos(\gamma-\alpha)$ is a geometrical factor. If we now make the plane wave approximations $f_B(\beta,k) = F_B(\beta,k)e^{i\theta_B(\beta,k)}$ and $f_C(\gamma,k) = F_C(\gamma,k)e^{i\theta_C(\gamma,k)}$ where $F_B(\beta,k)$, $F_C(\gamma,k)$ and $\theta_B(\beta,k)$, $\theta_C(\gamma,k)$ are the scattering amplitude and phase functions for scattering angles β,γ at atoms B, C, respectively, (2) can be rewritten as

$$\chi^{ABC}(k) = \frac{(-1)^l}{k}\Bigg\{\frac{F_C(\pi,k)}{r_{AC}^2}\, e^{-2r_{AC}/\lambda}\, e^{-2\sigma_{AC}^2 k^2}\sin[2kr_{AC}+2\delta_l'(k)+\theta_C(\pi,k)]$$

$$+\frac{2\cos(\gamma-\alpha)F_B(\beta,k)F_C(\gamma,k)}{r_{AB}r_{BC}r_{AC}}\, e^{-2\left[\frac{r_{AB}+r_{BC}+r_{AC}}{2}\right]/\lambda}\, e^{-(\sigma_{AB}^2+\sigma_{BC}^2+\sigma_{AC}^2)k^2}$$

$$\times\sin\left[2k\left(\frac{r_{AB}+r_{BC}+r_{AC}}{2}\right)+2\delta_l'(k)+\theta_B(\beta,k)+\theta_C(\gamma,k)\right]$$

$$+\frac{F_B(\beta,k)^2 F_C(\pi,k)}{r_{AB}^2 r_{BC}^2}\, e^{-2(r_{AB}+r_{BC})/\lambda}\, e^{-2(\sigma_{AB}^2+\sigma_{BC}^2)k^2}$$

$$\times\sin\left[2k(r_{AB}+r_{BC})+2\delta_l'(k)+2\theta_B(\beta,k)+\theta_C(\pi,k)\right]\Bigg\} \qquad (3)$$

It is apparent that these three terms correspond to the three pathways I, II, and III shown in Fig. 3 with effective distances r_{AC}, $(r_{AB}+r_{BC}+r_{AC})/2$, and $(r_{AB}+r_{BC})$, respectively. For a given pathway which can be detected and separated via Fourier filtering, it can be analyzed with the appropriate term in (2) or (3). However, for small β, the three pathways are generally not separable since $r_{AC} \lesssim r_{AB}+r_{BC}$. In such cases, we can assume $\gamma \approx \pi$ such that $\cos(\gamma-\alpha) \approx 1$ [11] and

$$f_C(\gamma,k) \approx f_C(\pi,k) \qquad (4)$$

(2) then becomes

$$\chi^{ABC}(k) = \frac{(-1)^l}{k} e^{-2\sigma_{AC}^2 k^2} \mathrm{Im}\, e^{i2\delta'_l(k)} \frac{f_C(\pi,k)}{r_{AC}^2} e^{-2r_{AC}/\lambda} e^{i2kr_{AC}}$$

$$\times \left\{ 1 + \frac{r_{AC}}{r_{AB} r_{BC}} f_B(\beta,k) e^{-\left(r_{AB}+r_{BC}-r_{AC}\right)/\lambda} e^{ik\left(r_{AB}+r_{BC}-r_{AC}\right)} \right\}^2 \tag{5}$$

Here we also assume that the root-mean-square displacements of the A-B and B-C bonds are uncorrelated and that $\sigma_{AC}^2 \approx \sigma_{AB}^2 + \sigma_{BC}^2$.

It has been shown [2] that (5) can be rewritten as

$$\chi^{ABC}(k) = \frac{1}{kr_{AC}^2} \Omega_B(\beta,k) F_C(k) e^{-2\sigma_{AC}^2 k^2} e^{-2r_{AC}/\lambda} \sin(2kr_{AC} + \phi_{AC}(k) + \omega_B(\beta,k)) \tag{6}$$

where $\phi_{AC}(k) = \phi_A^l(k) + \phi_C(k) - l\pi$.

It is apparent from (6) that the effect of multiple scattering via the intervening atom B is to multiply the amplitude $F_C(k)$ by $\Omega_B(\beta,k)$ and to add $\omega_B(\beta,k)$ to the phase $\phi_{AC}(k)$. That is, if one substitutes the modified amplitude $F_C(k)\Omega_B(\beta,k)$ for $F_C(k)$ and the corrected phase $\phi_C(k) + \omega_B(\beta,k)$ for $\phi_C(k)$, the EXAFS data can be analyzed in the usual way. Clearly, both $\Omega_B(\beta,k)$ and $\omega_B(\beta,k)$ are functions of the scattering angle β. Conversely, (6) can be used to determine the scattering angle β.

It should be noted that the approximation implicit in (4) becomes exact at $\beta = 0^\circ$ (i.e. linear case). However, at this extreme, one should in theory consider only the multiple scattering pathway III in Fig. 3.

Fig. 4 shows schematically the relative importance of the three scattering components in (2) or (3) and their peak positions in distance space at two scattering angles (β). The peaks are separated

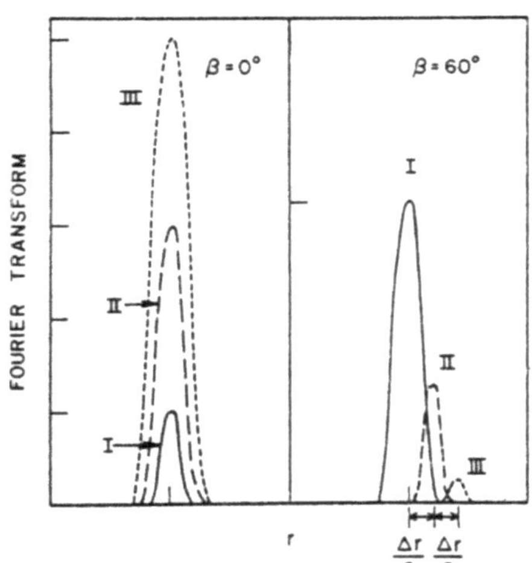

Fig. 4. Schematic illustrations of the relative importance of the three scattering components I, II, III (*cf.* Fig. 3) in distance r space in terms of their Fourier transform magnitudes. At $\beta = 0^\circ$, the three peaks merge at a distance corresponding to r_{AC} with relative intensities III > II > I. As β increases, III and II move to distances corresponding to $r_{AB} + r_{BC}$ and $(r_{AB} + r_{BC} + r_{AC})/2$, respectively, with a concomitant drastic decrease in intensity, resulting in a relative ordering of I > II > III. The spacing between the peaks corresponds to $\Delta r/2 = (r_{AB} + r_{BC} - r_{AC})/2$. The vertical axis of $\beta = 60^\circ$ case is enlarged for clarity

by $\frac{1}{2}(\Delta r) = \frac{1}{2}(r_{AB} + r_{BC} - r_{AC})$. At low β ($\lesssim 30°$), $\Delta r \approx 0$, the three peaks merge at a distance corresponding to r_{AC} with intensities III > II > I. As β increases, $\Delta r/2 > 0$, peaks III and II move to larger distances corresponding to $r_{AB} + r_{BC}$ and $\frac{1}{2}(r_{AB} + r_{BC} + r_{AC})$, respectively, with a concommitant dramatic decrease in intensities. At high β($\gtrsim 60°$) the peak magnitudes follow the order I > II > III. In this context, it should be cautioned that in Fourier filtering, one should not filter out the multiple scattering pathways II and III by choosing too narrow a filtering window for the *A*-*C* peak.

While focusing effect can, at low scattering angles (large bond angles), cause an order of magnitude enhancement in amplitude, its impact on the phase is, fortunately, less dramatic. In most cases, the corrections in distance determination due to multiple scattering processes amount to $\lesssim$0.10 Å for low *Z* and $\lesssim$0.20 Å for high *Z* intermediary atoms.

AB_1B_2C Systems

Multiple scatterings involving more than three atoms are likely to be of less importance due to the large effective total path lengths involved; except, perhaps, in cases where all the bond angles are close to linearity. Considering a four-atom framework AB_1B_2C shown in Fig. 5 and following the notations used in the previous section, it is obvious that there are two predominant groups of multiple scattering processes as shown in Fig. 6: those involving three atoms (a and b) and those involving four atoms (c, d, and e). The three-atom pathways (a,b) are in general more important than the four-atom pathways (c,d,e). Within each group, the relative importance of the multiple scattering pathways decreases with increasing number of path lengths involved. Hence, for the three-atom pathways a and b, I (2 distances) > II (3 distances) > III (4 distances) whereas for the four-atom pathways c, d, and e, I (2 distances) > II (4 distances) > III (6 distances). Obviously, the exact ordering also depends heavily on the scattering angles at each intervening atoms. In fact, when the atoms are arranged in a linear or nearly colinear fashion, the particular pathway(s) which have small scattering angle(s) will dominate. At small β values, the order within each group may be reversed: III > II > I.

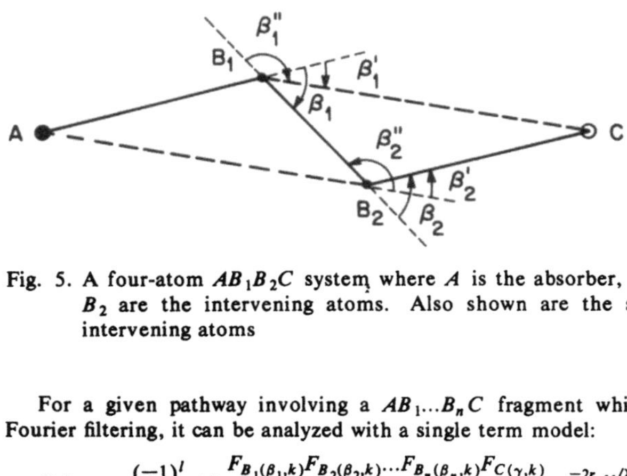

Fig. 5. A four-atom AB_1B_2C system where *A* is the absorber, *C* is the backscatterer, and B_1, B_2 are the intervening atoms. Also shown are the scattering angles β at the two intervening atoms

For a given pathway involving a $AB_1 \ldots B_nC$ fragment which is detectable and separable via Fourier filtering, it can be analyzed with a single term model:

$$\chi(k) = n\,\frac{(-1)^l}{k} \times \frac{F_{B_1}(\beta_1,k)F_{B_2}(\beta_2,k)\cdots F_{B_n}(\beta_n,k)F_C(\gamma,k)}{r_{AB_1}r_{B_1B_2}\cdots r_{B_nC}}\, e^{-2r_{eff}/\lambda}\, e^{-2\sigma_{eff}^2k^2}$$

$$\times \sin\left[2kr_{eff}+2\delta_l'(k)+\theta_{B_1}(\beta_1,k) + \theta_{B_2}(\beta_2,k)+\ldots+\theta_{B_n}(\beta_n,k) + \theta_C(\gamma,k)\right] \quad (7)$$

Here *n* is the geometrical factor and the photoelectron travels through distances r_{AB_1}, $r_{B_1B_2}$, $\ldots r_{B_nC}$, being scattered by atoms B_1, B_2, ... B_n, and *C* with scattering angles β_1, β_2, ... β_n, and γ, respectively. The effective pathlength, which corresponds to the distance observed in the Fourier transform after phase shift corrections, is $r_{eff} = (r_{AB_1}+r_{B_1B_2}+\ldots+r_{B_nC})/2$ and the effective Debye-Waller factor is $\sigma_{eff}^2 \approx (\sigma_{AB_1}^2+\sigma_{B_1B_2}^2+\ldots+\sigma_{B_nC}^2)/2$.

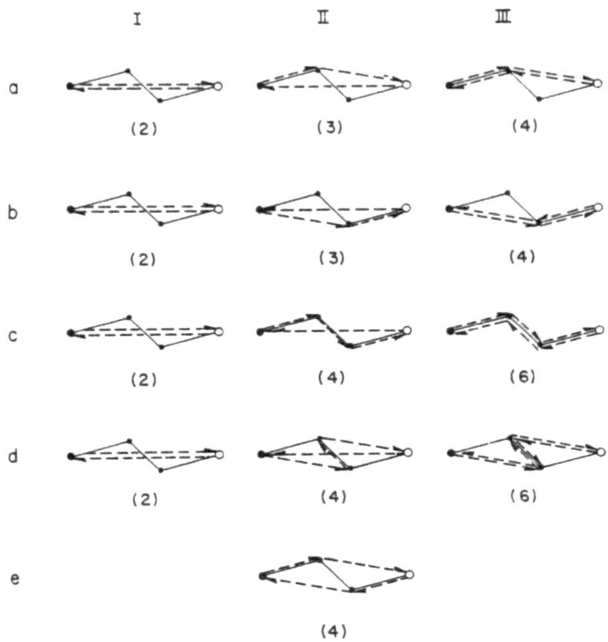

Fig. 6. Schematic representation of scattering pathways for the four-atom AB_1B_2C system shown in Fig. 5. The absorber A, the intervening atoms B_1, B_2 and the backscatterer C are represented by filled circles, dots, and open circles, respectively. Pathway I is the direct backscattering whereas pathways II and III are multiple scatterings involving three (a and b) or four (c, d, and e) atoms. The numbers in the parentheses indicate number of distances traveled. Pathways of type II should be counted twice (except e)

As an example, the multiple scattering pathway IIIc for the AB_1B_2C framework can be modeled by the following equation:

$$\chi(k) = \frac{(-1)^l}{k} \frac{F_{B_1}(\beta_1,k)^2 F_{B_2}(\beta_2,k)^2 F_C(\pi,k)}{r^2_{AB_1} r^2_{B_1B_2} r^2_{B_2C}} e^{-2\left(r_{AB_1}+r_{B_1B_2}+r_{B_2C}\right)/\lambda} e^{-2\left(\sigma^2_{AB_1}+\sigma^2_{B_1B_2}+\sigma^2_{B_2C}\right)k^2}$$

$$\times \sin\left[2k(r_{AB_1} + r_{B_1B_2} + r_{B_2C}) + 2\delta_l'(k) + 2\theta_{B_1}(\beta_1,k) + 2\theta_{B_2}(\beta_2,k) + \theta_C(\pi,k)\right] \quad (8)$$

For systems with small scattering angles β_1, β_2, ..., it is often not possible to separate out the individual pathways via Fourier filtering. Under these circumstances, it is necessary to take into account all the multiple scattering pathways involving the absorbing atom A, the end-atom C, and the intervening atoms B_1, ..., B_n. Following the treatment of the previous section and assuming $\gamma \approx \pi$, the multiple scattering processes depicted in Fig. 6 will give rise to an amplitude Ω and a phase ω modification factors where

$$\Omega e^{i\omega} = \sum_j \Omega_j e^{i\omega_j} - N \quad (9)$$

where j runs from a through e, each containing three pathways (except e which contains only one). In practice, the summation needs to run over only the *dominant* terms in the multiple

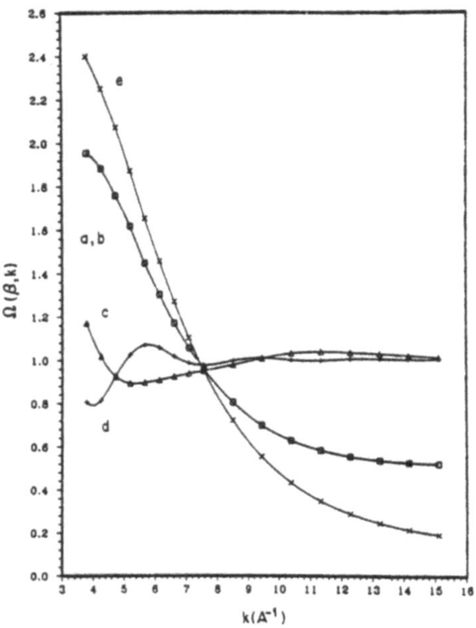

Fig. 7. The amplitude modification functions for various scattering pathways (a-e) for a zig-zag A-B_1-B_2-C system shown in Fig. 6. Here B_1, B_2 are oxygens and the structural parameters are given in the text

scattering processes and *included* in the Fourier filtering process. The scaler N is included to take care of the extra single-scattering processes included in each group (except pathway e). For example, if all five groups a to e are included, N=3, since the single scattering was overcounted three times. Depending upon the structure, some terms are orders of magnitude more important than others. The relative importance of each term depends on the total effective path length, the scattering angle(s) and the scattering power of the intervening atom(s). For peak separations of $\gtrsim$ 0.30Å, it is easier to analyze the component(s) individually via Fourier filtering.

Figure 7 illustrates the relative importance of the multiple scattering pathways for a zig-zag structure involving two oxygens as intermediary atoms at $A-B_1$, B_1-B_2, and B_2-C distances of 1.95, 1.31, and 1.95 Å, respectively. The $A-B_1-B_2$ and B_1-B_2-C angles are both 118°. It is apparent that three-atom pathways a and b (equivalent in this case) and the four-atom pathway e are most important, leading to amplitude enhancements at low k and amplitude reductions at high k. Pathways c and d can be ignored because they result in slight amplitude modification only at low k.

Linear or Nearly Linear $AB_1B_2...B_nC$ Systems

Generally speaking, the higher shells can be analyzed using the multiple scattering formalism developed so far for three- and four-atom systems. In most cases, however, the signals from the third and fourth shells would also have reached the detection limit of EXAFS due to factors such as Debye-Waller factor, inelastic electron mean free path and $1/r^2$ damping. The obvious exception to these considerations is the case where the atoms are arranged in a linear or nearly linear fashion. In such a case, significant amplitude enhancement due to forward scattering at small scattering angle *via* the intervening atoms can occur and EXAFS signals beyond the second or third shells can be observed. For these systems, the multiple scattering processes related to that involving the stepwise propagation of the photoelectron waves along the chemical bonds dominates. The latter can be described by (7) which can be simplified to a form similar to (6) for nearly linear systems. Details are given elsewhere [3].

Angle Determination

We shall now describe a general method for interatomic angle determination by EXAFS on unoriented (spherically averaged) materials based on the multiple-scattering theory and calculations described in the previous section.

ABC Systems

To determine the bond angle A-B-C ($\alpha = 180^\circ - \beta$), the distance r_{AC}, and the Debye-Waller factor σ_{AC}, one simply attempts to best fit the experimental curve with (6) by least-squares refining β, r_{AC}, σ_{AC}, ΔE_{o_C}, η_C, and S_C. The distance r_{AB} and the energy correction ΔE_{o_B} are taken from the AB fit (i.e., fitting the A-B peak with the single-scattering theory) and held constant throughout the curve fitting process. The distance r_{BC} can be calculated from r_{AB}, r_{AC}, and β in each cycle of refinement. The goodness of fit is then plotted as a function of β to determine the parameters for the best fit (minimum). The bond angle α is given by $180^\circ - \beta$. Two distances and one angle completely characterize a triangle.

AB_1B_2C Systems

To determine all the bond lengths and angles for the more complicated systems one needs only to carry the procedure to outer shells in a stepwise manner, taking into account only the most important multiple-scattering pathways. Thus, for the four-atom AB_1B_2C systems one can determine r_{AB_1}, $\Delta E_{o_{B_1}}$, σ_{AB_1}, λ_{B_1}, and N_{B_1} (coordination number) from the two-atom AB_1 fit. Carrying r_{AB_1} and $\Delta E_{o_{B_1}}$ into the three-atom AB_1B_2 fit, one obtains r_{AB_2}, β_1, $\Delta E_{o_{B_2}}$, σ_{AB_2}, as well as λ_{B_2} and N_{B_2}. In the curve-fitting process, $r_{B_1B_2}$ is calculated from r_{AB_1}, r_{AB_2}, and β_1. In fact, these latter three parameters completely define the triangle AB_1B_2. Carrying r_{AB_1}, r_{AB_2}, $r_{B_1B_2}$, β_1, $\Delta E_{o_{B_1}}$, and $\Delta E_{o_{B_2}}$ into the four-atom AB_1B_2C fit, one can obtain r_{AC}, β_2, ΔE_{o_C}, σ_{AC}, as well as λ_C, and N_C. In the process of curve-fitting, r_{B_2C}, β_1', β_2', β_1'', and β_2'' are calculated from other parameters. The two additional parameters r_{AC} and β_2 then completes the characterization of the four-atom AB_1B_2C structural framework.

$AB_1...B_nC$ Systems

To obtain structural parameters for the nearly linear $AB_1...B_nC$ system, the procedure outlined in the previous sections can be repeated for each of the preceding (shorter) fragments $AB_1...B_jC$ where j runs from 1 to n. In each step, one distance ($r_{B_jB_{j+1}}$) and one angle (β_j) will be sought (*i.e.*, refined) while the previously obtained structural ($r_{B_1B_2}, ... r_{B_{j-1}B_j}$, $\beta_1, ... \beta_{j-1}$) and energy ($\Delta E_{o_{B_1}}, ... \Delta E_{o_{B_{j-1}}}$) parameters will be fixed.

Conclusion

In this chapter, we have described an EXAFS formulation which takes into account the effect of multiple scatterings. It is shown that forward scattering is particularly strong within a cone (scattering angle) of 30° beyond which the effect is comparable to backscattering. The relative importance of various multiple scattering processes is assessed as a function of the scattering angle which helps unravel the "focusing" effect.

It is shown that the effect of multiple scattering on the outer shells can lead to orders of magnitude enhancement in EXAFS amplitude, rendering the single-scattering theory of EXAFS inapplicable. The effect on the outer-shell distances, however, are less dramatic, amounting to $\lesssim 0.1$ Å and $\lesssim 0.20$ Å corrections for low and high Z intermediary atom(s), respectively. It is also demonstrated that multiple scattering is most important for: (1) large bond angle ($\gtrsim 120^\circ$), or equivalently, small scattering angle ($\lesssim 60^\circ$); (2) scattering angles with large scattering amplitude $F(\beta,k)$; (3) low k region; (4) strong $A-B$, B_1-B_2, ... B_n-C bonds; (5) high Z intermediary atoms B_1, ...B_n; etc.

Based on these, a new method for bond angle determinations by EXAFS is devised. In most cases, the accuracy for angle determination is better than *ca* 6% for low Z ($Z \leq 10$) and *ca* 3% for high Z scatterers, which amounts to an accuracy of *ca* 5°. Since the B_1-B_2, ... B_n-C distances are determined indirectly, they have the least accuracy. In the case of low Z scatterers, the low accuracy arises from the shrinkage of outer shell distances. Nevertheless, if the shrinkage of the outer shell distances can be corrected theoretically or calibrated experimentally, the B_1-B_2, ... B_n-C distance, can be calculated with a better accuracy. Similarly, if model compounds with known structures are available, the accuracy of bond angle determinations can be improved.

The accuracy of bond angle determinations depends heavily on the quality of the EXAFS data, especially the amplitude. Equally important is the accuracy of theoretically calculated scattering amplitude and phase as functions of scattering angle (β) and photoelectron wavevector (k). Finally since forward scattering (low scattering angle) is affected by the outer electrons of the intermediary atom(s), bonding effects may become important in some cases.

It is interesting to note that at bond angles $\alpha \lesssim 120^\circ$ where the focusing effect is virtually disappearing, bond angle determination is still feasible. One reason is that, in general, the amplitude modification factor Ω exhibits characteristic shape for each α due to the Δr variation as bond angle changes. Another reason is that as α deviates far from the correct value, in either direction (viz., beyond the correlation limits of other parameters such as ΔE_0), the wrong distances give rise to progressively worsening phase mismatch which causes the quality of the fit to deteriorate rapidly.

It should also be cautioned that in Fourier filtering, one should not filter out the multiple scattering pathways which contain the angle information by choosing too narrow a filtering window. For an ABC system, the multiple scattering pathways correspond to the effective distances $(r_{AB}+r_{BC}+r_{AC})/2$ and $(r_{AB}+r_{BC})$ for pathways II and III, respectively. For systems with large $(r_{AB}+r_{BC}-r_{AC})/2$, each of these peaks may be analyzed separately using the corresponding term in (2) or (3). Fortunately, for most systems with large α, III > II > I and $r_{AB}+r_{BC} \approx r_{AC}$ whereas for most systems with small α, I > II > III and $(r_{AB}+r_{BC}+r_{AC})/2 \gtrsim r_{AC}$. That is, for both limits of α, the dominant term(s) are more likely than not to be buried under (or lie close to) the A-C peak. For more complicated systems, generally speaking, one needs only to consider only the most important scattering pathways.

In searching for a minimum in the goodness of fit as a function of the scattering angle (β), it is important to note that there is an upper bound as well as a lower bound of β beyond which the plot of the goodness of fit *vs* β is meaningless. For an ABC system, for example, the upper bound is generally determined either by the filtering window or the unreasonably long B-C distance (as β increases) whereas the lower bound is generally set by the unreasonably short B-C distance (as β decreases).

Extension of the application of the angle determination technique to a wide variety of systems is obvious. Examples include: stereochemical information beyond the first coordination shell for inorganic and biological systems, especially when the ligand atoms are arranged in a nearly colinear fashion; solution structural determinations such as those pertaining to the ligand fluxionality and kinetics; ligation of simple well-defined molecules to biological molecules in solution; etc. It is clear that such structural information involving the distant, normally nonbonding, shells can greatly facilitate either interpretation of the mode of binding of known ligands or identification of the unknown ligands.

References and Notes

1. This effect, which is analogous to an amplifying relay system, was first noticed in the fourth shell of copper and in nickel tetracarbonyl.
2. B. K. Teo, *J. Am. Chem. Soc.*, **103**, 3990 (1981).
3. B. K. Teo, submitted for publication.
4. The theory and calculations presented here and in Ref. 2 and 3 have been successfully applied to a number of systems, see, e.g., J. Goulon, *et al.* and D. Raouse, *et al.*, in this volume.
5. P. A. Lee and J. B. Pendry, *Phys. Rev. B*, **11**, 2795 (1975).
6. C. A. Ashley and S. Doniach, *ibid.*, **11**, 1279 (1975).
7. G. Beni, P. A. Lee, and P. M. Platzman, *Phys. Rev. B*, **13** 5170 (1976).
8. P. A. Lee and G. Beni, *Phys. Rev. B*, **15**, 2862 (1977).

9. J. J. Boland, S. E. Crane and J. D. Baldeschwieler, *J. Chem. Phys.*, **77**, 142 (1982).

10. B. K. Teo and P. A. Lee, *J. Am. Chem. Soc.*, **101**, 2815 (1979).

11. For scattering angles $\beta \lesssim 70^\circ$ where the multiple scattering effect is most important, one has $\cos(\gamma - \alpha) \gtrsim 0.8$. As an example, for $\beta = 60^\circ$ and $r_{AB} = r_{BC}$, $\cos(\gamma - \alpha) = \cos 30^\circ = 0.87$. Hence the assumption (*cf.* [2]) that for unpolarized X-ray source and unoriented samples, the geometrical factors can be ignored is, in general, a reasonable approximation (*cf.* [9]).

Role of Inelastic Effects in EXAFS

J.J. Rehr and S.-H. Chou

Department of Physics, University of Washington,
Seattle, WA 98195, USA

1. Introduction

The validity of the modern short-range-order theory of EXAFS (extended x-ray absorption fine structure) is largely a result of inelastic processes which give an energetic photoelectron a short mean free path. The need for including inelastic damping effects in calculations of EXAFS spectra $\chi(k)$ is graphically illustrated by a comparison of typical results [1] of one-electron theory and experiment [2] (Fig.1). Note that this theory, which has no inelastic losses, overestimates the observed EXAFS roughly by a factor of 2.

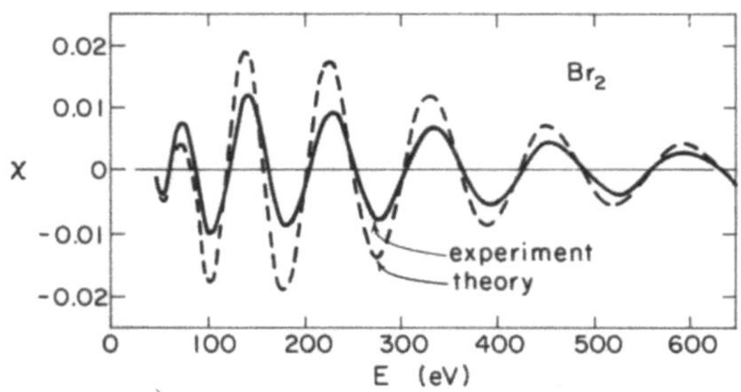

Fig.1. EXAFS spectra $\chi(k)$ based on a conventional one-particle, initial-state Hamiltonian and X_α exchange using the X_α scattered wave method [3]

In this paper the role of such inelastic effects is reviewed from a theoretical viewpoint based on ΔSCF Hartree-Fock formalism [4] and the treatment of inelastic losses developed by LEE and BENI [5]. This approach takes into account the most important many-body features of the problem. The principal inelastic processes are (a) losses in propagation (extrinsic processes) and (b) relaxation of the remaining N - 1 electrons in response to the creation of the core hole (intrinsic processes). The theory is illustrated by *ab initio* calculations of the EXAFS in Br_2 carried out by the authors [1]. It is convenient to begin with ΔSCF formalism, in which intrinsic losses appear naturally, and to deal with extrinsic losses subsequently.

2. ΔSCF Hartree-Fock Formalism

The starting point for the conventional treatment of x-ray absorption calculations is the Golden Rule expression for the transition rate in the dipole approximation

$$\mu \sim \sum_f |\langle\Psi_f^N|\hat{\varepsilon}\cdot\sum_i \vec{r}_i|\Psi_i^N\rangle|^2\delta(E_f - E_i - \hbar\omega) \ . \quad (1)$$

Here Ψ_i^N and Ψ_f^N are respectively N-particle wave functions for the initial and final states with respective energies E_i and E_f, and $\hbar\omega$ is the x-ray energy. In the ΔSCF approximation these wave functions are Slater determinants constructed from the one-particle eigenstates of effective one-electron Hamiltonians H and H' for these states, the prime denoting that the final state includes the core hole. In the sudden approximation valid in the EXAFS regime >200 eV from threshold, one can assume that the final state factorizes as $\Psi_f^N \simeq \phi_k' \Psi_0'^{N-1}$. Also, since one is primarily interested in the contribution to the absorption from a deep core state, the initial state is taken to have the form $\Psi_i^N \simeq \phi_{core}\Psi_0^{N-1}$. With these approximations the absorption coefficient from one-electron processes can be written as

$$\mu(\omega) \sim |\langle\phi_k'|\hat{\varepsilon}\cdot\vec{r}|\phi_{core}\rangle|^2 \; |\langle\Psi_0'^{N-1}|\Psi_0^{N-1}\rangle|^2 \quad , \tag{2}$$

i.e., as a one-particle transition rate $\mu^{(1)}(\omega)$ multiplied by $S_0^2 = |\langle\Psi_0'^{N-1}|\Psi_0^{N-1}\rangle|^2$. Since S_0^2 is the square of a many-body overlap integral, it is necessarily less than one. The reduction in cross section implied by this equation is, however, not observed. Instead, the missing oscillator strength in this "primary," one-electron channel is transferred to previously forbidden multielectron excitations, referred to as shake-up/shake-off processes.

3. Extrinsic Processes--Shake-up/Shake-off Excitations

Within the sudden approximation, the amplitude that the $N-1$ electron system finds itself in the nth excited state of the ion with a core hole is given by $S_n = \langle\Psi_n'^{N-1}|\Psi_0^{N-1}\rangle$. Thus the total absorption cross section, summed over all final states (or channels), is

$$\mu(\omega) \sim \sum_n \mu^{(1)}(\omega - \Delta E_n^{N-1})S_n^2 = \int \mu^{(1)}(\omega - \omega')A(\omega')d\omega' \quad , \tag{3}$$

i.e., a convolution of the one-electron absorption spectrum and the excitation spectrum $A(\omega) = \Sigma_n S_n^2 \delta(\omega - \Delta E_n^{N-1})$ of the system. Likewise, the net EXAFS spectrum, including the shake-up/shake-off excitations, is given by a convolution,

$$\chi(\omega) = \int_0^\omega \chi^{(1)}(\omega - \omega')A(\omega')d\omega' \Big/ \int_0^\omega A(\omega')d\omega' \quad . \tag{4}$$

Typically the excitation spectrum consists of discrete shake-up peaks at low energies, with the shake-off part distributed over a very long tail. Within the Hartree-Fock approximation the centroid of the excitation spectrum is the relaxation energy, $\Delta E_R = E^{N-1} - E'^{N-1}$; i.e., the difference in the total energy of the $N-1$ particle system with and without a core hole. For Br_2 the centroid is at 56 eV. Calculations of the many-body overlap integrals give $S_0^2 = 0.70$ and a total of 0.1 distributed over several shake-up peaks. These figures are in good agreement with earlier calculations [6].

At photoelectron energies smaller than 100 eV corrections to the sudden approximation become increasingly important. In particular the factorization of the final state assumed above fails, and one should consider additional terms in the full Slater determinant wave functions. For an attractive (positively charged) core hole these corrections enhance the primary channel (cf. the x-ray edge singularity [7]) and suppress the shake-up/shake-off excitations. Indeed, experimental studies of the near-edge structure indicate that the amplitude of multiple-electron excitations is quite small near threshold [8].

4. Energy-Dependent Exchange

As a simplification in the SCF procedure the non-local Hartree-Fock exchange operator is often replaced by a local potential proportional to $\rho^{1/3}$, where ρ is the local electron charge density. While this is appropriate for bound states and calculations of self-consistent potentials, this approximation fails completely when applied to a high-energy one-particle state. To see this, consider the behavior of the Hartree-Fock exchange operator, acting on a one-electron state of wave number k:

$$V_x\phi_k = -\int \sum_j^{\mathrm{occ}} \frac{\phi_j(r)\phi_j(r')}{|r-r'|}\,\phi_k(r')\,d^3r' \,. \tag{5}$$

For large k (small DeBroglie wavelength) one expects that only short distance behavior is important, so that the following approximations are valid:

$$\phi_k(r') \simeq \phi_k(r)e^{ik(r-r')} \,, \qquad r \simeq r' \,, \tag{6}$$

$$\sum_j^{\mathrm{occ}} \frac{\phi_j(r)\phi_j(r')}{|r-r'|} \simeq \frac{\rho(r)}{|r-r'|} \,, \qquad r \simeq r' \,. \tag{7}$$

Thus for large k one expects that $V_x \to 4\pi\rho/k^2$, so the local exchange operator decays as $\rho(r)$ rather than $\rho(r)^{1/3}$ and in addition is energy dependent, decaying with energy. In other words, at high energies the exchange hole turns off.

If one approximates the system locally as an electron gas of local density N/V, an expression for the exchange operator can be determined exactly [9]:

$$V_x^{\mathrm{EDX}} = -\frac{3}{2\pi}k_F\left(1 + \frac{k_F^2 - p^2}{2pk_F}\ln\left|\frac{p+k_F}{p-k_F}\right|\right) \tag{8}$$

where $p^2 = k_F^2 + k^2$. We refer to this approximation as energy dependent exchange (EDX).

5. Extrinsic Processes--Losses in Propagation

The addition of electron-electron interactions beyond the SCF approximation gives rise to additional loss mechanisms. In this case the Hartree-Fock states are no longer eigenstates and therefore decay. Two physical decay mechanisms are associated with this: electron-hole excitations and plasmons. HEDIN and LUNDQUIST [10] have incorporated these effects into a complex local potential (referred to as HL) based on the Sham-Kohn density functional formalism for excited states [11] and a one-pole approximation for the dielectric function of the electron gas. The potential has the form $V = V_{\mathrm{Hart}} + \Sigma$, where the self-energy Σ consists of two terms: a screened exchange potential and a polarization term arising from the correlation Coulomb hole. This potential has been implemented for calculations of EXAFS by LEE and BENI [5], and calculations of the scattering amplitudes and phase shifts for many elements have been tabulated by TEO and LEE [12]. A drawback to this potential is its complicated energy-dependent form, which requires a separate numerical integration at each k. The real part, for example, is given by [5]

$$\mathrm{Re}\,\Sigma = -\int \frac{d^3q}{(2\pi)^3}\frac{4\pi}{q^2}\left[\frac{f(\vec{k}+\vec{q})}{\varepsilon[q,\tfrac{1}{2}(k+q)^2-\omega]} - \frac{\omega_p^2}{2\omega_1(q)}\,\frac{1}{\omega_1(q)-\omega+\tfrac{1}{2}(k+q)^2}\right] . \tag{9}$$

We have found that a simple analytical interpolation formula can give a good fit to this equation. The real part is given by the energy-dependent exchange expression [8] together with an additional term from the screening of the exchange term and from the Coulomb hole:

$$\mathrm{Re}\,\Sigma \simeq V_x^{\mathrm{EDX}} - \frac{9k_F}{2\pi}\,\omega_p\,\frac{(p^4-k_F^4)}{p^5} . \tag{10}$$

The imaginary part is somewhat more complicated and is given by

$$\mathrm{Im}\,\Sigma \simeq \frac{\omega_p}{2p}\,\ln\left|\frac{q}{Q}\,\frac{[F(Q)+\omega_p^2+k_F^2Q/6]}{[F(q)+\omega_p^2+k_F^2q/6]}\right| \tag{11}$$

where $Q = 2k_F^2/3 + (4k_F^4/9 + k^4 - 4\omega_p^2)^{\frac{1}{2}}$, $q = [-2k_F^2/3 + p^2 - p(\omega_p/p^2 - k_F^2/3)^{\frac{1}{2}}]\omega_p^2$, and $F(q) = \omega_p^2(q^2/4 + k_F^2q/3 + \omega_p^2)^{\frac{1}{2}}$.

Since the scattering potential is now complex, the phase shifts have an imaginary part leading to the decay of propagating eigenstates. The amplitude reduction factors in EXAFS are due largely to losses in the central atom $\exp(-\mathrm{Im}\,2\delta_1)$, and to a lesser extent in the reduction in the back-scattering amplitude. In solids one has in addition a mean free path term depending on the value of the imaginary part of the potential at the muffin-tin radius (in this case the central atom and back-scattering phase shifts should be recalculated with $V - V_{\mathrm{mt}}$ [13]).

6. *Ab Initio* Calculation of EXAFS in Br_2

The theory outlined above will now be illustrated in calculations of EXAFS in Br_2. These calculations were carried out using X_α scattered wave programs [3]. For excited states we incorporated an exchange and correlation potential modified along the lines outlined above. As in the SCF formalism two sets of molecular orbitals were determined self-consistently, one set for the ground state and one for the ion with a 1s core hole. The self-consistency compensates for some of the arbitrariness in the choice of the muffin-tin radius; however, on comparison with results from atomic potentials it was found that molecular effects are of negligible importance in the EXAFS regime beyond 50 eV of threshold. These calculations are fully *ab initio* in the sense that there is no adjustable inner potential. The only arbitrariness is in the choice of potentials. In addition we have done the calculations both with and without the small-atom/plane-wave approximation.

The results for the EXAFS spectra as calculated both with HL and with EDX potentials are shown in Fig.2 and Fig.3, together with the experimental

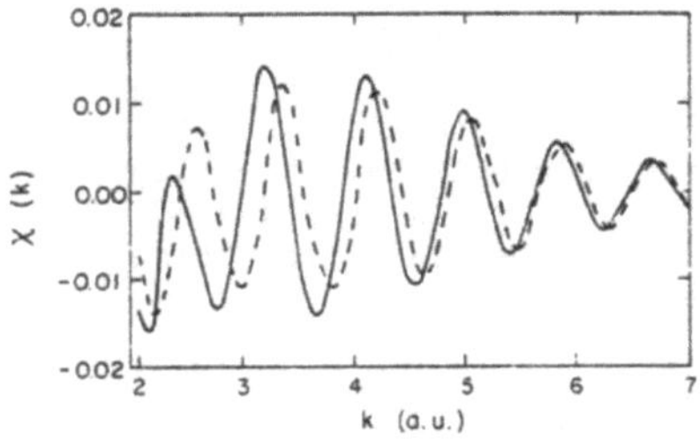

Fig. 3

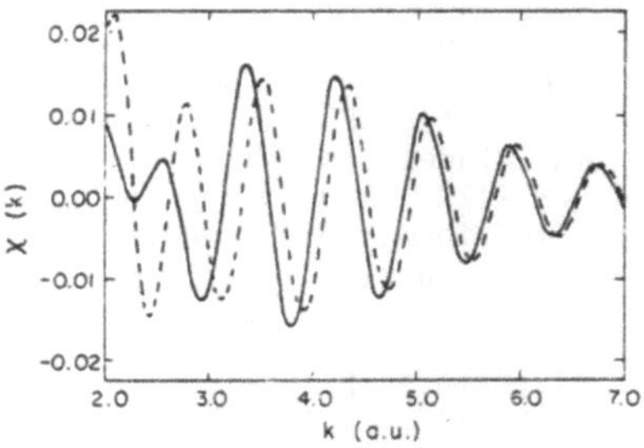

Fig. 2

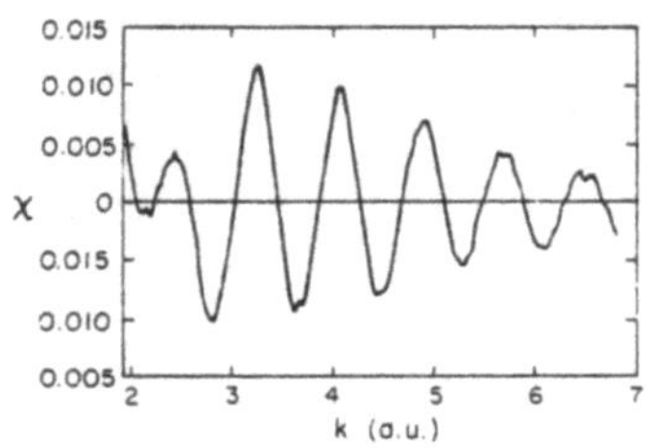

Fig.2. EXAFS spectra for Br_2 as calculated using the HL potential. The dashed (solid) line refers to calculations with (without) the small atom/plane-wave approximation

Fig.3. EXAFS spectra for Br_2 as calculated using the EDX potential. The dashed (solid) line refers to calculations with (without) the small atom/plane-wave approximation

Fig.4. Experimental results, from [2] and [14]

results in Fig.4. No correction for shake-up/shake-off excitations were included so their effect can be evaluated separately.

These results show that the small-atom/plane-wave approximation cannot be used to obtain the correct behavior of the phase and amplitude at wave numbers below k = 3 a.u. Instead, the full spherical wave calculation must be carried out. Remarkably, it is found that the simple EDX potential gives somewhat better results than the HL potential for the EXAFS phase, though the differences are not large (Fig.5). It will be interesting to see whether these findings hold true in other systems [15]. It is also interesting to note that an initial state Hamiltonian seems to give good agreement with the observed EXAFS phase (Fig.1); the reason seems to be that the correct final state potential has a core hole but lacks a full exchange hole in the EXAFS regime, so these differences tend to cancel.

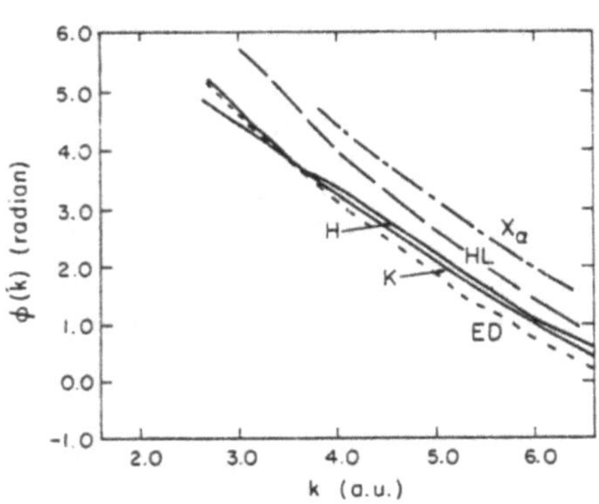

Fig.5. Overall EXAFS phase as calculated using X_α, HL, and EDX potentials, and as measured experimentally; K refers to [14], H to [2]

To assess the role of inelastic effects, we have calculated the EXAFS amplitudes in several ways. One is to take the HL potential both in the absorbing and in the back-scattering atom, which gives the results of Fig.4), multiplied by the amplitude reduction factor from the shake-up/ shake-off excitation spectrum. A second, more phenomenological approach is to take the EXAFS as calculated from the EDX potential multiplied by a mean free path term $\exp(-2R/\lambda)$, with λ being determined from a constant additional imaginary potential of 9 eV. The results are shown in Fig.6. Both calculations are found to give satisfactory agreement with the experimental results.

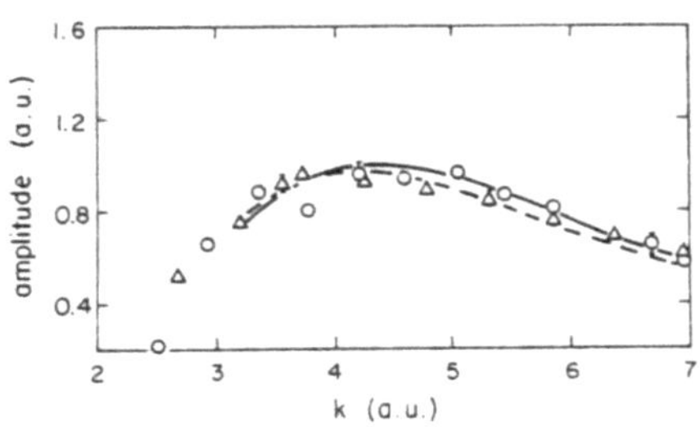

Fig.6. Overall EXAFS amplitude for Br_2 as calculated with the HL potential (solid line) and with an added constant imaginary potential (dashed line). Experimental data [2] are indicated by triangles and circles representing two different methods of analysis

7. Conclusions

We have found that it is essential to include inelastic effects to obtain a correct calculation of EXAFS amplitudes. Of the two types of processes, the *extrinsic* losses appear to be the more important. Such losses are well represented by the Hedin-Lundquist complex potential, as advocated by Lee and Beni. In addition, shake-up and shake-off excitations account for an additional reduction factor, which is roughly constant in energy in the EXAFS regime. When both of these loss mechanisms are combined, the standard EXAFS formula is recovered with modified phase and amplitude factors and a mean free path term. Although the HL potential yields an imaginary potential of about the right magnitude and energy dependence to explain the observed amplitude, it fits the EXAFS phase less well than the EDX potential. Morever, we have found that a simple, phenomenological mean free path term in addition to the EDX potential can yield good agreement for both amplitude and phase. The uncertainties in the magnitude of these many-body effects currently limit the accuracy of EXAFS amplitude determinations to about 10% in *ab initio* calculations. These results also suggest a need for further study of exchange and correlation potentials at high energies and for a better treatment of corrections to the sudden approximation.

8. Acknowledgments

The authors thank E. A. Stern, E. R. Davidson, J. Norman, and L. Noodleman for many helpful discussions. This work was supported in part by the National Science Foundation through grants DMR-79-07238 and DMR-82-207357.

References

1. S.H. Chou: Ph.D. Thesis, University of Washington (1982); S.H. Chou and J. J. Rehr (unpublished)
2. E.A. Stern, S.M. Heald, B. Bunker: Phys. Rev. Lett. 42, 1372 (1979), and private communication

3. K.H. Johnson: Adv. Quantum Chem. 7, 143 (1973)
4. J.J. Rehr, E.A. Stern, R.L. Martin, E.R. Davidson: Phys. Rev. B 17, 560 (1978)
5. P.A. Lee, G. Beni: Phys. Rev. B 15, 2862 (1977)
6. R.L. Martin, E.R. Davidson: Phys. Rev. A 16, 1341 (1977)
7. P. Nozieres, C.T. de Dominicis: Phys. Rev. 178, 1097 (1969)
8. E.A. Stern (unpublished)
9. P.A.M. Dirac: Proc. Cambridge Philos. Soc. 26, 376 (1930)
10. L. Hedin, S. Lundquist: Solid State Phys. 23, 1 (1969)
11. L.J. Sham, W. Kohn: Phys. Rev. 145, 561 (1966)
12. B.K. Teo, P.A. Lee: J. Am. Chem. Soc. 101, 2815 (1979)
13. D.B. Tran Thoai, W. Ekardt: Solid St. Commun. 40, 269 (1981)
14. B.M. Kincaid, P. Eisenberger: Phys. Rev. Lett. 34, 1361 (1976); P.H. Citrin, P.Eisenberger, B.M. Kincaid: Phys. Rev. Lett. 36, 1346 (1976)
15. See in this connection R.F. Pettifer: Proc. 4th EPS Gen. Conf. (1979), p. 522, and this volume

Relaxation Effects in Inner-Shell Photoemission and Absorption

Göran Wendin+

LURE, Bat 209C, Université Paris-Sud, F-91405 Orsay, France

1. Introduction

This paper will describe some characteristics of relaxation processes related to the depth and the symmetry of the inner-shell hole. The discussion will be based mainly on a particular element, lanthanum, comparing the response of atomic La with that of La complexes and La metal.

We shall emphasize the necessity of broad definitions of the concept of inner-shell excitation. It is well known that in a solid with a partially filled band, i.e. a metal, screening and correlation effects are extremely important. The same is true in a more general way, however, as long as not all of the subshells of an atomic main shell are filled. There can then be very large spatial overlap between empty and occupied levels and consequently very strong interaction effects, although the excitation energy may be high. In La systems this region of collective effects, giant dipole resonances [1,2] and giant Coster-Kronig relaxation and decay processes [2,3] involves the n=4,5,6 main shells, extending down through the 4s shell with a binding energy of almost 300 eV. The particular way in which many-electron effects are important depends critically on symmetry. Sometimes only a particle-hole excitation is affected in a dramatic manner, sometimes only an inner-shell hole, and occasionally also the final state electron, e.g., 5p-5d,εd and 4d-4f,εf will be strongly collective (giant dipole resonances [1,2]) due to large electron-hole overlap (exchange), while 4p-5d,εd transitions will be a many-electron process mainly in what concerns the 4p-hole (giant Coster-Kronig damping and shift of binding energy [2,3]).

In order to have a clear separation between deep core processes on the one hand, and shallow core and valence processes on the other, let us state that a deep core hole has to be created in a completely filled main shell. In the case of La this means 3d holes and deeper, starting in the 800-900 eV region. Good examples are the 3d-4f,εf and 2p-5d,εd transitions, in which cases the core-hole can be regarded as a classical charge distribution screened by the electronic medium, including the photoexcited electron. One then finds a "white line" absorption edge structure, a near-edge region which shows structure due to both multiple scattering and multiple electronic excitations (XANES), and at higher energies the usual EXAFS structure.

+On leave of absence from Institute of Theoretical Physics, Chalmers University of Technology, S-412 96 Göteborg, Sweden

2. Dynamics of photoionization processes

Figure 1 gives a condensed picture of various ways of testing the dynamics of a physical system (S). One may excite (ionize) the system with photons (ω) or electrons (e^-) and measure the energy and angle distributions of the final

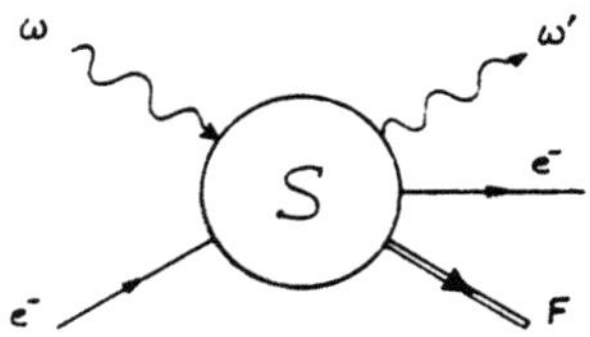

Fig.1. Schematic picture of excitation/emission processes

state products as functions of impact energies or transferred energies. In structure determination one usually wants to minimize any complications due to dynamics. One therefore (among other things) excites deep core levels, e.g. 1s,2s,2p in La, and uses photoabsorption or photoabsorption-related techniques, measuring yields of photons (ω'), electrons (e^-) or ions [4] and fragments (F), or measuring the intensities of individual Auger and fluorescence lines. One can also use electron impact excitation in an EELS [5] or APS [6] mode, or anomalous X-ray scattering [7]. In order to study dynamics, one has to consider all the partial channels that add up to the total absorption cross section. One would,e.g. study the intensity variation with photon energy of a main core line and all of its shake-up and shake-off structure in a photoelectron spectrum (PES). The structure information may now be more difficult to extract. Nevertheless, part of this dynamic information is necessary for a proper description of XANES and low-energy EXAFS (for a more extensive discussion of dynamic aspects, see refs. [8-14]).

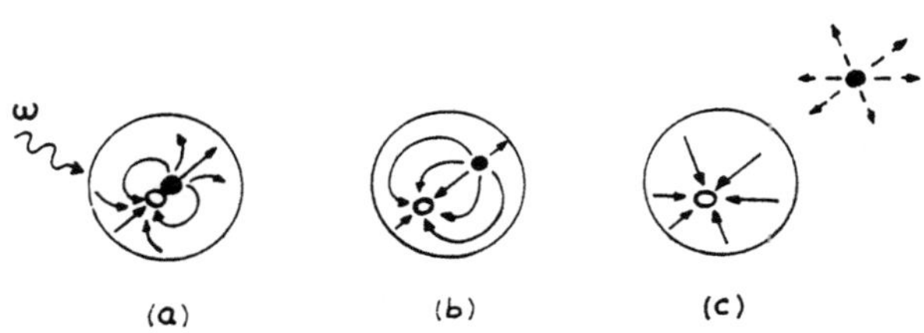

Fig.2. Schematic pictures of photoionization processes in real space and time, for short (a), intermediate (b) and long (c) times. The dashed arrows in (c) indicate that there is an accompanying screening cloud only when the electron moves in an electronic medium, not in vacuum

A qualitative picture of the dynamics of relaxation effects in photoionization in real space and time is shown in Fig.2. We distinguish between three different stages. At short times after absorption of a photon (Fig.2a) the electron and the hole overlap, and the system responds to a dipole perturbation (the flow of screening charge is suggested by arrows). At intermediate times (Fig.2b) the hole and the electron begin to separate, and individual screening clouds start to form around the electron and the hole. However, these screening clouds still share much of the screening charge: There is strong interference between the action of the hole and the action of the electron. Finally, at large times (Fig.2c) the photoelectron has completely separated from the hole and the two charges are screened separately, until the electron escapes into vacuum and loses its screening cloud.

The relative importance of the different stages of time development in Fig.2 will depend on the character of the initial and final one-electron levels. E.g., in transitions like 4d-4f,εf and 5p-5d,εd involving the collective giant dipole resonances, the stage in Fig.2a is of crucial importance, describing screening of the external dipole field by the atomic 4d and 5p shells. However, in the case of ionization of a deep shell, the photoelectron rapidly leaves the region of the core-hole, and the stages in Fig.2b,c become the important ones. As the photoelectron wave packet propagates out through the ion core, the shells become successively exposed to the potential of the core-hole and respond by relaxation, including excitation and ionization (shake-up and shake-off). The relaxation of the shells behind the photoelectron will set up a fairly long range polarization potential which will act back on the electron and ensure that it knows the dynamic state of the core. The polarization potential from relaxation around the photoelectron is velocity dependent and regulates the transition from the adiabatic to the sudden regime through its interference with the core-hole potential [8,10,12,13] (Fig.2b). It also provides a complex potential acting on the electron when this has separated from the core-hole region and propagates as a quasi-particle (Fig.2c) [9,10]. When the photoelectron finally escapes into vacuum it has to leave its correlation-hole behind. In an atom, the polarization potential then becomes a small correction to the ionic long-range Coulomb potential. However, in a metal this leads to a complex, dynamic image potential which may be very important due to the absence of any ionic potential.

Finally it should be noted that the shell structure of the ionic core introduces a spectrum of time scales for screening. Suppose that a photoelectron is ejected from the 2p shell in atomic La with 50 eV kinetic energy in the final state. All of the electrons in the n=3 and n=4 core shells will then relax elastically (adiabatically) because the lowest shake-up energy is more than 100 eV (4d shell). When the photoelectron penetrates the n=5 shell, the 5s shell (binding energy ~45 eV) and the 5p shell (binding energy ~30 eV) will show inelastic effects, allowing low lying shake-up excitations. 5s,5p relaxation is thus intermediate between adiabatic and frozen behaviour. When the electron finally reaches the valence region, the excess energy can be quite large. In a La atom the 6s shell will essentially have no time to respond (binding energy ~6 eV) (sudden limit); in a La complex or compound, the low energy charge transfer screening processes (ligand 4f) will not have time to respond, while the higher energy screening processes (ligand 5d, plasmons, interband transitions) will be in an intermediate situation.

3. How deep is a deep core-hole?

This question was addressed already in the Introduction. Here we shall expand the discussion by considering the illustrations in Fig.3. The boxes denote the open main shells n=4,5,6 in a La system. These have large radial overlap between filled and empty levels and therefore have very large polarizabilities. Core levels in this region will be referred to as shallow core levels. In order for a core-hole to be deep, it must be created in a completely filled main shell (n=3 or deeper). In such cases there is only weak overlap with empty levels and exchange effects become quite small. The most spectacular many-electron effects are associated with transitions within open main shells, e.g. the collective, giant dipole resonance in 4d absorption.

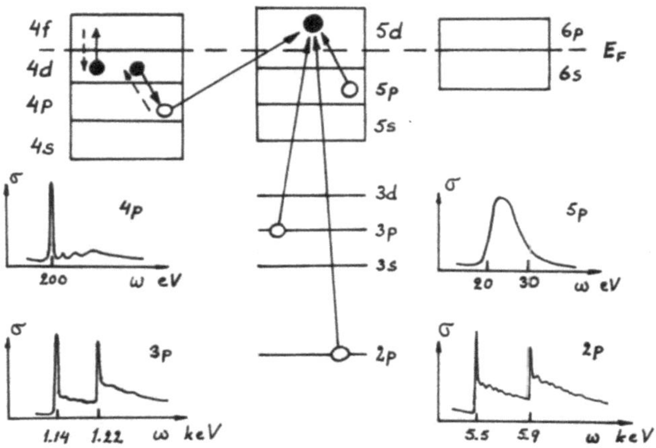

Fig.3. One-electron level structure of a La system. Open boxes denote open main shells and shallow core levels. Thin lines denote deep core levels in closed main shells. The inserts show schematic photoabsorption spectra for the various p shells

In such cases the atomic photoabsorption cross section or f-wave phase shifts cannot be calculated using ground state types of potentials, and XANES and EXAFS analysis will be much more difficult (cf. 4d-photoelectron diffraction [15] for Te on Ni; for experimental results on La compounds, see [16,17]; for theory, see [2,11,12]). In Fig.3 we choose however to discuss a sequence of p-shell absorption spectra. The 5p-5d,εd absorption spectrum [18,19] is dominated by the large 5p-5d exchange interaction, giving rise to strong collective behaviour [1,2]. The width of the spectrum is essentially an atomic property, and the solid state 5d bandwidth plays a secondary role. In the free atom a large part of the distribution will however consist of discrete structure. The major part of the oscillator strength of the 5p shell lies in the peak region, which makes it a giant dipole resonance. Proceeding now to the 4p-5d,εd spectrum, the transition itself takes place between different main shells (weak overlap) but the 4p core-hole presents complications due to the giant Coster-Kronig relaxation and decay processes [3] (involving $4p\text{-}4d^{2}4f,\varepsilon f$ configuration interaction; Fig.3), and the j=1/2 spin-orbit component will effectively appear to be lost (see section 4).

In order to find a reasonably simple spectrum, we have to go to a deep core-hole, starting with the 3p-5d,εd transitions. Due to the weak overlap, only a minor fraction of the oscillator strength is now in the 3p-5d edge peaks (which should therefore not be called giant dipole resonances!), and exchange effects should only weakly influence the continuum phase shift. From an EXAFS point of view there is still a complication: The 3p core-hole is quite broad (~5 eV) due to Coster-Kronig decay processes, in which the 3p hole is filled from the 3d shell and an outer-shell electron is ejected. In order to have smaller widths one has to go to the deeper 2p hole. There is now no Coster-Kronig decay possible (no 2d levels!) and the width is only a few electron volts. This is in fact a general rule: Holes in the highest angular momentum subshells will always be quite sharp (1s,2p,3d,4f etc.; the radiative broadening of a 1s hole in heavy elements is a different matter).

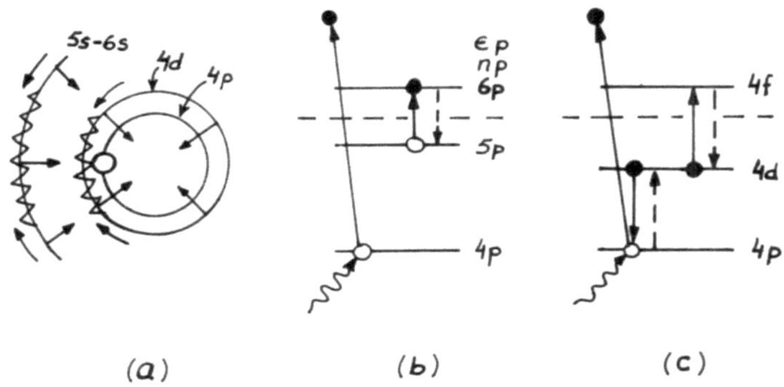

Fig.4. (a) Orbital pictures of the general relaxation process. /\/\/\ denotes accumulation of screening charge. Arrows indicate motion or relaxation and screeening charge. (b,c) One-electron level pictures of monopole relaxation and shake-up (b), and dipole (giant Coster-Kronig) relaxation and "shake-up" (c)

4. Dynamic effects in core-hole spectra

It is well known that in a solid, a hole in a valence band will propagate as a screened quasi-hole and that its wave function may be very different from a one-electron picture. The dynamic interactions may lead to symmetry breaking and localization, or to a number of intermediate situations. Here we shall briefly discuss an analogous case in a purely atomic environment, namely a 4p hole in La (4s and 5s holes are related cases [3]). In Fig.4a, the 4p hole is described as a wave packet in the 4p region, not as a 4p-symmetry orbital or a spherical shell. This is accomplished by the hopping of the core-hole between 4p and 4d levels (Fig.4c). The system can now relax by building up a screening cloud around the hole wave packet, and the entire object then moves around in the 4p-4d region as a quasi-hole. The relaxation process can be expanded with respect to a centre at the nucleus (multipole expansion), giving two main contributions: (i) Radial contraction (Fig.4a,b; monopole relaxation and shake-up) and (ii) angular contraction (Fig.4a,c; dipole relaxation). In La the radial contraction gives a relaxation shift of 7 eV

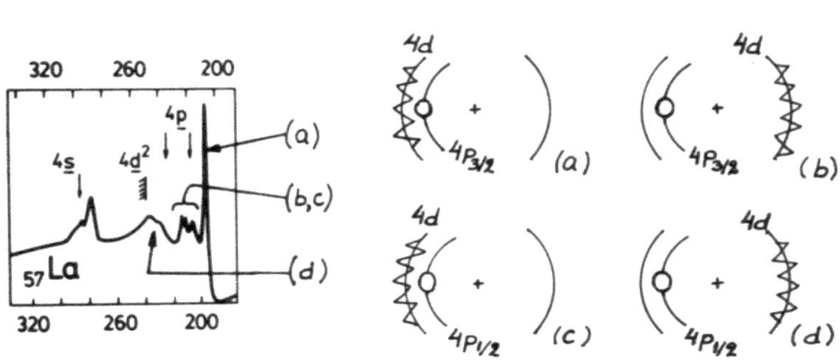

Fig.5. 4p X-ray photoelectron spectrum (XPS) of metallic La [21] together with an interpretation of the quasi-hole structure in real space. The lower (upper) scale gives the metallic (atomic) binding energy in eV. Vertical arrows denote ΔSCF energies (from ref. [3])

(can be obtained from a ΔSCF calculation), while the angular contraction gives a relaxation shift of 10 eV! This effect is thus comparable to the case of symmetry breaking and relaxation of deep holes in molecules like N_2 and O_2 [3,20] .

In this case of very strong interaction, the $4\underline{p}$ and $4\underline{d}^24f$ single-particle configurations lose their meaning. If we want to have a real space picture describing the quasi-holes that correspond to the various spectral peaks in Fig.5, then we might think of a 4p quasi-hole in terms of screened and "anti-screened" states, as suggested in Fig.5. In the case of a $^2P_{1/2}$ hole, the anti-screened state (Fig.5d) carries the main strength but the screening cloud is not stable. It has continuum components of f-wave type since the giant Coster-Kronig decay process is allowed at this energy, and the level becomes very broad. Therefore, the spectrum contains only a single strong, sharp level, of $^2P_{3/2}$ type.

5. Relaxation and shake-up in deep hole excitation spectra

We shall illustrate some aspects of the dynamics of deep core-hole excitation and screening processes using the highly simplified model level structure shown in Fig.6a. The v-v* transition denotes an intra-valence excitation that gives rise to a strong shake-up satellite ($\underline{c}\underline{v}v^*$) in the photoelectron spectrum (XPS). It might correspond to, e.g. a π-π^* transition in N_2 or C_2H_4, or to e-e* or t-t* excitations in transition metal complexes (see [3] for further references). With a strong v-v* shake-up excitation in XPS, one would expect to find double excitation structure in the photoabsorption spectrum (Fig.6b), although it should be weaker than in the photoelectron spectrum due to partial screening by the electron. Double excitations are well known and well established phenomena in core and valence photoabsorption spectra in atoms [22,23]. In molecules, including transition metal and rare-earth metal complexes and solids, double excitations should be much more prominent than in atoms, but it is difficult to assess their real importance due to problems in establishing even what is the one-electron photoabsorption spectrum. However, there is now fairly strong evidence for double excitations in core and inner-valence spectra in a number of cases (see, e.g. [4,24-28] and refs. therein).

Let us now discuss relaxation and shake-up starting from the absorption threshold (Fig.6b). As long as $\omega < \omega_S$, the valence charge will respond adia-

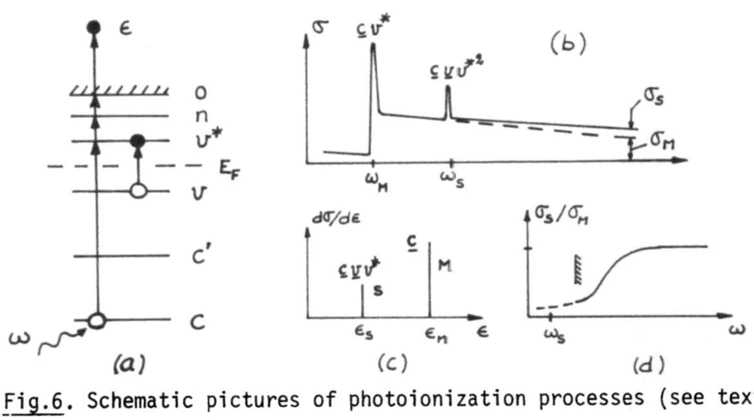

Fig.6. Schematic pictures of photoionization processes (see text)

batically (elastically), whether one excites the cv^* edge resonance (which is below the ionization limit in a molecule) or the $c\varepsilon$ continuum. In Fig.6c the satellite does not exist yet, and all intensity goes into the main line. However, note that the single excitation cross section depends on the presence of the double excitation channel: the screening is dynamic. This is of critical importance when one approaches the threshold for shake-up because it introduces a frequency-dependent interaction between different channels. There is intensity sharing between the cv^* main peak and the cvv^{*2} satellite, as well as intensity sharing between the $c\varepsilon$ and $cvv^*\varepsilon$ channels via continuum mixing. When the double excitation threshold ω_S is reached the valence charge can no longer respond entirely adiabatically: the electron-hole pair (Fig.2b) can now transfer its excess energy to an intra-valence v-v^* excitation. The intensity of this double excitation depends on how well the electron screens the hole, i.e. on the interference between scattering by the hole and by the electron. In this region, inelastic scattering of the photoelectron is associated with capture into v^* : there is not yet any satellite in the PES. The double excitation resonance will show up in the main line (c) intensity.

Increasing the photon energy beyond the threshold for ionization plus excitation, the photoelectron shake-up satellite (cvv^*) appears (Fig.6c,d). Nothing much should be seen in the photoabsorption continuum. The satellite cross section gains its intensity by stealing from the one-electron cross section. Moreover, there is no dramatic change in the main line intensity apart from the resonance phenomena in the threshold region (Fig.b,d). Note that there is no reason why the satellite intensity should go to zero at threshold (Fig.6d): it has to go smoothly over into the spectrum of double excitations. In an atom or molecule shake-up transitions cannot be turned off perfectly adiabatically.

The question of what potential the photoelectron is moving in [10] depends on whether we observe the final states or not. In absorption, we do not resolve the final states and do not have the problem (in principle) of how the screening process feeds intensity into the various photoelectron lines. The electron then moves in a dynamically screened potential, which at high energies tends towards a frozen, ionic potential: the electrons do not have time to screen. This point of view is relevant for EXAFS. The picture does not work, however, when the final states are resolved by observation of the intensity variations of the photoelectron lines (partial photoionization cross sections). A time argument no longer applies, and one has to consider the different potentials seen by the photoelectron corresponding to the main line and to each of the different satellites. For the main line, it should be an adiabatically relaxed single-hole potential (less attractive than the frozen potential) and for the satellites relaxed two-hole-one-electron potentials (more attractive). At high energies, the partial cross sections will nevertheless add up to the frozen core result for the total absorption cross section.

A few final comments: the satellite/main line ratio (Fig.6d) can actually increase when the energy is lowered towards threshold, due to correlation and inelastic scattering effects. This is the case in atomic He (1s-2s,2p satellite), which has recently been given much theoretical [29,30] and experimental [31,32] attention. Also, there is a strong trend in both theory and experiment towards investigating in detail various resonance phenomena in photoelectron emission from shallow and deep core levels ([23,33-38] and refs. therein). This should be the most direct way of revealing the dynamics of excitation and screening processes in core spectra.

1. G. Wendin, Vacuum Ultraviolet Radiation Physics, Eds. E.E. Koch, R. Haensel and C. Kunz (Pergamon/Vieweg, Braunschweig 1974), p. 225
2. G. Wendin, Proceedings of the Int. Conf. on X-ray and Atomic Inner Shell

Physics (X82), August 23-27 1982, Eugene, Oregon (to be published in the American Institute of Physics Conference Proceedings Series)
3. G. Wendin, Structure and Bonding 45, 1 (1981) (Springer Verlag)
4. R. Jaeger, R. Treichler and J. Stöhr, Surface Sci. 117, 553 (1982)
5. M. de Crescenzi, these Proceedings
6. T.L. Einstein, these Proceedings
7. A. Bienenstock and J. Kortright, these Proceedings
8. Many-Body Theory of Atomic Systems, Physica Scripta 21, no 3/4 (1980) (Proceedings of Nobel Symposium 46, Eds. I. Lindgren and S. Lundqvist); see in particular articles by L. Cederbaum, J.W. Gadzuk, O. Gunnarsson and K. Schönhammer, M. Sunjic and G. Wendin, and refs. therein; L. Hedin and S. Lundqvist, Solid State Phys. 23, 1 (1969)
9. J.J. Rehr, these Proceedings
10. C. Noguera and D. Spanjard, J. Phys. F 11, 1133 (1981); C. Noguera, X82 (see [2])
11. M.Ya. Amusia, Appl. Optics 19, 4042 (1980)
12. Photoionization and Other Probes of Many-Electron Interactions, Ed. F. Wuilleumier, NATO Advanced Study Institute Series (Plenum, New York 1976); see in particular articles by T. Åberg, H.P. Kelly and G. Wendin
13. J.J. Cheng and D. Langreth, Phys. Rev. B 8, 4638 (1974)
14. J. Kanski and G. Wendin, Phys. Rev. B 24, 4977 (1981)
15. G. Wendin, Solid State Commun. 38, 197 (1981)
16. P. Rabe, internal report DESY F41-74/2 (1974)
17. M. Aono, T.-C. Chiang, F.J. Himpsel and D.E. Eastman, Solid State Commun. 37, 471 (1981)
18. M. Cukier, B. Gauthé and C. Wehenkel, J. Physique (France) 41, 603 (1980)
19. D.H. Tracy, Proc. R. Soc. Lond. A357, 485 (1977)
20. P.S. Bagus and H.F. Schaeffer, J. Chem. Phys. 56, 224 (1972)
21. S.P. Kowalczyk, L. Ley, R.L. Martin, F.R. McFeely and D.A. Shirley, Faraday Discussions Chem. Soc. 60, 7 (1975)
22. J.P. Connerade, in New Trends in Atomic Physics, Les Houches Session 38, NATO Advanced Study Institute, June 28-July 29 1982
23. J.M. Bizau, F. Wuilleumier, P. Dhez and G. Wendin, Abstracts of the 8th Int. Conf. on Atomic Physics, August 2-6 1982, Göteborg, Sweden
24. G. Wendin, Int. J. Quant. Chem. S13, 659 (1979)
25. C. Brion, Proceedings of X82 (see [2])
26. E.A. Stern, these Proceedings
27. A. Bianconi, M. Campagna and S. Stizza, Phys. Rev. B 25, 2477 (1982)
28. B. Lengeler, G. Materlik and J. Müller, these Proceedings
29. T.N. Chang, J. Phys. B 13, L551 (1980)
30. K.A. Berrington, P.G. Burke, W.C. Fon and K.T. Taylor, J. Phys. B 15, L603 (1982)
31. J.M. Bizau, F. Wuilleumier, P. Dhez, D.L. Ederer, T.N. Chang, S. Krummacher and V. Schmidt, Phys. Rev. Lett. 48, 588 (1982)
32. V. Schmidt, H. Derenbach and R. Malutzki, J. Phys. B 15, L523 (1982)
33. G. Loubriel, T. Gustafsson, L.I. Johansson and S.J. Oh, Phys. Rev. Lett. 49, 571 (1982)
34. F. Gerken, Proceedings of X82 (see [2])
35. W. Eberhardt and A. Zangwill, to be published
36. D.A. Shirley, Proceedings of X82 (see [2])
37. D. Chandesris, J. Lecante and Y. Pétroff, Proceedings of X82 (see [2]), and Phys. Rev. B, to be published
38. G. Wendin, in Photoionization of Atoms and Molecules, Ed. B.D. Buckley, Daresbury Laboratory report DL/SCI/R11 (1978), p. 1
G. Wendin and K. Nuroh, to be published

Multiple Scattering Calculations of XANES

P.J. Durham

Science and Engineering Research Council, Daresbury Laboratory, Daresbury, Warrington WA4 4AD, UK

In this paper I will outline the multiple scattering approach developed at Daresbury for the calculation of X-ray absorption near edge structure (XANES), concentrating on the basic model used and its construction, and give two recent examples of promising applications. As stressed by BIANCONI [1], the near edge region of the spectrum contains information on a number of physical effects, and one of the most important is the co-ordination geometry around the excited atom. This is the aspect on which I will focus here.

In the independent particle model the electron states are calculated from a 1-electron Schrödinger equation containing an effective potential function. Our approach is to represent the potential around the X-ray excited atom by a finite cluster of non-overlapping muffin-tin potentials, each of whose scattering power is represented by a t matrix for angular momentum $L \equiv (\ell, m)$

$$t_\ell(E) = \frac{-1}{\sqrt{\varepsilon}} \sin \eta_\ell \, e^{i\eta_\ell}$$

where $\eta_\ell(E)$ is the ℓ-wave phase-shift. Within this model, the rate at which the system absorbs X-rays of energy $\omega(\hbar = 1)$ by excitation from a core state c is easily found to be

$$W_c(\omega) = -2 \sum_{LL'} m_L^i(\varepsilon) \, \mathrm{Im}\, \tau_{LL'}^{ii}(\varepsilon) \, m_{L'}^{i*}(\varepsilon) \, \theta(\varepsilon - \mu)$$

$$(\varepsilon = \varepsilon_c + \omega)$$

where $m_L^i(\varepsilon)$ is an atomic-like dipole matrix element linking the core state with high energy states (with angular momentum L) of the single muffin-tin potential at the excited atom (here given a site index i). Details are given elsewhere [2]. The scattering path matrix, $\tau_{LL'}^{ii}$, is a convenient way of summarising the multiple scattering paths relevant to X-ray spectra [3]. It sums the contributions of all multiple scattering paths which begin and end at the excited atom, i, and is given by

$$\sum_{k,L''} (t_{i,L}^{-1} \delta_{ik} \delta_{LL''} - g_{LL''}^{ik}) \tau_{L''L'}^{kj} = \delta_{ij} \delta_{LL'} .$$

Thus, to obtain the scattering path matrix one must invert a matrix which has inverse atomic t matrices on the diagonal and, off the diagonal, real space structure constants $g_{LL''}^{ik}$ which describe the propagation of the free spherical waves from site i to site k. This is a very general result; it is valid for any array of non-overlapping scatterers. In general, also, it is quite a difficult numerical task to perform the matrix inversion. The particular strategy we adopt is described elsewhere [4], and I do not intend to dwell on it here. Instead, I want to make some remarks about the construction of the muffin-tin potentials.

In the band theory of ordered metallic systems a simple and reasonably accurate method of obtaining potential functions, the so-called Mattheiss prescription [5], exists, and operates as follows.

1. Charge densities for neutral atoms are obtained from some Herman-Skillman-like self-consistent-field atomic calculation.

2. These atomic charge densities are placed on each site of the chosen geometry, and the superposed charge density is spherically averaged about the atom whose potential is required.

3. Using this spherical, superposed charge density, the Poisson equation is solved for the Coulomb part of the potential, and the exchange-correlation potential is found by means of a local approximation (we use an $X\alpha$-type formula with $\alpha = 1$).

This procedure lacks any formal justification, so far as I am aware, but it is very reliable in close packed metallic systems, at least on the scale of accuracy needed to account for XANES. We have used it quite successfully for a number of metals [6]. On the other hand, modern accurate band theory is done in a self-consistent way using the local density approximation (LDA) for the exchange-correlation potential [7], and good results for the XANES in small clusters have been achieved using self-consistent cluster calculations [8]. In metals, the XANES has been found to be essentially structural in content [9], and so self-consistency is unlikely to be very important for them. There is also the fundamental problem that self-consistent LDA calculations do not give excited states correctly, at least formally [10]; a self-energy correction (generally an unknown quantity) should, in principle, be applied in the calculation of excitation spectra. In the context of band theory it is known that good self-consistent calculations do not always give the best agreement with electron spectroscopic data, as in the case of photoemission from Cu [11]. Thus the rôle of self-consistency in XANES is not clear at present; it is certainly not guaranteed to improve agreement with experiment. The crude Mattheiss prescription outlined above presumably gives the gross changes in the potential of an atom when its environment changes, and the hope is that this is good enough for most XANES applications. It has, of course, the enormous advantage that it allows one to treat rather large and complex structures.

A possibly more serious difficulty has to do with the accuracy of any potentials (self-consistent or not) of the muffin-tin form. In band theory it is found that the more open the structure (e.g. in diamond structure semiconductors), the worse muffin-tins are, because the intersphere region is large. A commonly employed trick is to pack the interstitial region with empty spheres, and this might be useful in XANES calculations. A more elegant but also more complex alternative would be to use the extension of the multiple scattering theory for non-spherical potentials.

Thus, this problem of what to use for the potentials is, in principle, quite difficult. It often seems to be the case, however, that the geometrical arrangement of the atoms is considerably more important that the details of their potentials as regards XANES calculations, and I now want to give two examples of systems for which, somewhat surprisingly, the simple procedure I have outlined works quite well.

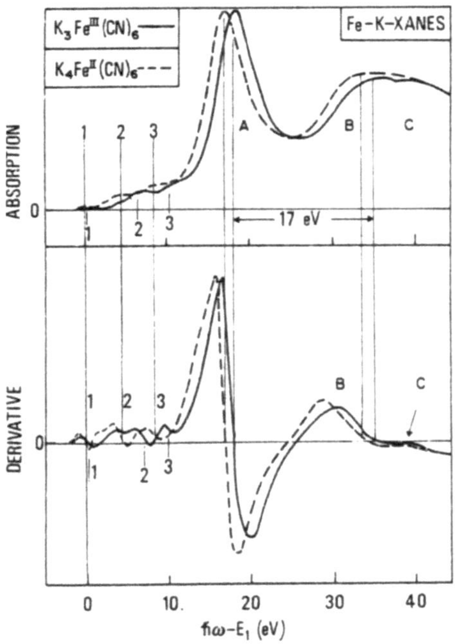

Fig.1. XANES (upper panel) and derivatives (lower panel) at the K edge of Fe in $Fe^{II}(CN)_6$ and $Fe^{III}(CN)_6$ complexes [12]

The first systems are the ferro- and ferricyanide complex ions, whose experimental near edge structures are shown in Fig.1 [12]. The XANES show strong oscillations, with an approximately 1 eV relative shift of the edge to lower energy in the ferrocyanide complex. Also the separation in energy of peaks A and B is 0.5 eV less in the Fe^{II} complex. XANES calculations using the procedure outlined above show the following important points [12]. First, the origin of the striking features A and B/C is the strong multiple scattering between the C and N atoms, which produces effects closely related to the shape resonances observed in diatomic molecules [13]. This multiple scattering make the XANES sensitive (mainly through the separation between peaks A and B) to the C-N distance, a parameter difficult to extract from the EXAFS because of the shadowing of the N atoms in the approximately octahedral co-ordination geometry. Figure 2 shows XANES calculations in which the atomic positions reported in previous diffraction studies were used. Satisfactory agreement with experiment is obtained using the atomic geometry of FIGGIS et al.[14] for $Fe^{III}(CN)_6$ and of TAYLOR et al.[15] for $Fe^{II}(CN)_6$. The diffraction data of KIRIYAMA et al. [16] for $Fe^{II}(CN)_6$ suggest that the cluster is considerably distorted away from octahedral symmetry, and when their geometry is used in a XANES calculation gross changes in the near edge structure result. No such changes are observed in the experimental data, and so the distortions suggested in ref.[16] seem doubtful. These XANES calculations do not include any effects due to the different charge distribution in the two formal oxidation states of Fe, and therefore indicate that the changes in the XANES of $Fe^{II}(CN)_6$ and $Fe^{III}(CN)_6$ have most to do with geometrical changes.

The second example is considerably more complicated - the XANES of Fe in haemoglobin (Hb) and its derivatives. Some experimental near edge data [17] are shown in Fig.3, and five features, labelled A-E, can be seen, es-

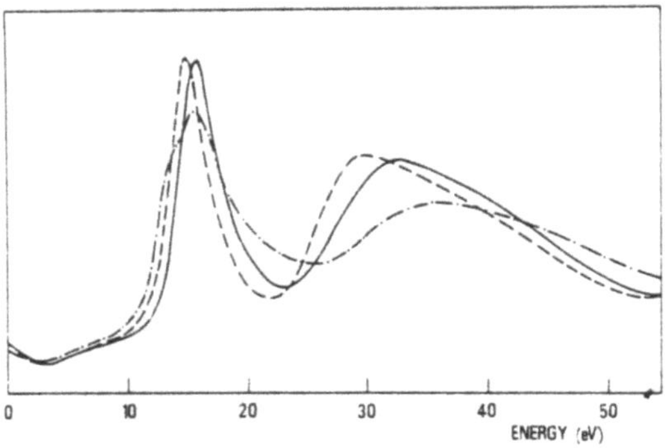

Fig.2. Calculated XANES for $Fe(CN)_6$ clusters using atomic co-ordinates taken from the diffraction studies of FIGGIS et al. [14] on $Fe^{III}(CN)_6$ (solid curve), TAYLOR et al. [15] on $Fe^{II}(CN)_6$ (dashed curve), and KIRIYAMA et al. [16] on $Fe^{II}(CN)_6$ (dash-dotted curve)

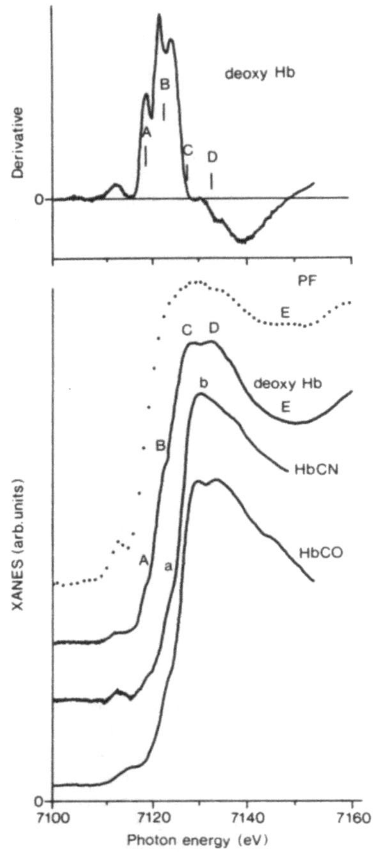

Fig.3. Experimental Fe K-edge XANES for the picket fence compound (PF), deoxy Hb, HbCN and HbCO [17]. The derivative spectrum is also shown for deoxy Hb

pecially in deoxy Hb and the picket fence compound, PF. The task of a XANES calculation is to be able to reproduce these features and to show how sensitive they are to the details of the bonding geometry of the Fe in its porphyrin environment. Preliminary calculations have now been done, and the details will be reported elsewhere [17]. Here I show in Fig.4 some calculations performed using clusters whose geometry is an idealised version of that in deoxy Hb or the very similar PF compound. The clusters vary in size from 1 shell (i.e. Fe + 5 nearest neighbour N atoms) to 3 shells (i.e. the full porphyrin system - see [17] for details), and it is clear that the XANES is strongly influenced by the outer parts of the porphyrin system, as is the EXAFS [18]. The calculations also predict a strong dependence of the XANES on the polarisation of the x-rays relative to the porphyrin plane; it would be interesting to try to observe this in experiments on single crystals. The "final" result, i.e. the 3-shell calculation averaged over polarisation, looks quite similar to the experimental data for the deoxy Hb, and in fact the relative peak positions (A-E) agree to within 1 eV. Also shown in Fig.4 is the effect of including a CN ligand coordinating vertically to the Fe (i.e. with the linear Fe-C-N bond normal to the porphyrin plane). The XANES changes in a way consistent with the experimental data for HbCN. At this stage, therefore, the results indicate

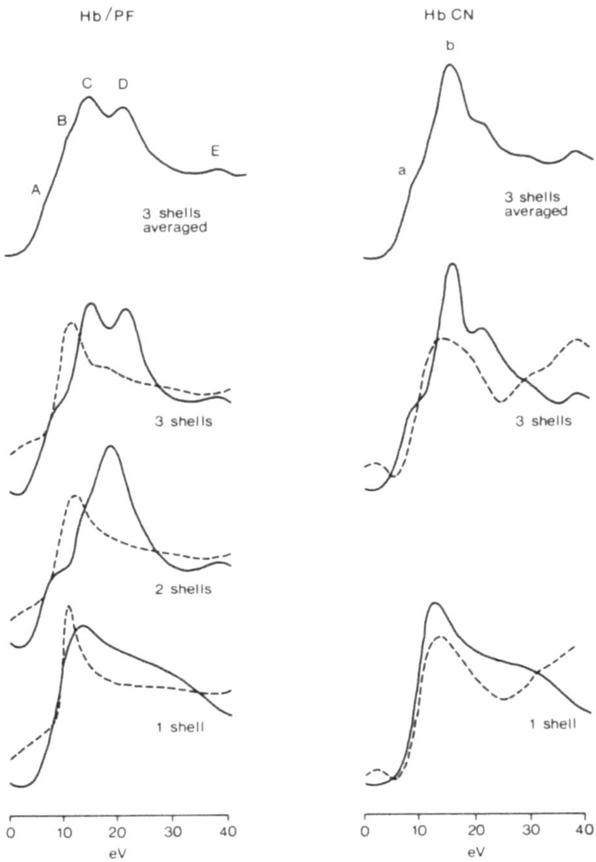

Fig.4. Calculated XANES for atomic geometries close to deoxy Hb and PF ("Hb/PF") and HbCN ("HbCN") [17]. Full curves: photon polarisation parallel to porphyrin plane. Dashed curves: polarisation normal to the plane. In the upper curves the polarisation is averaged

that it is feasible to attempt a quantitative analysis of the XANES in these complex systems, and the sensitivity to the bonding geometry around the Fe and to the charge distribution is currently being investigated.

References

1. A. Bianconi, Appl. Surf. Sci., 6, 392 (1980).

2. P.J. Durham, In: Extended X-ray Absorption Fine Structure. R. Prins and D. Koningsberger (Eds) New York: John Wiley (to be published).

3. B.L. Gyorffy and M.J. Stott, In: Band Structure Spectroscopy of Metals and Alloys. D.J. Fabian and L.M.Watson (Eds), London: Academic Press (1973).

4. P.J. Durham, J.B. Pendry and C.H. Hodges, Computer Phys. Commun., 25, 193 (1982).

5. L. Mattheiss, Phys. Rev., 134, A970 (1964).

6. Eg. G.N. Greaves, P.J. Durham, P. Quinn and G. Diakun, Nature, 294, 139 (1981).

7. A.R. Williams and U. von Barth, In: Theory of the Inhomogeneous Electron Gas. S. Lundqvist and N.H. March, (Eds), New York: Plenum (1982).

8. F.W. Kutzler, C.R. Natoli, D.K. Misemer, S. Doniach and K.O. Hodgson, J. Chem. Phys., 73, 3274 (1980).

9. J.E. Muller, O. Jepsen, O.K. Andersen and J.W. Wilkins, Phys. Rev. Lett., 40, 720, (1978).

10. U. von Barth, In: The Electronic Structure of Complex Systems. P. Phariseau and W. Temmerman (Eds), New York: Plenum (1983).

11. P. Thiry, D. Chandesris, J. Lecante, C. Guillot, R. Pinchaux and Y. Petroff, Phys. Rev. Lett., 43, 82 (1979)

12. A. Bianconi, M. Dell'Ariccia, P.J. Durham and J.B. Pendry, Phys. Rev. B. (in press).

13. J.L. Dehmer and D. Dill, J. Chem. Phys., 65, 5327 (1976).

14. B.N. Figgis, M. Gerloch and R. Mason, Proc. Roy. Soc., A309, 91, (1969.

15. J.C. Taylor, M.H. Mueller and R.L. Hitterman, Acta. Cryst., A26, 559, (1970)

16. R. Kiriyama, H. Kiriyama, T. Wada, N. Niizeki and H. Hiribayashi, J. Phys. Soc. Japan, 19, 540, (1964).

17. P.J. Durham, A. Bianconi, A. Congiu-Castellano, A. Giovannelli, S.S. Hasnain, L. Innoccia, S. Morante and J.B. Pendry, to be published.

18. M.F. Perutz, S.S. Hasnain, P.J. Duke, J.E. Hahn and J.L. Sessler, Nature, 295, 535, (1982).

Near Edge Absorption Structure in the Framework of the Multiple Scattering Model. Potential Resonance or Barrier Effects?

C.R. Natoli
Laboratori Nazionali di Frascati dell'INFN, I-00044 Frascati, Italy

1. Introduction

Near edge structures of X-ray absorption spectra from deep core levels, both in molecules and in extended media, have attracted, because of their prominence, the attention of the researchers in the field ever since they were experimentally observed.

Only recently [1, 2], however, has their interpretation met with some success. Indeed the theory of multiple scattering has provided the correct framework in which to interpret the experimental observation.

Most of the interpretation, however, has relied heavily on complicated computer programs with consequent lack of simple physical insight into the nature of the process under investigation.

A step toward a simple intuitive model for the interpretation of near edge absorption spectra, at least in molecules, has been taken by Nefedov and Dehmer [3, 4].

They introduced the concept of effective potential barrier (borrowed from the field of atomic physics), meaning a region of space where the potential is repulsive. This creates a sortof cage that traps the final state electron wave function, giving rise to a quasi-bound state decaying away with a lifetime connected with the tunneling probability through the barrier (Fig. 1).

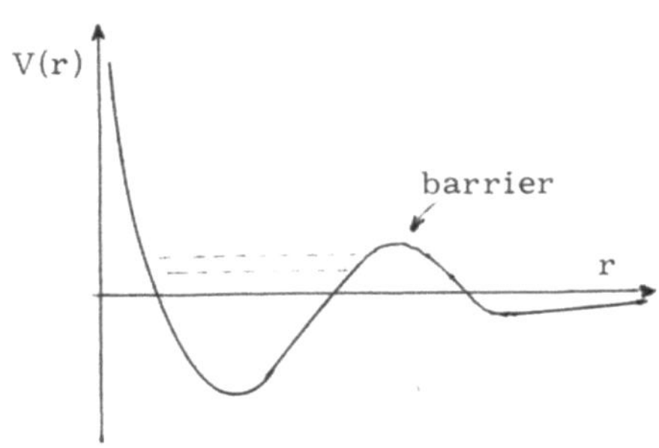

FIG. 1. The concept of effective potential barrier

In atomic physics the repulsive centrifugal potential provides the necessary barriers both around the origin and at some distance apart.

In molecules, Nefedov and Dehmer suggested that such a barrier could be provided by

(a) direct Coulomb repulsion due to an electron excess on the ligands;
(b) exchange interaction with the same electrons;
(c) repulsive forces derived from the Pauli exclusion principle;
(d) centrifugal effects.

However, interactions (a) and (b) are ruled out as a source of the potential barrier because in actual calculations (as in SF_6 or $GeCl_4$) the sum of the Coulomb potential (including nuclear attraction) and the exchange potential is everywhere negative.

Interaction (c) is a not too well-defined concept. It is due to the Pauli principle which prohibits the electron from moving freely in the space containing occupied molecular orbitals (the ligand region). The final state wave function must be orthogonal to all occupied orbitals; hence in the spirit of pseudopotential theory, Nefedov and Dehmer assume that one "may replace these orthogonality constraints by effective potential barriers in the heighborhood of the electronegative atoms in the molecule" [3].

However the concept itself of pseudopotential is not uniquely defined. It rests on the possibility of mocking up the relevant phase shifts of the true potential in the energy region of interest. But there are infinitely many pseudopotentials that can do the job, most of them everywhere negative.

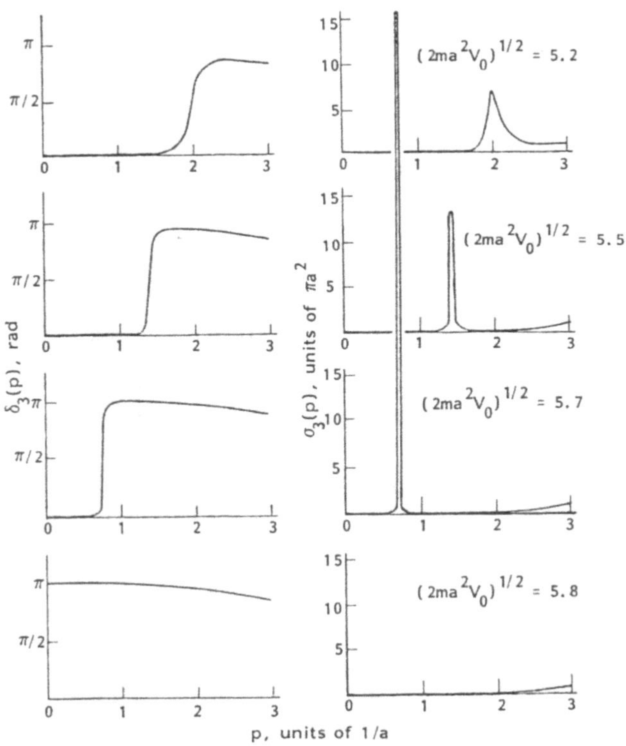

FIG. 2. The $l = 3$ phase shifts and cross sections for four successively deepar square wells. The top three wells are too shallow to support an $l = 3$ bound state - the fourth just can. Notice the $aV_0^{1/2}$, where a is the well radius and V_0 is the depth of the well, has varied only by 10% [5]

Centrifugal effects of point (d) are more real. They can actually provide a potential barrier whose height is $\sim l(l+1)/R^2$(Ryd), if R (measured in a. u.) is defined as the radius of the cluster beyond which the potential is negligible. In some molecules the concept seems to work, as is the case of the $l = 3$ resonances in N_2, C_2H_2 and in general diatomic molecules.

However sometimes one find resonances even for low l values ($l = 1, 2$) with bigger R (cfr. the case of $GeCl_4$ [2]), where the centrifugal barrier is decidedly insufficient to sustain a quasi bound state. Moreover, how to define R in the case of extended media (Cu, Ni crystals for example) where the same pronounced features in absorption spectra are observed?

Also there are instances where some other piece of information is needed as shown by the $l = 3$ resonance of a square potential well having about the necessary strenght to bind an $l = 3$ state (Fig. 2). When the strength is just enough to bind the state the resonance disappears, although the centrifugal barrier has not varied very much.

Clearly the concept of centrifugal barrier is insufficient to describe the variety of phenomena under investigation.

2. The language of potential scattering theory

Based on the above considerations, we feel the need for a more general framework capable of describing all situations from molecules to extended media, more mathematically founded and having more predicting power.

This framework is provided by the theory of potential scattering. Indeed it turns out that all concepts valid in the case of a single scattering center with spherical symmetry can be extended to the case of a collection of scattering centers (as are atoms in molecules or solids) through the theory of multiple scattering.

The key concept in potential scattering is the correspondence between bound states and resonances [5]. Resonances are quasi stationary states that would become bound if the strength of the potential could be varied to make it more attractive (Fig. 3)

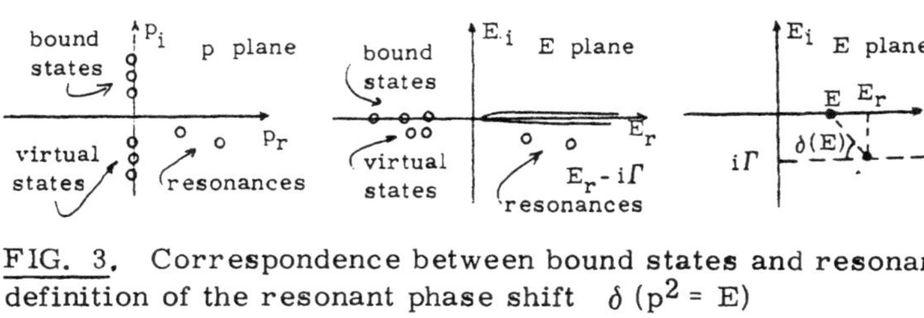

FIG. 3. Correspondence between bound states and resonances and definition of the resonant phase shift δ ($p^2 = E$)

The association of resonances with poles of the scattering matrix is not exact. Poles should be near the real axis so that the corresponding resonant phase shift varies rapidly through $\pi/2$. Also the background phase shift and the phase shift relative to other l-values should be small for the resonance to be well prominent.

Moreover there are examples of potentials exhibiting all the features of a resonance without a correspond pole in the scattering matrix. However in the case of atomic potentials (of Yukawa or exponential type) this latter behavior should not occur.

Nevertheless we shall assume a correspondence between at least the maxima of the scattering or absorption cross sections with the complex poles of the scattering matrix as a working hypothesis, leaving the exceptions to a more detailed investigation case by case, keeping in mind that the theory of potential scattering can also handle these pathological situations.

The properties of single scattering potential with spherical symmetry of interest for our future considerations are summarized in the following relations, where σ_t is the total scattering cross section:

$$\sigma \sim \sum_l \sigma_l(p) \qquad \sigma_l(p) = 4\pi(2l+1)\left|t_l(p)\right|^2 = 4\pi(2l+1)\,\frac{\sin^2\delta_l(p)}{p^2} \tag{2.1}$$

where

$$t_l(p) = \frac{e^{i2\delta_l(p)} - 1}{2ip} = e^{i\delta_l(p)}\,\frac{\sin\delta_l(p)}{p} \tag{2.2}$$

is the t matrix in the momentum space representation, diagonal in the angular momentum l and $\delta_l(p)$ are the phase shifts.

The t matrix can be also written in terms of the reactance matrix

$$k_l(p) = \frac{1}{p}\,\mathrm{tg}\,\delta_l(p) \tag{2.3}$$

as

$$t_l(p) = \frac{k_l(p)}{1 - ik_l(p)}\ . \tag{2.4}$$

The maxima of $|t_l(p)|$ coincide with the singularities of the reactance matrix $k_l(p)$. Near a singularity ($\delta_l \sim \pi/2$) $k_l(p) \sim (1/p)/(\Gamma/E-E_r)$ so that

$$t_l(p) = \frac{1}{p}\,\frac{\Gamma}{E - E_r + i\Gamma} \tag{2.5}$$

that is, to a singularity of $k_l(p)$, which occur for real p and $E = p^2$, there corresponds a pole of the t matrix in the fourth quadrant of the complex E plane. If the pole is sufficiently near the real axis ($\Gamma/E \ll 1$), the variation of the $\sigma_l(p)$ contribution is very rapid though its maximum and if all the other $\sigma_{l'}(p)$ are smooth functions of E, a sharp feature shows up in $\sigma = \sum_l \sigma_l \sim \frac{1}{p^2}\,\Gamma^2/\left[(E-E_r)^2 + \Gamma^2\right]$ + background.

Our task is then to define and calculate the corresponding quantities in the case of a non-spherically symmetric potential.

In such a case we define the $\underset{\sim}{S}$, $\underset{\sim}{T}$ and $\underset{\sim}{K}$ matrices through the asymptotic behavior of the wave function $\Psi(\vec{r}_o)$ referred to a given center (origin)

$$\Psi_L^S(\vec{r}_o) \rightarrow A_L\,\frac{1}{2}\left[4\pi i^l h_l^-(pr_o)Y_L(\hat{r}_o) + \sum_{L'} 4\pi i^{l'} h_{l'}^+(pr_o)Y_{L'}(\hat{r}_o)S_{L'L}\right] \tag{2.6a}$$

$$\varphi^{T}_{L}(\vec{r}_o) \rightarrow A_L \left[4\pi i^{l}\, j_l(pr_o) Y_L(\hat{r}_o) + ip \sum_{L'} 4\pi i^{l'} h^{+}_{l'}(pr_o) Y_{L'}(\hat{r}_o) T_{L'L}\right] \qquad (2.6b)$$

$$\varphi^{K}_{L}(\vec{r}_o) \rightarrow A_L \left[4\pi i^{l}\, j_l(pr_o) Y_L(\hat{r}_o) - p \sum_{L'} 4\pi i^{l'} h_{l'}(pr_o) Y_{L'}(\hat{r}_o) K_{L'L}\right] \qquad (2.6c)$$

where $h^{\pm}_{l}$, j_l, n_l are the usual Hankel, Bessel and Neuman functions.

These wave functions are all solution of the some Schrödinger or equivalent Lippman-Schwinger equation with different asymptotic normalization. Hence they are related by an amplitude matrix transform (e. g. $\varphi^{T}_{L} = \sum_{L'} (\underset{\sim}{I} - ip\underset{\sim}{K})^{-1}_{LL'}\, \varphi^{K}_{L'}$) whereas the matrices are related by

$$\underset{\sim}{S} = \underset{\sim}{I} + 2ip\underset{\sim}{T}\,, \quad \underset{\sim}{T} = \underset{\sim}{K}(\underset{\sim}{I} - ip\underset{\sim}{K})^{-1} \Rightarrow T_{LL'} = \sum_{L''} K_{LL''}(\underset{\sim}{I} - ip\underset{\sim}{K})^{-1}_{L''L'}\,. \qquad (2.7)$$

The total cross section is given in this case by

$$\sigma \sim 4\pi \sum_{LL'} \left|T_{LL'}\right|^2 .$$

We need a scheme to calculate actually either $\underset{\sim}{T}$ or $\underset{\sim}{K}$.

In the following we shall be concerned only with scattering from a static potential which may also include an absorptive imaginary part (Hedin-Lundquist type of potential [6]). In this latter case (complex potential) it is clear how to extend the preceeding considerations.

Many-body effects, as well as dynamically varying potentials due to relaxation effects interfering with the escaping electron should be introduced wherever necessary. Fortunately photoabsorption processes from very deep core states can be satisfactorily described in terms of a static potential. See G. Wendin's paper in these Proceedings.

3. The multiple scattering model

When there are many scatterers (as in any aggregate of atoms, whether extended or limited in size) we can partition the space into disjoint closed regions around the centers of the single scatterers (atoms), where the potential is strong, meaning that $|\vec{\nabla}V|/|V| \gtrsim 1$. In the region outside the potential is slowly varying $|\vec{\nabla}V|/|V| << 1$ so that a WKB approximation can be used.

If we can reasonably satisfy the condition

$$\int_{r}^{r'} \left[E - V(r(t))\right]^{1/2} dr(t) \sim \left[E - \bar{V}\right]^{1/2} (r - r') \qquad (3.1)$$

for all straight paths in the interstitial region, we can approximate the motion by free propagation, taking advantage of some reexpansion theorems valid for free motion which will be used later on.

The disjoint closed regions can instead be characterized by their phase shifts as scattering centers, without worrying about the interior solution of the Schrödinger equation, which is entirely determined by the phase shifts (boundary condition problem for second-order elliptic partial differential equations).

The model becomes therefore a multiple scattering model or phase shift model for several centers, with free propagation in the interstitial region and phase shift scattering from the various centers. Eq. (3.1) can be used as a "physical" definition of the "muffin-tin" constant $\bar{V}$.

Now, in the single scattering case, the Lippman-Schwinger equation $\Psi = \Phi_o + GV\Psi$ can be read as describing the process of an excitation by a plane wave (solution of the free equation) followed by a response, i. e. scattered wave that defines the t matrix through Eq. (2.6b) with $T_{LL'} = \delta_{LL'} t_l$.

In close analogy, in the many-center case (see ref. [7] for more details) the total wave function in the interstitial region can be written as the sum of an incident wave (regular when continued to the center) referred to a given origin $\vec{o}$, with wave vector $x = [E - \bar{V}]^{1/2}$

$$\Psi_{inc}(\vec{r}_o) = \sum_L A_L(\vec{o})\, 4\pi i^l j_l(xr_o)\, Y_L(\hat{r}_o) \tag{3.2}$$

defining the excitation amplitudes $A_L(\vec{o})$, plus scattered or radiated waves (irregular at the center), each one referred to a given center $\vec{n}$ ($\vec{r}_n = \vec{r} - \vec{n}$)

$$\Psi_{scat}(\vec{r}) = ix \sum_{\vec{n}} \sum_{LL'} 4\pi i^l h_l^+(xr_n)\, Y_L(\hat{r}_n)\, t_{LL'}(\vec{n})\, A_{L'}(\vec{n}) \tag{3.3}$$

defining the excitation amplitudes relative to the various centers $\vec{n}$.

The equation satisfied by the amplitudes $A_L(\vec{n})$ can be easily found if one uses the following reexpansion theorem.

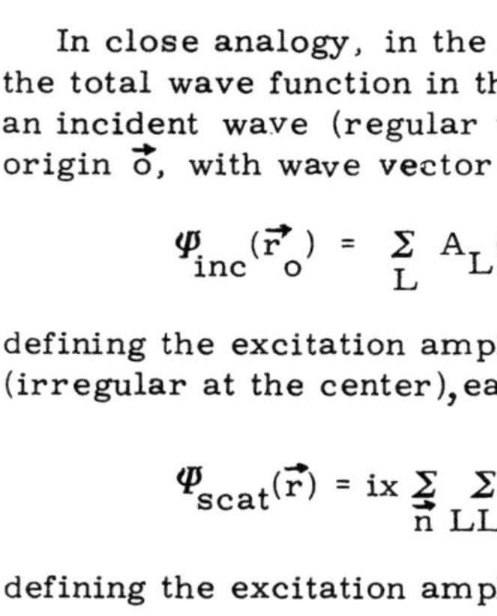

FIG. 4. Definition of the various vectors intervening in the reexpansion theorem

With reference to Fig. 4, the theorem reads :

$$ix\, 4\pi i^l h_l^+(xr_n) Y_L(\hat{r}_n) = \sum_{L'} 4\pi i^{l'} j_{l'}(x\, r_{n'}) Y_{L'}(\hat{r}_{n'}) G_{L'L}(\vec{n}' - \vec{n}) \tag{3.4a}$$

$$\text{if} \quad |\vec{n} - \vec{n}'| > |\vec{r}_n|$$

$$= ix \sum_{L'} 4\pi i^{l'} h_{l'}^+(xr_{n'}) Y_{L'}(\hat{r}_{n'})\, \Delta_{L'L}(\vec{n}' - \vec{n}) \tag{3.4b}$$

$$\text{if} \quad |\vec{r}_n| > |\vec{n}-\vec{n}'|$$

where

$$\Delta_{L'L}(R) = \sum_{L_1} C_{L'L}^{\;L_1}\, 4\pi i^{l_1} j_{l_1}(xR)\, Y_{L_1}(\hat{R}) , \tag{3.5a}$$

$$G_{L'L}(R) = ix \sum_{L_1} C_{L'L}^{\;L_1}\, 4\pi i^{l_1} h^+_{l_1}(xR)\, Y_{L_1}(\hat{R}) , \tag{3.5b}$$

$$C_{L'L}^{\;L_1} = \int Y_{L'}(\Omega)\, Y^*_L(\Omega)\, Y_{L_1}(\Omega)\, d\Omega \quad . \tag{3.5c}$$

Indeed, using Eqs. (3.2), (3.3) and (3.4a) we find that the excitation amplitude $A_L(\vec{n})$ at site $\vec{n}$ due to scattered waves which resulted from an excitation amplitude $A_L(\vec{n}')$ on site $\vec{n}'$ is given by

$$A_L(\vec{n}) = \sum_{L'L''} G_{LL'}(\vec{n}-\vec{n}')\, t_{L'L''}(\vec{n}')\, A_{L''}(\vec{n}') \, . \tag{3.6}$$

This shows that if we have an incident excitation amplitude $A^o_L(n)$ on site $\vec{n}$ with scattering factor $t_{LL'}(\vec{n})$, then the actual excitation amplitudes due to both the incident wave and the scattered waves from all other sites satisfy the equation

$$A_L(\vec{n}) = A^o_L(\vec{n}) + \sum_{\vec{n}\neq\vec{n}'} \sum_{L'L''} G_{LL'}(\vec{n}-\vec{n}')\, t_{L'L''}(\vec{n}')\, A_{L''}(\vec{n}') \, . \tag{3.7}$$

Since we must refer the outgoing wave to a given origin $\vec{o}$, we must transform the incident wave at $\vec{o}$ given by (3.2) into an incident wave on the scattering site n. The relation is

$$\Psi_{scat}(\vec{r}_n) = \sum_{LL'} 4\pi i^{l'} j_{l'}(xr_n)\, Y_{L'}(\hat{r}_n)\, \Delta_{L'L}(\vec{n}-\vec{o})\, A_L(\vec{o}) \tag{3.8}$$

and is obtained by taking the imaginary part of the Eq. (3.4b), remembering that $h^{\pm}_l = j_l \pm i n_l$.

Hence the incident excitation amplitude on site $\vec{n}$ is simply $\Delta_{L'L}(\vec{n}-\vec{o})A_L(\vec{o})$. Eq. (3.7) must then be rewritten as

$$A_L(\vec{n}) = \Delta_{L\underline{L}}(\vec{n}-\vec{o})\, A_{\underline{L}}(\vec{o}) + \sum_{n'\neq n} \sum_{L'L''} G_{LL'}(\vec{n}-\vec{n}')\, t_{L'L''}(\vec{n}')\, A_{L''}(\vec{n}') \; . \tag{3.7a}$$

In terms of the quantities $B_L(\vec{n}) = \sum_{L'} t_{LL'}(\vec{n}) A_{L'}(\vec{n})$ Eq. (3.7a) can also be written as

$$B^{\underline{L}}_L(\vec{n}) = \sum_{L'} t_{LL'}(\vec{n})\, \Delta_{L'\underline{L}}(\vec{n}-\vec{o})\, A_{\underline{L}}(\vec{o}) + \sum_{\vec{n}'\neq\vec{n}} \sum_{L'L''} t_{LL'}(\vec{n})\, G_{L'L''}(\vec{n}-\vec{n}')\, B^{\underline{L}}_{L}(\vec{n}') \tag{3.7b}$$

where the extra index $\underline{L}$ in $B_L^{\underline{L}}(n)$ comes from choosing a particular excitation amplitude $A_{\underline{L}}(o)$ in the inhomogeneous term.

By defining the following matrices

$$\underset{\sim}{G} \equiv G_{LL'}(\vec{n}-\vec{n}') = G_{LL}^{\vec{n}\vec{n}'} \ , \qquad \underset{\sim}{T}_a = T_{LL'}^{\vec{n}\vec{n}'} = t_{LL'}(\vec{n})\, \delta_{\vec{n}\vec{n}'}$$

and the vectors $B(\underline{L}) \equiv B_L^{\vec{n}}(\underline{L}) = B_L^{\underline{L}}(\vec{n})$, and $\Delta(\underline{L}) \equiv \Delta_L^{\vec{n}}(\underline{L}) = \Delta_{L\underline{L}}(\vec{n}-\vec{o})$, one for each value of L we can write (3.7b) in the compact form

$$B(\underline{L}) = \underset{\sim}{T}_a\, \Delta(\underline{L}) + \underset{\sim}{T}_a \underset{\sim}{G}\, B(\underline{L}) \ , \tag{3.7c}$$

the formal solution of which is

$$B(\underline{L}) = \left[I - \underset{\sim}{T}_a \underset{\sim}{G}\right]^{-1} \underset{\sim}{T}_a\, \Delta(\underline{L}) = \left[\underset{\sim}{T}_a^{-1} - \underset{\sim}{G}\right]^{-1} \Delta(\underline{L}) \ . \tag{3.9}$$

One can also show that the full T matrix of the cluster $T_{\underline{L}\underline{L}'}$ is given by

$$T_{\underline{L}\underline{L}'} = \Delta(\underline{L}) \left[\underset{\sim}{T}_a^{-1} - \underset{\sim}{G}\right]^{-1} \Delta(\underline{L}') = \Delta(\underline{L}) B(\underline{L}') \ . \tag{3.10}$$

From Eq. (3.3) and the definition of $B_L^{\underline{L}}(\vec{n})$ one easily finds the cross section for photoabsorption of a core electron at site $\vec{n}_c$ to be

$$\sigma \propto \sum_{LL_f} \left| B_{L_f}^{L}(\vec{n}_c) \right|^2 \left| \langle \varphi_{L_f}(\vec{n}_c) | \vec{r} | \Phi_{in} \rangle \right|^2 \tag{3.11}$$

(where $\varphi_{L_f}(\vec{n}_c)$ is the L_f component of the final state w.f. centered on the atom at site $\vec{n}_c$)

which can be shown to be $\sim (d/dE)\ \varrho_{L_f}(r_{\vec{n}_c}, E)$, that is the energy differential of the charge density projected on site $\vec{n}_c$ and angular momentum L_f. In practice all the structural information around the photoionized atom at site $\vec{n}_c$ is carried through the amplitude $B_L(\vec{n}_c)$. The other factor is simply an atomic matrix element giving the broad background of the absorption spectrum.

For electron scattering off the multicenter cluster we find as anticipated

$$\sigma \propto \sum_{LL'} \left| T_{LL'} \right|^2 \ . \tag{3.12}$$

In both cases the singularities of $\left[\underset{\sim}{T}_a^{-1} - \underset{\sim}{G}\right]^{-1}$ in the complex E plane provide the sharp features of the spectra, which are superimposed to the atomic background.

Now a practical way to calculate the $\underset{\sim}{T}$ matrix is through the calculation of the $\underset{\sim}{K}$ matrix and use of the relation (2.7). The advantage is that $\underset{\sim}{K}$ is an hermitian matrix, actually real symmetric if one uses a real basis for spherical harmonics. Moreover the singularities of K for real E are connected to the singularities of $\underset{\sim}{T}$ for complex E (cfr. Eq. (2.5)).

The equation for the amplitudes in $\underset{\sim}{K}$-matrix normalization are exactly the same as those for the amplitude in $\underset{\sim}{T}$-matrix normalization except for the replacement

$$\underset{\sim}{T}_a \to \underset{\sim}{K}_a \equiv K_{aLL'}(n)\delta_{\vec{n}\vec{n}'}\ , \qquad \text{and} \tag{3.13a}$$

$$\underset{\sim}{G} \to \underset{\sim}{G}^o \equiv G^o_{LL'}(\vec{n}-\vec{n}') = x \sum_{L_1} C_{LL'}^{L_1}\, 4\pi i^{l_1} n_{l_1}(x|\vec{n}-\vec{n}'|)\, Y_{L_1}(\widehat{\vec{n}-\vec{n}'}). \tag{3.13b}$$

Hence the solution is, in analogy with (3.10)

$$\underset{\sim}{K} \equiv K_{\underline{L}\underline{L}'} = \Delta(\underline{L})\left[\underset{\sim}{K}_a^{-1} - \underset{\sim}{G}^o\right]^{-1}\Delta(\underline{L}')\ . \tag{3.14}$$

The relation (2.7)

$$\underset{\sim}{T} = \underset{\sim}{K}\left[\underset{\sim}{I} - ix\underset{\sim}{K}\right]^{-1}$$

between the $\underset{\sim}{T}$ matrix and the $\underset{\sim}{K}$ matrix provides a way to associate the singularities of $\underset{\sim}{K}$, with the poles of $\underset{\sim}{T}$, in analogy with the single scattering case. To see this it is sufficient to diagonalize the K matrix. Since this letter is obtained by a matrix inversion ($\underset{\sim}{K} \sim \underset{\sim}{M}^{-1} = \left[\underset{\sim}{K}_a^{-1} - \underset{\sim}{G}^o\right]^{-1}$), $\underset{\sim}{K}$ is singular wherever

$$\text{Det}\,\underset{\sim}{M} = \text{Det}\left|\underset{\sim}{K}_a^{-1} - \underset{\sim}{G}^o\right| = \text{Det}\left|K^{-1}_{aLL'}(\vec{n})\,\delta_{\vec{n}\vec{n}'} - G^o_{LL'}(\vec{n}-\vec{n}')\right| = 0\ . \tag{3.15}$$

Notice that with the definitions (3.13) the multiple scattering matrix is real symmetric if one uses real spherical harmonics and multiplies the rows by i^l and the columns by $i^{l'}$.

For spherically symmetric single scatterers Eq. (3.15) simplifies to

$$\text{Det}\,\underset{\sim}{M} = \left|\text{Det}\left|x\,\text{cotg}\,\delta_l^{\vec{n}}\ \delta_{ll'}\ \delta_{\vec{n}\vec{n}'} - G^o_{LL'}(\vec{n}-\vec{n}')\right|\right| = 0\ . \tag{3.16}$$

4. Some consequences and conclusions

Remembering our assumption about the correspondence between maxima in the absorption or scattering cross section and poles of the $\underset{\sim}{T}$ matrix, Eq. (3.16) can therefore be interpreted as relating the wave vector of the maxima $x_r = \left[E_r - \overline{V}\right]^{1/2}$ to the distances $\vec{n} - \vec{n}' = \vec{R}_{\vec{n}\vec{n}'}$ between the scattering centers.

Because of (3.13b), (3.16) is an equation of the form

$$\mathcal{F}\left[x_r \vec{R}_{\vec{n}\vec{n}'}\ ,\ \delta_l^{\vec{n}}(x_r)\right] = 0 \qquad . \tag{4.1}$$

From (4.1) we can immediately read off some consequences :

a) If we have two systems of scatterers (e. g. Cu and Ni) situated at lattice points differing only by an overall scaling factor (different lattice parameters), with quite similar phase shifts, smooth functions of x, then the maxima of absorption x_r in the two cases scale with the inverse factor, since

$$\mathscr{F}\left[x\vec{R}_{\vec{n}\vec{n}'},\ \delta_1^{\vec{n}}(x)\right] \simeq \mathscr{F}\left[\frac{x}{\alpha}\ \alpha\,\vec{R}_{\vec{n}\vec{n}'},\ \delta_1^{\vec{n}}\left(\frac{x}{\alpha}\right)\right] = 0 \quad .$$

Calculations have been performed by <u>Müller</u> et al. [8] on X-ray absorption spectra from K edges of 3d transition metals, L edges of 3d and 4d metals, and M edges of Palladium. They found that the spectra of the elements with the same crystal structure show a striking similarity and the maxima line up when plotted versus an energy scale proportional to the inverse of the lattice constant squared. A thorough investigation of the experimental spectra is needed to assess the validity of this theoretical conclusion.

b) Differentiating (4.1) with respect to $\vec{R}_{\vec{n}}$ gives the relation

$$\left[\frac{1}{2}\sum_{\vec{n}\vec{n}'}\frac{\partial\mathscr{F}}{\partial R_{\vec{n}\vec{n}'}}R_{\vec{n}\vec{n}'} - \sum_{\vec{n}l}\frac{\partial\mathscr{F}}{\partial\delta_1^{\vec{n}}}\frac{d\delta_1^{\vec{n}}}{dx}\right]\frac{dx_r}{d\vec{R}_{\vec{n}}} = -\sum_{\vec{n}'}\frac{\partial\mathscr{F}}{\partial R_{\vec{n}\vec{n}'}}\frac{\vec{R}_{\vec{n}\vec{n}'}}{R_{\vec{n}\vec{n}'}}x_r \tag{4.2}$$

linking the variation of given maximum at x_r to the variation of $R_{\vec{n}}$.

c) Specialization of Eq. (4.1) to the case where there is only one distance and putting in an electron damping for completeness gives

$$\mathscr{F}\left[x_r R,\ R,\ \delta(x_r)\right] \equiv \mathscr{F}\left[x,\ y,\ z\right] = 0 \quad .$$

Differentiation with respect to R, assuming $x_r = x_r(R)$, gives ($\mathscr{F}_x = \frac{\partial\mathscr{F}}{\partial x}$)

$$-\frac{\Delta x_r}{x_r} = \frac{\Delta R}{R}\frac{\left[1 + x_r^{-1}\mathscr{F}_y/\mathscr{F}_x\right]}{\left[1 + R^{-1}\mathscr{F}_z/\mathscr{F}_x\frac{d\delta}{x_r}\right]} \tag{4.3}$$

which is a special case of (4.2) when $\mathscr{F}_y = 0$.

There are two cases to be discussed:

c_1) In a region of energy when $d\delta/dx \gg 1$ (atomic resonance) there is little dependence, to a first approximation, of the position of the maximum on the distance R. This is also true in the general case (4.2). This behaviour is reminiscent of the behaviour of 3d bands in transition metals. The KKR equation determining E as a function of k (E(k)) for 3d states is similar to Eq. (3.16), namely

$$\operatorname{Det}\underset{\sim}{M}' = \operatorname{Det}\left|\,x(\operatorname{cotg}\delta_1^{\vec{n}} - i)\ \delta_{\vec{n}\vec{n}'}\delta_{11'} - G_{LL'}(k)\right| = 0$$

where $G_{LL'}(k) = \sum_{\vec{R}_{\vec{n}\vec{n}'}} e^{i\vec{k}\vec{R}_{\vec{n}\vec{n}'}}G_{LL'}(\vec{R}_{\vec{n}\vec{n}'})$ and the atomic phase shifts $\delta_{l=2}^{\vec{n}}$ are resonating. Hence the solution E as a function of k, E(k), shows little dependence on the structure factors G(k), and hence on k, so that 3d bands are mostly flat.

Transition from L edges to empty d states just above the Fermi level in rare-earth metals or to empty antibonding d states in oxides or clorides of the same elements also provide a striking confirmation of this behaviour.

c_2) When $d\delta/dx \ll 1$, then Eq. (4.3) implies (if $\mathscr{F}_y/\mathscr{F}_x \sim 1$, which is usually the case)

$$-\frac{\Delta x_r}{x_r} = \frac{\Delta R}{R} \qquad \text{or} \qquad x_r R = \text{const.}$$

It is tempting to interpret this relation as arising from the constraints imposed to the electron wavefunction by the singularities of the potential. This latter becomes infinite of the nuclear sites, where the wave function must vanish or have a cusp.

Since the electron moves in a dispersive medium, the constant will in general be different from the free electron case with the same boundary conditions.

Such a relation could be used, with some caveat , to determine bond lengths of adsorbates on surfaces, if the same bond distance is known for the free molecule.

For a more detailed discussion on this point, the reader is referred to the papers by the author et al. in this Section of the Proceedings.

All the previous considerations presuppose that the condition Det $\underset{\sim}{M} = 0$ can actually be satisfied. However if all phase shift are small, as in the limit $x \to \infty$, the condition can never be satisfied. Indeed in such a case we have

$$\underset{\sim}{M} = \underset{\sim}{\Lambda} - \underset{\sim}{G}^o$$

where Λ is a diagonal matrix with fairly large elements $\Lambda_{nn} \sim \text{cotg}\,\delta_1^n$ and G^o contains only off-diagonal elements at most of the order of 10 or less.

If, as is plausible under these conditions,

$$\|\underset{\sim}{\Lambda}^{-1}\underset{\sim}{G}^o\| < 1$$

where $\|\underset{\sim}{A}\|$ denotes the maximum eigenvalue of the matrix $\underset{\sim}{A}$, then the theorems on linear operators tell us that

$$\underset{\sim}{M}^{-1} = \left[\underset{\sim}{\Lambda}(\underset{\sim}{I} - \underset{\sim}{\Lambda}^{-1}\underset{\sim}{G}^o)\right]^{-1} = (\underset{\sim}{I} - \underset{\sim}{\Lambda}^{-1}\underset{\sim}{G}^o)^{-1}\underset{\sim}{\Lambda}^{-1} = \sum_{n=0}^{\infty}(\underset{\sim}{\Lambda}^{-1}\underset{\sim}{G}^o)^n\underset{\sim}{\Lambda}^{-1}$$

which is equivalent to say that the condition Det $\underset{\sim}{M} = 0$ cannot be satisfied. In these circumstances, we also have $\|\underset{\sim}{K}\| = \|\Delta\underset{\sim}{M}^{-1}\Delta\| < 1$ so that

$$\underset{\sim}{T} = \underset{\sim}{K}\left[\underset{\sim}{I} - ix\underset{\sim}{K}\right]^{-1} = \sum_{n=0}^{\infty}(ix\underset{\sim}{K})^n\underset{\sim}{K}\,.$$

In practice it is better to start from the expression (3.10) and derive

$$\underset{\sim}{T} = \Delta \underset{\sim}{T}_a \left[\underset{\sim}{I} - \underset{\sim}{G}\underset{\sim}{T}_a\right]^{-1}\Delta = \Delta \underset{\sim}{T}_a \sum_{n=0}^{\infty} (\underset{\sim}{G}\,\underset{\sim}{T}_a)^n \Delta = \tag{4.4}$$

$$= \Delta \underset{\sim}{T}_a \Delta + \Delta \underset{\sim}{T}_a \underset{\sim}{G}\underset{\sim}{T}_a \Delta + \Delta \underset{\sim}{T}_a \underset{\sim}{G}\underset{\sim}{T}_a \underset{\sim}{G}\underset{\sim}{T}_a \Delta + \cdots\cdots$$

(EXAFS)

since again $\| \underset{\sim}{G}\underset{\sim}{T}_a \| < 1$. We have also indicated the terms that contribute to the EXAFS signal in the expansion for the $\underset{\sim}{T}$ matrix. It is interesting to notice the presence of an extra term $\Delta \underset{\sim}{T}_a \underset{\sim}{G}\underset{\sim}{T}_a \Delta$ of the same order of magnitude as the EXAFS term , but not propagating, also found by Lee [9]. We are carefully investigating its meaning, unless some sort of cancellation occurrs as suggested by Lee in his paper, although without proof.

Summarizing, in a photoabsorption spectrum three regions of energy ($\bar{E} = E - \bar{V}$) can be distinguished:

a) $\bar{E} \lesssim 1$ Ryd. Resonant atomic phase shifts may be present. The resonances in the spectra can be described in terms of molecular orbitals by analytic continuation. The position of the maxima are fairly insensitive to the relative distances of the constituents atoms.

b) 1 Ryd $\leq \bar{E} \leq$ 3-12 Ryd. Some phase shifts, although not resonating, may still cross $\pi/2$. The Born series (4.4) for the amplitudes $B(\underline{L})$ or for the $\underset{\sim}{T}$ matrix is not convergent. One can speak about infinite multiple scattering process. Some sort of renormalized perturbation theory must be devised. The energies at which maxima occur scale with inverse squared distances of atomic constituents.

c) $\bar{E} \gtrsim 12$ Ryd. The condition Det $\underset{\sim}{M} = 0$ cannot be satisfied. In this case the Born series (4.4) converges. How many terms to retain in the series (whether single or finite multiple scattering contributions) depends on the size of $\| \underset{\sim}{G}\underset{\sim}{T}_a \|$. More exactly, $\| \underset{\sim}{G}\underset{\sim}{T}_a \| \ll 1$ for single scattering approximation to be valid (EXAFS). Again the maxima scale with inverse distances squared in energy scale.

Finally, we remark that in the case where the condition $\| \underset{\sim}{G}\underset{\sim}{T}_a \| < 1$ cannot be satisfied, it is still possible to set up a sort of renormalized perturbation theory by eliminating the "pathological" phase shifts from the problem through a submatrix inversion. Indeed by interchanging rows and columns in the multiple scattering matrix $\underset{\sim}{M}$ we can isolate a diagonal submatrix, corresponding to the large ($\sim \pi/2$) phase shifts, which is responsible for the fact that $\| \underset{\sim}{G}T_a \| > 1$. Hence the amplitude equations can be formulated as follow

$$\left[\begin{array}{c|c} M_s & M_r \\ \hline M_\ell & M_p \end{array}\right] \left[\begin{array}{c} B_s \\ \hline B_p \end{array}\right] = \left[\begin{array}{c} \Delta_s \\ \hline \Delta_p \end{array}\right] \Rightarrow \begin{array}{ll} \underset{\sim}{M}_s B_s + \underset{\sim}{M}_r B_p = \Delta_s & (4.5a) \\ \underset{\sim}{M}_\ell B_s + \underset{\sim}{M}_p B_p = \Delta_p & (4.5b) \end{array}$$

where $M_s = I_s - G_s T a_s$ is the singular submatrix ($\|G_s T a_s\| > 1$) and $M_p = I_p - G_p T a_p$ can be inverted by perturbation theory ($\|G_p T a_p\| < 1$).

In such a case, from (4.5a) we derive

$$B_s = M_s^{-1}(\Delta_s - M_r B_p)$$

which inserted in (4.5b) gives

$$M_\ell M_s^{-1}(\Delta_s - M_r B_p) + M_p B_p = \Delta_p$$

or

$$(M_p - M_\ell M_s^{-1} M_r) B_p = \Delta_p - M_\ell M_s^{-1} \Delta_s$$

writing

$$M_p - M_\ell M_s^{-1} M_r = I - G_p T a_p - M_\ell M_s^{-1} M_r = I - R$$

it is very likely that the condition $\|R\| < 1$ is satisfied since

$$\|R\| = \|G_p T a_p - M_\ell M_s^{-1} M_r\| \leq \|G_p T a_p\| + \|M_\ell M_s^{-1} M_r\|$$

and M_ℓ has matrix elements proportional to the small quantity $\|T a_p\|$. Hence

$$B_p = \sum_{n=0}^{\infty} R^n (\Delta_p - M_\ell M_s^{-1} \Delta_s)$$

so that

$$B_s = M_s^{-1}\left[\Delta_s - M_r \sum_{n=0}^{\infty} R^n (\Delta_p - M_\ell M_s^{-1} \Delta_s)\right]$$

defining the renormalized perturbation expansion.

Acknowledgments.

It is a pleasure to acknowledge illuminating discussions on the subject with Prof. S. Doniach of the Dept. Appl. Phys. Stanford University, Stanford, Ca. in the framework of the longstanding collaboration persued in this field since 1976.

References.

1. J. L. Dehmer, Phys. Rev. Letters 35, 213 (1975); C. R. Natoli, D. K. Misemer, S. Doniach and F. W. Kutzler, Phys. Rev. A22, 1104 (1980); F. W. Kutzler, C. R. Natoli, D. K. Misemer, S. Doniach and K. O. Hodgson, J. Chem. Phys. 73, 3274 (1980).
2. P. J. Durham, J. B. Pendry and C. H. Hodges, Sol. St. Commun. 38, 159 (1981); Comp. Phys. Commun. 25, 193 (1982); A. Bianconi, M. Dell'Ariccia, P. J. Durham and J. B. Pendry, Phys. Rev. B (to be published). A.

A. Pavlychev, A. S. Vinogradov, T. M. Zimkina, Proceedings of the VII Int. Conf. on VUV Radiation Physics USA, II-95, II-97 (1980).
3. V. I. Nefedov, J. Struct. Chem. 11, 277 (1970); J. L. Dehmer, J. Chem. Phys. 56, 4496 (1972).
4. J. L. Dehmer, J. Chem. Phys. 65, 5327 (1976).
5. J. R. Taylor, Scattering Theory (Wiley and Sons, 1972).
6. L. Hedin and B. I. Lundquist, J. Phys. C4, 2064 (1971) and refs. therein.
7. P. Lloyd and P. V. Smith, Adv. in Phys. 21, 69 (1972).
8. J. E. Müller, O. Jepsen and J. W. Wilkins, Solid State Com. 42, 365 (1982).
9. P. A. Lee, Phys. Rev. B13, 5261 (1976).

Bond Length Determination Using XANES

A. Bianconi and M. Dell'Ariccia
Dipartimento di Fisica, Università di Roma, La Sapienza, P. Aldo Moro 2, I-00185 Roma, Italy

A. Gargano and C.R. Natoli
Laboratori Nazionali di Frascati dell'INFN, I-00044 Frascati, Italy

1. Introduction

The use of XANES (X-ray Absorption Near Edge Structures) as a tool for determining coordination geometry, symmetry of the unoccupied valence electronic states or effective charge on the absorbing atom has been suggested by several authors [1], [2], [3].

Also in the C K-shell absorption spectra of simple molecules (C_2H_2, C_2H_4) an empirical relation has been observed between the wave vector k_r of the only resonance in the continuum ($k_r = \sqrt{E_r}$, E_r being the photoelectron energy at which the maximum in the absorption cross section occurs) and the C-C distance R, of the type $k_r R$ = const, implicitly suggesting the use of this relation for bond length determination [2].

However, apart from relying on complex computer calculations of near edge structures or on this empirical relation, with no hints as to its limitation, no simple quantitative methods exist at the moment to relate the properties of unknown system under study to those of known system. Indeed up to now the kind of information extracted from the observation of near edge structures has been mainly of qualitative character.

2. Theoretical background

We are now in a position to state that the above empirical relation has a theoretical justification. Being aware of its limitations, we can confidently use it to extract information about bond length, complementary or even substitutional to that extracted by EXAFS techniques.

Indeed the multiple scattering theory approach to calculation of absorption cross sections requires that at a resonance, defined as a maximum of the absorption, the determinant of the multiple scattering matrix in K-matrix asymptotic normalization

$$M^{jj'}_{LL'} = K^{aj}_{LL'}(x)\,\delta_{jj'} - G^{ojj'}_{LL'}(x\vec{R}_{ij}) \qquad (1)$$

must vanish. We refer to the paper by one of us (C. R. Natoli [4]) in these Proceedings for a thorough discussion of this point and definitions of the symbols used. We only recall here that $x = [E - \bar{V}]^{1/2}$, where E is the photoelectron energy $h\nu$-I measured from the ionization threshold and $\bar{V}$ is some sort of "muffin-tin" constant interstitial potential defined so that the WKB phase

$$\int_{r}^{r'} \left[E - V(r(t))\right]^{1/2} \dot{r}(t)dt \sim \left[E - \bar{V}\right]^{1/2} (r - r') \tag{2}$$

along all straight paths in the interstitial region.

The vanishing of Det M at a resonance defines an implicit relation of the type

$$\mathscr{F}\left[x_r \vec{R}_{ij},\ \delta^j_\ell(x_r)\right] = 0 \tag{3}$$

between the reduced wave vector of the resonance $x_r = \left[E_r - \bar{V}\right]^{1/2}$ and the distances $\vec{R}_{ij} = \vec{R}_i - \vec{R}_j$ of the various atomic scatterers in the cluster with phase shifts $\delta^j_\ell(x)$.

A very useful and simple application of eq. (3) occurs when at a particular energy the zero of $\mathscr{F}$ is sensitive mainly to one particular distance R_{ij} (for example the first strong absorption maximum in any cluster is mainly related to the nearest neighbours distance from the absorbing atom) or when there is effectively only one distance in the cluster (diatomic molecules, octahedral or tetrahedral clusters, etc.).

In such a case eq. (3) reduces to

$$\mathscr{F}\left[x_r R,\ \delta^j_\ell(x_r)\right] = 0\ . \tag{4}$$

Now if we want to apply this relation to different systems for bond length determination, we must further assume that the atomic phase shifts $\delta^j_\ell(x)$ are transferable, in the sense that the functional dependence of δ^j_ℓ on x must be the same or very nearly the same in the different systems under study.

Moreover, if we stay away from the energy region where atomic resonances are localed (in practice the energy region below the first strong absorption maximum) the dependence of the phase shifts on energy is very smooth, so that we can neglect $\frac{d\delta}{dx}$ with respect to $R(\frac{1}{R}\frac{d\delta}{dx} \ll 1)$. In such a case eq. (4) reduces to [4]

$$\mathscr{F}(x_r R) = 0 \qquad \text{or} \qquad x_r R = \text{const.} \tag{5}$$

We notice that a relation of this type under the above assumptions was to be expected in the framework of the multiple scattering model, on the basis that the resonant behaviour reduces to a kinetic energy effect, with constraints imposed by the boundary conditions (zeros of the wave function at the singularities of the atomic potentials), associated with the free propagation of the photoelectron from one site to another.

We have tested eq. (5) on the linear molecule C_2H_2 by performing several C-K shell multiple scattering calculations (with incident light polarized along the C-C bond) with different values of the C-C distance, noting each time the position of the maximum of the absorption cross section. Fig. 1 shows some of the spectra so obtained. The results are summarized in Table I and Fig. 2.

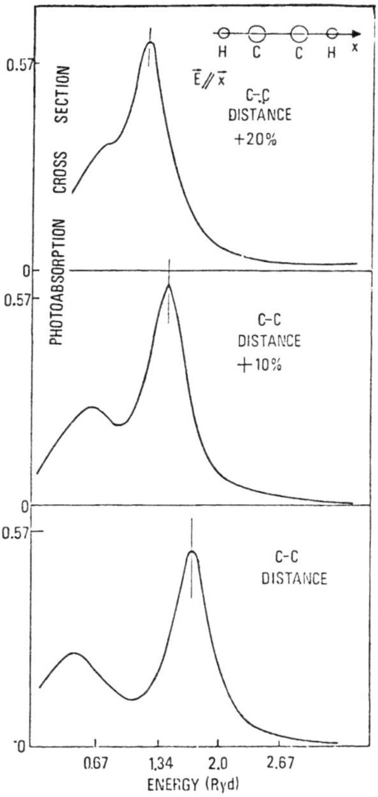

FIG. 1. Calculated spectra for C_2H_2. Only the continuum part is shown

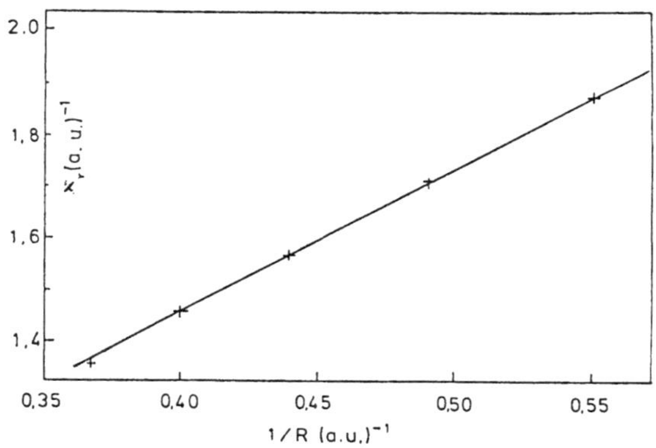

FIG. 2. Relation between x_r versus R^{-1} in C_2H_2

$\Delta R/R$	R(a. u.)	x_r(a. u. $^{-1}$)	x_rR
0	2.28	1.565	3.56
+10%	2.50	1.458	3.64
+20%	2.73	1.360	3.70
-10%	2.05	1.710	3.50
-20%	1.82	1.869	3.40

TABLE I. The slight variation of x_rR with R can be attributed to the slight dependence of the phase shifts on x

In these calculations $\bar{V}$ has been taken to be the volume average of the potential in the interstitial region between the atomic spheres and an outer sphere tangent to the outermost of them ($\bar{V} \sim -11$ eV).

The correct value for $\bar{V}$ to use would be the one obtained through eq. (2), which is not readily estimated. Hence we propose to determine it as a parameter to be fitted on the experimental data.

3. Application to chemisorption processes of diatomic molecules

We exemplify the kind of procedure to follow in the case of light diatomic molecules chemisorbed on a metallic surface. It is known, in fact, that for such moleculesthe EXAFS signal is not observable, so that eq. (5) offers an alternative way to measure bond lengths.

From its definition, $\bar{V}$ is a quantity that is peculiar to the particular bond under investigation, so it would seem that its determination be impossible on the basis of eq. (5) and experimental data only.

In fact if for two different molecules we know bond lengths and resonance energies, measured from the ionization threshold, relation (5) would entail

$$(E_{r1} - \bar{V}_1) R_1^2 = (E_{r2} - \bar{V}_2) R_2^2 \qquad (6)$$

which is seen to be indetermined.

However there are cases, like for example K-shell absorption from C and N in CO and N_2 moleculas, for which the final excited state, using the Z+1 rule, is the same and equal to NO^+. In such cases eq. (6) is sufficient to determine $\bar{V}_1 = \bar{V}_2$.

If we further suppose that the interaction with the substrate is not strong enough to modify $\bar{V}$ we have a way to measure modifications of the bond length of a molecule upon chemisorption. In fact, as soon as the constant $\bar{V}$ has been defined for a particular bond, all the others, within the range of validity of eq. (6), can be determined through eq. (6) itself.

Table II illustrates the results that can be obtained by using the above procedure in case of CO, N_2, NO chemisorbed on Ni and C_2H_2, C_2H_4 in the gas phase.

TABLE II. Calculated bond lengths for CO, N_2, NO chemisorbed on Ni and determination of V through the procedure described in the text

	Gas phase			Chemisorbed on Ni		
Molecule	R(Å)	$\bar{V}$ (ev)	E_r(ev)	E_r(ev)	R(Å)	ΔR(Å)
N_2	1.094	- 11.73	9.0[a]	8.6[a]	1.104	0.01
CO	1.130	- 11.73	7.7[a]	6.6[a]	1.163	0.03
NO	1.151	- 14.73	4.0[a]	4.0[a]	1.151	0.0
C_2H_2	1.204	2.49	19.6[b]	--	--	--
C_2H_4	1.334	1.96	15.6[b]	--	--	--

a - Data taken from ref. [5]; b - Data taken from ref. [2].

The $\bar{V}$ values so derived for N_2, CO, NO are very near (within 1-2 eV) to those calculated in our multiple scattering program using the definition of the volume average given above. That it should be so is understandable on the basis of eq. (2), if the potential is slowly varying with limited variation in the interstitial region.

In the case of C_2H_2 and C_2H_4, however, the values found for V, although very similar, which is expected, are even slightly positive, far away from the value found in m. s. calculations ($\sim$ - 11 ev). Even assuming that $\bar{V}$ is equal for the two molecules and using eq. (6) with $\bar{V}_1 = \bar{V}_2 = \bar{V}$ one would derive a value $\bar{V} = - 0.35$ ev, very near to zero and to the previous values.

Something different, it must be concluded, happens upon excitations in these two last molecules from what happens in the first three.

It might be that polarization effects in the final state play a different role in the two sets of molecules. This point is currently under investigation.

Summarizing, eq. (6), within its limitations, can be used to predict bond lengths. The optness of the prediction should be checked experimentally, in some cases, however, in order to able to use it with some confidence.

Moreover a novel feature that emerges from the preceding analysis is the role played by the constant $\bar{V}$. Without it the naive formula $\sqrt{E_r} \cdot R =$ = const would not work in nearly all cases. We think that a physical meaning should be attached to $\bar{V}$ and it would be of extreme importance to devise easy theoretical methods able to estimate this quantity. In such a way the powerfulness of eq. (6) would be greatly improved.

1. C. R. Natoli, D. K. Misemer, S. Doniach and F. W. Kutzler, Phys. Rev. A22, 1104 (1980).
2. A. Bianconi, XANES and their application to local structure determination, Daresbury Lab. Report DL/SCI/R17E, and these Proceedings.
3. F. W. Kutzler, C. R. Natoli, D. K. Misemer, S. Doniach and K. O. Hodgson, J. Chem. Phys. 73, 3274 (1981).
4. C. R. Natoli, Near Edge Absorption Structures in the framework of multiple scattering model, This Section of the Proceedings.
5. J. Stöhr and R. Jaeger, Phys. Rev. B26, 4111 (1982).

Double Ionization in Inner Shells

E. Bernieri, E. Burattini, N. Cavallo, M.R. Masullo,
A. Morone, and R. Rinizivillo

INFN, Frascati National Laboratory and University of Naples,

C. Natoli, L. Palladino, and A. Reale[x]

INFN, Frascati National Laboratory, I-00044 Frascati, Italy

1. Introduction

There is a great interest in studying the radiative decay scheme of atoms doubly ionized in the same inner shell because it gives direct information on interaction of the vacancies involved, on the use of intermediate coupling in various atoms and, indirectly, on the probability of double ionization in heavy ion collisions and in photoionization. It can happen, in fact, that correlation effects change the probability by more than one order of magnitude.
Transitions involving more than one electron with emission (or absorption) of one photon have been considered many years ago [1], [2], [3], [4], but only recently has it been possible to observe fluorescence lines of this nature, when very intense heavy ion beams became available.
This paper is devoted to the study of 'two electrons-one photon' radiative decay of atoms doubly ionized in deep inner shells (K and L).
Theoretical results, obtained in a given approximation, are presented.

2. Decay schemes

A double ionization in the K shell implies either the emission of hypersatellites (and satellites) lines or a decay as the one we are considering here. For these two competitive processes we indicate with $P_{\alpha\alpha}$ the double K-vacancy transition rate (t.r.) for a 'two electrons-one photon' process and $K_{\alpha\alpha}$ the photon energy (Fig. 1); $P_{\alpha}^{(h)}$ the double K-vacancy t.r. for a 'one electron-one photon' process (competitive process); photon energy: hypersatellite K_{α}^{h} and satellite K'_{α} cascade (Fig. 2).
The single K-vacancy decay rate, corresponding to the usual lines K_{α}, K_{β}, is indicated by P_{α}.
The transition rate of a doubly ionized atom decaying radiatively with a simultaneous filling of the two K-shell vacancies results from the interaction between the 2 electrons involved and the radiation field: it can be imagined as a two-step process (as shown in Fig. 1) corresponding to the transition matrix element:

$$P_{\alpha\alpha} \sim \left| \left(\int_{\text{cont}} + \sum_{i} \right) \frac{\langle f \mid H \mid i \rangle \langle i \mid V \mid o \rangle}{E_o - E_i} \right|^2$$

where V is the Coulomb interaction between the two jumping L electrons; H is the usual e-m interaction of the electrons with the radiation field.

[x]Present Address: University of L'Aquila, L'Aquila

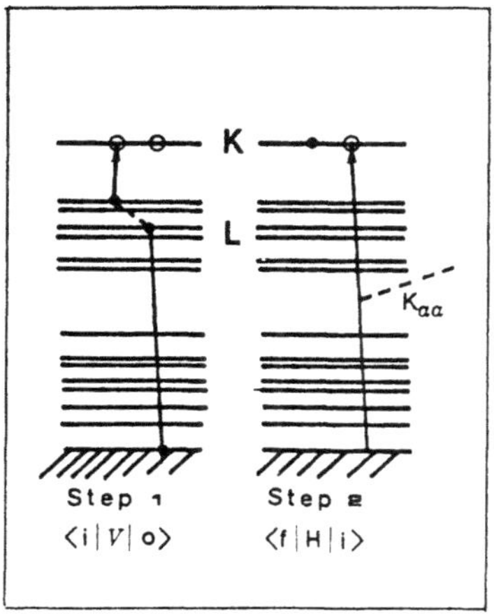

Fig. 1. The two-step process 'two electrons-one photon'

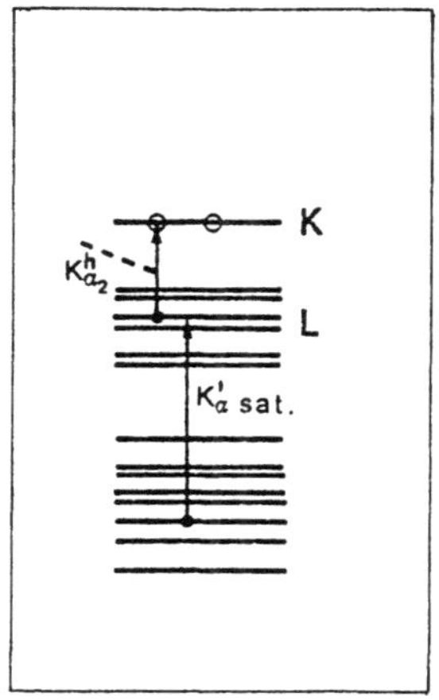

Fig. 2. Hypersatellites (and satellites) emission

The calculation has been performed with the following assumptions:

.electric dipole approximation

.appropriate hydrogen-like two particles wave function in pure LS representation:

$|o\rangle$: $L_I \otimes L_{II}$ and $L_I \otimes L_{III}$ two particles w.f.

$|i\rangle$: Sommerfeld's function [16]

$|f\rangle$: S waves

.'frozen' orbitals approximation (no rearrangement taken into account).

3. Results

Within the given approximations (for low and intermediate Z) our result is:

$$P_{\alpha\alpha} = 0{,}078\ \alpha^3\ \omega_{\alpha\alpha} \qquad \alpha = 1/137 .$$

Our simple approach, where the calculation of the transition rate involves eventually only single particle wavefunctions, can be compared with more sophisticated approaches [8], [9], [10], [11] based on complex multiparameter variational wave functions.

A comparison of the results is given in Table 1.

Table 1. Comparison of the calculated values with other theoretical and experimental results

Reference	$P_{\alpha\alpha}$ [sec^{-1}]	$P_\alpha^h / P_{\alpha\alpha}$	Ni (Z=28) $\Delta K_{\alpha\alpha}$ [eV]	ΔK_α^h [eV]
Present Work	6.9 10^{11}	1800	180	350
[11]	5.8 "	3060	125	198
[9]	9.2 "	2120	-	-
[10]	0.81 "	4700	-	-
[15]	-	-	-	292
[7] exp. (Ni, Z=28)	-	5000±600	154±22	211±26

We give also the energy shifts $\Delta K_{\alpha\alpha}$ and ΔK_{α}^{h} of the expected transition lines which have been calculated using the well-known formulas for the total energy of helium-like ions in first-order perturbation theory, and calculating the effective screening parameter ΔZ for K^{-2} and $K^{-1} L^{-1}$ states:

$$\Delta K_{\alpha\alpha} = (K_{\alpha\alpha} - 2\, K_{\alpha}); \qquad \Delta K^{h} = (K_{\alpha}^{h} - K_{\alpha}) \;.$$

These predictions could be checked in heavy-ion collision processes, where double K ionizations are quite frequent ($\sigma \simeq$ mb). A suitable apparatus for accurate measurement of fluorescent yields and energies is, of course, needed.
With synchrotron light there is already evidence of K-shell ionization accompanied by excitation of another electron on to an excitation level (σ^{+*}). There is also experimental evidence of double inner shell ionization process (σ^{++}) with an X-ray source [14] .
A program for more systematic theoretical experimental investigation of double ionization processes is in progress in Frascati: we are looking for double ionization threshold effects in photon energy regions far from EXAFS region.

REFERENCES

1. Richtmeyer et al. - J.Franklin Inst. 208 (1929) 325.
2. Richtmeyer et al. - Ph.Rev. 49 (1963) 1.
3. Richtmeyer et al. - Rev.Mod.Phys. (1937) 391.
4. Bloch et al. - Phys. Rev. 48 (1935) 187.
5. Bearden - Rev.Mod.Phys. 39 (1967) 78.
6. Richard et al. - Phys. Rev. Lett. 29 (1972) 373.
7. Wolfli et al. - Phys. Rev.Lett. 35 (1975) 656; Phys.Lett. A58 (1976).
8. Åberg et al. - Phys.Rev. Lett. 37 (1976) 63.
9. Gravila et al. - J.Phys. B 11 (1978) 1353.
10. Khristenko - Phys.Lett. A 59 (1976) 202.
11. Jalbert et al. - Proc.Int.Conf. Stirling (1980) 281.
12. Wuilleumier et al. - C.R.Acad.Sci. Paris 270 (1970) 1229.
13. Madden et al. - Phys.Rev. (1965); Lengeler et al. - DESY S.R./82/02.
14. Salem et al. - Proc.Int.Conf. on X-Ray - Stirling (1980) 793.
15. Briand et al. - J.Phys. B 9 (1976) 1055.
16. Handbuch der Ph. XXIV 289.

Part II

EXAFS Data Analysis

The Reliability of ab initio Calculations in Extracting Structural Information from EXAFS

R.F. Pettifer and A.D. Cox
Department of Physics, University of Warwick,
Coventry, CV4 7AL, England

1. Introduction

Ab initio calculations for extracting structural information are of importance for several reasons. Firstly, we need calculated amplitudes and phases for those situations where reliable standards are not available. Secondly, the calculations should provide absolute parameters rather than relative values. Thirdly, they continue to give insight into the important physical processes involved in EXAFS. Prior to using ab initio calculations in structural determinations, it is important to have some guidelines within which we can trust the results.

The starting point in EXAFS calculations is the construction of the atomic potential from which the phase-shifts can be established. From these the phase and amplitude of back-scattering along with the emitter phase can be extracted. LEE and BENI [1] constructed an atomic-like Hartree muffin-tin potential to which was added a complex energy dependent potential which accounted for exchange and correlation effects within the local density approximation.

From these phase factors shell radii were determined which were in impressive agreement with the known values, i.e. better than 0.015Å.

More recent work, however has suggested that other effects may be present which influence structural data obtained from ab initio calculations.

NOGUERA and SPANJAARD [2] have examined the dynamical screening of the core hole in aluminium and have calculated that this effect should produce errors in the determined bond length which depend on the maximum photo-electron momentum used in the interpretation and also the shell radii under study. In particular, the slope of the dynamically screened emitter phase-shift is less negative than the two extreme cases and thus a shell radius determination based on the fully screened emitter (say) should provide a shorter bond-length than the true value. Their calculations are for a low Z metal and the effects on other materials remain unclear at present.

TRAN THOAI and EKARDT [3] buried their Hartree potential in a homogeneous electron gas, a situation more appropriate to solids than that of LEE and BENI [1]. This produced differences in phase and amplitude for the case of copper. The authors also point out that the k dependence of the real potential in the inter muffin-tin region should give phase perturbations in particular to distant shells. Later the same authors [4] applied many-body theory to include lifetime effects which again produced phase and amplitude effects.

The experimental amplitude of the fine structure is always less than that calculated theoretically. This has been explained on the basis of multiple excitation of the emitter atom by REHR et al.[5] and STERN et al.[6]. A search has been made in the latter paper for energy dependence of these effects. These should occur in the low energy region of the spectrum, but it is important that the approximations involved in calculating the back-scattering amplitude are appreciated. The correct treatment of the back-scattering amplitude can be found in the paper by LEE and PENDRY [7]. The formula commonly used

$$f(\theta) = \frac{1}{k} \sum_l (2l + 1)e^{i\delta_l} \sin(\delta_l) P_l(\cos \theta) \qquad (1)$$

is appropriate for an incident plane wave whereas it must be remembered that the emitted wave has the form $h_l^{(1)}(kr)Y_{lm}(\Omega)$, i.e. a spherical wave. Two levels of approximation are necessary to give the standard EXAFS equation. Firstly, the Hankel function is replaced by its asymptotic form and then the wave is assumed to be planar at the scatterer. One expects that as the kinetic energy of the electron increases, the effective ion core for scattering reduces and equation 1 will be a reasonable approximation. On the other hand, as the size of the scatterer increases, the approximation should be more unreliable. The effects of the above approximations can be seen in Ref [7] for copper and more drastic effects of this approximation for the case of tellurium scattering are seen in Refs [8,9]. It must be noted that this is a geometrical effect and can be calculated with no adjustable parameters. Its neglect can give large errors especially in coordination numbers for high Z scatterers.

LEE & BENI's [1] work is theoretically an improvement over a Slater type exchange potential, however WOOLFSON, GURMAN and HOLLAND [10] have recently examined electron scattering for energies $\lesssim$ 100 eV from rare gas atoms. They compared the experimental results with four types of exchange. $X\alpha$ with $\alpha = 0.7$ and 0.4, the HARA [11] exchange, and the LEE & BENI [1] treatment of exchange and correlation. In this work the potentials are well defined in that there are not ambiguities in the exact momentum of the electron, or of the radius of the scattering atom. These authors find that above 100 eV all of the potentials give reasonable agreement but in the low energy region only the Hara and $X\alpha$ ($\alpha = 0.4$) results match experiment. The authors attribute this to a failure of the local density approximation.

In the rest of this paper we present some calculation of back-scattering and emitter phases using these potentials, together with an illustration of the effects of wave curvature on the results. Finally, data is analysed using these results and an assessment is made of the reliability of structural parameters. A single spectrum is analysed, being that of niobium at liquid nitrogen temperature.

2. Phase-shift Calculations

All of the calculations were performed with variations of a programme called MUFPOT [12]. The Hartree potentials used were all identical and were calculated with the radius of the muffin-tin potential set at one half of the interatomic distance. Contributions from the surrounding atoms was included using a procedure given by LOUCKS [13]. To this potential was added the energy-independent Slater-type exchange, or the energy-dependent Hara exchange, (after WOOLFSON [14]), or the complex exchange correlation potential of LEE & BENI (after GURMAN [15]). Thirty phase-shifts were then calculated by numerical intergration. The HARA [11] exchange is written in atomic units as

$$V_x = -\frac{2}{\pi} k_F \left(\frac{1}{2} + \frac{1-\eta^2}{4\eta} \ln \left| \frac{1+\eta}{1-\eta} \right| \right) \qquad (2)$$

where k_F is the local Fermi wavevector and $\eta = k/k_F$. The potential becomes weaker as k increases.

3. Calculation of Structural Parameters from EXAFS

For efficient comparison of theory to experiment we write an EXAFS equation as

$$\chi(k) = \sum_j A(k,R_j) \frac{N_j}{R_j^2} \exp(-2\sigma_j^2 k^2) \, \mathrm{Cos}(2kR_i + \Phi(k,R_j) + 2\varepsilon(k)) \qquad (3)$$

where the amplitudes A and phases Φ are weakly radius dependent. A refinement can be made of the shell radii using least squares techniques ensuring that local minima are avoided. We also include a polynomial in E to account for residual background errors and multiply the $\chi(k)$ function by a weighting factor e^{sE}. The program is called EXAFSFT and further details can be found in the articles by COX [16] and PETTIFER and COX [17].

4. Experimental

The spectrum of niobium was taken using the synchrotron DESY in an EMBL bunker. The resolution was estimated to be 2eV. Niobium is a b.c.c. structure so multiple scattering effects should not be large out to the second shell.

5. Results

5.1 Phase-shifts Calculations

Figure 1 shows the shell radius dependence of the amplitude and phase of Φ backscattering. Also shown for comparison are the results of using the asymptotic plane wave approximation. Both the amplitude and phase are affected. In particular the amplitude has a noticably sharper minimum at lower energies. This is typical of the behaviour of this row of the periodic table.

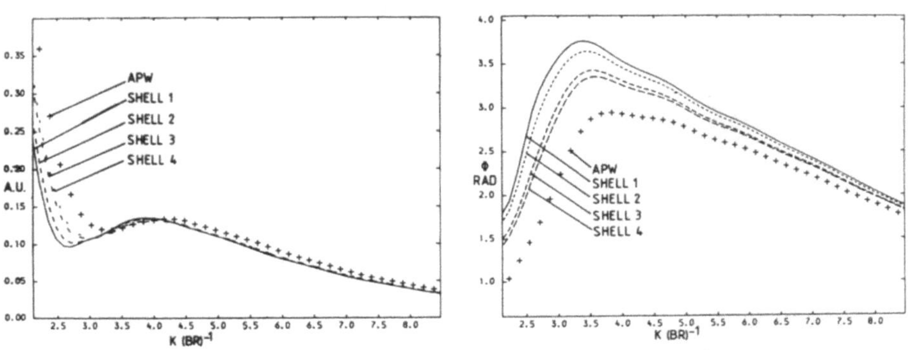

Fig.1. The shell radius dependence of the amplitude A and phase Φ of back-scattering for niobium. A and Φ are as defined in Eqt (3). The limits of the k range is $4 \rightarrow 16$ Å^{-1}. The crosses show the results using Eqt(1)

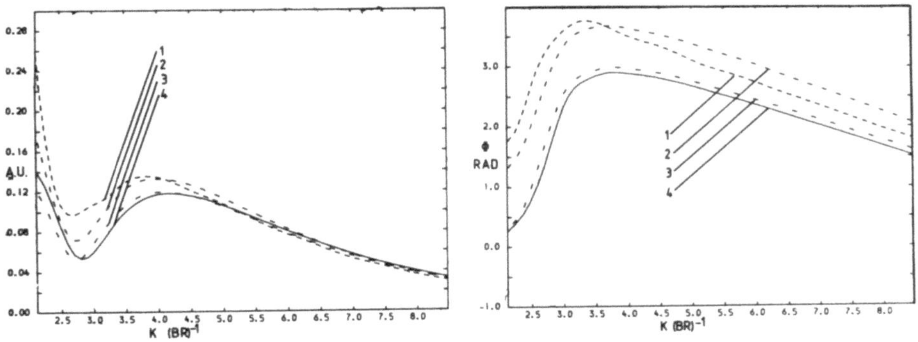

Fig.2. The first shell amplitude A and phase Φ for varying exchange potentials. Inelastic effects are included implicitly in the LEE & BENI results. For the others, an allowance is made for inelastic effects by including a constant imaging potential of -4eV. (1) LEE-BENI, (2) Xα α = 1.0, (3) Xα α = 0.4 and (4) HARA

Figure 2 shows the dependence of amplitude and phase on the exchange-correlation treatment. The major differences appear in the region below 5 BR^{-1}.

Figure 3 is a plot of the emitter phases calculated using the Z+1 fully relaxed potential except for the LEE-BENI case which corresponds to neutral atom. Three of the four potentials have approximately the same slope. The exception is for the Xα (α=1) curve, this shows a more negative slope and should thus result in a reduced shell radius. The same slope results from the fact that except for α=1 the exchange makes a **negligible** contribution to the phase shifts at these energies and is dominated principally by the Hartree potential.

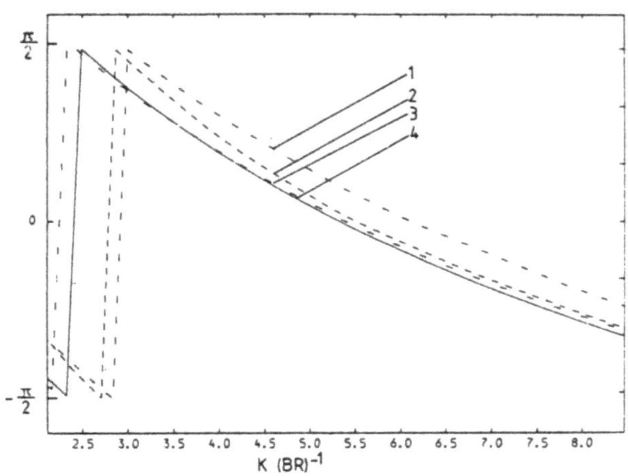

Fig.3. The emitter potential phase shift using the Z+1 fully screened potential for (2) Xα α=1.0 (3) Xα α=0.4 and (4) HARA. For the LEE-BENI curve 1 the neutral atom phase shift is plotted

5.2 Results of Fitting

Four shells were fitted to the data over two energy ranges 61 → 1490eV ($4Å^{-1}$ → $20Å^{-1}$) and 61 → 548eV ($4Å^{-1}$ → $12Å^{-1}$). An example of the fits are shown in Figure 4 and the results are summarised in table 1 for the first two shells in niobium. Over the extended energy range the shell radii predicted theoretically agree with the true value within 0.005Å and 0.013Å for the first and second shells respectively except for the Xα (α=1) potential. The first shell σ^2 also agree with calculated values to within 7% and the second to within 30%. The mean first coordination number is effectively 0.65 of the true value which is consistent with estimates of S_0^2 made by others [5], [6]. The consistency despite the variation in potential is merely due to the extended energy range of the data which makes the results relatively insensitive to the amplitudes, phases and models of inelastic loss used.

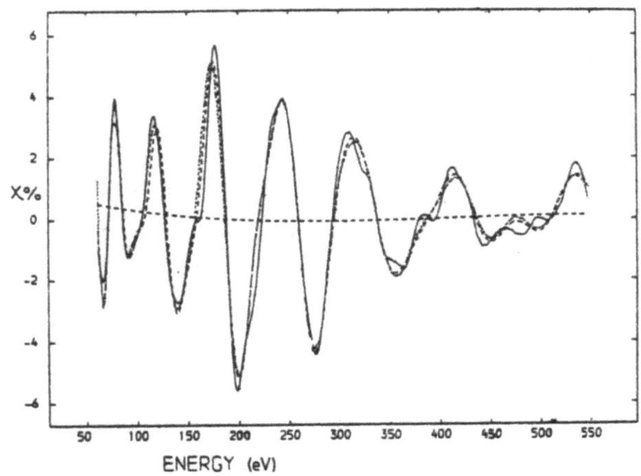

Fig.4. Results of fits to the experimental data over the reduced range. Solid curve is the experiment. Fine dashed line the LEE-BENI potential and long dashed line the HARA potential

When the range of data is reduced, however, the situation changes dramatically. The shell radii are preserved for the Hara and α = 0.4 phases which indicates the correct phase behaviour in the low energy region. The results for Lee-Beni are artificially worse owing to the incorrect absolute phase of the emitter. An accommodation has been made for this by varying E_0. The correlation coefficient for E_0 and R_1 is 0.93 for this region and the coefficient of linearity is 0.006Å/eV. The situation for σ_1^2 and N_1 over the reduced range indicates that the errors in σ_1^2 are ~ 25% and those in coordination number are ~ 10% when corrected for a constant S_0^2. The exception is the result of LEE and BENI which appears to underestimate inelastic loss and compensate for this by adjusting N_1 and σ_1^2 (correlation coefficient 0.94). The situation for shell 2 is bad for all models and again is probably associated with problems involving inelastic loss processes.

Finally, although the fits produced poor results for N and σ_j^2 for shells other than the first, their radii were fitted. Over the reduced

Table 1. The results of fitting the niobium spectrum over two energy ranges. The calculated mean square relative displacements are found from SCHMIDT [18]. Allowance has been made for thermal expansion in the shell radii. The units are R(Å), σ^2(Å^2 x 10^3), and E and Range (eV)

Potential	R_1	R_2	σ_1^2	σ_2^2	N_1	N_2	E_o	Range
Xα α=1	2.835	3.269	2.8±0.1	2.3±0.2	4.9±0.2	2.4±0.2	- 0.7	61-1490
Xα α=0.4	2.850	3.280	3.0±0.1	2.8±0.2	5.6±0.2	3.3±0.2	-11.7	"
Hara	2.857	3.289	3.1±0.1	3.0±0.2	5.8±0.2	3.4±0.2	-11.9	"
Lee Beni	2.854	3.290	2.7±0.1	2.8±0.2	4.6±0.2	3.1±0.2	- 1.8	"
Xα α=1	2.800	3.220	2.6±0.5	3.0±0.6	4.1±0.3	3.6±0.4	- 6.6	61-548
Xα α=0.4	2.851	3.279	2.5±0.5	0.0±0.6	5.2±0.4	2.3±0.3	-11.5	"
Hara α=1	2.857	3.285	2.2±0.5	0.0±0.6	5.0±0.3	2.4±0.3	-11.8	"
Lee Beni	2.835	3.257	0.0±0.5	2.8±0.6	2.6±0.2	4.1±0.3	- 5.5	"
Meas. or Calc.	2.852	3.293	2.9	3.2	8	6	-	-

Table 1.(Continued) The correlation matrix for the reduced energy range 61-548eV using Hara phase-shifts

	E_o	R_1	R_2	σ_1^2	σ_2^2	N_1	N_2
E_o	1	0.96	0.92	0.45	-0.63	0.64	-0.72
R_1		1	0.92	0.46	-0.62	0.65	-0.72
R_2			1	0.39	-0.64	0.60	-0.71
σ_1^2				1	-0.05	0.93	-0.24
σ_2^2					1	0.34	0.96
N_1						1	0.51
N_2							1

energy range, the Hara and Xα (α=0.4) potentials gave agreement to within 0.02Å for all four shells, although they were in general shorter than the true value.

Conclusions

In determining distances in niobium, it is found that a reduced Slater exchange or a Hara exchange potential produces shell radii which are comparable to those obtained using the Lee-Beni potentials.

The major discrepancies in comparing potentials occurs in the low energy region where most uncertainty exists in the behaviour of inelastic and many electron effects. However, it appears that LEE & BENI underestimate the magnitude of inelastic processes.

The calculation of back-scattering amplitudes and phases from large ion cores should involve removing the plane-wave approximations.

The existence of a good fit to the data is a necessary but not sufficient condition to insure that the correct structural information has been obtained. In particular N, σ^2 and R, E_o are highly correlated over a limited energy range.

Acknowledgements

We have great pleasure in acknowledging the help given in this work by S.J. Gurman, T. Clark, B.W. Holland, J.M. Taylor and M. Woolfson. I also acknowledge EMBL for access to synchrotron radiation and SRC for financial help.

References

1. P.A. Lee & G. Beni, Phys. Rev. B 15 p.2862 (1977)

2. C. Noguera & D. Spanjaard, J. Phys. F. 11 1133 (1981)

3. D.B. Tran Thoai & W. Ekardt, Solid State Comm. 40 p.269 (1981)

4. W. Ekardt & D.B. Tran Thoai, Solid State Comm. 40 p.939 (1981)

5. J.J. Rehr, E.A. Stern, R.L. Martin and E.R. Davidson, Phys. Rev. B 17 p.560 (1978)

6. E.A. Stern, B.A. Bunker and S.M. Heald, Phys. Rev. B 21 p.5521 (1980)

7. P.A. Lee & J.B. Pendry, Phys. Rev. B 11 p.2795 (1975)

8. R.F. Pettifer, P.W. McMillan and S.J. Gurman, The Structure of Non-Crystalline Materials ed. P.H. Gaskell p.63 Taylor and Francis (1977)

9. R.F. Pettifer, Inner-Shell and X-ray Physics of Atoms and Solids Ed. D.J. Fabian, H. Klienpopper and L.M. Watson p.653 (1981)

10. M.S. Woolfson, S.J. Gurman and B.W. Holland, Surf. Sci 117, p.450 (1982)

11. S. Hara, J. Phys. Soc. Japan 22 p.710 (1967)

12. J.B. Pendry, unpublished.

13. T. Loucks, Augmented Plane Wave Method p.54, Benjamin (1969)

14. M.S. Woolfson, unpublished.

15. S.J. Gurman, private communication

16. A.D. Cox, EXAFS for Inorganic Systems, Daresbury Laboratory, DL/SCI/R17 p.51 (1981)

17. R.F. Pettifer and A.D. Cox, to be published.

18. V.V. Schmidt, Bull Acad. Sci. USSR, Phys. Ser. 27 p.392 (1963)

EXAFS Analysis in Anisotropic Systems

Peter Rabe

Institut für Experimentalphysik, Universität Kiel,
D-2300 Kiel, Fed. Rep. of Germany

Nonrelativistic Dipole Approximation

The anisotropic emission of photoelectrons from atoms in principle leads to an anisotropy in EXAFS. In the nonrelativistic dipole (NRD) approximation the differential cross section for exciting a photoelectron from a free atom with polarized X-radiation is described by [1]

$$\frac{d\sigma}{d\Omega} = \frac{\sigma_o}{4\pi} \left(1 + \frac{\beta}{2} (3 \cos^2\Theta - 1)\right) \quad . \tag{1}$$

Here σ_o denotes the total cross section and Θ the angle between the direction of polarization and the momentum vector of the photoelectrons. The asymmetry parameter β depends on the subshell to be excited and in general on the electron kinetic energy. For s-wave bound states $\beta=2$ so that (1) reduces to

$$\frac{d\sigma}{d\Omega} = \frac{3\,\sigma_o}{4\,\pi} \cos^2\Theta \quad . \tag{2}$$

Using this approximation the EXAFS showing up beyond K edges is described by [2]

$$\chi(k) = - \frac{3\cos^2\Theta}{kR^2} |f| \sin\,(2kR + 2\,\delta_1 + \arg f)\ . \tag{3}$$

Here k is the photoelectron wave number, f the complex scattering amplitude, R the interatomic distance, δ_1 the central atom scattering phase and Θ the angle between electric vector and the vector which connects the absorbing and the scattering atom. In the case of L_2 and L_3 shell absorption excitation to s- and d-symmetric final states are possible. Both channels contribute to the fine structure resulting in [3]

$$\begin{aligned}\chi(k) = \frac{|f|}{kR^2} \Big\{ & \frac{1}{2} (1+3\cos^2\Theta) M_2^2 \sin(2kR + 2\delta_2 + \arg f) \\ & + \frac{1}{2} M_o^2 \sin(2kR + 2\,\delta_o + \arg f) \\ & + (1-3\cos^2\Theta) M_o M_2 \sin(2kR + \delta_o + \delta_2 + \arg f) \Big\} (M_2 + \tfrac{1}{2} M_o)^{-1}\ . \end{aligned} \tag{4}$$

Here M_o and M_2 are the radial dipole matrix elements for transitions to s- and d-symmetric final states and δ_0 and δ_2 the corresponding central atom phase shifts.

It has been noted that for heavy elements (Z>50) which are the only interesting objects in view of a structural analysis the ratio M_2/M_o takes values between 4 and 5 [3,4,5]. This means that in the case of unoriented ($\langle 3 \cos\Theta \rangle = 1$) samples transitions to s states contribute only 2%-3% to the EXAFS amplitudes. In most cases these contributions can be neglected if one remembers that typical uncertainties for amplitudes are of the order of 10%. In oriented samples, however, the mixed term contributes up to 20% and therefore in a careful data analysis may not be neglected.

It should be noted that also for unpolarized radiation an anisotropy can be observed. The angular distribution of photoelectrons from free atoms is given by [1]

$$\frac{d\sigma}{d\Omega} = \frac{\sigma_0}{4\pi} \left(1 - \frac{\beta}{4} \left(3\cos^2 \Theta' - 1 \right) \right) \quad . \tag{5}$$

Here Θ' is the angle between photon and electron momentum. The formal description of EXAFS above K edges in this case is given by (3) where $\cos^2\Theta$ is simply replaced by $\sin^2\Theta'$.

Relativistic and Multipole Corrections

All anisotropic EXAFS spectra up to now have been interpreted in the NRD approximation and generally good agreement is observed. On the other hand it is well known that relativistic effects and higher multipoles may influence the angular distribution of photoelectrons so that it seems to be apt to examine the effect of these corrections on the anisotropy of EXAFS.

For unpolarized X-rays the more general form of the photoelectron angular distribution is a Legendre expansion of the differential cross section [6]

$$\frac{d\sigma}{d\Omega} = \frac{\sigma_0}{4\pi} \sum B_n P_n \left(\cos\Theta' \right) \quad . \tag{6}$$

In this expansion always $B_0=1$ and $2B_2$ is identical with β in (5), i.e., $B=-1$ in the NRD approximation. The two corrections (relativistic and multipole) influence EXAFS in different ways.
A relativistic correction retains the general form of (5). However the coefficient $|B_2| < 1$ ($\beta < 2$) and varies with energy [7,8]. Numerical calculations show that in most cases of considerable interest B_2 is negative so that the angular distribution is still peaked in the direction of the electric vector and the EXAFS amplitudes are maximal for X-rays propagating perpendicular to the line connecting absorbing and scattering atom (for unpolarized light). Deviations from the NRD approximation are less

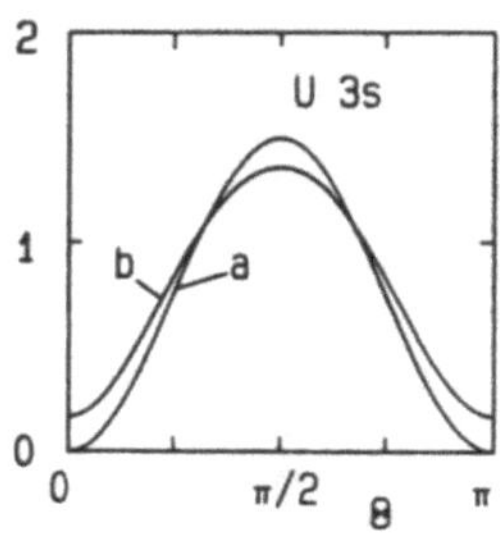

Fig.1. Angular distribution of photoelectrons (E_{kin}=0.1keV) excited from the M_1 level of uranium with unpolarized X-rays calculated within the dipole approximation. a) Nonrelativistic result; b) Relativistic result with parameter B_2 in (6) taken from [8]. Θ' is the angle between photon and electron momentum

pronounced for inner shells of light atoms. For inner shells of heavy atoms ($Z > 70$) the relativistic corrections are significant. This is demonstrated in Fig.1 for the 3s shell of uranium in a hypothetical oriented diatomic heteronuclear molecule. The NRD case predicts zero EXAFS amplitudes when the molecular axis is oriented parallel to the direction of propagation of X-rays. In the relativistic case the angular distribution is superimposed on a constant background, the amplitude is reduced by 25%. An analysis of this distribution within the NRD approximation would assume 20% of the molecules to be unoriented.

The inclusion of higher multipoles leads to a nonsymmetric distortion of the angular distribution (Fig.2). The distributions have been calculated from published relativistic multipole expansion coefficients [7,8]. As expected, for light elements the angular distribution is close to the NRD approximation. For carbon, e.g., significant deviations from $\sin^2\Theta$ do not occur below 0.5keV kinetic energy of the photoelectrons. For demonstration the distribution at 1 keV is shown in Fig.2. The distribution at 0.1keV is virtually identical to the $\sin^2\Theta$ curve. For heavier elements the distributions are asymmetric and the shape of the distribution changes more drastically with photoelectron energy. For tin the peaks are shifted to larger angles, i.e., photoelectrons are preferably emitted in backward direction. In contrast to this electrons from the L_1 and M_1 shell of uranium are emitted in forward direction. The deviations from the $\sin^2\Theta$ distributions amount up to 20 degrees, a value which could easily be detected in an EXAFS experiment. Although such an experiment certainly will not be performed at the uranium K edge it is interesting to note that the distribution in this case would peak backwards. All these effects show the sensitivity of the angular distributions to relativistic effects and multipole excitations. For many experiments especially on not too heavy elements the NRD approximation is accurate enough with regard to other experimental uncertainties. Significant corrections have only to be made for heavy elements. On the other hand the sensitivity of EXAFS to the angular distribution of photoelectrons could be used to measure relativistic and multipole corrections. These experiments seem to be much easier to perform than angular resolved photoemission in the X-ray region.

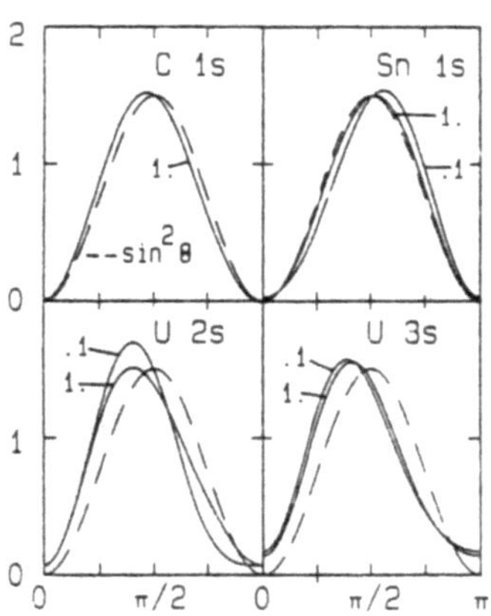

Fig.2. Angular distribution of photoelectrons calculated within the nonrelativistic multipole approximation for E_{kin}=0.1keV and 1keV. The B_n in (6) have been taken from [7,8]. The distribution obtained in nonrelativistic dipole approximation has been included for comparison

Applications

Although an anisotropy in the EXAFS amplitudes should be observable with unpolarized X-rays, as the bremsstrahlung from conventional sources, to our knowledge no reliable data have been published. X-ray sources as high-energy accelerators combine a high degree of polarization with a high photon intensity. According to the finite source height the degree of polarization is slightly less than 100% even when narrow exit slits are used [9]. The degree of polarization is increased by the Bragg reflecting crystals. The width of the single crystal diffraction pattern for p-polarized X-rays varies as $\cos^2\alpha$ with α the Bragg angle. With a proper choice of the crystal one can go as close as possible to $\alpha = 45$ degrees where only the s-polarized component is reflected.

Synchrotron radiation studies of the anisotropic absorption of polarized X-rays have been carried out on bulk single crystals and surfaces. Experiments on bulk crystals compete with the more precise diffraction experiments and therefore have added little to structure information. They have rather been used to test the theoretical model underlying (3) and (4). HEALD and STERN [3] have studied the anisotropic absorption in the layered compounds 2H-WSe and 1T-TaS above the Se K edge and the L_2 and L_3 edges. They found a good agreement with theory. From the cross term in (4) they concluded that the contribution of the final s state is 0.02 of that of the final d states to the total absorption of the W L_3 edge. The anisotropy of the K edge of EXAFS of a zinc single crystal was measured by BROWN et al. [10]. At a temperature of T = 20K, they determined the difference in nearest-neighbour distance $R_1^{\varepsilon /\!/ c} - R_1^{\varepsilon /\!/ a} = 0.0225 \pm 0.0005$ nm which is consistent with the X-ray diffraction result, 0.0222nm. RABE et al. [11] investigated the EXAFS of a GeS single crystal at the Ge K edge. From the orientation-dependent radial structure function the three-dimensional arrangement of atoms in the crystal was evaluated. Again a good agreement with X-ray diffration data was observed.

Bond length and site determination of adsorbates on a single crystal surface by means of EXAFS competes with the much less accurate low-energy electron diffraction (LEED). Typical accuracies for distances between adsorbed atoms and surface atoms of the substrate extracted from LEED data are 0.01-0.02 nm [12]. Moreover the analysis of LEED requires extensive multipole scattering calculations. Compared to this structure parameters can be calculated from EXAFS with comparatively little computational effort and a precision which is one order of magnitude better than LEED data.

According to the extreme dilution of an adsorbate / substrate system simple absorption experiments are unfeasible. Therefore the yield of nonradiative Auger or secondary electrons has been measured as a function of the photon energy. The yield of Auger electrons is directly proportional to the absorption coefficient. This experimental technique however necessitates an energy analysis of the emitted electrons which in general narrows the solid angle where electrons can be detected and therefore suffers from an extremely low signal. Moreover coincidences of Auger electron and photoelectron energies at certain photon energies dramatically distort the yield spectra [13] so that this technique can only be applied to heavier elements where the Auger electron energies are larger than the EXAFS range above the absorption edge.

A larger signal results from the secondary electrons. These electrons result from a cascade of scattered Auger and photoelectrons. In principle no energy analysis of the electrons is necessary. All electrons which are emitted from the sample contribute to the signal. Several experiments have shown that total yield spectra show a close similarity to the absorption coefficient [14,15]. This fact is by no means obvious as far as EXAFS is concerned. Including the scattering terms shown in Fig.3, a direct measurement of the photoelectron yield would result in a normalized fine structure which is described by [2]

$$\chi(k) = 6(\hat{\varepsilon}\,\hat{R})(\hat{\varepsilon}\,\hat{\kappa})\,\mathrm{Re}\big(-e^{ikr(1-\hat{\varepsilon}\hat{R})} f(\Theta)/R + e^{2ikR} f_o(\pi-\Theta)\, f(\Theta)/R^2\big) \quad . \qquad (7)$$

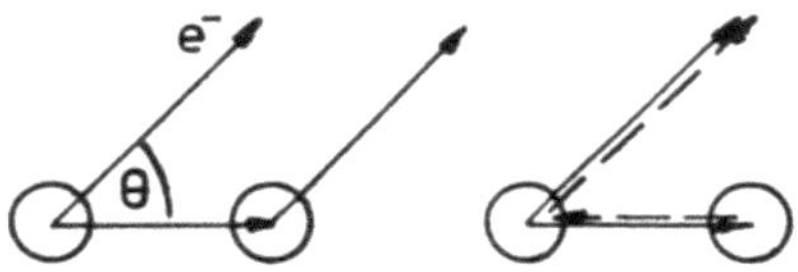

Fig.3. Scattering processes which lead to the fine structure (7)

Here $\hat{\varepsilon}$, $\hat{R}$ and $\hat{\kappa}$ are unit vectors in the direction of polarization, the line connecting the central and the neighbouring atom and the direction of propagation of the photoelectron. f_o is the complex scattering amplitude of the central atom. As can be seen from (7) for a particular direction Θ the periodicity of the photoelectron yield spectra is different from that of EXAFS. FUJIKAWA [16] has calculated the fine structure which would be observed in photoelectron yield spectra by exciting the 1s level of an oxygen atom adsorbed on an Al(110) surface (Fig.4). As expected, these spectra differ completely from a calculated EXAFS spectrum. The question arises, why the experimental total electron yield spectra reproduce EXAFS. Theoretically the EXAFS formalism is exactly recovered by integrating (7) over 4π . In the experiment there must be a mechanism which is working as an effective averaging process. The first quantitative comparison between EXAFS in absorption and yield spectra have been performed on bulk samples [15]. In that cases the measured signal originates from atoms located within the escape depth of the electrons. This depth can be large compared to the mean free path of the primary electrons. Due to multiple scattering the electrons loose the information about the primary directional anisotropy. For adsorbates the total yield obviously contains a significant fraction of electrons which originate from electrons primarily emitted into the substrate hemisphere. Monte Carlo calculations have confirmed this assumption [17]: For instance, one 10keV electron primarily emitted from the surface into the substrate produces a cascade of scattered electrons. From these

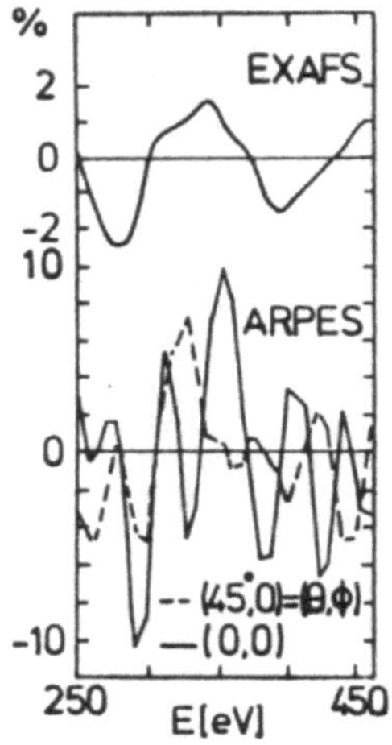

Fig.4. Comparison between EXAFS and angular resolved X-ray photoelectron spectra. The spectra have been calculated for photoionization from the 1s orbital of O adsorbed on top of an Al atom of a Al(110) surface [16]. The Al-O bond length was assumed to be 0.193 nm. X-rays are polarized normal to the surface. Θ and Φ are the polar and azimuthal angles

more than twenty almost simultaneously are emitted into the vacuum hemisphere.

The dependence of the EXAFS amplitudes on the direction of polarization is very useful to determine the site of atoms adsorbed on single crystal surfaces. This has been shown for iodine adsorbed on copper (111) and (100) surfaces [18]. From the anisotropy of I L_3 spectra 3-fold and 4-fold hollow sites could be unambiguously identified. In the soft X-ray region a monolayer oxygen coverage on Al (111) has been studied [19]. Total electron yield spectra taken at different direction of polarization allowed the discrimination of the chemisorbed phase where the oxygen covers 3-fold hollow sites on top of the surface against the oxide phase where the oxygen penetrates through the first surface layer and occupies either tetrahedral or octahedral interstitial positions. Furthermore the O-O contribution could be identified which shows up only when the polarization vector has a component parallel to the surface. An extensive study has been crried out on the adsorption of Br_2 and Kr on Grafoil [20, 21]. In this case the problem of low signal in surface experiments was circumvented by the extremely high surface area ($22m^2/g$) of Grafoil. From the polarization dependence of K absorption spectra it was concluded that at a coverage between 0.6 and 0.9 monolayers the Br_2 molecules lie flat in the basal plane. From the fact that at lower coverages no anisotropy has been observed it was concluded that only one atom is bound to the surface and the molecular axis is free to rotate. For Kr on Grafoil the adsorption site was determined to be the center of the carbon hexagons.

These few examples should show that the analysis of anisotropic effects in EXAFS yield additional information which may be necessary for the completion of a structural model. The most promising perspectives certainly can be expected in surface science. One should, however, consider experiments on bulk single crystals especially with a fair degree of imperfection where diffraction experiments loose their high potential accuracy.

References

1. J.W.Cooper and S.T.Manson, Phys.Rev.177,157 (1969)
2. P.A.Lee, Phys.Rev.B 13,5261 (1976)
3. S.M.Heald and E.A.Stern,Phys.Rev.B 16,5549 (1977)
4. B.-K.Teo and P.A.Lee,J.Am.Chem.Soc.101,2815 (1979)
5. P.Rabe, G.Tolkiehn, and A.Werner,J.Phys.C 12, 899 (1979)
6. R.H.Pratt, A.Ron, and H.K.Tseng,Rev.Mod.Phys.45,273 (1973)
7. H.K.Tseng, R.H.Pratt, S.Yu, and A.Ron,Phys.Rev.A 17,1061 (1978)
8. Y.S.Kim, R.H.Pratt, A.Ron, and H.K.Tseng,Phys.Rev.A 22,567 (1980)
9. P.Rabe, G.Tolkiehn, and A.Werner,Nucl.Instr.Methods 171,329 (1980)
10. G.S.Brown, P.Eisenberger, and P.Schmidt,Solid State Commun. 24,201 (1977)
11. P.Rabe, G.Tolkiehn, and A.Werner,J.Phys.C 13,1857 (1980)
12. F.Jona,Surf.Sci 68,204 (1977)
13. J.Stöhr,Jap.J.Appl.Phys.17-2,217 (1978)
14. W.Gudat and C.Kunz,Phys.Rev.Letters 29,169 (1972)
15. G.Martens, P.Rabe, N.Schwentner, and A.Werner,J.Phys.C 11, 3125 (1978)
16. T.Fujikawa,J.Electr.Spectr.22,353 (1981)
17. G.Martens, W.Pronkow, and P.Rabe, to be published
18. P.H.Citrin, P.Eisenberger, and R.C.Hewitt,Phys.Rev.Letters 45,1948 (1980)
19. L.I.Johansson, J.Stöhr, and S.Brennan,Appl.Surf.Science 6, 419 (1980)
20. S.M.Heald and E.A.Stern,Phys.Rev.B 17,4609 (1978)
21. C.Bouldin and E.A.Stern,Phys.Rev.B 25,3462 (1982)

EXAFS in Metallurgy

E. Dartyge[1], A. Fontaine[1], and J. Mimault[2]

LURE, CNRS, Bât. 209c, F-91405 Orsay Cedex, France

[1]Lab. Physique des Solides, F-91405 Orsay Cedex, France

[2]Lab. Met. Physique, 40 Av. Rect. Pineau, F-86022 Poitiers, France

1. Introduction

Focused on metallurgy this paper addresses crystalline systems only, since amorphous metallic alloys have been discussed elsewhere [1]. The outline of this short review will be to show typical metallurgical problems where EXAFS has proved its efficiency. Some problems previously reviewed in the Daresbury meeting will not be recalled [2].

The first section is a short account of how a host matrix accommodates an isolated solute atom. Although very simple this measurement was only possible by indirect experiment before EXAFS became available.

The study described in section II concerns one-spin glass system (Ag-Mn) where atomic short-range order has been observed by X-ray diffuse scattering. The two aims of this study were to measure the first Cowley parameter, and to check the lack of distortion of the host matrix about the solute magnetic atom, which has been assumed for the analysis of diffuse scattering intensities.

As a last point the Al-Zn system provides a nice system for which the identification of the backscatters proceeds essentially from the difference in the delays of phase of the partial backcoming electronic waves. It has been thus possible to investigate the chemical composition of the GP zone which is almost spherical clusters, embedded coherently in the host matrix which is not completely depleted.

Such studies of the environment of the dilute species [4,5,6] are almost impossible (or very indirect in specific cases) using standard diffraction techniques since these sum over all atomic pairs whatever the species are. Hence the contribution from the solute atoms to the signal is swamped by that from the host atoms.

Besides these experiments where point defects or extended defects are involved, a very exciting challenge would be linear defects. We mean dislocations in a binary III-V system for example. Does the edge-dislocation end the half-plane preferentially on one atomic species or on the other? Can we measure the distortion in the core of this linear defect? Using fluorescence detection it would be possible to do that if a third atom, isomorphous of one of the two components, can be preferentially substituted... but today this project is speculative...

2. Point defects : solute atom in random solid solutions

The determination of the radius of the first shell around a solute atom is the most simple measurement, very suited to EXAFS since it requires only the knowledge of accurate phase shifts which are possible to extract from well-defined standards.

As it was stressed during this conference [3] the transferability of the phase shifts of an atomic pair between two chemically close systems is really well settled , whereas the transferability of amplitudes requires more care.

Such measurement is not possible using diffraction methods since these sum, over all distances of the crystal the largest ones being more numerous than the shortest ones. The positions of the Bragg peak yield an average of the structure over distances which are large compared to the next neighbour distance.

Nevertheless, because of the lack of available results, the assumption that the solute-host distance is equal to the host-host distance has been often used even for local properties which may depend drastically on distance. Hence the isolated atom can be used to probe the crystal field through the commonly called b_4 coefficient [7] which follows a rapid power law with respect to the first-neighbour distance. This cannot be checked if the elastic core effect is large as AZOULAY et al. [8] underlined.
The estimation of the amount of distortion is needed too to analyse small-angle scattering data. It may be very difficult to conclude on the rate of the atomic short-range order without any information about the relaxation of the host matrix.

In that field the contribution of EXAFS has been really original and will become larger and larger during the next years. An extensive review of EXAFS studies of dilute alloys including a scope in materials science has been given by S.J.GURMAN [9], so we shall only emphasize here as a typical physical result, the large dicrepancy of the EXAFS measurement of the elastic core effect in dilute aluminium-based alloys [10] with the values derived from VEGARD's law via an elastic theory which treats the host matrix as a *continuous medium*. This last calculation tells us how the local change in the interatomic distance due to point defects (the solute atoms)creates the macroscopic effect which is the variation of the lattice parameter of a substitutional solid solution. This continuum analysis gets the sign of the change of the first shell radius correct (negative for Cu and Zn in Al, positive for Mg in Al as it is expected from their values of atomic radii) but underestimates the magnitude of the change. The discussion to know whether it is due to the discrete atomic nature of the crystal or to the electronic forces rising from the screeming effect over the first shell will be found in this book [11].

Experimental phase shifts extraction

Finally we would mention briefly the problem of quasi-experimental extraction of accurate phase shifts of a heterogeneous pair built by elements whose Z numbers differ only by a few units. Starting from the phase shifts of the Al-Al pair or the Mg-Mg pair provided by EXAFS spectra on elemental metals it has been possible to simulate the total phase shifts of the Mg-Al pair using either a theoretical correction [12] for change of the central atom or a theoretical correction for change of the backscatterer [10]. Since the two Z numbers differ only by one the theoretical correction is small and the inaccuracy in the final results may be expected of the second order of magnitude.

But we have to point out that each phase shift for homogeneous pairs is not unique since it depends upon the choice of the origin of the kinetic energy (E_0) which determines the k scale. However the two proposed ways to build these phase shifts (Mg-Al) serves as a consistency check since it has to lead to a unique set of values.

It can be seen on Fig.1 that the trial when the origin of k is put on the inflection point of each absorption edges the discreapancy between the two set is large even if their k dependences show very similar slopes. Indeed E_0 should not be taken at the Fermi level, but a lower energy near the bottom of the conduction band of these free electron-like metals. The correct value should be the extrapolation of the parabolous-like curve observed for fast free photo-electrons. Since Al and Mg are very simple metals it is

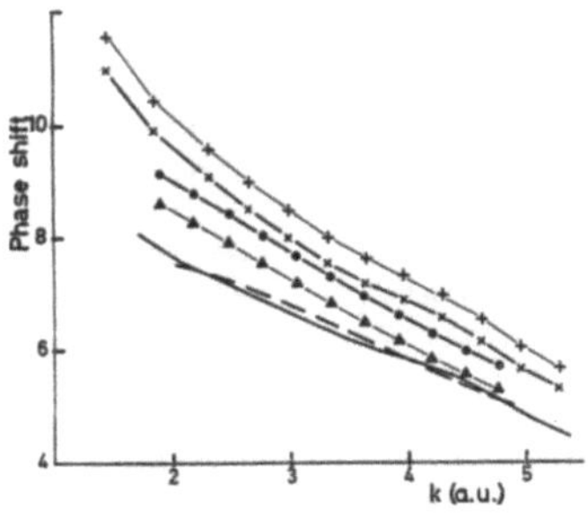

Fig.1. Experimental determinations of Al-Al (+) and Mg-Mg (Δ) phase shifts, the k origin at the inflection point of the K edges.
- corrected phase shifts which simulate a Mg-Al pair. (X) result from the Al-Al data corrected for the central atom and (.) from the Mg-Mg data corrected for the backscatterer, the k origin at the inflection point of the edges.
- the two curves below show the same phase shifts with the kinetic energy origins taken at the bottom of the conduction bands, i.e. 7.1 eV below the Mg edge (---) and 11.7 eV below the Al one (——)

rather good to choose for E_0 the bottom of the conduction band : 11.7 eV and 7.1 below the Fermi level for Al and Mg respectively. The two determinations for the Mg-Al phase shifts are then the same within experimental errors.

The same method is currently used to determine the phase shifts of the Cu-Zn pair, for the two orientations of the single crystal of Zn doped with 1 at. % Cu, in order to study the anisotropy of the elastic core effect in an anisotropic matrix [13]. The Fourier transforms of the Fig.2a show clearly the expected orientation dependence of the EXAFS spectra. They reproduces almost exactly the Fourier transforms of the two spectra recorded for both orientations of the pure zinc crystal except a slight shift of the distances. For comparison Fig.2b represents the F.T. of the powder-like spectrum which yields the expected radius (3.92 Å) for the third shell using theoretical phase shifts [6] and origin of the kinetic energy located 5eV below the edge.

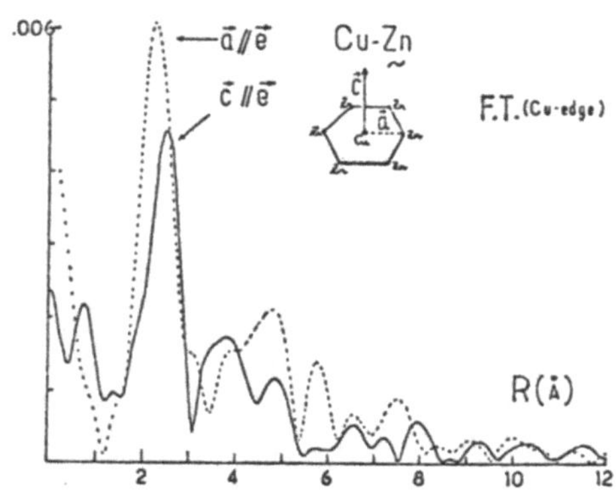

Fig.2a. Orientation dependence of the Fourier transforms of Cu-edge spectra of the single crystal Zn 1 at % Cu

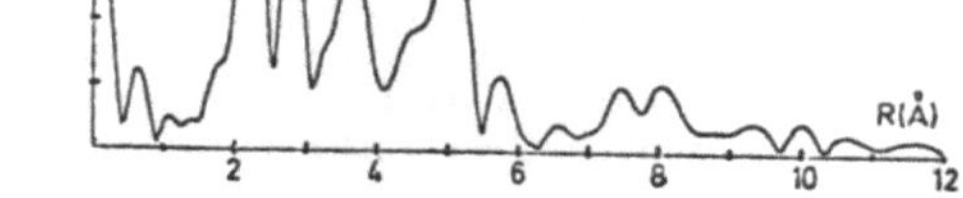

Fig.2b. Fourier transform for polycrystal Zn

3. Atomic short-range order in Ag-Mn spin glass

In order to correlate changes in magnetic properties in the Ag-Mn spin glass alloys with changes in short range order (s.r.o.), EXAFS experiments have been performed on Ag-Mn quenched alloys as well as pure Ag and α-Mn samples. These measurements follow X-ray diffuse scattering experiments carried out on Ag-Mn single crystals [14]. The X-ray scattering experiments showed that s.r.o. exists in Ag-Mn quenched alloys of composition larger than 6 at. %.

Two models of s.r.o. were compatible with this earlier diffuse scattering experiments :

i) a heterogeneous model, with Mn atoms in microdomains of the 50 at. % stoichiometry and $I4_1/amd$ symmetry.

ii) a statistical homogeneous model which can be described with the Cowley short-range order parameter. Small angle X-ray scattering experiments suggested us to eliminate the first model and to adopt the homogeneous one.

However, when handling the diffuse scattering data the scattered intensity profiles are assumed to be Lorentzian and centred at (1 1/2 0) and equivalent reciprocal spots. No distortion effect is taken into account. The α_i values yielded by this idealized model of the diffuse scattering have to be checked by an other independent measurement.

The estimation of the average number of Mn in next-neighbour position of another Mn atom, and the measurement of the core relaxation of the matrix is specifically in the EXAFS field.

The magnitude of the local variation of the first shell radius about Mn atoms and the value of the Cowley s.r.o. parameter (α_1) were measured on five alloys from Ag-6 at. % Mn alloy up to the 24 at. % one. For every sample the Mn-Ag distance differs only by 2/100 Å from the Ag-Ag distance measured at the same temperature in pure silver. Thus only a very weak distortion effect exists in these alloys. The Cowley parameter (α_1) are in good agreement with what the analysis of the X-ray diffuse scattering experiments provides. That confirms that the number of Mn atoms first neighbours of one Mn atom is less than the statistical value (Table 1).

Table 1 [15]

C Mn at. %	N(Mn) EXAFS	N(Mn) diff.scattering	N(Mn) random alloy
6	0		0.7
10	0	0	1.2
15	0	0	2
20	1.5	0.9	2.2
24	2	1.3	2.9

α-Mn

The theoretical phase shifts and amplitude required in the EXAFS analysis have been checked on elemental silver and α-Mn. This last metal provided

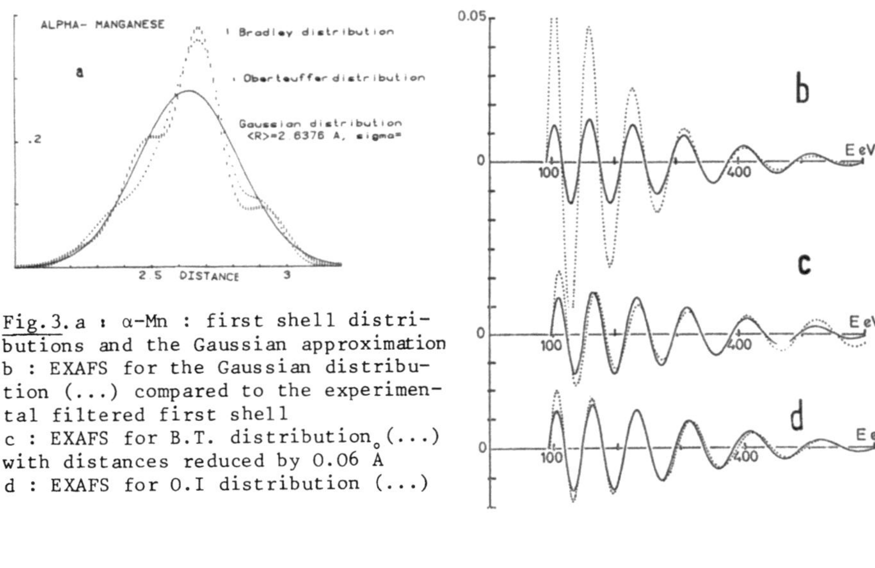

Fig.3.a : α-Mn : first shell distributions and the Gaussian approximation
b : EXAFS for the Gaussian distribution (...) compared to the experimental filtered first shell
c : EXAFS for B.T. distribution (...) with distances reduced by 0.06 Å
d : EXAFS for O.I distribution (...)

a very good example to check the power of EXAFS to discriminate even broad asymetrical distributions in amorphous systems. Since 58 atoms of Mn were present in the unit cell the first determination of the structure given by [16] in 1927 was not as accurate as the last one [17] Fig.3a. Only slight differences in the shape of the distribution are recorded.

On Fig.3b,c,d, three theoretical first shell spectra are compared to the experimental filtered first shell. The discrepancy is well shown for the amplitude when using the Gaussian distribution built with the same $\bar{R}$ value and standard deviation as the Oberteuffer one. With the Bradley distribution a better fit is obtained but with distances reduced by .06 Å. Conversely the Oberteuffer distribution gives a good fit without any other reduction of the distance but the dilatation effect.

4. G.P zone composition in Al-Zn

We have previously emphasized the extreme sensitivity of the EXAFS spectrum to progressive Al substitution by Zn atoms which allows to follow accurately the birth of clusters in the AlZn alloy [2]. This suitability to EXAFS studies is intimately related to

i) the difference in phase shifts which is near π between the two possible backscaterrers, Al or Zn atoms.

ii) the largest magnitude of the backscattering which is for the solute atom Zn dispersed in a light matrix.

Hence for the unannealed solid solution (Al-0.83 at. % Zn) and for the ultimate annealed Al-6.8 at. % Zn alloy - where large GP zones are coherently embedded in the host matrix - the two spectra are in phase opposition in almost the whole energy range. The annealing-induced progressive change of the first shell composition between these extreme metallurgical states provides a rapid fall of the magnitude of the Fourier transform (F.T.) followed by a deeply splitted first peak which can be interpreted as a composition of 7 Al atoms plus 5 Zn atoms for the average first shell of each Zn atom (Fig.4).

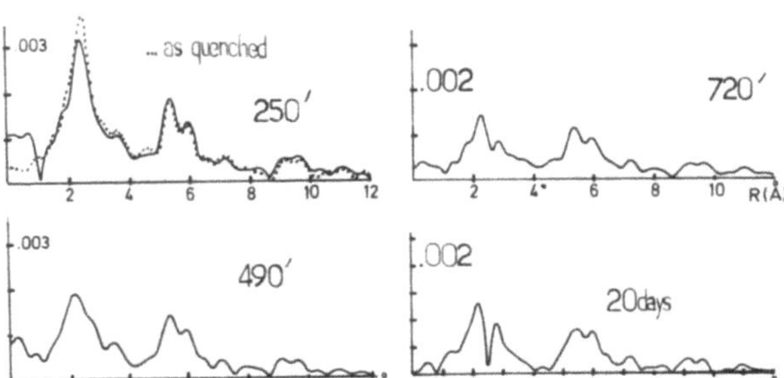

Fig.4. Progressive changes of the Fourier transforms of annealed Al-4.4 at. % alloy at room temperature with increasing time

This sequence of shape changes of the first peak is observed with the same features (including the splitted first peak) for larger concentrations but at shorter annealing times (20 mn for 6.8 at. %, just as quenched for 9.6 at.% instead of 20 days for 4.4 at. %).

Nevertheless the topological situation of Zn atoms involved in middle-sized clusters (let us say a few ten atoms) is too complex - because of possible large size distributions - to give more information than an average environment. Hence we focus our study of large GP zones formed at room temperature whose composition is only known by rather indirect measurements reviewed in [18]. The Zn concentration of the remaining solid solution at room temperature has been estimated mainly by X-ray small angle scattering to be 1.8 at. % - whereas the GP is claimed to be 70 at. %.

EXAFS data has been collected for four atomic compositions (4.4, 6.8, 9.6, 15 at. % Zn) in order to change the weight of the GP zones spectrum with respect to that of the solid solution [19]. For the three last concen-

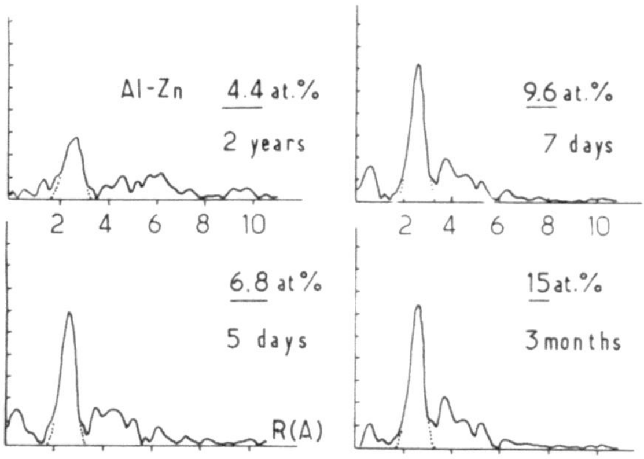

Fig.5. Fourier transforms of EXAFS spectra of 4 Al-Zn alloys annealed at room temperature

trations the Fourier transform looks alike except small changes in magnitude. The size of the GP zone has been calculated using the Guinier plot to display the small-angle scattering intensities. That leads to the following Guinier diameters 30 Å, 50 Å, 54 Å, 52 Å respectively for the four alloys. The annealing times were 2 years for the 4.4 at. % alloy, 5 days, 50 and 90 days for 6.8 at. %, 7 days, 3 and 18 months for 9.6 at. %, 3 and 18 months for 15 at. %.

EXAFS analysis were performed using the Fourier - filtering method. The first shell EXAFS were fitted by theoretical spectra using phase shifts and amplitudes checked in the previous study [6]. Details will be available elsewhere (19). Since for GP zones, lattice parameters are

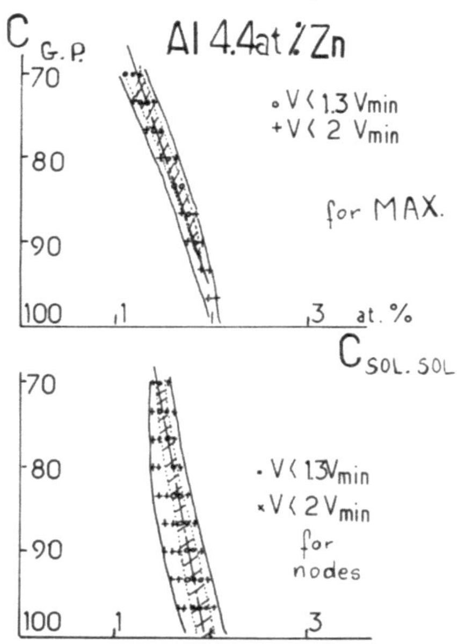

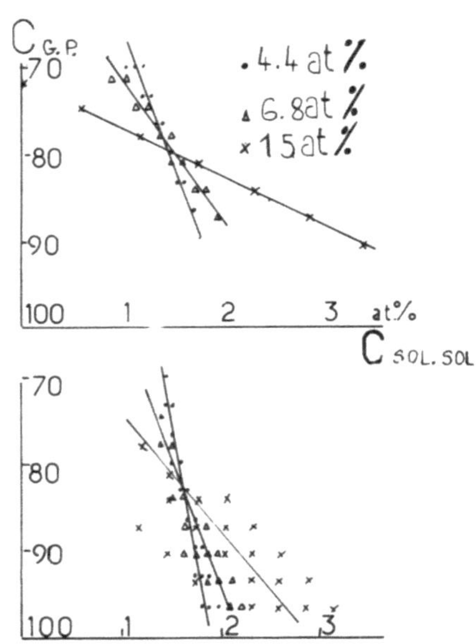

Fig.6. Al-4.4 at. % Zn annealed at r.t. for 2 years.
a) Variance of the magnitude of the extrema in a C(GP zone), C(sol.sol.) diagram
b) id. of the zero locations

Fig.7. The same plot for the 3 concentrations

known, the independent parameters can be reduced to the GP zone concentration and the solid solution concentration. In planes defined by these two coordinates, it is possible to plot the variation of the set of variances calculated from the standard deviation of the zero locations of the theoretical fits compared to those of the filtered first shell (Fig.6b). The same calculations were performed for the magnitude of the extrema of the EXAFS oscillations (Fig.6a). That yields narrow loops defined by a maximum variance at the border equal to twice the best one. The bottom of the valleys is recalled by an almost straight line which is more sloped for the magnitude diagram than for the nodes one. That means that the nodes locations are less sensitive to small change in the repartition of Zn atoms between GP zone and solid solution than the magnitude of the spectrum for larger compositions as both loops are more inclined. Fig.7 gathers these calculations of variance. All valleys intersect each other for the 80 at. % value for the GP zone contraction and 1.5 at. % for the solid solution one.

5. Conclusion

The ability of EXAFS to investigate defects correlated to solute atoms diluted in a host matrix has been demonstrated. The contrast of EXAFS spectrum is improved if the dilute species has a Z number larger than the one of the host matrix. The α-Mn structure which provides an amorphous-like first shell viewed in terms of asymmetric broad distribution yields a good example of the sensitivity of EXAFS to slight changes in the atomic distribution. We finally emphasize the power of this structural tool to investigate all problems of metastable alloys in which segregation often occurs in form of small clusters which are, virtually, impossible to study using conventional diffraction methods.

References

1 D.Raoux, published in this book
2 A.M.Flank, A.Fontaine, P.Lagarde, M.Lemonnier, J.Mimault, D.Raoux, A.Sadoc, EXAFS for Inorganic Systems ed. by C.D.Gardner and S.S.Hasnain (1981) p.70
3 P.Eisenberger, published in this book
4 A.Fontaine, P.Lagarde, A.Naudon, D.Raoux, D.Spanjaard, Phil.Mag. B40, 17 (1979)
5 H.Maeda, T.Tanimoto, H.Terauchi, M.Hida, Phys.Stat.Sol. 58, 629 (1980)
6 J.Mimault, A.Fontaine, P.Lagarde, D.Raoux, A.Sadoc and D.Spanjaard, J.Phys. F. 11, 1311-26 (1981)
7 Urban, Seipler, J.Phys.F. 78, 1589 (1977)
8 Azoulay, E.Stern, Shaltiel, Grayevski, Phys.Rev. 25, 9, 5627 (1982)
9 S.J.Gurman, J. of Materials Science, 17, 1541-1570 (1982)
10 D.Raoux, A.Fontaine, P.Lagarde, A.Sadoc, Phys.Rev. B24, 10, 5547 (1981)
11 A.Craievich, E.Dartyge, A.Fontaine, D.Raoux, published in this book
12 B.K.Teo, P.A.Lee, J.of Am.Chem.Soc. 101, 2815 (1979)
13 E.Dartyge, A.Fontaine, to be published
14 H.Bouchiat, E.Dartyge, P.Monod, M.Lambert, Phys.Rev. B23, 1375 (1981)
15 E.Dartyge, A.Fontaine, submitted to J. of Physics F
16 J.Bradley, J.T.Tewlis, Proc.Roy.Soc. A115, 456 (1927)
17 J.A.Oberteuffer, J.A.Ibers, Act.Cryst. B26, 1499 (1970)
18 A.Junqua (thesis) Poitiers (197)
19 J.Mimault, A.Fontaine, to be published.

The Effect of the Noise on the Structural Parameters Deduced by EXAFS

Settimio Mobilio and Lucia Incoccia
PULS-CNR Laboratori Nazionali di Frascati, I-00044 Frascati, Italy

1. Introduction and method.

The accurate determination of structural parameters from an EXAFS spectrum requires many precautions both in the experimental data collection and in the data analysis. Indeed amplitude non-transferability [1], defects in the samples [2] higher harmonic presence in the X-ray beam are well-known causes of poor accuracy, principally in the determination of coordination number and of Debye-Waller factors.

Also the noise of the measurements plays an important role in limiting the accuracy of the physical quantities measured by EXAFS [3]. This point turns out to be especially important whenever it is very difficult to get good S/N ratio, as is the case, for example, of highly dilute samples or of SEXAFS spectra.

We evaluated the influence and the limitations to the accuracy of each structural parameter obtained from EXAFS, caused to the noise in the measurement. The method adopted was the following:

a) Generation of the ideal spectra for several materials of known structure at T = 77 K and T = 300 K using calculated phases, amplitudes [4] and D. W. factors [5], and the relation (Fig. 1):

$$\chi(k) = \sum_i \frac{N_i e^{-2\sigma_i^2 k^2}}{kR_i^2} \sin(2kR_i + \varphi_i) A_i(k) .$$

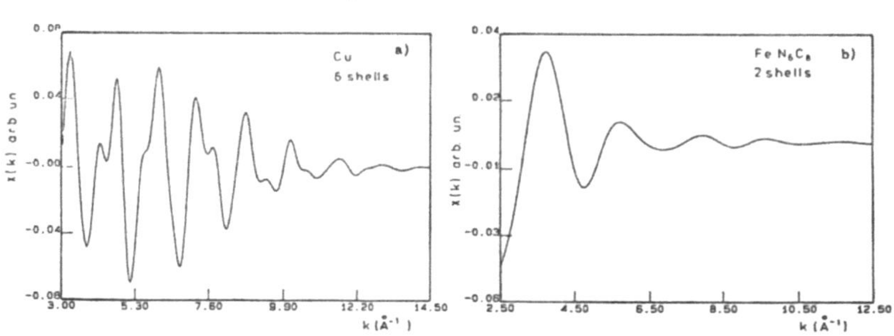

FIG. 1. Theoretical $\chi(k)$ for 6 shells of fcc Cu and for the FeN_6C_8 molecule at T = 300 K

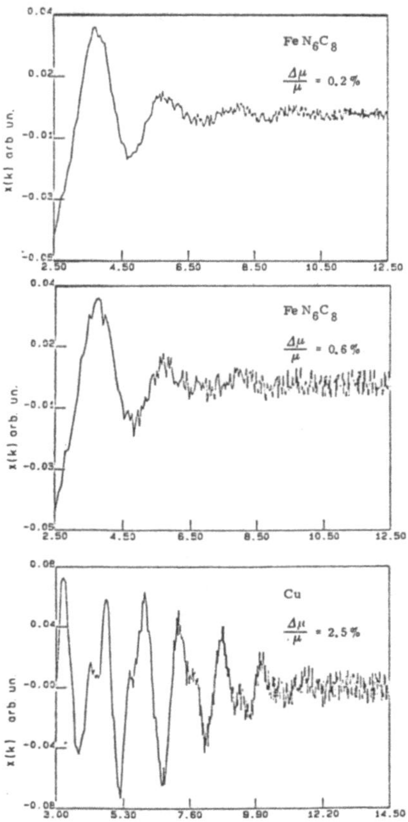

FIG. 2. Examples of the real $\chi(k)$'s obtained for different noise value

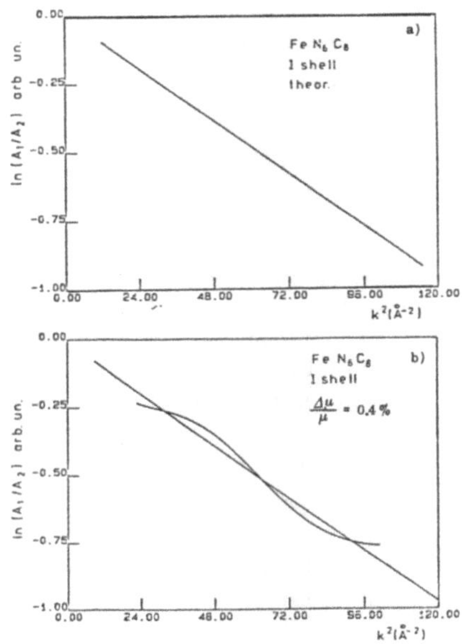

FIG. 3. Plot of $\ln(A/A_M)$ for the 1st shell of FeN_6C_8. a) from the ideal spectrum; b) from the a real spectrum with 0.4 % error on μ; the straight line is the result of a fit in the range $k^2 = 20\text{-}100\ A^{-2}$

b) Generation of the real spectra at T = 300 K by adding a given noise to the ideal ones. The noise was obtained using a random number generator with both uniform and Gaussian distributions (Fig. 2).

c) Analysis of the real spectra at 300 K using the relative LNT ideal spectra as models. From the Fourier transform of the spectra the position and intensity of the peaks (R, y) were obtained; from the inverse Fourier transform of a given peak the relative coordination number and D. W. factors were evaluated by fitting a straight line to the plot of $\ln A_i(k)/A_M(k)$ vs k^2. Here $A_M(k)$ is the envelope function of the inverse F. T. of the model (Fig. 3).

This procedure was repeated 1000 times for each material for several values of the noise magnitude. The distributions of the results obtained were statistically analyzed evaluating the mean value and the functional standard deviation $d = \sqrt{V^2/\bar{x}}$ where V^2 is the variance of the distribution.

2. Results and Conclusions.

In tables I and II we report the results obtained for the two shells in the case of metallic Cu and the fictive molecule FeN_6C_8 (6 N atoms at 2.05 and 8 C atoms at 2.78).

TABLE I - a) Errors on R and y evaluated from the distributions for the first and third shell of Cu; b) errors on N and $\Delta\sigma^2$.

Cu	(a)				(b)		
	I shell $\bar{R}$ = 2.24 Å $\bar{y}$ = 0.71		III shell $\bar{R}$ = 4.10 Å $\bar{y}$ = 2.28		Cu I shell	$\overline{N_1/N_2}$ = 1	$\overline{\Delta\sigma^2}$ = 0.0042 Å^2
$\frac{\Delta\mu}{\mu}$	ΔR	dy	ΔR	dy	$\frac{\Delta\mu}{\mu}$	d_N	d_{σ^2}
1.0 %	0.007 Å	1.4 %	0.012 Å	3.2 %	1.0 %	3.2 %	7.1 %
2.5 %	0.01 Å	2.4 %	0.02 Å	5.8 %	2.5 %	6.3 %	13.8 %
5.0 %	0.015 Å	4.6 %	0.04 Å	10.5 %	5.0 %	12.3 %	26.4 %
6.5 %	0.022 Å	6.7 %	0.06 Å	14.0 %			

TABLE II - Same as Table I for the 1st and 2nd shell of FeN_6C_8.

	(a)				(b)		
	I shell $\bar{R}$ = 1.57 A $\bar{y}$ = 2.54		II shell $\bar{R}$ = 2.27 A $\bar{y}$ = 0.690		FeN_6C_8 I shell	$\overline{N_1/N_2}$ = 1	$\overline{\Delta\sigma^2}$ = 0.004 Å^2
$\frac{\Delta\mu}{\mu}$	ΔR	dy	ΔR	dy	$\frac{\Delta\mu}{\mu}$	d_N	d_{σ^2}
0.4 %	0.03 Å	6.0 %	0.06 Å	22.3 %	0.2 %	10 %	30 %
0.8 %	0.06 Å	11.0 %			0.4 %	18 %	54 %
1.2 %	0.09 Å	14.0 %					

The differences found in the two cases reported are mainly due to the different k range available for the analysis : in the case of a heavy backscatterer, indeed, the large k range for which $\chi(k)$ is greater than the noise allows a good averaging of the noise itself so that its influence is, at last for the first shell, negligible both on R and on N and σ^2 even for highly noisy spectra. In such a case the limitations to the accuracy of EXAFS data are due to other factors. As for the third shell it can be observed that a good accuracy on R determination ($\delta R \simeq 0.05$) can still be achieved at high values of the noise. For the amplitude parameters an R-space determination is still accurate (error of 14% on peak intensity for 65% on the spectra), while a K-space analysis is not reliable due to the loss of resolution of the peak with respect both to the noise peaks and to the other shells.

In the case of light backscatterers, the situation is much worse. An accurate analysis of Table II shows that in such case an error of less than 1%

on μx is enough to lose the usually claimed accuracy of 0.01 A on R determination only as a consequence of the experimental noise. The results on the amplitude parameters clearly indicate that for a light backscatterer extremely good spectra are required to get good accuracy on N and σ also for the first shell.

The main conclusion of this work is that often the noise in a spectra is the main limitation to the accuracy of the EXAFS structural determination principally when dealing with second or third shell and with light atoms as backscatterers : in any case the table here reported are to be consulted to get an upper limit to the accuracy one can get from an EXAFS spectrum.

1. E. A. Stern, B. A. Bunker and S. M. Heald, Phys. Rev. B21, 5521 (1980); P. Eisenberger and B. Lengeler, Phys. Rev. B22, 3551 (1980).
2. E. A. Stern and K. Kim, Phys. Rev. B23, 3781 (1980); J. Goulon, C. Goulon-Ginet, R. Cortes and J. M. Dubois, to be published; P. Rabe, G. Tolkiehn and A. Werner, Nuclear Instr. and Meth. 171, 329 (1980).
3. L. Incoccia and S. Mobilio, Frascati Report LNF-82/ (1982); to appear in Riv. Nuovo Cimento.
4. B. K. Teo and P. A. Lee, J. Am. Chem. Soc. 101, 2015 (1979).
5. E. Sevillano, H. Meuth and J. J. Rehr, Phys. Rev. B20, 4908 (1979).

Debye-Waller Factors in *R*-Space Analysis

Lucia Incoccia and Settimio Mobilio

PULS-CNR Laboratori Nazionali di Frascati, I-00044 Frascati, Italy

It has been recently underlined that the knowledge of the EXAFS Debye-Waller factors σ^2_{DW} with high accuracy provides information on the interacting potential between atom pairs in the condensed matter [1, 2]. The evaluation of σ^2_{DW} is usually done in k space, following the now standard procedure suggested by Stern et al. [3].

This procedure is affected by conceptual difficulties such as amplitude non-transferability or mean-free-path effects [4, 5] as well as by experimental factors like the noise in the measurements and the distortion due to the data analysis.

We have shown [6] that the effect of the noise in R space is much smaller than in k space.

Thus from this point of view an R-space approach to calculate Debye-Waller factors and coordination numbers should be more accurate. Furthermore the influence of factors like amplitude non-transferability or many-body effects is probably less important in an R-space analysis.

The evaluation of N and σ in R space has been attempted before [7, 8], the conclusion being that the presence of side lobes in the FT and of correlation effects (whose origin is in the convolution of the RDF with the FT of the backscattering amplitude) prevent the use of this technique. In our approach, we define the reduced function:

$$\chi'(k) = \frac{\chi(k)\,k}{f(k,\pi)} \quad ,$$

where $\chi(k)$ is the component of the experimental spectrum due to a given shell, and $f(k,\pi)$ is the corresponding backscattering amplitude. Working on $\chi'(k)$ instead of $\chi(k)$ eliminates the correlation effect [9].

Assuming a Gaussian RDF around the atomic site R and a square window function, the Fourier transform of $\chi'(k)$ in the range $k_{min} - k_{max}$ is given by [11]:

$$FT\,\chi'(k) = \frac{i N e^{-(R-r)^2/2\sigma^2}}{8\sigma R^2} \left\{ \mathrm{erf}\left(\sqrt{2\sigma^2}\,k_{max} - \frac{i(R-r)}{\sqrt{2\sigma^2}}\right) - \right.$$

$$\left. - \,\mathrm{erf}\left(\sqrt{2\sigma^2}\,k_{min} - \frac{i(R-r)}{\sqrt{2\sigma^2}}\right) \right\} .$$

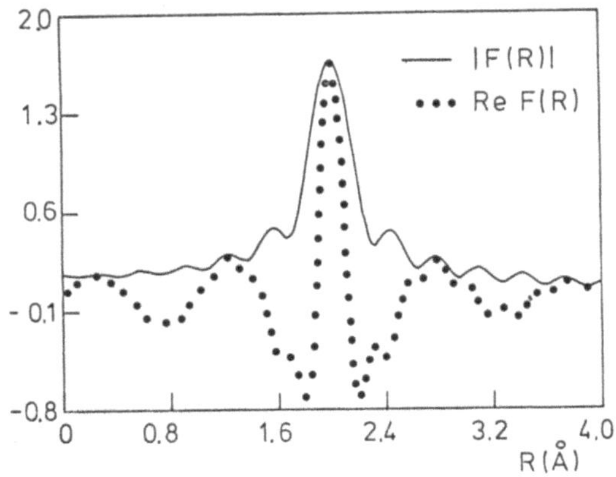

Fig. 1. Fourier transform of the function $k\chi'(k)$ built with the following parameters: N = 1, R = 2 $Å^{-1}$, σ^2 = 0.004 $Å^{-1}$. The transformation range was: k_{min} = 3 $Å^{-1}$, k_{max} = 13 $Å^{-1}$

Since the function erf (x + iy) exhibits oscillations whose main frequency is 2xy, neither the real nor the imaginary part of the FT of $\chi'(k)$ can be used as a picture of the radial distribution function of the system. Moreover the magnitude of the FT shows side lobes (Fig. 1). Such undesirable distortions to the lineshape of the |FT| are removed if good spectra are collected up to k_{max} values such that $e^{-2k^2_{max}\sigma^2} \simeq 0$.

In this case the function:

$$\mathrm{erf}\left(\sqrt{2\sigma^2}\, k_{max} - \frac{i(R-r)}{\sqrt{2\sigma^2}} \right)$$

can be approximated by its limiting value 1. Still the real and imaginary parts of the FT exhibit oscillations (Fig. 2), whose main frequency is $2Rk_{min}$, but due to the $\pi/2$ shift between them the magnitude no longer shows side lobes (Fig. 2).

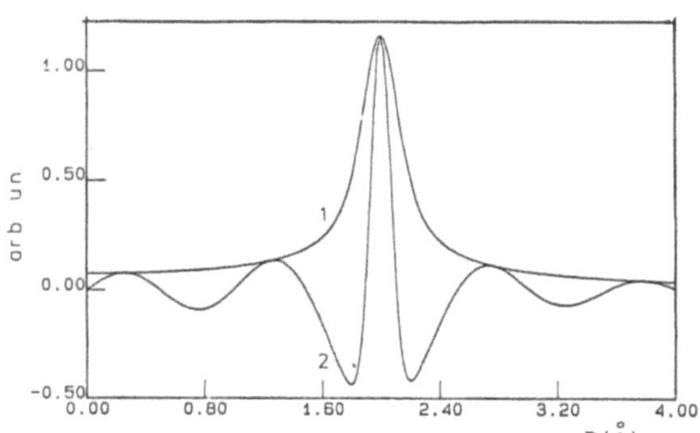

Fig. 2. Magnitude (1) and imaginary part (2) of the FT of $\chi'(k)$ in the case k_{min} = 3 $Å^{-1}$ and k_{max} -- ∞

The maximum M of the |FT| is now related to both N and σ^2 by the relation:

$$N = \frac{8\sigma R^2 M}{(1 - \operatorname{erf}(\sqrt{2\sigma^2}\, k_{min}))} .$$

The correlation between N and σ prevents a general use of the above relation. Nevertheless it is possible to correlate the width of the |FT| with the σ^2_{DW} only. This relation is graphically shown in Fig. 3, for several values of k_{min} [11].

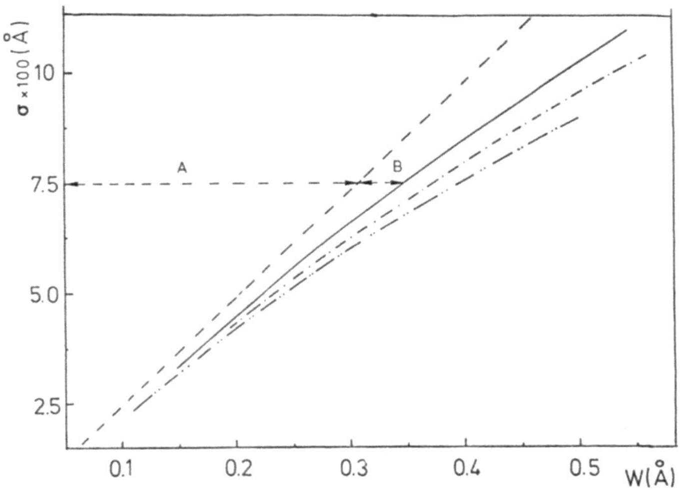

Fig. 3. Behavior of χ as a function of the FWHM of the magnitude of the Fourier transform of $\chi'(k)$ for different values of k_{min} ($Å^{-1}$): 0 (---), 2 (——-), 3 (· -· -), 4 (·· -· ·-)

From the figure it is possible to evaluate the two contributionsto the width of the |FT| due to Debye-Waller factors (A) and by the finite range of transformation (B). It comes out that for typical values of σ the contribution of the Debye-Waller factors to the width is around 80%.

We tested the applicability of this R-space approach in some crystalline systems, measuring the variation of σ^2 in going from liquid nitrogen temperature to room temperature.

The spectra at the two temperatures were taken at the facility PULS in Frascati. The reduced function $\chi'(k)$ was build by dividing the $k\chi(k)$ of each RT spectrum by the envelope function of the inverse Fourier transform of the first shell of the LNT spectrum, i. e.

$$\chi'(k) = \frac{k\chi_{300K}(k)}{A_{77K}(k)} .$$

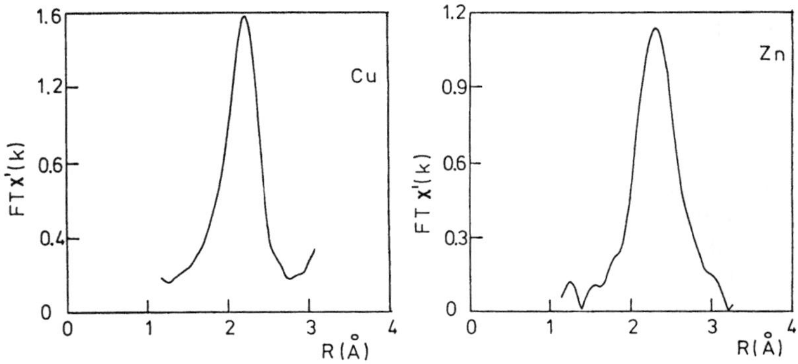

Fig. 4. Magnitude of the FT of $\chi'(k)$ in the case of Cu (a) and Zu (b)

In this way $\chi'(k)$ carries the information about $\Delta\sigma^2$, that is the difference between σ^2_{DW} of the 1st shell at the two temperatures.

In Fig. 4 we show the FT of the reduced $\chi'(k)$ in the case of Cu and Zn. From these FT's the $\Delta\sigma^2$ values were obtained by means of the plot of Fig. 3. Besides this R-space evaluation, we also determined $\Delta\sigma^2$ following the more usual procedure in k space, from the slope of the function

$$\ln \frac{A_{300k}(k)}{A_{77k}(k)} \quad \text{vs.} \quad k^2 .$$

In Table I we report the results so obtained together with some theoretical values when available.

We want to stress two points : a) there is a very good agreement between the results obtained by the usual k-space procedure and the ones obtained by our approach, thus demonstrating that the method is applicable ; b) our results are in closer agreement with theoretical values, suggesting that this R-space method is more accurate.

Table I Values of $\Delta\sigma^2$ obtained in R space compared with the ones obtained in k space and with the values of the literature, both experimental and theoretical

Sample	$\Delta\sigma^2$ R space	$\Delta\sigma^2$ k space	$\Delta\sigma^2$ calculated	$\Delta\sigma^2$ other authors
Ni	0. 31	0. 33		
Cu	0. 43	0. 49	0. 44(2)	0. 48(5, 10) 0. 46(8)
Co	0. 34	0. 37	0. 34	0. 25(10)
Ti	0. 44	0. 48		
Zn	0. 74	0. 75		
Ge	0. 17	0. 20	0. 15(5)	0. 04

1. G. Beni and P. M. Platzman, Phys. Rev. B14, 1514 (1976).
2. E. Sevillano, H. Meuth and J. J. Rehr, Phys. Rev. B20, 4908 (1979).
3. E. A. Stern, D. E. Sayers and F. W. Lytle, Phys. Rev. B11, 4836 (1975).
4. P. Eisenberger and B. Lengeler, Phys. Rev. B22, 3551 (1980).
5. E. A. Stern, B. A. Bunker and S. M. Heald, Phys. Rev. B21, 5521 (1980).
6. L. Incoccia and S. Mobilio, Riv. Nuovo Cimento (in press); also Frascati report LNF-82/52 (1982).
7. D. E. Sayers, F. W. Lytle and E. A. Stern, Jour. of non Cryst. Solids 8, 401 (1972).
8. R. B. Greegor and F. W. Lytle, Phys. Rev. B20, 4902 (1979).
9. P. A. Lee and G. Beni, Phys. Rev. B15, 2862 (1977).
10. W. Böhmer and P. Rabe, J. Phys. C12, 2465 (1979).
11. S. Mobilio and L. Incoccia, Riv. Nuovo Cimento (in press); also Frascati report LNF-82/44 (1982).

Thickness Effects on the Determination of EXAFS Amplitudes

J. Goulon[1,2], C. Goulon-Ginet[1], R. Cortes[3], and J.M. Dubois[4]

We wish to highlight here some artefactual limitations in the current EXAFS analysis which are to be encountered for thick or inhomogeneous samples. We shall also suggest a curative approach to the problem of holes and harmonics.

1. Beam instabilities

It is especially crucial to prepare highly homogeneous samples if either the vertical or the horizontal stability of the X-ray beam is not properly monitored. We experienced recently this difficulty at L.U.R.E. for the EXAFS-II station when we tried to obtain a lateral deflection of the beam by slightly tilting ($\Delta\psi_z < 1°$) one of the two crystals of the monochromator around an axis parallel to the beam [1]. If θ_B denotes the energy-dependent Bragg angle, the lateral shift x_L measured at the distance $L = 5$ m downstream of the monochromator is indeed energy dependent too : $x_L = L\,\Delta\psi_z \sin\theta_B$ and can give rise to detectable horizontal shifts of the beam during an EXAFS scan. This effect is insensitive for large sized homogeneous samples or has to be compensated by moving the sample. This problem is also to be considered for sagittal focusing with a triangular shaped bent crystal [1].

2. Holes and harmonics ([2] and references therein)

The transmission of harmonics by the monochromator and any leakage of radiation around or through the sample (e.g. holes) are affecting the apparent absorption coefficient $\mu_A\bar{d}$ through a perturbation $\xi(\mu_1)$ according to :

$$\mu_A\,\bar{d} = -\ln\frac{I}{I_o} = \mu_1\,\bar{d} - \frac{1}{2}\,\mu_1^{\,2}\,\sigma_d^{\,2} - \ln(1+\xi) - \ln A_o \qquad (1)$$

with : $\xi(\mu_1) = \exp\left\{\mu_1\bar{d} - \frac{1}{2}\,\mu_1^{\,2}\sigma_d^{\,2}\right\}\sum\eta_p$,

[1] Laboratoire de Chimie Théorique, E.R.A. 22 au C.N.R.S., Université de Nancy I, B.P. 239, F-54506 Vandoeuvre-les-Nancy, France

[2] L.U.R.E., Laboratoire Propre du C.N.R.S., Université de Paris-Sud, Bât. 209C, F-91405 Orsay, France

[3] G.R. 4 du C.N.R.S., "Physique des Liquides et Electrochimie", Université de Paris VI, 4 Place Jussieu, F-75230 Paris Cedex 05, France

[4] Laboratoire de Métallurgie, L.A. 159 au C.N.R.S., ENSMIN, Parc de Saurupt, F-54042 Nancy Cedex, France

I, I_o referring to the signal of the ion chambers, $\mu_1 = \mu_1^b + \mu_1^o(1+\chi)$ to the absorption coefficient of the fundamental, $\bar{d}$ and σ_d^2 to the average sample thickness and its variance, A_o being a scaling term figuring the linearity of the ion chambers, η_p describing the level of harmonics or the proportion of holes [2]. We may expand next $\mu_A(\chi)$ as a function of the EXAFS modulation χ in Taylor series :

$$\mu_A(\chi) = \mu_A(0) + \chi\,\mu_A' + \frac{1}{2}\chi^2\,\mu_A'' + \ldots \qquad (2)$$

where the partial derivatives are easily calculated analytically and are thickness dependent. Thus, Fig. 1 summarizes the results : the coordination numbers have to be corrected by a factor $T_1(E_o) = \mu_A'/\mu_A^o$ and the Debye-Waller factor by the quantity $\Delta\sigma_j^2$, whereas the edge jump $\Delta\mu_A^o$ drops rapidly. Spectral harmonics are also to be predicted due to the non-linear term : $T_2\chi^2 = \frac{1}{2}\left[\mu_A''/\mu_A^o\right]\chi^2$. This fairly general approach has been extended to REFLEXAFS and emission EXAFS measurements [2].

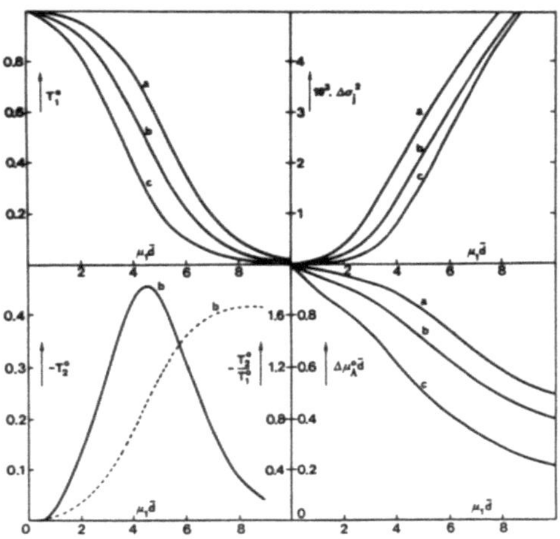

Fig.1. Typical thickness effects on T_1, T_2, $\Delta[\sigma^j]^2$ and the edge jump $\Delta\mu_A^o\,\bar{d}$ for simulated spectra of Co_3B under variable proportions of holes and harmonics

In order to reduce the effects of holes and harmonics, we suggested to replace μ_A by a different function of the experimental quantity $J(E) = I/I_o$:

$$\mu_B = -\ln\left\{1 - \Delta_o G(E)\right\} \qquad \text{with :} \qquad G(E) = \frac{J(E) - J_B(E_o)}{J_o(E_o) - J_B(E_o)}$$

J_B being the extrapolated preedge contribution to J whereas $J_o(E) = J(E,\chi\equiv 0)$. The corrective factor : $\Delta_o = 1 - \exp\left\{-\mu_1^o\bar{d} + \frac{1}{2}\left[\mu_1^2(E_o) - \mu_1^{b2}(E_o)\right]\sigma_d^2\right\}$ requires only a rough estimation of $\bar{d}$, σ_d and the quantities $\mu_1^{o,b}$ available from the McMasters tables. X-ray leakage effects are thus virtually eliminated as demonstrated in reference [2].

1. J. Goulon, M. Lemonnier, R. Cortes, A. Retournard and D. Raoux, To be published in the Proc. of the SRI Conf. (Hambourg, 7-13 Aug. 1982)
2. J. Goulon, C. Goulon-Ginet, R. Cortes and J.M. Dubois, J. Physique 43, (1982), 539.

Thickness Effects in X-Ray Absorption Measurements

S.M. Heald

Brookhaven National Laboratory, Upton, NY 11973, USA

The most common source of error in an x-ray absorption measurement is the thickness effect. It is caused by the detection of x rays which are not absorbed by the sample in the usual fashion. The primary cause is leakage of x rays around or through the sample. Leakage can be due to an inhomogeneous sample (e.g. cracks or pinholes), radiation scattered around the sample, or a sample smaller than the beam. Another type of leakage present even for an ideal sample is contamination of the incident beam by other energies. In all of these cases the leakage signal is not affected by the absorption changes in the XANES or EXAFS structure and thus reduces the measured structure.

Thickness effects were first discussed by Parratt et al. [1], and later shown to also be an important source of error at synchrotron radiation facilities by Heald and Stern [2]. Since then the general theory of thickness effects has been fully worked out [3-5]. Two parameters are important in calculating the thickness effect: the percentage of the incident beam which is leakage and the x-ray thickness of the sample. The effect becomes larger as both parameters are increased. Figure 1 shows the results of Stern and Kim [3] to illustrate this point. Plotted are the reduction in μx and the EXAFS $\chi(k)$ as a function of sample thickness for various amounts of leakage. For comparison it is found at SSRL that a well-tuned monochromator can easily have a leakage from harmonics of several percent. This would give unacceptable errors if $\mu x=2.6$, which is the statistically optimum value. However, the harmonic contamination can be reduced either by detuning the monochromator or using a total external reflecting mirror in the case of synchrotron radiation facilities, and by operating the x-ray tube below the harmonic excitation energy for a laboratory facility.

Another type of leakage comes from the sample itself. It is due to the fluorescence emitted by the sample as the incident beam is absorbed. This is primarily a problem for the second or I detector, and depends on the quantity $\Omega e/4\pi$ where Ω is the solid angle subtended by the detector and e is the fluorescence efficiency [5]. Figure 2 shows the magnitude of the error for various values of $\Omega e/4\pi$. Again there are unacceptable errors for thick samples. It is a particular problem for laboratory machines where the tight focusing geometry may require the detectors to be quite close to the sample.

As seen in the figures the thickness effect is small for thin samples. Empirically it is found that if reasonable precautions are taken to avoid leakage then the effect can be ignored for samples with $\mu x<1.5$. However, if a thickness effect is suspected and accurate amplitudes are required, the only sure procedure is to measure various thicknesses of sample and

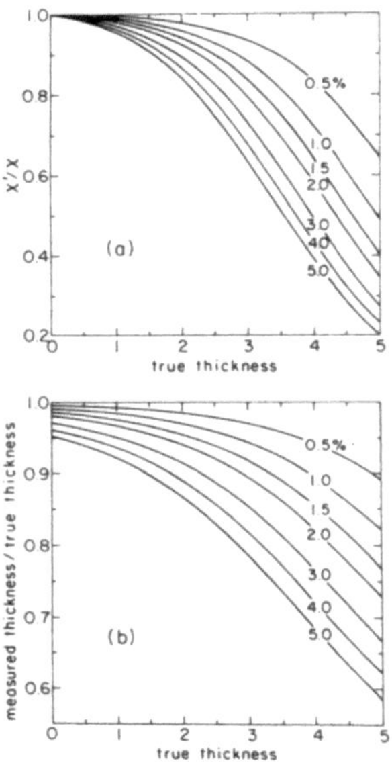

Fig. 1. The reduction of χ(k) and μx for various levels of leakage as a function of the true sample thickness

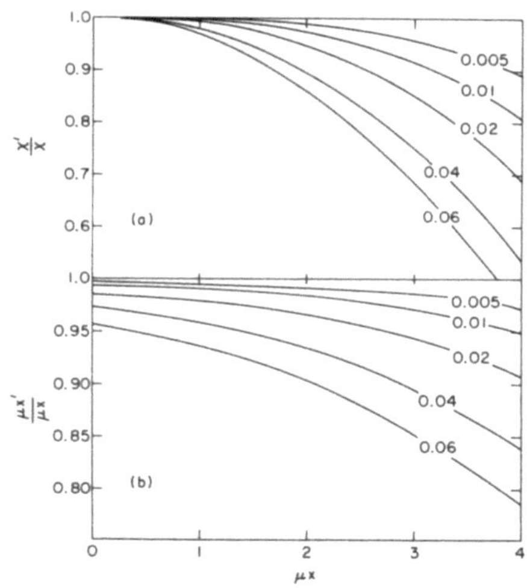

Fig. 2. The reduction of χ(k) and μx for various values of Ωe/4π as a function of the true sample thickness

to extrapolate the results to zero thickness. Finally it should be mentioned that in some cases the thickness effect can also distort the EXAFS phase [5].

Acknowledgment

This work was supported in part by the Material Sciences Division of the U.S. Department of Energy under contracts DE-AS05-ER10742 and DE-AC02-76CH00016.

References

[1] L. G. Parratt, C. F. Hempstead, and E. L. Jossem, Phys. Rev. 105, 1228 (1957).
[2] S. M. Heald and E. A. Stern, Phys. Rev. B16, 5549 (1977).
[3] E. A. Stern and K. Kim, Phys. Rev. B23, 3781 (1981).
[4] K-q Lu and E. A. Stern, to be published
[5] J. Goulon, C. Goulon-Ginet, R. Cortes, and J. M. Dubois, J. Physique 43, 539 (1982).

Interference and Multiple Scattering Effects in Porphyrinic Complexes

J. Goulon[1,2], P. Friant[1], J.L. Poncet[3], R. Guilard[3], J. Fischer[4], and L. Ricard[4]

1. Introduction

Synthesis and structural studies of dioxygen complexes (peroxo, superoxo, μ-peroxo...) of metalloporphyrins stem from their potential interest in the activation of molecular oxygen. In this paper, we shall report the EXAFS spectrum of the peroxotitanium(IV)tetraphenylporphyrin (TiO_2:TPP) to be compared to the spectrum of the oxo complex (TiO:TPP). However, in order to back up our discussion concerning an apparent effect of destructive interferences, we shall present here the spectra of the vanadyl and thiovanadyl(IV)2,3,7,8,12,13,17,18,octaethylporphyrin (V=O:OEP, V=S:OEP). As a typical illustration of the so-called focusing effect in the colinear arrangement of two scatterers we shall also reproduce the EXAFS spectrum of the dibromovanadium(IV) OEP complex. It should be noticed that the FT spectra $\tilde{\chi}(R)$ reproduced in this paper are all corrected for the amplitudes $F_j(k)$ and phase shifts $\psi_j(k)$ of a selected shell j (e.g. j=1: M*...N or Br*...V) according to :

$$\tilde{\chi}_j(R) = \int_0^\infty dk\; g(k)\; \chi(k)\; \frac{k\, R_j^2}{N_j |F_j| A_j}\; \exp\left\{2\sigma_j^2 k^2 - 2ikR - i\psi_j(k)\right\} \quad , \qquad (1)$$

where g(k) is a Kaiser window function and $A_j(R_j,k)$ accounts for the loss of coherence of the photoelectron [1]. $F_j(k)$ and $\psi_j(k)$ were numerically calculated using the parametrization suggested by TEO et al. [2], E_0 being adjusted according to the LEE-BENI criterion of coincidence of the maxima of $|\tilde{\chi}(R)|$ and Im $\tilde{\chi}(R)$ [3]. N_j, R_j and σ_j were fixed in a self-consistent way, according to the results of a preliminary fit.

2. Destructive Interference in (VS:OEP) [4]

Since the pioneering work of CRAMER et al. [5], it is well known that the FT spectra of metalloporphyrins exhibit a typical pattern due to the highly sym-

[1]Laboratoire de Chimie Théorique, E.R.A. 22 au C.N.R.S., Université de Nancy I, B.P. 239, F-54506 Vandoeuvre-les-Nancy, France

[2]L.U.R.E., Laboratoire Propre du C.N.R.S., Université de Paris-Sud, Bât. 209C, F-91405 Orsay, France

[3]Laboratoire de Synthèse et d'Electrosynthèse, L.A. 33 au C.N.R.S., Faculté des Sciences, 6 Boulevard Gabriel, F-21100 Dijon, France

[4]E.R.A. 8 au C.N.R.S., "Cristallographie et Chimie Structurale", Université de Strasbourg I, 4 rue B. Pascal, F-67070 Strasbourg Cedex, France

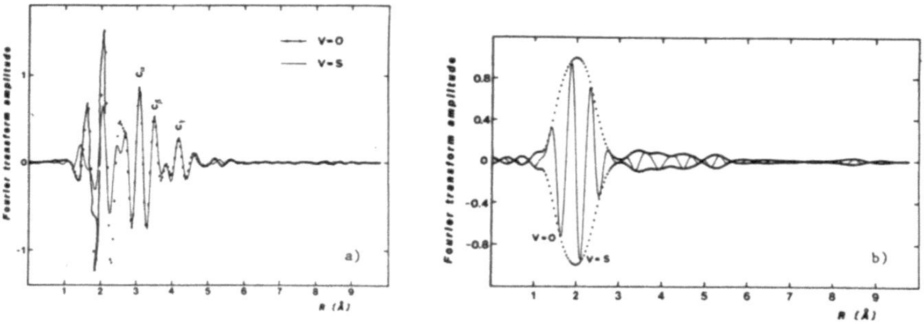

Fig.1. a) Comparison of the real part of $\check{\chi}_1(R)$ for VO:OEP (dotted line) and VS:OEP (full line). b) Difference spectrum showing the phase inversed V=O and V=S signals

metrical porphyrinic ring, making it possible to identify the contributions of the 4N, $8C_\alpha$, $4C_\beta$, $8C_\gamma$ and $4C_\phi$ (for TPP only) shells [6]. For the VO:OEP compound, all the corresponding distances fit nicely with the crystal structure [7]. As illustrated by the comparative figure 1a, a destructive interference effect is clearly detected in the case of the VS:OEP complex, as a direct consequence of the scattering phase shift difference between S and N. As the geometry of the porphyrinic ring remains identical for both compounds, a difference FT spectrum gives (Fig.1b) the negative contribution of the V=O shell (1.6_1 Å) together with the V=S signal. A fitting procedure can thus give a refined V=S distance ($2.0_7 \pm 0.0_2$ Å) in excellent agreement with the V=S distance recently reported for another compound [8].

3. Focusing effect in (Br-V-Br:OEP) [9]

From the EXAFS spectrum reproduced in Fig.2, we could probe the trans-configuration of the dibromo complex studied at the Br(*) edge. The linear Br...V...Br sequence results in a typical enhancement of the Br...Br shell due to a now classical focusing effect. An additional phase shift explains why the observed Br...Br distance ($\sim 4.7_0$ Å) is definitively smaller than twice the Br...V bond (2.41_7 Å).

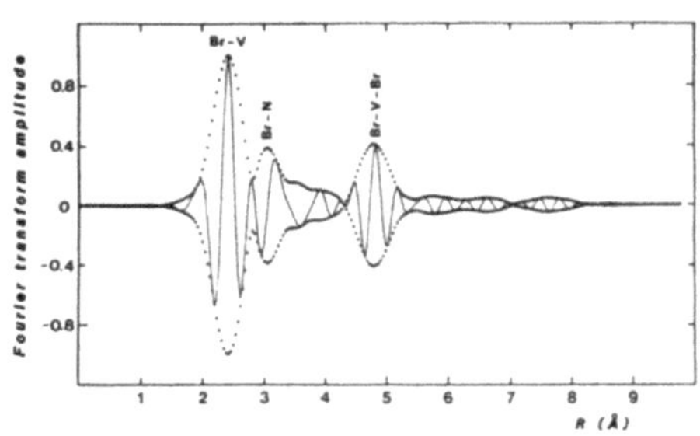

Fig.2. FT spectrum $\tilde{\chi}_1(R)$ of Br-V-Br:OEP, corrected for the Br*...V shell

4. TiO_2:TPP and TiO:TPP complexes

A few years ago, GUILARD et al. have reported [10] the first synthesis of the peroxo complexes TiO_2:TPP and TiO_2:OEP, the crystal structure of the lat-

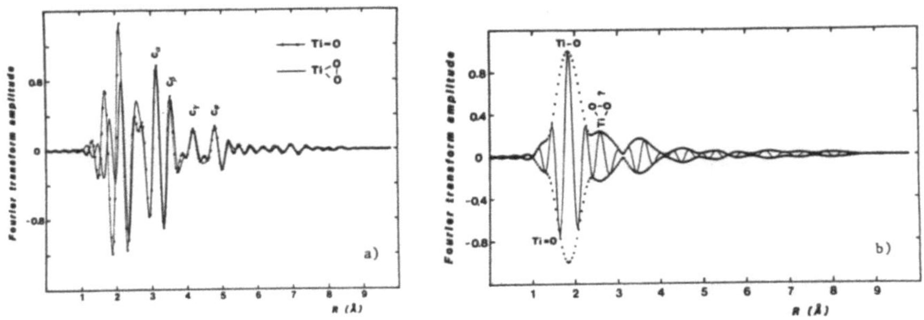

Fig.3. a) Comparison of the real part of $\tilde{\chi}_1(R)$ for TiO:TPP(dotted line) and TiO_2:TPP (full line). b) Difference spectrum TiO_2:TPP minus TiO:TPP

ter compound giving a definitive proof of the dioxygen bonding to the metal ($Ti...O_1 = 1.82_7$ Å, $Ti...O_2 = 1.82_2$ Å, $O...O = 1.44_5$ Å). However the comparative Fig.3a again seems to indicate an unexpected destructive interference of the Ti...N signal. We ensured by IR spectroscopy that the complex did not decompose during the EXAFS measurements carried out at low temperature ($T \simeq 80$ K). Furthermore, a decomposition would have led to the TiO:TPP complex. Thus a first idea was to add the triangular path ($TiO_1-O_1O_2-O_2Ti = 5.04$ Å), taking into account the additional phase shift recently calculated by TEO [11]. However it did not help much because, as illustrated by the difference spectrum (Fig.3b), the interfering path is expected at a shorter distance than the one calculated. We are presently investigating the hypothesis of a change of the phase shift of the single path $Ti...O_1$ perturbated by the close oxygen atom O_2. Another effect which might perhaps be involved is the fact that the potential barrier hindering the rotation of the oxygen adduct around the porphyrin axis is rather low [10], thus making a slight asymmetrical distribution of the Ti...O, Ti...N distances not impossible.

References

1. J. Goulon, C. Goulon-Ginet and M. Chabanel, J. Sol. Chem. 10, (1981), 9.
2. B.K. Teo, P.A. Lee, A.L. Simmons, P. Eisenberger and B.M. Kincaid, J. Am. Chem. Soc. 99, (1977), 3854.
3. P.A. Lee and G. Beni, Phys. Rev. B15, (1977), 2862.
4. J.L. Poncet, R. Guilard, P. Friant and J. Goulon, submitted to Polyhedron.
5. S.P. Cramer, T.K. Eccles, F. Kutzler, K. Hodgson and S. Doniach, J. Am. Chem. Soc. 98, (1976), 8059.
6. J. Goulon, C. Goulon, F. Niedercorn, C. Selve and B. Castro, Tetrahedron 37, (1981), 3707.
7. F.S. Molinaro and J.A. Ibers, Inorg. Chem. 15, (1976), 2278.
8. M. Sato, K.M. Miller, J.H. Enemark, C.E. Strouse and K.P. Callahan, Inorg. Chem. 20, (1981), 3571.
9. P. Richard, J.L. Poncet, J.M. Barbe, R. Guilard, J. Goulon, D. Rinaldi, A. Cartier and P. Tola, J. Chem. Soc. Dalton Trans., (1982), 1451.
10. R. Guilard, J.M. Latour, C. Lecomte, J.C. Marchon, J. Protas and D. Ripoll, Inorg. Chem. 17, (1978), 1228.
11. B.K. Teo, J. Am. Chem. Soc. 103, (1981), 3990.

Multiple Scattering Effects in the EXAFS of Fe and TiFe

N. Motta Scuola Normale Superiore, Pisa, Italy

M. De Crescenzi Dipartimento di Fisica, Università della Calabria, Cosenza, Italy

A. Balzarotti Istituto di Fisica, Università di Roma II, I-00185 Roma, Italy

TiFe is an Asano alloy of great technological interest for its ability to incorporate large amounts of atomic hydrogen. It has a very stable CsCl lattice whose ordered phase is made of Ti and Fe atoms located in the bcc lattice sites (1).

To test the local ordering around each atomic constituent we have studied the EXAFS and near-edge regions beyond Ti and Fe K edges. Since large multiple scattering contributions have to be expected in bcc compounds (2), we have first considered the K-edge EXAFS spectrum of crystalline Fe, which has a lattice parameter close to that of TiFe and the same bcc structure.

EXAFS spectra collected at the PULS X-ray line have been reported elsewhere (3). In Fig.1 a) we compare the radial distribution function F(R) for pure Fe, measured at 80 K, with a model EXAFS calculation of $k\chi(k)$ in the single scattering approximation. The amplitudes and phases used are those of Teo and Lee (4) and the Debye-Waller factors for the first two shells are taken from (5). All relevant parameters are listed in Table I. The zero of the energy scale has been taken at the inflection point of the K edge at 7112 eV. To improve the fitting in both momentum and real space, a nearly constant phase-shift factor $\Delta\phi = -1.6$

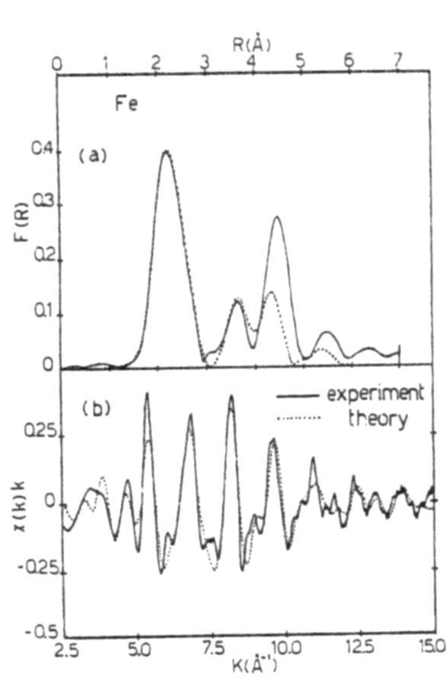

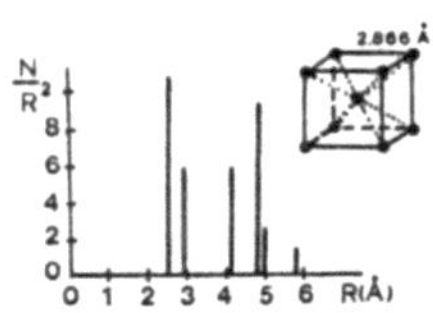

Fig.1. Comparison between experimental spectra at 80 K of Fe (solid line) and model calculation (dotted line): a) radial distribution function; b) $\chi(k)k$. The calculation is made within the single scattering theory (4) and with an additional constant phase factor $\Delta\phi=-1.6$ rad. The structural parameters used are shown in Table I

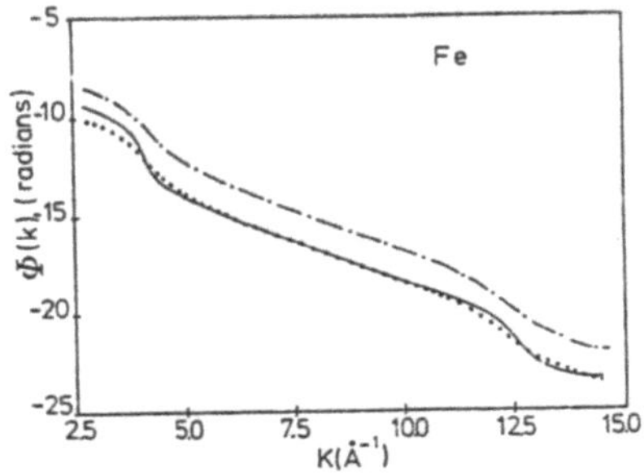

Fig.2.Phase shift for the first two shells of Fe: solid(experimental); line-dotted (theoretical from data of ref. (2); dotted (theoretical with a constant shift $\Delta\phi = -1.6$ rad)

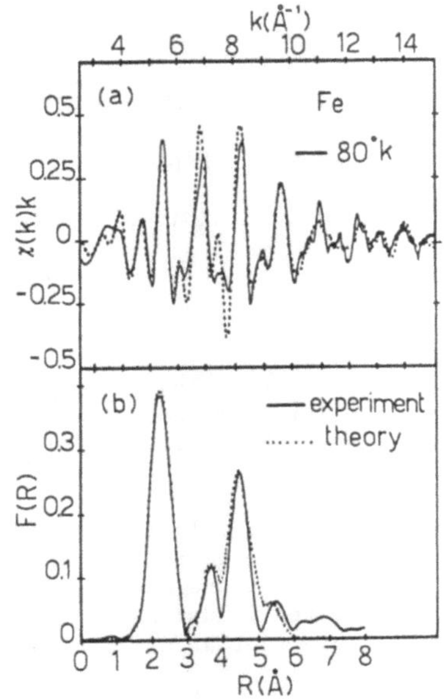

Fig.3.Comparison between experimental spectra at 80 K (solid) and a model calculation with multiple scattering corrections for the 5th and 6th shell (dotted). The forward amplitude and phase factors are those of (7).
a) $\chi(k)k$; b) Radial distribution function

rad is required. Attempts to compensate for this shift by changing E_o introduce a mismatch of the peaks' position of F(R) (6).Taking the effective radius of the backscattering potential into account, one gets from simple geometrical arguments a correction $\Delta\phi_j$ to the plane-wave phase shift for the shell j:

$$\Delta\phi_j = \frac{2k\,|F(\pi,k)|^2}{R_j} \tag{1}$$

where R_j is the distance of the j-th coordination shell. This correction has to be applied to the tabulated phases (4) which are independent of the shell radii. For the first two scattering shells of Fe the constant correction $\Delta\phi = -1.6$ rad brings the experimental and model $\chi(k)$ in good agreement. In Fig.2 we show the experimental and theoretical phases for these two shells of Fe before and after the correction.

Table 1.Structural parameters used in the simulation of bcc Fe

Shell	N_j	σ_j^2 x10^{-3}(Å^2)	R_j(Å)
1	8	2.5	2.482
2	6	3.	2.866
3	12	4.	4.053
4	24	4.	4.752
5	8	6	4.971
6	6	8	5.74
E_o(eV)		7112	
S_o^2		0.4	

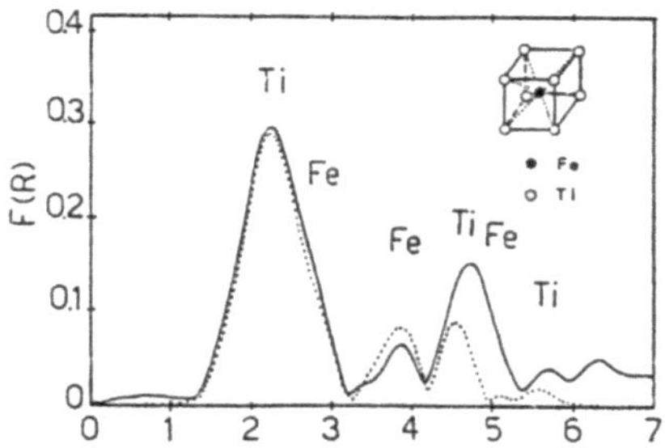

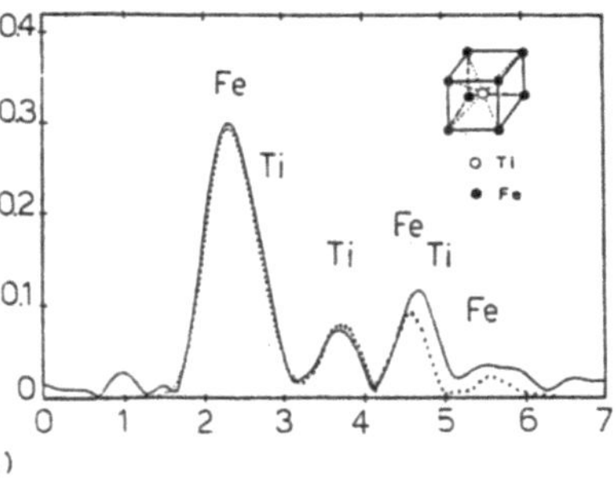

Fig.4. Radial distribution function of TiFe around the iron atom. The dotted curve is a single-scattering model calculation

Fig.5. Radial distribution function of TiFe around the Ti atom. The dotted curve is a single-scattering model calculation

Multiple scattering enhancement in the amplitude of the 5th and 6th shell are substantial, as shown in Fig.3. The improvement to $\chi(k)$ is particularly evident at low k values, as shown in Fig.3 a). Multiple scattering effects modify the amplitudes and phases through corrective terms $\Omega_B(o,k)$ and $\omega_B(o,k)$, in presence of the intervening atom B, as discussed in (2). Shadowing effects to the fifth and sixth shell come from the first and second shell atoms aligned along the cube diagonals and along the perpendicular to the cube faces, respectively.

These large multiple scattering effects exist also in TiFe where the single scattering EXAFS formula is inadequate to explain the F(R) for the outer shells. In Fig.4 and Fig.5 a comparison is made, for both Fe and Ti edges, between the experimental and the theoretical F(R) calculated without multiple scattering corrections. The intensity of the 5th shell in Fig.4 and Fig.5 is smaller when the intermediate first shell atom is Fe, because its forward scattering power is lower than that of Ti (7).

In Fig.6 the Ti and Fe edge structures of TiFe are compared with calculated d-like and p-like partial DOS (8). It is apparent that only the final density of states of p-like symmetry at the excited atom site describes accurately the near K-edge line shape, at least within 10 eV above the edge (9).

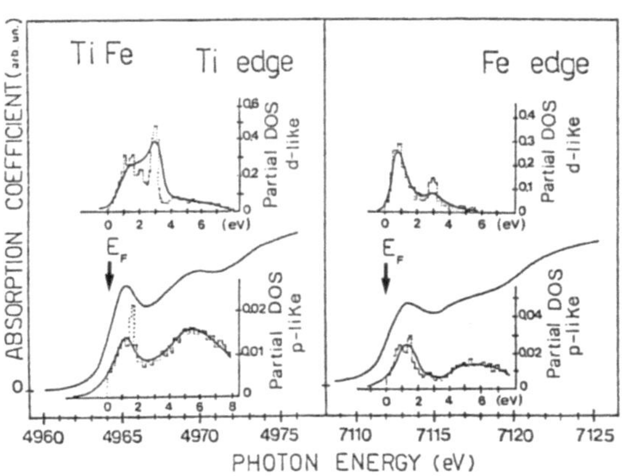

Fig.6. Partial d- and p-like densities of the conduction band states of TiFe at the Ti and Fe sites, compared with the experimental near-edge structures

References

1 - J. Yamashita and S. Asano, Progr. Theor. Phys. 48, 2119 (1972)
2 - B.K. Teo,J. Am. Chem. Soc. 103, 3990, (1981)
3 - N. Motta, M. De Crescenzi and A. Balzarotti, to be published
4 - B.K. Teo and P.A. Lee, J. Am. Chem. Soc. 101, 2815 (1979)
5 - E. Sevillano, J.J. Rehr, H. Meuth, Phys. Rev. B 11, 2795 (1975)
6 - M.B. Stearns, Phys. Rev. B 25, 2382 (1982)
7 - M. Fink and J. Ingram, Atomic Data 4, 129 (1972)
8 - D.A. Papaconstantopoulos, Phys. Rev. Lett. 31, 1050 (1973)
9 - A. Balzarotti, M. De Crescenzi, L. Incoccia, Phys. Rev. B 25, 6349 (1982)

A Model Comparision Between EXAFS and X-Ray Scattering

R. Frahm, R. Haensel, and P. Rabe

Institut für Experimentalphysik, Universität Kiel,
D-2300 Kiel, Fed. Rep. of Germany

Introduction

For crystalline samples with Gaussian pair distribution functions the structure data obtained from X-ray diffraction (XRD) and EXAFS are generally in good agreement. In disordered systems like liquids and amorphous solids, however, differences in the interatomic distances and coordination numbers have been reported |1|. They were assumed to originate from the respective regions of momentum space, which contribute to the Fourier transform (FT). In XRD, data are available in typical wave vector ranges $1Å^{-1} \leq q \leq 18Å^{-1}$, whereas the analogous EXAFS range of 2k extends approximately from 6 to $22Å^{-1}$. The loss of low k data makes EXAFS sensitive to the broadening of pair correlation peaks and removes long-range order information .

Details of Calculations

To compare the distances obtained by both techniques we have calculated the XRD structure factor S(q) and the EXAFS $\chi(k)$ for a simple model pair distribution function given by

$$\rho(r) = \begin{cases} \frac{1}{al^2} e^{(R-r)/a} & \text{for } r \geq R \\ 0 & \text{elsewhere} \end{cases} \qquad (1)$$

where $\rho(r)$ is the number of atoms per unit volume in the distance r from the central atom and l is the unit length. The factor $1/(a \cdot l^2)$ was included to normalize the peak area to unity. This model with its hard core distance R represents a simple approximation to an unrelaxed dense random packing of hard spheres.

The EXAFS at K edges for an arbitrary number density $\rho(r)$ is given by

$$\chi(k) = -4\pi |f(\pi,k)| \int_0^\infty dr \frac{\rho(r)\, e^{-2r/\lambda(k)}}{k} \sin(2kr+\Phi) \qquad (2)$$

where $|f(\pi,k)|$ is the backscattering amplitude of the scattering atom, $\lambda(k)$ is the mean free path of the outgoing photoelectron and Φ is the total phase shift for the considered atom pair. The atomic distances are usually taken from the peak positions of the modulus of the complex FT

$$F(r)=\frac{1}{\sqrt{2\pi}}\int_0^\infty k^n \chi(k)\, e^{2irk}\, d(2k) \quad . \tag{3}$$

The distances obtained in this way have to be corrected for the phase shift Φ, known from reference materials. In the following the exponent n of the weighting function is chosen to be n=1 to allow full comparison with the FT of XRD data.

The structure factor S(q) for XRD is given by

$$S(q)= 1 + 4\pi \int_0^\infty dr \; \frac{r(\rho(r)-\rho_0)}{q} \sin(qr) \quad , \tag{4}$$

where ρ_0 is the average atomic density. Atomic distances are normally determined from the reduced distribution function calculated by the sine FT of q(S(q)-1)

$$G(r)=4\pi r(\rho(r)-\rho_0) = \frac{2}{\pi}\int_0^\infty q(S(q)-1)\sin(qr)\, dq \quad . \tag{5}$$

Using the pair distribution (1) and assuming for simplicity a constant backscattering amplitude, the phase shift Φ=O and an infinite mean free path of the photoelectrons, the structure functions calculated from (2) and (4) are given by

$$\chi(k)=-\frac{4\pi}{kl}\,\frac{1}{\sqrt{1+(2ak)^2}}\,\sin(2kR+\tan^{-1}(2ak)) \tag{6}$$

and

$$S(q)-1= -\frac{R}{2l}\,\chi(q/2) + \frac{4\pi}{ql^2}\,\frac{a}{(1+a^2q^2)}\,\sin(qR+\tan^{-1}(\frac{2aq}{1-a^2q^2})) \quad . \tag{7}$$

The XRD structure factor is similar to $\chi(k)$ if k is substituted by q/2 and contains only a small correction term.

A numerical FT was performed in the respective data regions mentioned above of the simulated spectra with a hard core radius of R=2.4Å for different values of a. The complex transform was calculated for both data sets. In Fig.1 we show the variation of peak position $\Delta r/R$ in the modulus and the imaginary part of the FT with respect to the hard core radius R as a function of the asymmetry parameter a. The magnitude of the FT is rather insensitive to different values of a. It peaks near R with a maximum deviation of 1.3%. This is due to the fact that the structural phase term $\tan^{-1}(2ak)$ is nearly constant ($\pi/2$) over the transformed k range for larger asymmetries and shows effect only for a$\lesssim$0.2Å. The imaginary parts of the FTs however are influenced by the constant phase shift $\tan^{-1}(2ak)$. In this case peaks show up at larger distances compared to R. Both the EXAFS and XRD results however differ by less than 2%. The peak shift of the sine transform is especially pronounced in the XRD transform at

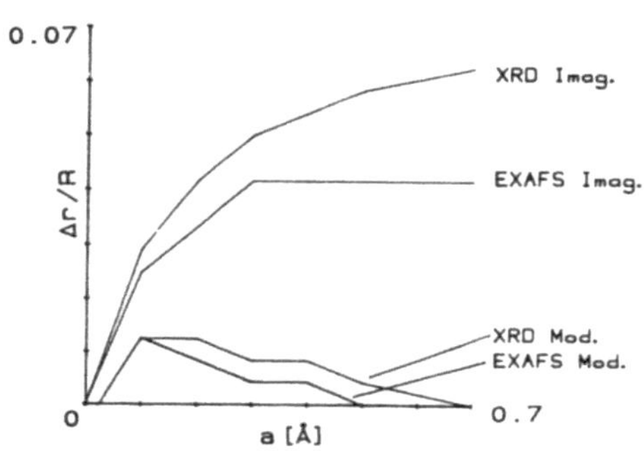

Fig.1. Deviations in the peak position from the hard core radius R as a function of the parameter a in (1)

large values of a, where the second term in (7) is effective. This shift, however, does not reach R+a, as has been stated previously |2|.

It has been incorrectly stated that both EXAFS and XRD must be used to determine the degree of asymmetry. We have calculated the parameter a from the tabulated XRD data of WASEDA |3| for liquid Zn at 450°C by taking the complex FT as shown in Fig.2. In this case we observe a peak shift between magnitude and imaginary part of 0.14Å. Such a deviation is expected from our simulations for an asymmetry of a=0.5Å. The XRD atomic density function can be simulated by convolution of such an exponential with a Gaussian function with mean square fluctuation of $\sigma^2=.025Å^2$. For the EXAFS data a shift of 0.1Å is expected from our simulation for such an asymmetry. Both values agree well with the experimental EXAFS results of CROZIER and SEARY |4|, who measured a shift between magnitude and imaginary part of the FT of 0.09Å and $\sigma^2=.03Å^2$ for liquid Zn.

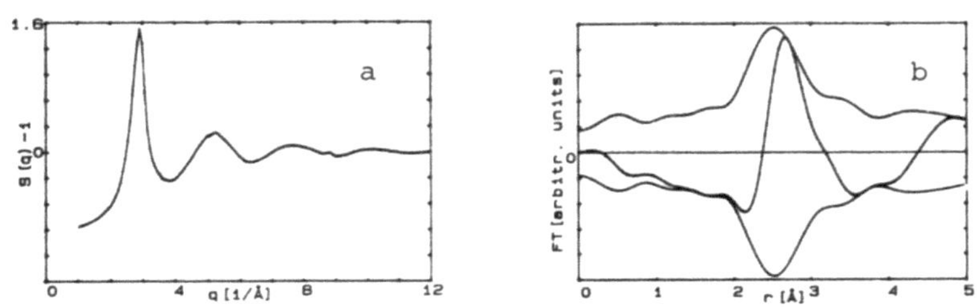

Fig.2. Structure factor S(q)-1 for liquid Zn at 450°C as given by WASEDA |3| (a) and complex FT (imaginary part and modulus) (b)

References

1. R.Haensel, P.Rabe, G.Tolkiehn and A.Werner, in: Liquid and Amorphous Metals, ed. E.Lüscher and H.Coufal (Sijthoff&Nordhoff Intern. Publ. 1980)
2. M.De Crescenzi, A.Balzarotti, F.Comin, L.Incoccia, S.Mobilio, and N.Motta, Solid State Comm., 37, 921 (1981)
3. Y.Waseda: The structure of non-crystalline materials, McGraw Hill, New York, 1980
4. E.D.Crozier and A.J.Seary, Can. J. Phys, 58, 1388 (1980)

EXAFS Database and Program Library

E. Pantos and D. Firth

Science and Engineering Research Council, Daresbury Laboratory,
Warrington WA4 4AD, UK

1. Introduction

Daresbury Laboratory provides central support for the computing needs of various research projects at the SRS. EXAFS is one of the most active ones. Data reduction and analysis needs are covered by a number of programmed utilities that form part of the SRS program library [1]. In addition a database of reduced EXAFS or data files needed by analysis programs has been formed.

2. Program library

Data produced at the EXAFS experimental stations are forwarded by the data acquisition system to the laboratory's AS/7000 mainframe computer for cataloguing in the SRS Database System and for subsequent analysis.

Initial raw data reduction may be carried out using program EXCALIB. It corrects for detector and monochromator characteristics and produces a normalized absorbance spectrum. It can also be used to remove glitches and discontinuities. EXBACK may then be used for background subtraction and normalization of EXAFS. Both programs use interactive graphics on a Tektronix terminal.

Computation of structural parameters is carried out by programs EXINT (interactively) and EXAFS (batch) using the plane wave or spherical wave approximation [2] based on Lee and Pendry's theory. Input structural parameters may be altered and at each iteration the experimental and calculated spectra or their Fourier transforms may be inspected. The phase shift values required by these programs may be calculated by program MUFPOT [3]. Program EXANES [4] may be used to calculate near edge structure.

Additional utilities for general spectroscopic manipulation, averaging, smoothing, plotting, etc., are available through the SRSPL [1]. All programs are centrally maintained and are executed through a command that takes care of file allocation and job control language syntax. They are executable by anyone with access to the Daresbury computer through the country-wide SERC network. Users need not have detailed knowledge of the local computing environment or even of proramming.

3. Databases

3.1 Atomic wave functions, Crystal structures and Phase shifts

Clementi wave functions for neutral and excited atoms are available in machine readable form for all elements from lithium to uranium. The excited atom wave functions are modelled using the wave function of the neutral atom one place higher in the periodic table but with a hole in the 1s electron shell.

Crystal structure datasets for all metals and many other elements and compounds are also available. These datasets asign closed-packed structures to metals, diamond/zincblende to covalent element compounds and NaCl and CaF_2 structures to most others. All structures use classical metallic, covalent or ionic radii as the muffin-tin radii.

The atomic wave function and crystal structure datasets have been used to compute the phase shifts of several systems using MUFPOT. The naming convention and method of access by EXINT and EXAFS is described in detail in the SRSPL documentatation [1] (command EXDBS). Magnetic tape copies of these files are available on request.

3.2 Reduced data Database

This Database contains EXAFS datasets after instrumental corrections and includes details of sample and experimental conditions and other information supplied by authors of publications where the analysis and interpretation is described. A Database Management System (Rapport) allows on-line or off-line retrieval, report generation and searches on key fields identifying the data, such as atom type and edge, experiment type, etc. Utilities are provided for inserting new entries or dispatching data in machine readable form on request.

Acknowledgements

The atomic wave function, crystal structure and phase shift datasets were prepared by K. Gaskel under the supervision of Dr. G.N. Greaves. We would also like to thank Dr. S.S. Hasnain for his cooperation in designing the reduced data Database.

References

[1] E. Pantos "The SRS Program Library Documentation", Daresbury Laboratory, June, 1982

[2] S.J. Gurman Daresbury Laboratory Internal Report DL/SCI/TM 21T (April 1980).

[3] J.B. Pendry "Low Energy Electron Diffraction", Academic Press, London, 1974.

[4] P.J. Durham, J.B. Pendry and C.H. Hodges Com. Phys. Commun. 25, 193 (1982).

Regular Methods in Real-Space Analysis of EXAFS

Yu. A. Babanov and A.L. Ageev
Institute of Metal Physics, Ural Science Research Center of the USSR, Academy of Sciences, Sverdlovsk, USSR

V.V. Vasin and N.V. Ershov
Institute of Mathematics and Mechanics, Ural Science Research Center of the USSR, Academy of Sciences, Sverdlovsk, USSR

The EXAFS technique is employed to determine the short-range order in atomic arrangements from oscillations of the X-ray absorption coefficient. This is one of a family of ill-defined inverse problems [1,2]. The errors arising in the registration of the absorption spectrum and during the preliminary processing of experimental data are apt to lead to arbitrarily large variations in the solutions. The resulting ambiguity causes great difficulties in the interpretation of results.

For a single-species problem the EXAFS equation reduces to the Fredholm integral equation of the first kind

$$\chi(k) = 4\pi\rho \frac{|f(k,\pi)|}{k} \int_0^\infty g(r) \exp[-2r/\lambda(k)] \sin[2kr + \psi(k)]\, dr \quad , \qquad (1)$$

where $\chi(k)$ stands for the normalized oscillating part of the X-ray absorption coefficient; $f(k,\pi)$ is the backscattering amplitude for a photoelectron with a momentum k; $\psi(k)$ is the total phase shift in the scattering; $\lambda(k)$ is the mean free path of an electron; and ρ is the mean atomic density. The function of radial atomic distribution $g(r)$ has the following properties:

(1) $g(r) \geq 0$ for any r ,

(2) $g(r) \to 1$ when $r \to \infty$,

(3) $\frac{1}{V} \int_0^\infty g(r) 4\pi r^2\, dr = 1$.

The regular methods of solving Eq. (1) which were proposed and realized in [1,2] make it possible to select solutions from an infinite set of possible ones according to physical information (2).

To ascertain the effectiveness of the regular methods of deriving $g(r)$ from the EXAFS data, numerical experiments were carried out. The function $\chi^{th}(k)$ was calculated according to (1) for the well-known model function

$g_m(r)$ (crystal or amorphous). By simulating the errors arising at different stages of EXAFS spectrum processing on $\chi^{th}(k)$, we could trace the effect of those errors on the solution of g(r) obtained with the help of two methods, viz., regularization and the Fourier method. The results of these experiments allowed us to draw the following conclusions:

1) Restriction of the interval in k space to 3.5-15 $Å^{-1}$ prescribing $\chi(k)$ is no essential error for the regular methods as opposed to the Fourier procedure.

2) The stability of the regular methods is not worse than that of the Fourier transformation method with respect to errors such as statistical noise and errors in determining the background of atomic absorption and of the parameter E_0.

3) In contrast to the Fourier transformation method, regularization is not associated with a particular kind of integral equation kernel. This circumstance permits one, in principle, to reject a number of model restrictions such as neglect of the electron wave-front curvature and the approximation $kr \gg 1$ for any r.

4) The application of regular methods involving the use of physical requirements yields a solution which practically coincides with the model function $g_m(r)$, not only for crystal but also for amorphous solids.

5) The regular methods permit a strict formulation of the problem of determining the partial distribution function for a binary system.

References

1. Yu.A. Babanov, V.V. Vasin, A.L. Ageev, N.V. Ershov: Phys. Status Solidi B *105*, 747 (1981)
2. N.V. Ershov, A.L. Ageev, V.V. Vasin, Yu.A. Babanov: Phys. Status Solidi B *108*, 103 (1981)

Regular Methods of Solving Inverse Problems in Spectroscopy

V.V. Vasin

Institute of Mathematics and Mechanics, Ural Science Research Center
of the USSR Academy of Sciences, Sverdlovsk, USSR

A problem is called ill-defined (unstable) if small variations in input data may cause large variations in the solution sought [1,2]. Errors in input data are apt to stem from inaccuracy of measurement, inadequacy of the model involved, and computation errors.

Mathematically ill-defined problems are usually formulated as the solution of a linear (or nonlinear) equation of the first kind [1-3]:

$$Au = f \tag{1}$$

with a completely continuous operator $A:U \to F$, where U, F are metric spaces, for example, $L_2[a,b]$ with $\rho(u,v) = \|u - v\| = (\int_a^b |u(s) - v(s)|^2 ds)^{1/2}$. Problem (1) is ill-defined (unstable) because the inverse operator A^{-1} is discontinuous in this case, and therefore the convergence of approximate solutions $\rho_U(u,u_\delta) \to 0$, $\delta \to 0$, where u_δ is the solution of the equation $Au_\delta = f_\delta$, does not follow from that of the right-hand sides of (1) $\rho_F(f,f_\delta) \to 0$, $\delta \to 0$.

This circumstance makes every reasonable interpretation of the approximate solution and its traditional treatment difficult and sometimes impossible. The integral equations:

$$Au \equiv \int_a^b K(t,s)u(s)ds = f(t) \tag{2}$$

relate to problems of type (1). These equations describe a wide range of what are known as inverse problems. The latter arise in geophysics, astrophysics, automatic control, spectroscopy [1-3], and particularly when processing EXAFS data [4].

Mathematicians have now developed regular (i.e., stable with respect to perturbations of input data) methods of solving ill-defined problems. The concept of a regularizing operator and regularized solution introduced by TIKHONOV [5] has been instrumental in constructing regular methods.

Let us describe a problem (2) solution algorithm which has made a good showing in the processing of EXAFS data [4,6]. To a simplest single-scattering approximation for K absorption spectra, the equation connecting the normalized oscillating part of the X absorption coefficient $\chi(k)$ with the radial atomic distribution function may be written as

$$Ag \equiv 4\pi\sigma \int_0^b [\psi(k)/k]\exp(2r/\lambda)\sin(2kr + 2\beta + \varphi)g(r)dr = \chi(k) \quad . \qquad (3)$$

The additional information about the solution g(r) used in the algorithm is determined by the relation of normalization and positiveness of g(r):

$$(V,g) \equiv (3\pi b^3/4) \int_0^b 4\pi r^2 g(r)dr = 1 \quad , \quad g(r) \geq 0 \quad . \qquad (4)$$

Three stages in the construction of the algorithm must be noted.

1) Regularization of the equation by Tikhonov's method (allowing for the normalization condition by means of a penalty function):

$$\min\left\{\|Ag - \chi\|^2_{L_2} + \alpha\|g\|^2_{L_2} + x|(v,g) - 1|^2 : g \in L_2\right\} \quad . \qquad (5)$$

2) Discretization of problem (5) by the quadrature or collocation method [4,7]:

$$\min\left\{\|A_n g_n - \chi_n\|^2_{l_2^n} + \alpha\|g_n\|^2_{l_2^n} + x|(v_n,g_n) - 1|^2 : g_n \in l_2^n\right\} \quad . \qquad (6)$$

3) Numerical solution of problem (6) and iterative correction of the formula [8]:

$$g_n^{(m)} = P\left(A_n^* A_n + xv_n^* v_n + \bar{\alpha}E\right)^{-1}\left(A_n^* f_n + xv_n^* + \alpha g_n^{(m-1)}\right) \quad ,$$

$$\bar{\alpha} > \alpha > 0 \quad , \quad (m = 1,2,\ldots) \quad ,$$

where P is a metric projection on a positive orthant.

A large number of numerical experiments have shown that the method proposed is highly efficient compared to the Fourier transformation method and other techniques. The described algorithm can be applied to solve the problem of determining partial functions of radial atomic distribution in multicomponent systems [6].

1. A.N. Tikhonov, V.Ya. Arsenin: *Metody resheniya nekorrektnyh zadach* (Nauka, Moscow 1979)
2. M.M. Lavrent'ev: *O nekotorykh nekorrektnyh zadachakh matematicheskoi fiziki* (Sibirsk. Otdel. Akad. Nauk SSSR, Novosibirsk 1962)
3. V.K. Ivanov, V.V. Vasin, V.P. Tanana: *Teoriya lineinykh nekorrektnyhk zadach i eye prilozheniya* (Nauka, Moscow 1979)
4. A.L. Ageev, Yu.A. Babanov, V.V. Vasin, N.V. Ershov: Trudy Instituta Matematiki i Mekhaniki UNC, Akad. Nauk, SSSR, 1981, pp.3-18
5. A.N. Tikhonov: Dokl. Akad. Nauk SSSR *151*, 501-504 (1963)
6. Yu.A. Babanov, V.V. Vasin, A.L. Ageev, N.V. Ershov: Phys. Status Solidi *105*, 747-754 (1981); *108*, 103-111 (1981)
7. V.V. Vasin: Dokl. Akad. Nauk SSSR *258*, 271-275 (1981)
8. V.V. Vasin: *Trudy konferentsii: Chislennoe reshenie kraevykh zadach i integralnykh uravnenii* (Tartus. Gos. Univ., Tartu 1981) pp.76-78

Part III

XANES

XANES Spectroscopy for Local Structures in Complex Systems

A. Bianconi

Dipartimento di Fisica, Università di Roma, La Sapienza, P. Aldo Moro 2, I-00185 Roma, Italy

1. Introduction

The microscopic local order plays a key role on the properties of complex systems such as the biological function of proteins, the activity of catalysts, the chemisorption of molecules on solid surfaces and electronic properties of amorphous materials. Local structure determination is the first step toward a quantitative understanding of these systems. The expanding interest of physics from simple systems (such as molecules and crystals) toward complex systems requires new experimental methods for local structure determination.

X-ray absorption spectroscopy using synchrotron radiation is a powerful tool for investigation of complex systems because of its capability to probe the local structure near a selected atom in non-crystalline systems. Two energy regions in X-ray absorption spectra have been distinguished since the early days (1931) of X-ray spectroscopy: the high energy Kronig structure probing the local structure and the low energy Kossel structure probing the empty electronic states. The renewal of interest in the Kronig structure began with the work of Sayers et al. [1] who first pointed out the potential of EXAFS (Extended X-ray Absorption Fine Structure) for determination of local structure in non-crystalline materials. In these last ten years EXAFS has become, with the use of synchrotron radiation, a well-established method to measure the pair-correlation function of atomic distribution, i. e. interatom ic distances and coordination numbers in a large variety of complex systems of biological, chemical or material science interest.

The Kossel structure of crystals has been assigned to the "partial local density of states" of the conduction band complicated by final state effects such as excitons and many-body effects [2]. The difficulties faced by the theory of electronic structure both to calculate high energy empty levels and final state effects in the excited state were such that only the features in the first ~ 8 eV energy range have been explained in metal and insulator crystals. In complex systems the low energy Kossel structures in the first ~ 8 eV energy range have been explained for a long time using a molecular orbital approach [3]. In this approach the final states are the calculated empty valence orbitals of a cluster formed by the absorbing atom and its first neighbours.

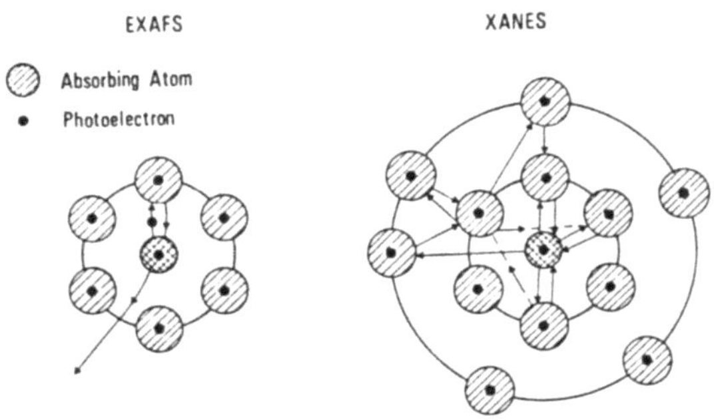

FIG. 1. Pictorial view of the scattering processes of the excited photoelectron in the single scattering regime (EXAFS) and in the full multiple scattering regime (XANES)

From extended investigation on biological molecules [4-6], surfaces [7--11] and simple systems [11, 5] we have pointed out that the full series of structures extending over a ~40 eV energy range near the edge defined by us [5, 11, 13] XANES (X-ray Absorption Near Edge Structure) are directly determined by the local structure. The potential of XANES for determination of local geometry of atomic arrangments giving unique information complementary to EXAFS in complex systems was shown.

The XANES peaks extending from ~8 eV to ~40 eV above the threshold (which escape EXAFS interpretation) have been identified as multiple scattering resonances (MSR) [4, 11, 13] of the excited photoelectron scattered by neighbour atoms in the continuum part of the spectrum.

In the photoionization process the low kinetic energy (10-40 eV) excited photoelectron is strongly backscattered, generating a multiple scattering process. It is for such multiple-scattering involving several atoms that XANES is informative on the relative positions of neighbour atoms. As it is shown in Fig. 1 the EXAFS oscillations are due to a different scattering process where the high kinetic energy photoelectron emitted by the central atom is weakly scattered by only one neighbour atom in a single scattering process. Therefore like EXAFS also XANES is a technique probing the local structure through the electron scattering by neighbour atoms of the emitted electron at the central atom. The advantageous simple characteristics of EXAFS are also its limitations. It gives information about local structures only in terms of atomic radial distribution (interatomic distances and coordination numbers) and no information on coordination geometry can generally be extracted. XANES contains information on stereochemical details (coordination geometry and bond angles) which are particularly important for complex systems, like proteins, characterized by weak order and low symmetry.

2. Site geometry and distance variations by theoretical analysis of XANES spectra.

2.1. XANES of N_2, C_2H_2 and C_2H_4: energy shift of multiple scattering resonances and distance contraction.

Historically the XANES has been first assigned to "inner well resonances" induced by an electronegative potential barrier on the first neighbour atoms by Dehmer [14]. Later, in the XANES of the covalent N_2 molecule shown in Fig. 2 [12], the resonance G, above the continuum photoionization threshold, has

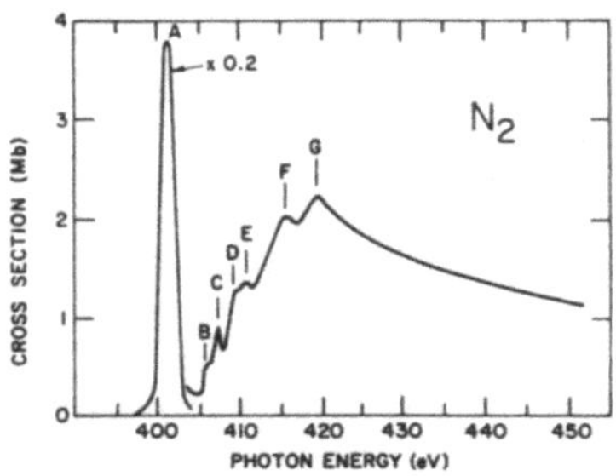

FIG. 2 . N_2 K-edge absorption spectrum. The continuum threshold is at 410 eV, G is the multiple scattering resonance σ_u and F is a "shake-up" or two-electron transition

been identified, by X_α multiple scattering calculations, as a "shape resonance" [15]. The wave function of the excited photoelectron is strongly enhanced and confined inside the molecule of the resonance energy as it is shown in Fig. 3 [16]. The hypothesis that this resonance is due to the molecular centrifugal barrier acting on the higher-l components of the final states (partial wave expansion) was pointed out. These higher-l components arise from the multiple scattering in the molecular potential of the excited p wave by the photoionization of the 1s shell.

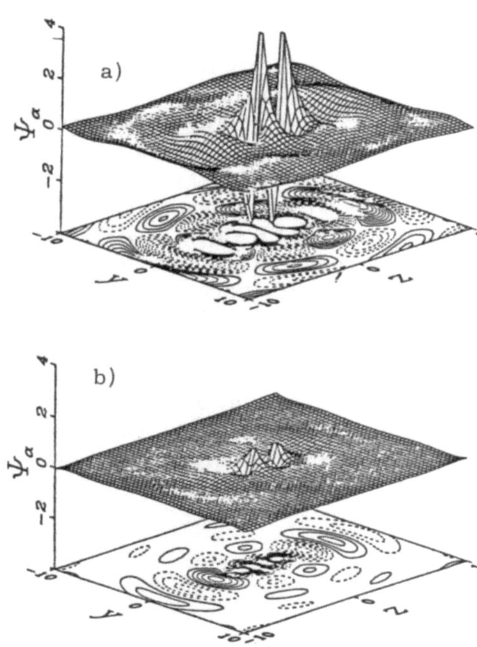

FIG. 3 Wave function of the excited photoelectron in the continuum at the resonance energy of peak G in the N_2 absorption spectrum in Fig. 2 (a), and out of resonance at higher energy (b) (Ref. [16])

In a linear molecule like N_2, CO, CN characterized by very short interatomic distance, two strong shape resonances appear, one in the continuum part (peak G in Fig. 2) and one in the discrete part of the spectrum (peak A in Fig. 2), which is a bound state because of the electron-hole interaction, arising from the π_g and σ_u photoionization channels respectively. The K-XANES spectra of C_2H_2 and C_2H_4 shown in Fig. 4 shows the characteristic two resonances like biatomic molecules because of the very weak electron backscattering amplitude of H atoms. Therefore the two multiple scattering resonances can be assigned to the electron scattering from the two carbon atoms. Decreasing the C-C distance the energy shift of the σ_u resonance follows the $k \cdot R$ = const rule where k is the photoelectron wave vector and R the interatomic distance.

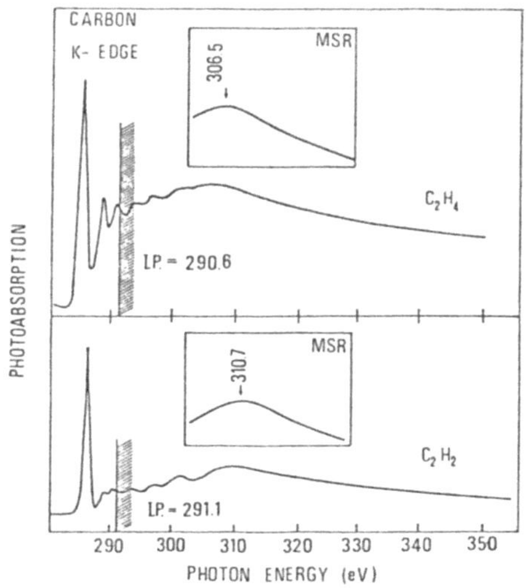

FIG. 4. K XANES of C_2H_2 and C_2H_4. The energy shift of the multiple scattering resonance (MRS) at wave vector k is a measure of the variation of the carbon-carbon distance R following the rule KR = = const

The comparison between the theory and experiment shows that it is not possible to predict by XANES calculations [14, 17] the exact energy position of the continuum resonance due mainly to the muffin-tin approximation for the molecular potential, but it is possible to extract interatomic variations by the energy shift of the multiple scattering resonance. This approach has been shown to work in the case of C_2H_2 and C_2H_4 where the resonance shift follows the C-C contraction [18, 19]. This seems to be a general rule for XANES following from the fact that in the electron scattering between the same type of atoms the phase shifts remain the same and neglecting, for small energy variation, their energy dependence, the energy position of resonances for small interatomic distance variations should follow the $k \cdot R$ = const rule like EXAFS peaks [18, 20]. Recently angular resolved XANES of CO or N_2 molecules chemisorbed on a surface of an oriented single crystal have been measured [21]. This measurements using the polarization of synchrotron radiation allows the determination of the molecular orientation. Moreover the multiple scattering resonance in the continuum σ_u shows an energy shift toward lower energy in comparison with the gas phase indicating anincreased interatomic distance when the molecule is chemisorbed on the surface.

2. 2. K XANES of iron hexacyanide complexes.

One of the most important problems of XANES interpretation was the determination of the size of the cluster determining the XANES. As we have seen, the molecular approach for the edge region was to consider the cluster formed only by the first shell, also in the first interpretation of the XANES as "inner well resonances" the first shell was considered to play the main role. In fact experimental evidence that the XANES spectra of quartz, of amorphous silica, and of the ~5 Å oxide layer on silicon surface were similar supported the idea that the silicon $L_{2,3}$ XANES where determined by only the SiO_4 microscopic cluster [9, 11]. The extension to many systems of the investigation of the effect of long range on XANES has shown [5] that for non-tetrahedrally coordinated metal atoms the XANES are strongly determined by the second and third neighbour shells. This seems now well established from analysis of many systems that for metal ions in tetrahedral coordination, perhaps both due to the symmetry and to the generally shorter interatomic distances, the multiple scattering resonances are confined mostly within the first coordination shell while for other coordination symmetries the second and third shells play an important role. The observed effect of the second and third shells is clearly more important in ordered atomic arrangements and the successive inclusion of shells induces a splitting of XANES features [4].

The approach based on the multiple scattering of the excited photoelectron by successive neighbour shell proposed by Durham et al. [22] has the important advantage in comparison with other approaches that a large number of atoms can be included in the calculation. Fig. 5 shows the measured iron K XANES of $K_3Fe(CN_6)$ and $K_4Fe(CN_6)$ and the calculated XANES [23].

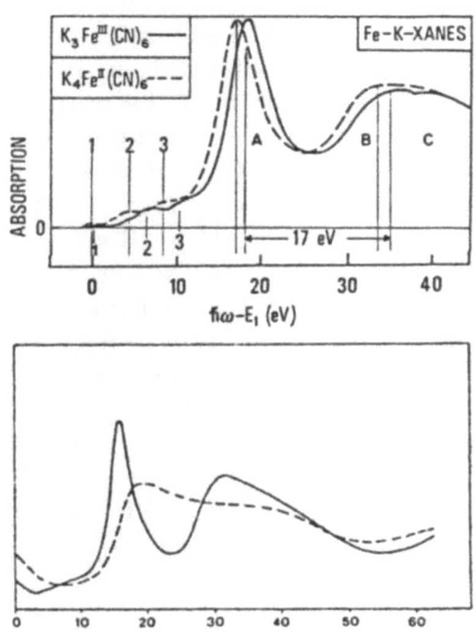

FIG. 5. Iron K-edge XANES of iron hexacyanide complexes $[Fe(CN)_6]^{2+}$ and $[Fe(CN)_6]^{3+}$. The important role of the second neighbour shell is shown by XANES calculations in the lower panel for a cluster $[Fe(CN)_6]^{3+}$ with both C and N shells included (full curves) and with the C first shell only included (dashed curve). The energy shift of peak A follows the Fe-C distance variation, the shape and intensity ratio between peaks A, B and C follow the geometric distortions induced by different charge state of the cluster

The XANES of the FeC_6 cluster (only of the first coordination shell) is quite different from the experimental spectrum. Including the second shell the XANES of the $Fe(CN)_6$ cluster gives account of the experimental XANES features of these systems.

The XANES theory has been used to determine the local geometry variation between the $[Fe(CN)_6]^{3+}$ and $[Fe(CN)_6]^{2+}$ clusters. A very subtle structural change is expected to occur because of the effect of electron "back donation" between Fe and CN groups. In fact the EXAFS gives the same ($\pm$ 0.01 Å) Fe-C distance for both charge states. The XANES shows a shift of peak A in Fig. 5 and a change of shape of the features B and C. We have first calculated the XANES of the $[Fe(CN)_6]^{3+}$ cluster finding a good agreement with the experimental spectrum. Then we have calculated the effect of changing atomic positions on XANES. A contraction of the Fe-C distance induces a shift of peak A; a contraction of the C-N distance induces an increase of the A-B energy splitting in the XANES. Introducing distortions in the geometry induces a change in the shape of the B and C peaks, as well as of peak A. Therefore the different XANES spectra indicate structural differences not detected by EXAFS like different C-N distances and distortion of octahedral symmetry induced by the extra electronic charge in $[Fe(CN)_6]^{2+}$ cluster. Because only three structures are observed the best procedure for XANES analysis is to calculate the spectra of different proposed strucutres to choose between models in competition. We have introduced in our calculations two different structures one determined by X-ray diffraction and the other by neutron diffraction and we have found a good agreement with experimental XANES for the last proposed structure.

2. 3. K XANES of hemoglobin and mioglobin.

We have measured and calculated the K XANES of iron in the heme-proteins like hemoglobin and mioglobin [30]. It comes out from experiments that the XANES of heme-proteins should be determined by a large number of atoms in several shells. In fact a good agreement between theory and experiment, shown in Fig. 6, has been obtained by including nearly 30 atoms in the cluster for XANES calculation. The effect of binding of a biatomic molecule to Fe has a strong effect on the XANES as can be seen in the experimental Hb-CN spectrum in agreement with calculations. This large effect can be explained by a trapping of the excited photoelectron in the CN molecule which has a very short interatomic distance and therefore exhibits a strong shape resonance in the continuum where the electron wave function is confined along the CN molecule axis. Rotation of the CN molecule should induce a large variation of XANES. Fig. 7 shows low resolution XANES of mioglobin (Mb) measured in collaboration with B. Chance and L. Powers on the focused beam line II at the Stanford synchrotron radiation facility SSRL. The fash-photolysis of mioglobin CO at low temperature (4 K) induces a large variation of the XANES spectrum. For time resolved studies of structural variations XANES analysis offers the advantage, in comparison with EXAFS, of strong spectral structures and more sensitivity to local geometry variations. High resolution XANES is necessary for a quantitative analysis of XANES; however the

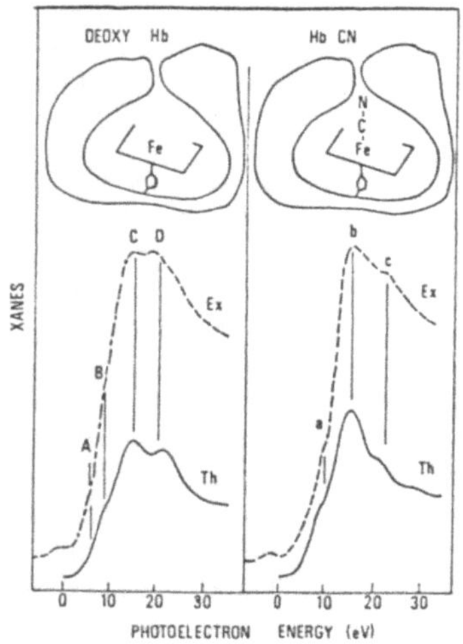

FIG. 6. XANES spectra of Fe in hemoglobin (Hb), experiment (Ex) and theory (Th). The agreement between theory and experiment is obtained only by including ~ 30 neighbour atoms showing the full porphyrin plane contribute to the XANES. The effect of a molecule (CN) bound to Fe induces important changes in the XANES. The good agreement between theory and experiments is obtained with a vertical position of the CN molecule relative the porphyrin plane. The rotation of the ligand can be determined mainly by change in peak b of Hb-CN spectrum [30]

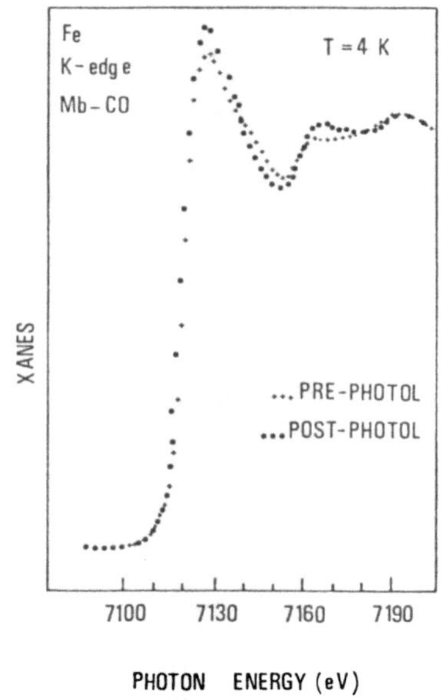

FIG. 7. Low resolution XANES spectra of mioglobin CO at low temperature (4 K) (+++ curve), and the modified spectrum after flash photolysis (••• curve). The CO movement induces large effects on XANES

observed changes in the XANES spectrum are in agreement with a possible rotation of the CO molecule from a tilted configuration to a normal configuration related to the porphyrin plane after flash photolysis.

3. The site structure determination by XANES using model compounds.

We first proposed an experimental approach for structural determination by XANES using model compounds [4, 11, 13]. The idea follows the identification of XANES as determined only by the local structure; therefore the XANES spectrum of an unknown system should be the same as a model compound with similar local structure but with a different long-range order. This approach has been successfully applied to several systems [13]. The main difficulty of this approach concerns the size of the cluster determining

the XANES. Only in the case of tetrahedrally coordinated atoms does the similarity between the first coordination shell in different systems induce similar XANES spectrum. In the study of complexes and proteins the right application of this method requires the use of model compounds with the same type of coordinated molecular groups, to measure the effect of their different geometrical orientation in known systems and then the comparison with the XANES spectrum of the system where the atom is expected to be coordinated by the same type of molecular groups. For example, the calcium phosphates, with $Ca(PO_4)_n$ local clusters, are good model compounds for calcium bound to ATP or membranes; calcium formate or acetate, formed by $Ca(COO^-)_n$ local cluster, are good model compound for calcium in proteins when Ca^{2+} is coordinated by carboxyl groups. Using this approach we have studied calcium binding proteins: calmodulin [24] and tropinin-C [25]. From the identity between the XANES of different calcium sites in troponin-C we have deduced the similarity between local structures. Fig. 8 shows that the calcium XANES are affected by binding of Mg^{2+} in some low affinity sites perhaps on the external part of the protein. This is the only method that gives direct information on the local structure change induced by magnesium. The shift of peak A induced by Mg^{2+} indicates a decreasing of the Ca-O nearest distance within the first shell and the variation of peak B should be related with a rotation of the coordinated carboxyl groups.
This shows that Mg^{2+} induces important structural modifications on this protein which controls the muscle contraction by binding and releasing two calcium ions [26].

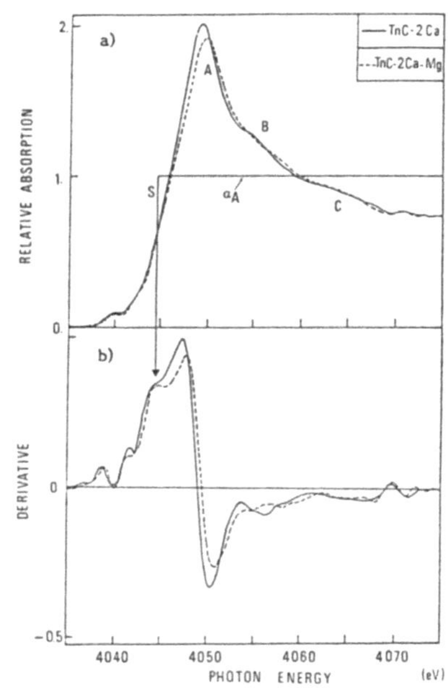

FIG. 8. The XANES spectra of calcium in the calcium binding protein troponin-C. The multiple scattering resonances A, B and C show energy shift and intensity variation due to change in the local geometry (movement of CO and COO^- molecular groups bound to calcium) induced by binding of Mg in external low affinity sites of the protein

4. Surface XANES studies of clean surfaces and of chemisorption processes.

Surface XANES of clean surfaces can be measured by intermediate energy Auger yield detection mode [13]. Collecting the intensity low energy Auger electrons of kinetic energy E_f = 30-80 eV, and scanning the photon energy, a surface photoabsorption spectrum with high surface contrast can be measured over a energy range above threshold $\Delta E = E_f + \Phi$, where Φ is the work function. Fig. 9 shows the surface XANES of clean aluminium surfaces (111) and (100) compared with the bulk Al XANES of the $L_{2,3}$ edge [8]. The high energy structure at 97 eV can be explained by EXAFS theory; however, for its low energy it should be considered a XANES structure. The different electronic structure on the different

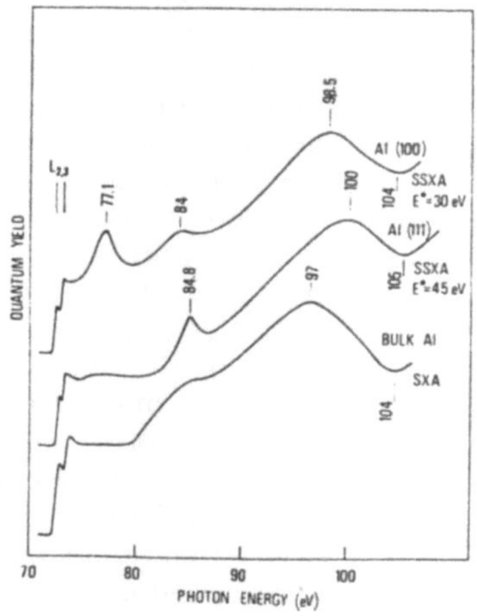

FIG. 9. Surface XANES of clean aluminium surfaces (111) and (100) selected by intermediate energy (30-40 eV) Auger yield to have a high surface contrast compared with bulk XANES. The differences between the XANES structures maxima and minima at ~104 eV are related with interatomic distance variations following the k· R = const rule

surfaces can account for the differences in the low energy part of the spectrum. The maxima at 97 eV and the minimum at ~104 eV should be directly related with atomic arrangements at the surface. Using the rule $\Delta E/2E = -\Delta R/R$ where the energy variation ΔE of multiple scattering resonance is related with small distance variation ΔR between the same type of atoms, we find a contraction $\Delta R = -0.04 \pm 0.02$ A for the Al-Al distance on the

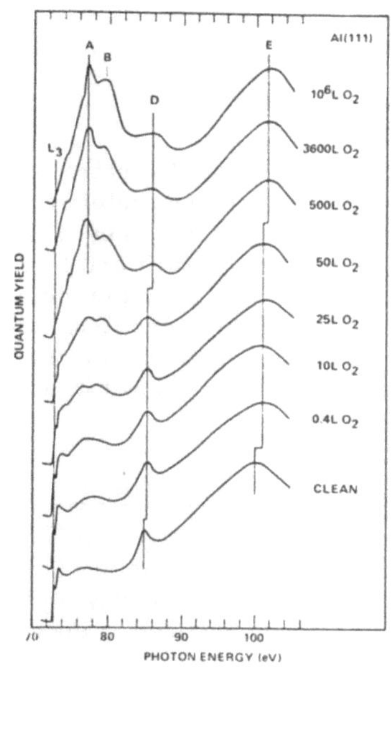

FIG.10. Site structure variation near Al atoms on the surface are detected with high sensitivity by XANES during the exposure of the surface to oxygen. The clean change of the local site structure going from the atomic chemisorption phase (below 100 Langmouir exposure) on the Al(111) surface to the oxide phase is observed. By comparison with model compounds the XANES spectrum of the omorphous oxide phase shows that Al is coordinated by six oxygens

Al(111) surface (between one Al atom of the first and one of the second surface monolayer because of light polarization). This contraction is smaller than that extracted by EXAFS analysis [27].

The chemisorption process can be followed by surface XANES studies of the local structure near a surface metal atom of the crystal interacting with the chemisorbed atom. The Al atom interacting with oxygen on an oxygen exposed surface can be discriminated by a bulk Al atom by the collecting interatomic (Al-O) Auger electrons [8]. Fig. 10 shows the aluminum $L_{2,3}$ XANES of the Al(111) surface at different oxygen exposure. The transition from the chemisorption phase of oxygen atoms to the oxidation phase above 100 Langmouir exposure can be clearly observed. These spectra show that XANES is a very sensitive probe for local structure variations.

5. Studies of metal-insulator phase transition: edge studies.

We have shown that the multiple scattering responances from ~8 to ~40 eV above threshold are mainly structure dependent while the edge region at the absorption threshold is sensitive to the electronic structure near the Fermi level. We have recently shown that the metal-insulator transition in VO_2 due to temperature variation (with the opening of a 0.5 eV energy gap in the insulator and with very little structural variation) is in fact a large effect in the edge region with a very weak effect on the XANES at higher energy [28].

Fig. 11 shows the VO_2 spectrum compared with V_2O_5. The edge and XANES of the two compounds are quite different as their local structure is

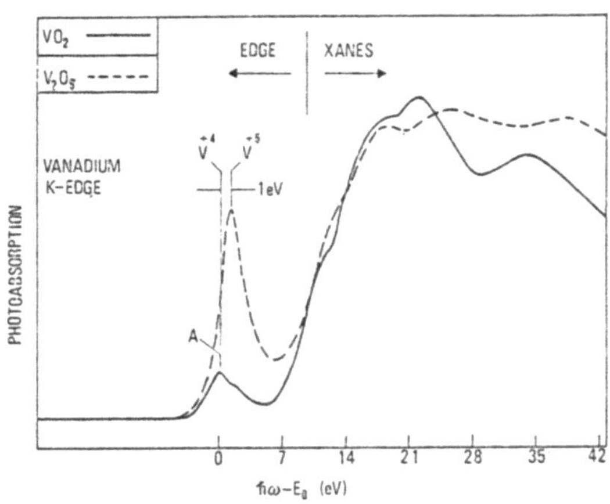

FIG. 11. Vanadium K edge of VO_2 and V_2O_5. The XANES region dominated by multiple scattering resonances in the continuum is distinguished from low energy near threshold "edge region" dominated in V_2O_5 by a bound valence state giving a withe line "A". The chemical shift of peak A between different oxides is related with the effective charge of vanadium ion going from V^{4+} to V^{5+} formal states, while the rising edge of multiple scattering resonances at ~10 eV is related also with metal-ligand covalency and structural differences

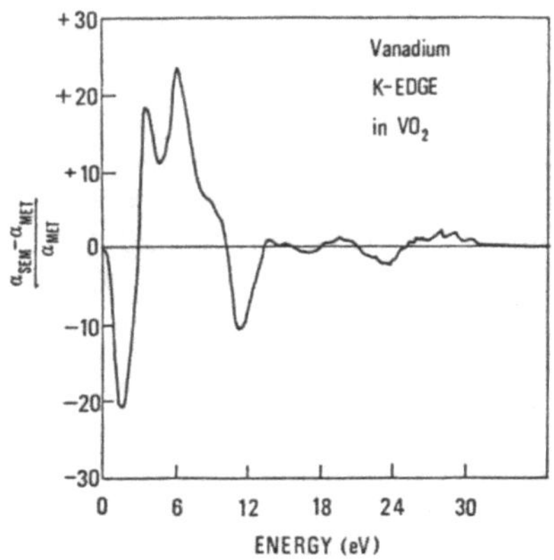

FIG. 12. Relative difference between the semiconducting (SEM) and metal (MET) VO_2 XANES

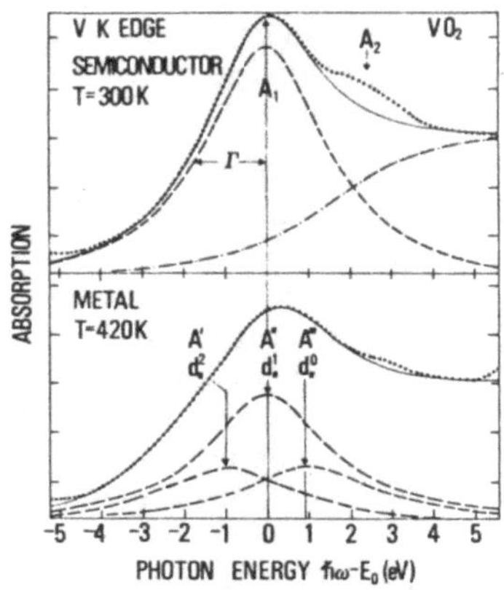

FIG. 13. Curve fitting of the V 1s threshold in metal and semiconductor VO_2

different. In V_2O_5 the fivefold coordination with a double V-0 bond gives a p-like empty valence state in the energy region of the 3d'-derived orbitals. This gives the white line at the K edge. The energy shift $\Delta E = 1$ eV between the first excited state at the edge is related with the effective charge of vanadium ions [13]. The relative variation of the photoabsorption between the semiconducting and insulating phases of VO_2 is plotted in Fig. 12. Up to 25% variation is observed in the edge region, up to ~ 15 eV due to electronic structure variation close to the Fermi level. Only less than 2-3% variation is observed in the XANES region determined by the local structure.

The large difference between spectra at the edge arises from a broadening of the edge absorption peak in the metal phase. High resolution measurements and curve-fitting procedures of the edge show in Fig. 13 the presence of multiplet splitting in the metal phase. In VO_2 the Fermi level crosses the threefold T_{2y} band which can be separated between a localized $d_{\parallel}$ band and a delocalized π band. There is a single electron in the conduction band. While in the excited final state of insulating phase only the $2p^5 d_{\parallel}^1 \pi^1$ configuration is possible, in the metal phase several final state configurations are possible: $2p^5 d_{\parallel}^0 \pi^1$, $2p^5 d_{\parallel}^1 \pi^1$ and $2p^5 d_{\parallel}^2 \pi^0$. The different energies of these configurations arise from the different electron hole interaction due to the large Coulomb attraction of the electron in the localized $d_{\parallel}$ band in comparison with that of electrons in the delocalized band.

This final state effect in VO_2 is characteristic of crystals with localized bands. The final state effects in these systems have not yet been well studied theoretically. Understanding these final state effects is very important for a quantitative interpretation of X-ray absorption edge spectra of intermediate valence crystals like rare-earth compounds [29].

1. D. E. Sayers, E. A. Stern and F. W. Lyttle, Phys. Rev. Letters 27, 1204 (1971); E. A. Stern, D. E. Sayers and F. W. Little, Phys. Rev. B11, 4836 (1975).

2. F. C. Brown, in "Synchrotron Radiation Research", ed. by S. Doniach and H. Winick (Plenum Press, 1980), p. 61.
3. L. V. Azaroff and D. M. Pease, in "X-ray Spectroscopy", ed. by L. V. Azaroff (McGraw Hill, 1974), p. 296.
4. A. Bianconi, S. Doniach and D. Lublin, Chem. Phys. Letters 59, 121 (1978).
5. M. Belli, A. Scafati, A. Bianconi, S. Mobilio, L. Palladino, A. Reale and E. Burattini, Solid State Commun. 35, 355 (1980).
6. A. Bianconi, S. Biocca, L. Castellani, P. Fasella, F. Comin and S. Mobilio, Italian J. of Biochem. 29, 71 (1980).
7. A. Bianconi, R. Z. Bachrach and S. A. Flodström, Solid State Commun. 24, 539 (1977).
8. A. Bianconi, R. Z. Bachrach and S. A. Flodström, Phys. Rev. B19, 3879 (1979); B19, 2837 (1979).
9. A. Bianconi, Surface Sci. 89, 41 (1979).
10. A. Bianconi and R. S. Bauer, Surface Sci. 99, 76 (1980).
11. A. Bianconi, Appl. of Surface Sci. 6, 392 (1980).
12. A. Bianconi, H. Petersen, F. C. Brown and R. Z. Bachrach, Phys. Rev. A17, 1907 (1978).
13. A. Bianconi, in "EXAFS for Inorganic Systems", ed. by C. D. Garner and S. S. Hasnain, Daresbury Laboratory SERC DL/SCI/R17 (1981), p. 13.
14. J. L. Dehmer, J. Chem. Phys. 56, 4496 (1972).
15. J. L. Dehmer and D. Dill, J. Chem. Phys. 65, 5327 (1976).
16. M. G. Lynch, D. Dill, J. Siegel and J. L. Dehmer, J. Chem. Phys. 71, 4249 (1979).
17. F. W. Kutzler, C. R. Natoli, D. K. Misemer, S. Doniach and K. O. Hodgson, J. Chem. Phys. 73, 3274 (1980).
18. A. Bianconi, M. Dell'Ariccia, C. Gargano and C. R. Natoli, this volume.
19. A. Bianconi, M. Dell'Ariccia, C. R. Natoli, R. Z. Bachrach and F. C. Brown, to be published.
20. C. R. Natoli, this volume.
21. J. Stöhr, K. Baberschke, R. Jaeger, R. Treichler and S. Brennan, Phys. Rev. Letters 47, 381 (1981); J. Stöhr and R. Jaeger, Phys. Rev. B26, 4111 (1982).
22. P. J. Durham, J. B. Pendry and C. H. Hodges, Solid State Commun. 38, 159 (1981).
23. A. Bianconi, M. Dell'Ariccia, P. Durham and J. B. Pendry, Phys. Rev. B, to be published.
24. A. Bianconi et al., this volume.
25. A. Bianconi, A. Giovannelli, L. Castellani, S. Alemà, P. Fasella, B. Oesh and S. Mobilio, J. Molecular Biology, to be published.
26. S. Alemà, A. Bianconi, L. Castellani and P. Fasella, in "Cell Function and Differentiation" (Alan Lyss, 1982).
27. A. Bianconi and R. Z. Bachrach, Phys. Rev. Letters 42, 104 (1979).
28. A. Bianconi, Phys. Rev. B26, 2741 (1982).
29. A. Bianconi, M. Campagna and S. Stizza, Phys. Rev. B25, 2479 (1982).
30. P. Durham, A. Bianconi, A. Congiu Castellano et al., to be published.

Experimental and Theoretical XANES Studies of Selected Vanadium Compounds

J. Wong, R.P. Messmer, and D.H. Maylotte

General Electric Corporate Research and Development,
Schenectady, NY 12301, USA

F.W. Lytle

The Boeing Company, P.O. Box 3999 2T-05, Seattle, WA 98214, USA

1. Introduction

As the region near an absorption edge is scanned in energy the ejected photoelectron sequentially probes the empty electronic levels of the material. The resulting spectra have long been known to be rich in chemical and structural information [1]. An attempt to correlate a large number of spectra, including vanadium compounds, was made by WHITE and McKINSTRY [2]. Reconciliation of molecular orbital theory with x-ray near-edge spectra as made by BEST [3] for tetra-oxo-bonded transition metals, by SEKA and HANSON [4] for octahedral complexes and by FISCHER and BAUM [5] for Ti compounds. All of these earlier efforts suffered from the (varying) poor resolution of their experiments. With the availability of intense synchrotron radiation sources, these spectra can now be gathered more quickly, simply and with greater resolution than ever before. A number of interesting papers have recently been published on K-edge and L-edge spectra [6] measured with synchrotron sources. For transition metals there are generally weak but distinct features in the spectra just before the onset of absorption. The so-called near-edge features have been attributed by SHULMAN et al. [7] to electronic transitions from 1s to some d,s (non-dipole), and p-like (dipole-allowed) empty states. In this paper we report high resolution XANES spectra of vanadium in a series of selected compounds of well-defined coordination geometry and known bond distances. The intensity and position of these near-edge spectral features are analyzed within a quantum molecular orbital framework and a simple coordination charge concept.

2. Results and Discussion

The vanadium K-edge spectra for a series of selected compounds are shown in Fig. 1-4. In each figure, the zero of energy is taken at the first inflection point of vanadium metal (at 5465 eV), the spectrum of which is also given for comparison. Table 1 summarizes the structure of the first coordination shell in these model compounds, and Table 2 lists the energy values of the more prominent features of their spectra in the vicinity of the vanadium K-absorption edge. Two different types of near-edge spectra are observed: (i) the octahedrally coordinated compounds, such as V_2O_3, etc. (Fig. 1), which exhibit a weak pre-edge absorption peak, and (ii) the tetrahedrally coordinated compounds, such as NH_4VO_3, etc. (Fig. 2), or those compounds with a short vanadyl bond in a square pyramid (Fig. 3 and 4), all of which exhibit a rather strong pre-edge peak.

The transition of interest in the pre-edge region of the XANES spectrum is vanadium 1s to the total empty manifold of the compound. For a transition metal atom in octahedral symmetry, the 1s to 3d transition is dipole forbidden, but is an allowed quadrupole transition. This is a weak transition but is still observable. The spectra of V_2O_3, V_2S_3, and Roscoelite illustrate this type of spectrum (Fig. 1). For

Fig. 1

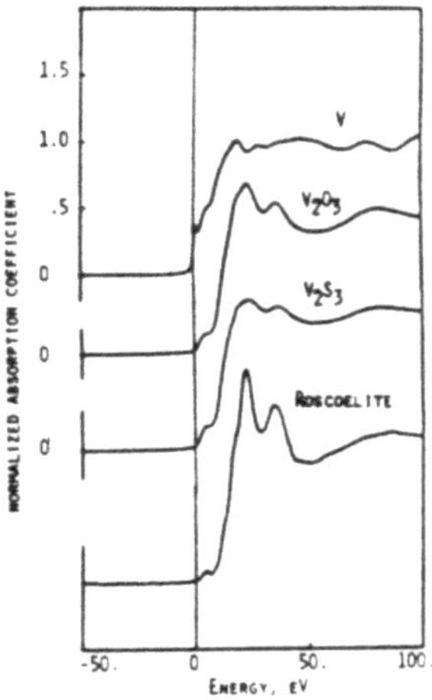

Fig. 2

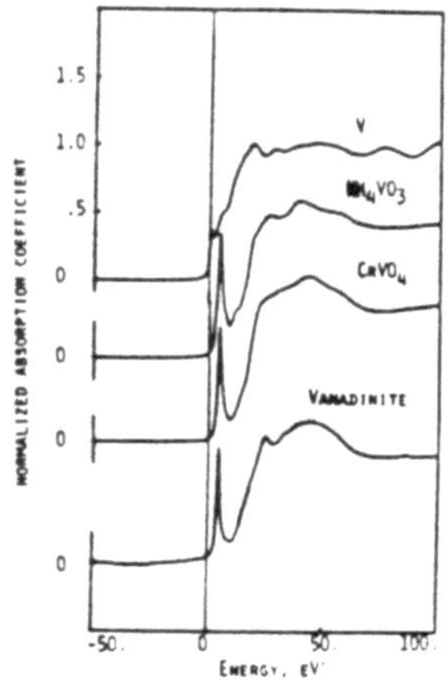

Fig. 3

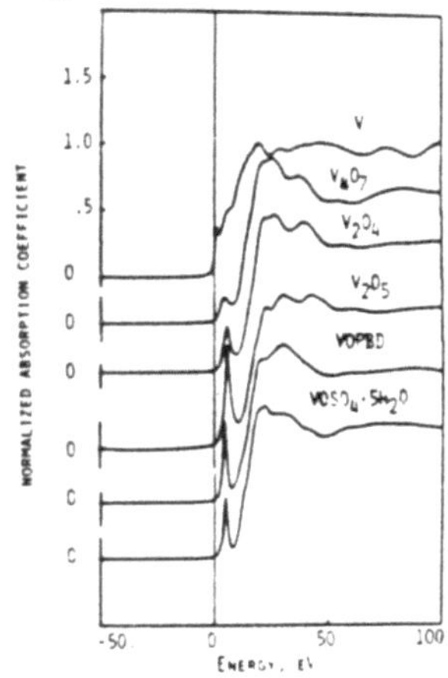

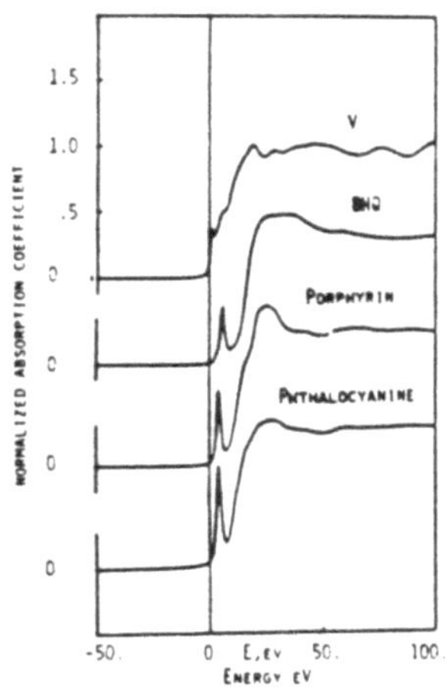

Fig. 4

Figure 1. Vanadium K-edge absorption spectra for vanadium in ectahedral coordination

Figure 2. Vanadium K-edge absorption spectra for vanadium in tetrahedral oxygen coordination

Figure 3. Vanadium K-edge absorption spectra for vanadium compounds with a vanadyl plus other oxygen ligands

Figure 4. Vanadium K-edge absorption spectra for compounds with a vanadyl oxygen bond plus (some or all) nitrogen ligands

tetrahedral complexes, the lack of a center of inversion permits a dipole transition 1s to the T_2 orbitals which can contain much metal 3d character. This gives rise to the strong pre-edge peak shown in Fig. 2 for NH_4VO_3, $CrVO_4$, and Vanadinite. KUTZLER et al. [8] have pointed out the importance of the molecular "cage" defined by the first-shell ligands: (i) a cage which breaks inversion symmetry at the metal atom site can enable dipole 1s to nd transitions; (ii) the confining effect of the cage potential to "inner well" unoccupied orbitals is to enhance the oscillator strength of transitions to these orbitals.

These points indicate not only the reason for the strong pre-edge transition in tetrahedral compounds but in any compound lacking a center of inversion (e.g., all the compounds with the short vanadyl bond given in Table 1). Furthermore, oscillator strength should correlate with cage size [8]. Within one type of coordination geometry, this would appear to be true. In Fig. 5, we have plotted our model compound results grouped according to their geometries. It is interesting to note that vanadium 8-hydroxy quinoline, which contains three oxygen and two nitrogens in its first coordination shell, falls midway between the line for a purely oxygen environment and the line for the porphyrin-type environment.

Table 1

LOCAL STRUCTURE PARAMETERS FOR VANADIUM REFERENCE COMPOUNDS

Compound	Vanadium Formal Valence	Bond Type	Number of Bonds, *N*	Bond Distances (Å)						References
V Metal	0	V-V	8	2.622						62
V_2O_3	3	V-O	3	1.96						63
			3	2.06						
V_4O_7	3.5	V-O	6	1.883	1.897	1.937	2.032	2.042	2.101	64
V_2O_4	4	V-O	6	1.76	1.86	1.87	2.01	2.03	2.05	65
V_2O_5	5	V-O	5	1.585	1.780	1.878	1.878	2.021		66
NH_4VO_3	5	V-O	4	1.66	1.66	1.81	1.81			67
$CrVO_4$	5	V-O	4	1.72	1.72	1.80	1.80			68
$VOSO_4.nH_2O$	5	V-O	1	1.591						69
		$V-OSO_3$	1	1.983						
		$V-OH_2$	3	2.035	2.037	2.048				
		V-O	1	2.23						
Roscoelite $KAlV_2Si_3O_{10}(OH)_2$	3	V-O	6	2.0						70
Vanadinite $Pb(VO_4)_3Cl$	5	V-O	4	1.7						71
Vanadyl bis 1-phenyl-1,3 butane dionate	5	V-O	5	1.612	1.946	1.952	1.986	1.982		72
Vanadyl 8-hydroxy quinoline	4	V-O	3	1.60	1.921	1.921				73
		V-N	2		2.136	2.136				
Vanadyl Phathalocyanine	4	V-O	1	1.58						
		V-N	4	2.008	2.024	2.034	2.038			74
Vanadyl Tetra-phenyl prophyrin	4	V-O	1	1.62						75
		V-N	4	2.097	2.099	2.102	2.110			
V_2S_3	3	V-S	6	2.08 (average)						*N* and *R* Determined by EXAFS

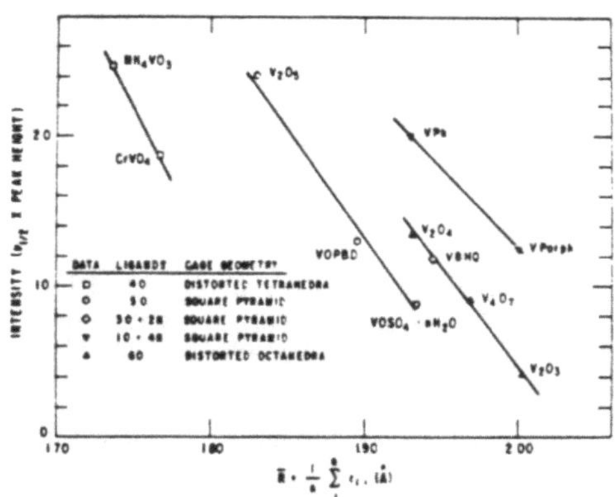

Figure 5. Intensity of pre-edge peak versus average bond length R of first shell ligands in vanadium model compounds

The integrated intensity A of the prepeak is found to vary according to $A \sim [\sum c_i R_i^2 \exp(-\alpha_i R_i)]^2$, where c_i are the coefficients of LCAO making up the final state MO and of the V complex, R_i is the bond distance of the ith ligand and α_i is the orbital exponent in the Slater-type orbitals (STO) of the ith type ligand. With $i = 1,2$ and R_1 being the short V=0 bond, (see Table 1) self-consistent iterations yield a slope of 2 in the plot of ℓnA vs. $\ln (c_i R_i^2 \exp(-\alpha_i R_i))$ only for $c_1 = .55$ (Fig. 6). In other words, the strength of the prepeak transition is strongly dependent on the size of the "molecular cage" defined by the short V=0 bond.

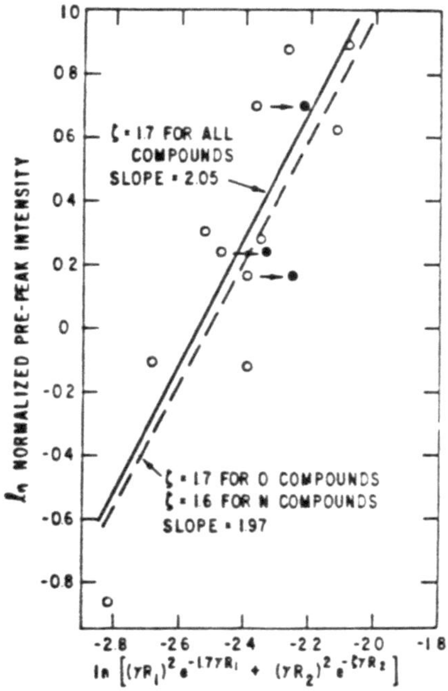

Fig. 6

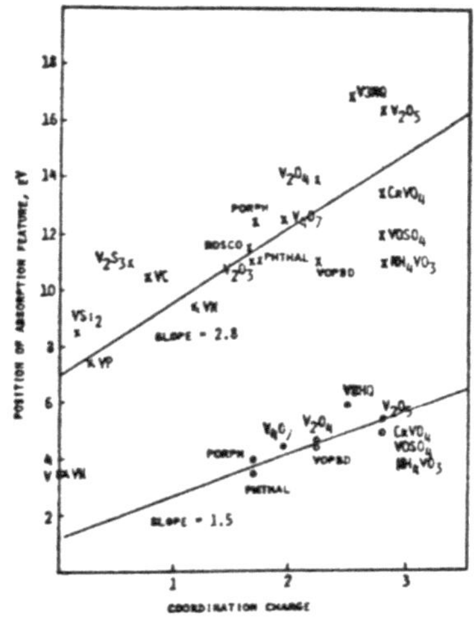

Fig. 7

Table 2. Pre-edge peak position and intensity (normalized height x half-width) of V model compounds

Vanadium Compound	Position (± 0.2 eV)	Intensity
V Metal	0.0 at first inflexion point	–
V_2O_3 (small quadruple peak)	4 (shoulder)	0.4
V_4O_7	4.5	0.90
V_2O_4	4.7	1.40
V_2O_5	5.5	2.40
NH_4VO_3	5.0	2.45
$CrVO_4$	5.0	1.86
$VOSO_4.nH_2O$	5.0	0.88
Roscoelite	(small quadruple peak)	
Vanadanite	5.0	1.45
Vanadyl bis-l-phenyl-1,3 butane dionate	4.5	1.20
Vanadium 8-hydroxyl Quinoline	6.0	1.19
Vanadyl phthalocyanine	3.5	2.01
Vanadyl tetra-phenyl porphyrin	4.0	1.26
V_2S_3	(small quadruple peak)	

The position (chemical shift) of the pre-edge peak can be approximatley accounted for by the concept of coordination charge [9]. As the valence electrons are pulled away from the metal atom by the electronegativity, X, of the coordinating ligands, all of the other electrons become more tightly bound in order to shield the unchanging nuclear charge. Hence, a K-shell transition must increase in energy with an increase in the product of valence (m) and ionicity (I); i.e., its coordination charge η given by $\eta = mI$. Here, I is the Pauling single bond ionicity given by $I = 1 - \exp[-1/4(X_A - X_B)^2]$.

In Fig. 7 the data for the energy position of the pre-edge sharp peak (lower line) and for the main absorption edge (inflection point) are plotted versus coordination charge. Although there is considerable scatter, particularly for the main edge where a variety of spectral features contained in the edge make identification of the edge quite subjective, the simple coordination charge theory allows much better organization of the data than does valence alone. In the lower line, where a specific orbital feature is easily identified, the scatter of data is much less. The lines were drawn using linear regression analysis and the difference in slopes is statistically significant. The difference is also chemically reasonable in that the pre-edge peak (primarily metal 3d electron character) is at lower energy and is expected to be more tightly bound, hence less sensitive to changes in valence and ionicity than is the more diffuse, high energy "np" potential barrier. In general terms, the sharply defined 3d level is "hard". The np continuum is "softer".

References

1. L.V. Azaroff and D.M. Pease, in X-ray Spectroscopy (L.V. Azaroff, ed.) McGraw-Hill, NY 1974 p. 184.

2. E.W. White and H.A. McKinstry, Adv. in X-ray Analysis 9, 376 (1966).

3. P.E. Best, J. Chem. Phys. 44, 3248 (1966).

4. W. Seka and H.P. Hanson, J. Chem. Phys. 50, 344 (1969).

5. D.W. Fischer and W.L. Baum, J. Appl. Phys. 41, 3561 (1970).

6. R.G. Shulman, Y. Yafet, P. Eisenberger, W.E. Blumberg, Proc. Nat. Acad. Sci. USA 73, 1384 (1976); J.E. Miller, O. Jepsen, O.K. Anderson, J.W. Wilkins, Phys. Rev. Lett. 40, 720 (1978); V.O. Kostroun,, R.W. Fairchild, C.A. Kukkonen, J.W. Wilkins, Phys. Rev. B13, 3268 (1976); V.W. Hu, S.I. Chan, G.S. Brown, Proc. Nat. Acad. Sci. USA 74, 3821 (1977); S.P. Cramer, T.K. Eccles, F. Kutzler, K.O. Hodgson, S.J. Doniach, Amer. Chem. Soc. 98, 8059 (1976); S.P. Cramer, K.O. Hodgson, E.I. Stiefel, W.E. Newton, J. Amer. Chem. Soc. 100, 2748 (1978); P.S.P. Wei and F.W. Lytle, Phys. Rev. B19, 679 (1979).

7. R.G. Shulman, Y. Yafet, P. Eisenberger, W.E. Blumberg, Proc. Nat. Acad. Sci. USA 73, 1384 (1976).

8. F.W. Kutzler, D.K. Misemer, C.R. Natoli, S. Doniach, K.O. Hodgson, J. Chem. Phys. 73, 3274 (1980).

9. S.S. Batsanov, Bull. Acad. Sci., USSR, Phys. Ser. 31 936 (1967) (English Translation).

X-Ray Absorption of Crystalline Titanium Compounds: Evidence of Electronic Relaxation Effects

Adalberto Balzarotti
Istituto di Fisica, Università di Roma II, I-00185 Roma, Italy
PULS-CNR Laboratori Nazionali di Frascati, I-00044 Frascati, Italy

1. Introduction

In distinction to XANES which involve strong multiple scattering processes occurring within a cluster surrounding the central atom [1], the near-edge structures are mostly related to the lowest unoccupied crystal levels and then reflect band structure details. As a matter of fact, at the onset the mean free path is long enough that the excited electron will travel several unit cells and its wave function will resemble a stationary Bloch state.

We present experimental evidence that the local partial density of empty conduction band states (PDOS) can be effectively used to reproduce the metal near K-edge features of a few crystalline metallic and insulating titanium compounds. It will be also shown that in metal carbides and nitrides the core hole induces relaxation of the outer electrons which shows up as a shift of the inner 2p levels of Ti with respect to the Fermi level. A strict correlation is found with the electron density-of-states distribution at the Fermi surface.

The experiments have been performed at the PULS facility (National Lab. of Frascati) using the X-rays emitted by the ADONE storage ring (E=1.5 GeV, I=50 mA). Details of the experimental apparatus have been reported elsewhere [2]. The samples were polycristalline powders sandwiched between two kapton tapes. The carbides and nitrides were 99.9 % pure and stoichiometric samples refined and vacuum melted with an oxygen content less than a few ppm.

2. Edge features and local partial density of states

Core levels E_l have negligible energy dispersion through the Brillouin zone Ω and a well-defined angular momentum l. In the single-electron approximation the absorption coefficient from an inner level (i) to a final unoccupied band state (f) is given by the expression [3]:

$$\mu(\omega)=\left(\frac{2\pi e}{m}\right)^2 \frac{1}{nc\omega} \sum_{\nu} |P_{f,\nu}|^2 \int_{\Omega} \frac{d^3\underline{k}}{4\pi^3} |A_{f,\nu}(\underline{k})|^2 \, \delta(E_f(\underline{k})-E_l-\hbar\omega), \tag{1}$$

where $A_{f,\nu}(\underline{k})$ is the contribution of the atomic orbital of symmetry $\nu \equiv \{(l,m)\}$ forming the final crystal wavefunction and $P_{f,\nu}$ is the transition matrix element between the atomic functions $\Phi_l(\underline{r})$ and $\Phi_\nu(\underline{r}-\underline{R}_j)$ centered on the lattice sites $\underline{R}_j$. The dipole approximation picks out the final states of symmetry $l\pm1$ which are coupled to the initial state of symmetry l. Since $A_{f,\nu}(\underline{k})$ are slowly varying functions of $\underline{k}$, the integral in (1) is the weighted final DOS for the ν-type symmetry orbital on the j-th atom and for the band f. This PDOS function gives the number of electronic states per unit cell and unit energy associated with the ν-type orbital. In the augmented-plane-wave (APW) method of band calculations, suitable for metals and semiconductors, the PDOS are evaluated by expanding the electron wave functions in spherical harmonics inside muffin-tin type potentials centered on the atomic sites. The PDOS has the same critical behaviour (Van Hove singularities) as the final DOS function, unless $A_{f,\nu}(\underline{k})=0$. Sharp peaks of μ arise mostly from regions of Ω where the energy gradient vanishes.

Many-body corrections to the single-electron picture have to be considered particularly when the core-hole potential perturbs heavily the excited electronic states, as in insulating solids. For NiO, an antiferromagnet oxide which exhibits narrow 3d bands embedded in broader conduction bands (c.b.), the DOS picture is invalid and the threshold spectral peaks are due to core excitons [4]. In V oxides the localization of the excited wave-function causes excitonic effects to prevail and, accordingly, a sharp prepeak - the white line - is observed below the K edge whose strength depends on the size and the geometry of the molecular cage surrounding the V atom [5]. Although conclusions of general validity cannot be drawn, it seems reasonable that excitonic effects are weaker the larger the bond covalence is. This seems to be the case at least for TiO_2, one of the insulating compounds considered below [4].

2.1 Specific examples

D-band cubic perovskites have simple c.b.'s π and σ derived from the antibonding 3d-like states of Ti splitted by the octahedral field of the ligand O^{--} ions. The π band has a predominant 2D character [6] and its bandwidth is determined by the two-center transfer integral (pdπ) between nearest-neighbours O 2p orbitals and Ti t_{2g} nd orbitals. The σ band lies little higher in energy and its bandwidth is controlled by (pdσ). In Fig. 1 we compare the measured Ti K-edge absorption of $SrTiO_3$ and $BaTiO_3$ with that of rutile TiO_2 where Ti is 6-fold coordinated to O as in perovskites but has a tetragonal lattice structure. The spectral features of the three compounds differ markedly in the edge region, a signal that solid-state effects are important. For rutile a molecular- orbital (MO) picture applied to the $(TiO_6)^{8-}$ cluster seems adequate for interpreting the peaks at the edge [11]. Furthermore, an extended Hückel tight-binding DOS calculation [4] predicts peaks which are broadened versions of the MO t_{2g} and e_g levels.

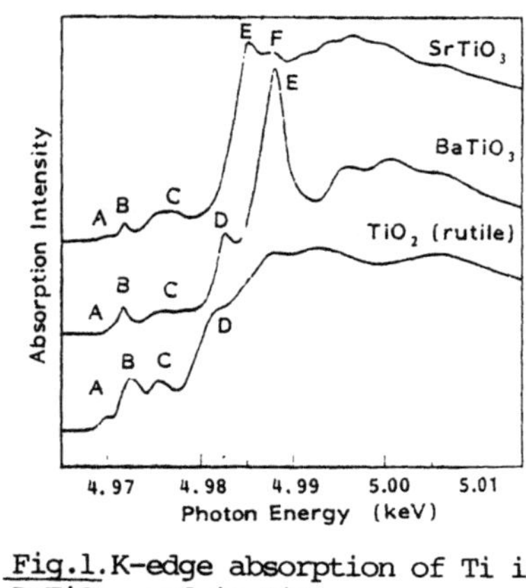

Fig.1. K-edge absorption of Ti in $SrTiO_3$, $BaTiO_3$ and in TiO_2

For the titanates we have computed [7] d- and p-like PDOS for these bands using the analytical expressions recently derived by WOLFRAM et al. [8]. The results for $BaTiO_3$ (Fig. 2) give evidence of a strong logarithmic singularity P_1 due to the flatness of the π band along the XM direction

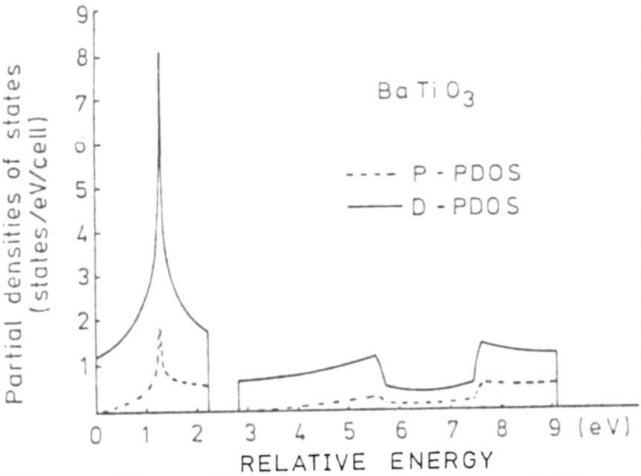

Fig. 2. Theoretical PDOS for $BaTiO_3$ using the parameters of Ref. 9

of the BZ as well as of P_0 and P_2 singularities occurring at the band edges. In the p-PDOS reported in Fig. 3 we have used a Lorentzian broadening factor γ=1.5 eV to account for the finite lifetime h/τ_{hole} of the 1s hole due to the Auger decay and to the instrumental resolution. The overall agreement between theory and experiment is good for what concerns the structural features and the width of both π and σ bands, suggesting that the (pdπ)=1.8795 eV and (pdσ)=-3.7454 eV integrals used in the computation of the energy bands [9] are basically correct [10]. A similar analysis applied to $SrTiO_3$ yields an average $t_{2g}-e_g$ splitting of the c.b. smaller than the experimental one. This indicates that the p-d hybridization has been underestimated in previous band-structure calculations [11].

Metallic carbides and nitrides TiC and TiN provide another good test of the PDOS of the lowest c.b.'s. A comparison is made in Fig. 4 between the theoretical total (mainly d-like) and the p-like PDOS and the experimental K near edge structure of TiC. The p-PDOS describes rather well the region extending approximately 5.5 eV and 8.0 eV above E_F in TiN and TiC, respectively, in close agreeement with APW calculations [12].

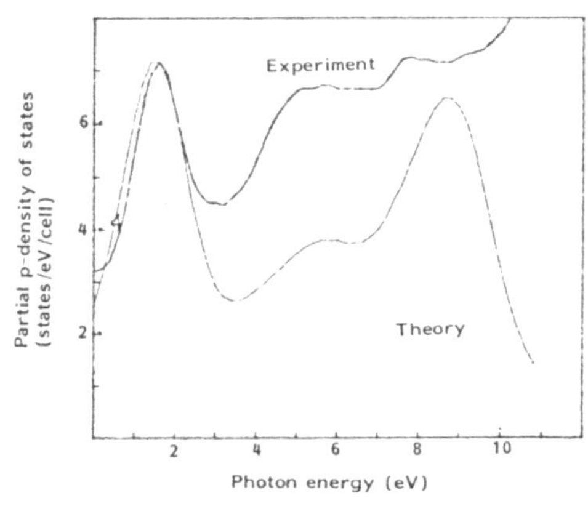

Fig.3. Comparison between the p-PDOS features and the experimental near-edge structure for $BaTiO_3$

As a further example we consider TiFe, one of the Asano-type transition metal alloys, which has a c.b. derived mainly from Ti orbitals. The K absorption edge of Ti (4964 eV) and Fe(7112 eV) are similar to those of the Ti metal and reflect the p component of the PDOS of the c.b. of Ti [13]. Fig. 5 gives the comparison with accurate computations of p-like and d-like PDOS inside Ti and Fe muffin-tin spheres [14].

3. Relaxation effects

When a core electron is ejected, the remaining one-electron orbitals relax adiabatically around the positive core hole. At high kinetic energies, appropriate to the X-ray photoemission (XPS) regime, the outgoing electron is fast and leaves behind a bare core hole. In the opposite regime of X-ray absorption (XAS) and X-ray emission (XES), the electronic system has time enough to readjust itself around the core hole which remains screened and thus the distribution of valence electrons resembles that of the ground state. The screening is larger in the metals, where the valence charge is effective to screen, than in insulators where the core hole is poorly screened.

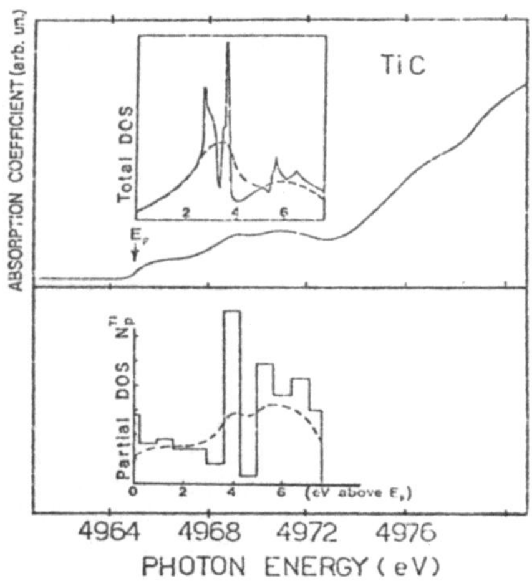

Fig. 4. Near-edge structure of TiC. The upper and lower insets are the total and partial p-PDOS (after Neckel et al. Ref. 12)

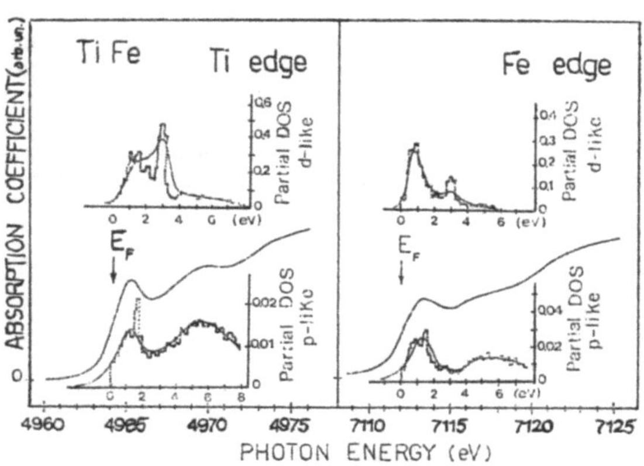

Fig. 5. p- and d-PDOS of the conduction band states of TiFe and Ti and Fe sites, compared with the experimental line shapes

In metallic titanium carbide and nitride as well as in Ti the XPS binding energy of the $2p_{3/2}$ level (L_3 edge) exceeds that determined by XAS and XES by a quantity Δ which for TiC is about 2 eV [2], as shown in Fig. 6. In the sudden limit, the XPS peak energy of the $2p_{3/2}$ level is shifted towards higher binding energies with respect to the adiabatic edge by [15]:

$$\Delta = - \gamma \cot (\pi/(2 - \alpha)), \quad (2)$$

where α, the asymmetry threshold index, is connected to the s-wave scattering phase shift of the plane-wave electrons at E_F and to the lifetime γ of the core hole. Thus the "apparent" binding energy of the XPS threshold depends strongly on the scattering potential of the hole. According to GADZUK and ŠUNJIĆ [15], the finite γ produces a core level broadening which forces to average over the shake-up's energy distribution. This shifts the centroid of the spectrum towards higher binding energies. We remark that for Ti the Δ value computed from (2) with γ =0.33 eV [16] and α =0.24 [17] agrees with the measured shift of XAS and XPS edges. Since the exponent α is inversely proportional to the DOS at the Fermi level [18], the decrease of Δ with increasing the density of states (Fig. 7) can be directly correlated to the larger availability of movable charge at E_F.

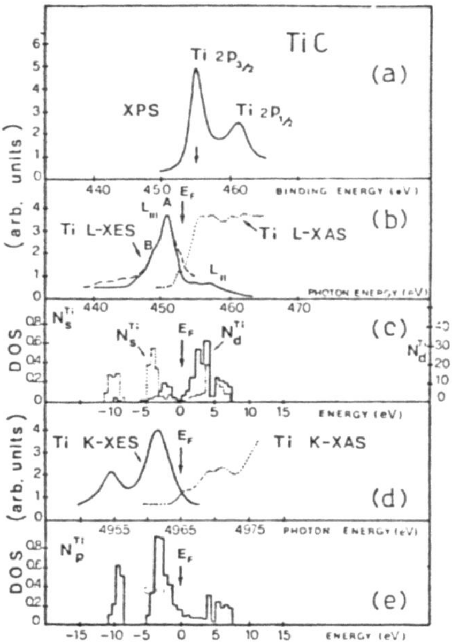

Fig. 6. Juxtaposition of L-XAS, L-XES, K-XAS and K-XES from several sources. The hystograms are the PDOS (states/Ry/unit cell) after Ref. 12

Final-state interactions in metals are more pronounced at the $L_{2,3}$ edges than at the K edges [19]. In the latter case the lower repulsive centrifugal barrier potential due to the charge readjustement should reduce the effect of the final-state interactions, in agreement with the experimental evidence that K edges are little altered by changes in the local electronic environment.

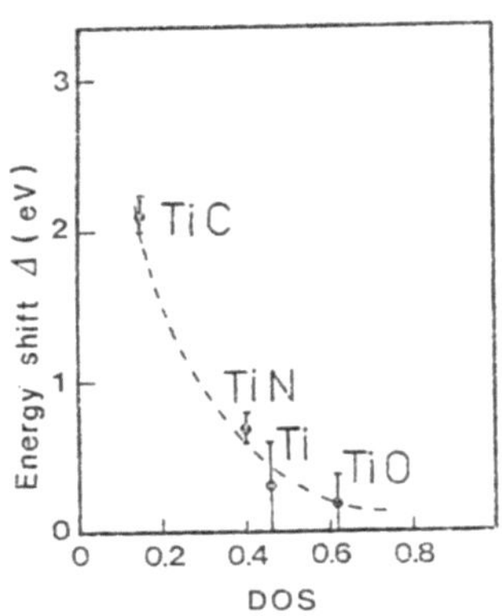

Fig. 7. Difference between the XPS line and the Fermi energy position for Ti and some titanium metal compounds vs the theoretical DOS

References

1. P.J. Durham, J.B. Pendry, and C.H. Hodges, Solid. State Commun. 38, 159 (1981).
2. A. Balzarotti, M. De Crescenzi, and L. Incoccia, Phys. Rev. B 25, 6349 (1982).
3. F. Antonangeli, M.L. Apicella, A. Balzarotti, L. Incoccia, and M. Piacentini, Physica 105 B, 25 (1981).
4. L.A. Grunes, R.D. Leapman, C.N. Wilker, R. Hoffmann, and A.B. Kunz, Phys. Rev. B 25, 7157 (1982).
5. J. Wong, F.W. Lytle, R.P. Messmer, and D.H. Maylotte, this volume.
6. T. Wolfram, Phys. Rev. Letters 29, 1383 (1972).
7. A. Balzarotti, to be published.
8. T. Wolfram and S. Ellialtioglu, Phys. Rev. B 25, 2697 (1982).
9. P. Pertosa and F.M. Michel-Calendini, Phys. Rev. B 17, 2011 (1978).
10. Better agreement could be achieved by including the contribution of the upper Ti 4p(t_{1u}) levels and of the matrix element $P_{f,\nu}$ of (1).
11. A.Balzarotti, F.Comin, L.Incoccia, M.Piacentini, S.Mobilio,and A. Savoia,Solid State Commun. 35, 145 (1980).
12. A. Neckel, P. Rastl, R. Eibler, P. Weinberger, and K. Schwarz, J. Phys. C 9, 579 (1976).
13. N. Motta, M. De Crescenzi, and A. Balzarotti, to be published.
14. D.A. Papaconstantopoulos, Phys. Rev. B 11, 4801 (1975).
15. J.W. Gadzuk and M. Šunjić, Phys. Rev. B 12, 524 (1975).
16. M.O. Krause, J. Phys. Chem. Ref. Data 8, 307 (1979).
17. S. Hufner, K. Wertheim, Phys. Lett. 51 A, 301 (1975).
18. F. Antonangeli, A. Balzarotti, A. Bianconi, P. Perfetti, P. Ascarelli, and N. Nistico', Solid State Commun. 21, 201 (1977).
19. E. Mehreteab and J. Dow, Solid State Commun. 43, 837 (1982).

Application of XANES to the Study of AsF_5-Graphite Intercalation Compounds

S.M. Heald

Brookhaven National Laboratory, Upton, NY 11973, USA

H.A. Goldberg and I.L. Kalnin

Celanese Research Co., Summit, NJ 07901, USA

1. Introduction

The As-F compounds AsF_3, AsF_5 and AsF_6^- have very large absorption peaks at the onset of the As K absorption edge. These peaks are due to a large density of unfilled 4p states associated with the As-F anti-bonding levels. As shown in Fig. 1(a) the position and shape of the peaks can be used to identify the absorbing species. In addition, since the 4p states are associated with the As-F bonds the strength of the transition depends on the orientation of the molecule with respect to the x-ray polarization direction. In the similar case of the Br_2 molecule this polarization dependence has already been demonstrated to have the same $\cos^2\theta$ dependence as the EXAFS [1]. Therefore, measurements of the As XANES in AsF_5-graphite intercalation compounds can be used to identify the As species present, and in some cases the orientation of the molecules.

X-ray absorption measurements on these compounds were first carried out by Bartlett et al. [2] using powdered HOPG as a starting material. On the basis of their results they proposed that intercalation is accompanied by the reaction

$$3AsF_5 + 2e^- \quad 2AsF_6^- + AsF_3. \qquad (1)$$

Little AsF_5 was found, which indicated that each intercalated AsF_5 resulted in a charge transfer from the graphite of 2/3 e. However, all other charge transfer measurements [3-10] found smaller values for the charge transfer with an average of ~0.3 e. In the present work, new measurements were made using both HOPG and graphite fibers are used as a starting material in an attempt to clarify the situation.

2. Experimental

Results are reported for two samples: HOPG and Celion^R GY-70 fibers with an average Young's modulus of 600 GPa. The samples were sealed into rectangular pyrex tubes into which thin (0.2 mm) x-ray windows were etched. A weighed amount of AsF_5 gas was admitted prior to sealing. The fiber sample had 1.9% AsF_5 as measured by weight uptake and from the x-ray absorption was measured to have 1.4%. The discrepancy is reasonable since invariably some of the reactive AsF_5 gas is lost during the filling and sealing process. On the other hand, a much larger discrepancy was found for the HOPG sample for which the weight increase gave 7.4% while the x rays gave 0.8%. This sample was decidedly inhomogeneous and it appeared that intercalation had taken place mainly at the surface. However, the $\Delta\mu x$ at the edge was only 0.078, and no thickness effects were observed. All measurements were made at CHESS line C-2 with two 0.25 mm slits used to give ~2 eV resolution.

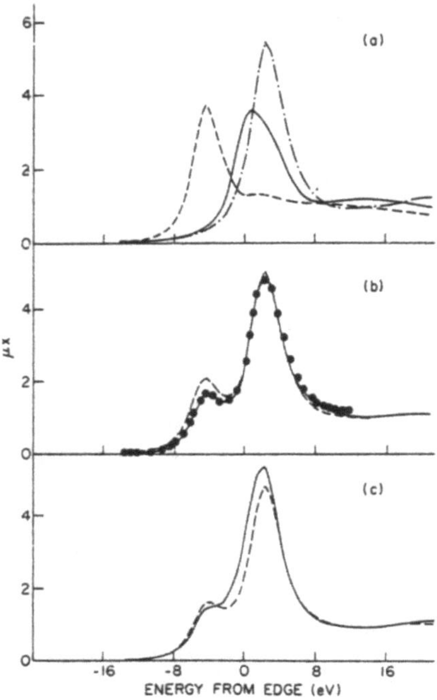

Fig. 1. X-ray absorption spectra of: a) AsF_3 (dashed-line) AsF_5 (solid line) and $Na^+AsF_6^-$ (dotted-dashed line) standards; b) AsF_5 - intercalated Celion[R] GY70 fibers with the x-ray polarization parallel (dashed line) and perpendicular (solid line) to the fiber axis (A fit to the data using the above standards is shown by the dots); c) AsF_5 - intercalated HOPG (solid line) and fibers (dashed line) with polarization perpendicular to the fiber axis. The data has been scaled so that the change in absorption from above to below the As edge equals one. The zero of energy is taken to be 11.868 KeV

3. Results

Figure 1(b) shows the data obtained on the fiber sample for two polarization directions: e perpendicular and parallel to the fiber axis. To quantitatively analyze the edge a fit was made to the data using a linear combination of the three standards shown in 1(a). The points show the fit results for the perpendicular data, and it is seen to agree quite well with the data. The HOPG data is shown in Fig. 1(c) and is somewhat different from the fiber results. Fits were also made for this case and all the results are tabulated in Table 1.

In all cases the ratio of AsF_3/AsF_6^- was ~1/2 in agreement with (1). However, for the HOPG sample only about half of the AsF_5 has reacted. This would give a charge transfer of ~1/3, but it should be mentioned that most charge transfer measurements were made on much more concentrated samples (Stage 2 or 3 which is ~75 or 50 wt%). It might be expected that the more dilute samples would react more completely, although because of the inhomogeneity our measurements may be representative of higher concentrations than the average. More measurements are planned to answer these questions.

For the fiber sample the reaction goes to completion, possibly because the large surface area allowed a much more homogeneous reaction. Also observed is a distinct orientation dependence of the AsF_3 peak. Since the graphite c axis is perpendicular to the fiber axis, the AsF_3 pyramidal molecules sit on the graphite basal planes on the tripod formed by the three fluorines. The magnitude of the orientation dependence is somewhat larger than that expected for the free molecule bond angles. This implies

Table 1. The relative concentrations of AsF_3, AsF_5 and AsF_6^- obtained by fitting the data shown in Figs. 1(b) and 1(c) with the data obtained from the standards (shown in Fig. 1(a)). Note the approximate 2:1 ratio of AsF_6^- to AsF_3 in both fibers and HOPG in agreement with (1)

Graphite:	Species: AsF_3	AsF_5	AsF_6^-
Fibers - perpendicular orientation	0.30	0.08	0.62
Fibers - parallel orientation	0.36	0.09	0.55
HOPG	0.19	0.45	0.36

either a slight flattening of the molecule upon intercalation or the possibility of a second As^{+3} species which has a larger orientation dependence. More details will be published elsewhere [11].

Acknowledgments

We thank G. Breckenridge for his help in taking the data, Dr. A. W. Moore for the HOPG, and the CHESS staff for their assistance. This work was supported in part by the Army Research Office and the U.S. Department of Energy under Contract No. DE-AC02-76CH00016.

References

[1] S. M. Heald and E. A. Stern, Phys. Rev. B17, 4069 (1978).
[2] N. Bartlett, R. N. Bianconi, B. W. McQuillan, A. S. Robertson, and A. C. Thompson, J. Chem. Soc. Chem. Comm., 200 (1978).
[3] R. S. Markiewicz, H. R. Hart, Jr., L. V. Interrante, and J. S. Kasper, Synthetic Metals 2, 331 (1980).
[4] R. S. Markiewicz, J. S. Kasper, and L. V. Interrante, Synthetic Metals 2, 363 (1980).
[5] S. Tanuma, Y. Iye, O. Takahashi, and Y. Koike, Synthetic Metals 2, 341 (1980)
[6] B. R. Weinberger, J. Kaufer, A. J. Heeger, J. E. Fischer, M. Moran, and N. A. W. Holzworth, Phys. Rev. Lett. 41, 1417 (1978).
[7] L. R. Hanlon, E. R. Falardeau, and J. E. Fischer, Solid State Comm. 24, 377 (1977).
[8] J. Blinowski, N. Hy Hau, C. Rigaux, J. R. Vieren, R. Letooles, G. Furdin, A. Herold, and J. Melin, J. Phys. (Paris) 41, 47 (1980).
[9] M. J. Moran, J. E. Fischer, and W. R. Salaneck, J. Chem. Phys. 73, 629 (1980).
[10] C. Zeller, L. A. Pendrys, and F. L. Vogel, J. of Mat. Science 14, 2241 (1980).
[11] S. M. Heald, H. A. Goldberg, I. L. Kalnin, and J. B. Stamotoff, submitted to Phys. Rev. Lett.

XANES of 3*d* Elements: Effects of Local Symmetry Evidenced Using High-Resolution Spectra

J. Petiau[1] and G. Calas

Laboratoire de Minéralogie-Cristallographie, LA CNRS 09, F-75230 Paris, France

and LURE, CNRS, Bât. 209C, F-91405 Orsay Cedex, France

1. Introduction

During the study of amorphous and imperfectly crystallized compounds (gels, clay minerals...) we take advantage of the complementarity of information obtained from EXAFS and XANES. Both studies are improved by the systematic comparison with well-known crystalline compounds. In the specific case of XANES which is discussed in this paper, the study of the same element in various environments permits one to decipher the relative influence of the various crystallochemical parameters which affect the edge position and structure [1, 2]. Recently a multiple scattering approach [3, 4] gave a complementary model to the more classical treatment by molecular orbital calculations [5, 2]. Consequently we must now distinguish between various regions in the XANES overall structure (Fig. 1) a "pre-edge" structure, the main edge region (up to the absorption maximum) and the "intermediate range" region, (which extends beyond the maximum and makes transition with the EXAFS). Only the two first regions will be discussed here in oxide and sulphide crystalline compounds, and briefly in oxide glasses. Although we present here the results concerning mainly Fe compounds, similar observations can be made about other 3d elements.

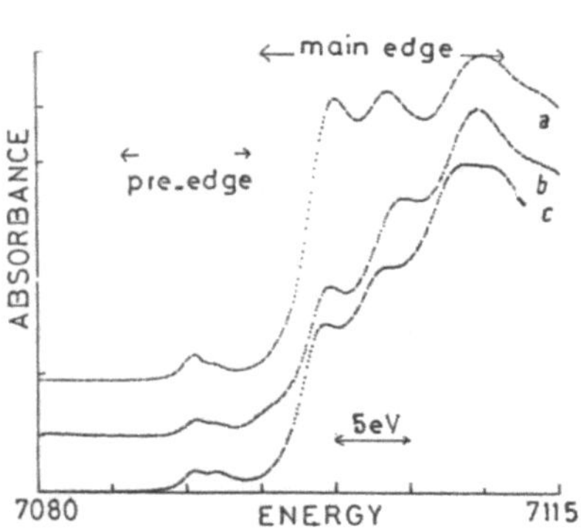

Fig.1.XANES of tetrahedral Fe(II) in various matrices : a- Staurolite ($Fe_2Al_9O_7(SiO_4)_4(OH)$); b- Chromite ($FeCr_2O_4$) ; c- Ulvospinel (Fe_2TiO_4)

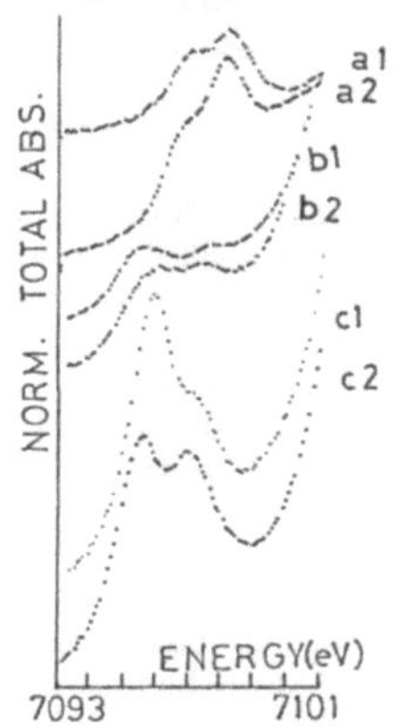

Fig.2.Pre-edge structure of K-edge Fe: a- octahedral Fe(III) : 1- acmite ($NaFeSi_2O_6$); 2-goethite (FeO OH); b- octahedral Fe(II) : 1- siderite ($FeCO_3$) ; 2- fayalite (Fe_2SiO_4) ; c- tetrahedral Fe(II) : 1 - staurolite ; 2 - ulvospinel

[1] Ecole Normale Supérieure de Montrouge

2. Experimental

The measurements were made by the transmission method at LURE-DCI. We have chosen a Si400 channel-cut monochromator in order to get a good spectral resolution. The angular width of the rocking curve and the angular acceptance of the beam lead to an experimental resolution which just matches the intrinsic limitation due to the finite lifetime of the core-hole. This gives an overall resolution of about 1.4 eV at Fe K-edge. The observed pre-edge splittings (Fig.1 and 2) confirm that this resolution is attained. The energy scanning is made by 0.1 eV steps. Special care is taken to assure a reliability of about 0.1 eV between the various spectra.

3. Oxide compounds

Coordination, local symmetry and metal-ligand covalency have already been recognized to affect the edge structure and position. As an example Figure 1 shows three minerals containing tetrahedral Fe(II). The similarity between the number and positions of the observed features is apparent in spite of a different Fe(II)-site distortion.

Figure 2 compares pre-edge features between compounds having various Fe crystallochemistry. Two peaks are always observed and are relative to 1s to mainly 3d transitions, partly allowed by O-2p mixing. These peaks are enhanced in the tetrahedral symmetry case, due to a greater mixing of metal and ligand orbitals (absence of inversion centre). The relative intensity of both components is not related to the respective number of empty 3d states, but is rather governed by the respective orbital symmetries which affects the mixing efficiency. The almost constant value of the splitting (about 1.7eV), apart from siderite where it is 2.0 eV, is still not interpreted and does not obey the simple crystal-field considerations.

The main part of the edge depends strongly of the symmetry of the site. It rises to higher energies as the metal-oxygen distance decreases [1]. It is also affected by site distortions [6] which give an apparent splitting of the absorption maximum and a sharper beginning of the edge. This last point is probably related to an easier mixing of metal and ligand orbitals. Few calculations were made on empty molecular levels in the molecular orbital theory. The published models [7] are in good agreement with our data, explaining the gap of 5 to 7 eV between the 3d levels and the further levels of dominantly p character. The molecular orbital approach of this part of the edge is at least partly justified by the similarity of the observed features, whatever the differences in crystalline environment beyond the first shell.

XANES of oxides are rather different from those observed in other minerals : they exhibit a continuously rising prepeak region to the absorption maximum which results in an apparently shifted beginning of the edge towards the lower energies. Covalency effects and Fe-Fe interactions [8] could explain this behaviour.

Oxide glasses present some particularities [6]. A prepeak enhancement is always observed and is related to a greater p covalency of metal states, and the consequently shorter metal-oxygen distances. The main-edge features are partially smeared out but allow a comparison with crystalline compounds, particularly concerning the distortion of the site.

4. Sulfide compounds

Two kinds of systems are compared in Figure 3. Dilute systems give results similar to those seen in oxide compounds : tetrahedral Fe(II) surrounded by sulphur (curve a) exhibits a similar structure of the main

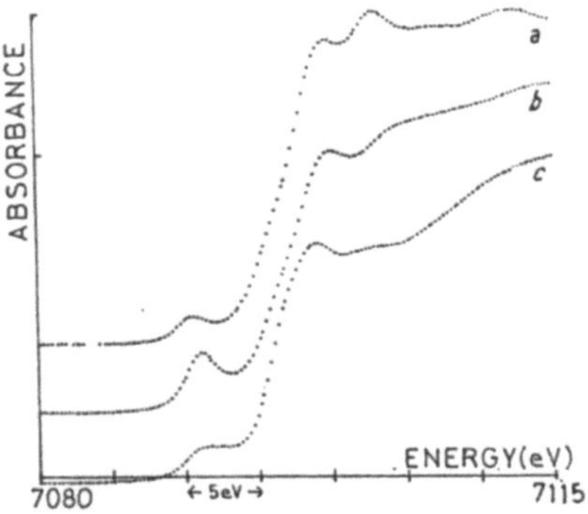

Fig. 3 XANES of Fe surrounded by sulphur : a - ZnS:Fe (tetrahedral Fe(II)) - b - chalcopyrite, $CuFeS_2$ (tetrahedral Fe(III), c- pyrite, FeS_2 (octahedral low-spin Fe(II))

edge as tetrahedral Fe(II) surrounded by oxygens (Figure 1). The same shift towards lower energies of the first part of the edge is observed, as in oxide compounds, and the same explanation by greater covalency effects can be made. Iron sulphides are semiconductors and band theory can be used to interpret the observed features. In $CuFeS_2$ the presence of an intense prepeak is due to tetrahedral Fe(III). The first maxima are well explained using the band theory model [9]. However the beginning of the edge does not exhibit the classical chemical shift due to the trivalent state of iron. In FeS_2 (pyrite), the pre-edge is rather smooth which is explained by the low spin state of Fe(II) in this compound (only the e_G-type Fe3d orbitals are empty) and by the absence of localized states. The absorption maximum is shifted towards higher energies : the edge spreads out 30 eV wide, which can be related to very short Fe-S distances.

References

[1] J. PETIAU, G. CALAS : in "EXAFS for inorganic systems", (Proceedings of the Daresbury Conf., March 1981), 127-129.

[2] A. BIANCONI - id., 13-22.

[3] G.N. GREAVES, P.J. DURHAM, G. DIAKUN, P. QUINN : Nature 294, 139-142 (1981).

[4] A. BIANCONI, M. DELL'ARICCIA, P.J. DURHAM, J. PENDRY : (to be published).

[5] G.L. GLEN, C.G. DODD : J. of Appl. Phys., 39, 5372-5384 (1968).

[6] G. CALAS, J. PETIAU : in "Structure of Non Crystalline Materials II" (to be published).

[7] D.J. VAUGHAN, J.A. TOSSEL : Geochim. Cosmochim. Acta, 38, 993-1005 (1974).

[8] J.B. GOODENOUCH : in "Metallic oxides" (ed. by Pergamon Press) (1971).

[9] T. HAMAJIMA, T. KAMBARA, K.I. GONDAIRA, T. OGUCHI : Phys. Rev. B, 24, 3349-3353 (1981).

The Adsorption of Oxygen on Nickel (001) Studied by XANES

D. Norman, P.J. Durham, and J.B. Pendry

Science and Engineering Research Council, Daresbury Laboratory, Daresbury, Warrington WA4 4AD, UK

J. Stöhr and R. Jaeger

Corporate Research Science Laboratories, EXXON Research and Engineering Company, Linden, NJ 07036, USA

Abstract

We report the results of multiple scattering calculations for the X-ray absorption near-edge structure (XANES) for oxygen adsorbed on Ni(001), and compare them with experimental results. The theory shows considerable sensitivity to changes in adsorption site or distance, while there is no change in the experimental spectra for a wide range of oxygen coverages. We conclude that there is no significant difference between the local coordination sites for adsorption in the p(2×2) or c(2×2) configurations, contrary to recent speculation. It is shown that XANES is a promising technique for studying adsorbates on surfaces.

X-ray absorption spectroscopy, in the form of EXAFS, is increasingly being applied to the study of adsorbates on surfaces, and is now producing very precise results for short-range structures which can offer unusual insights into bonding mechanisms [1]. The XANES spectrum is, of course, collected in the same experiment as the EXAFS, but to date, little use has been made of the near-edge part of the spectrum for surface structural analysis. The multiple scattering in the XANES region means that analysis cannot proceed by Fourier transformation, as for EXAFS, and must be achieved by comparison with the results of a full multiple scattering calculation. A computational scheme has been developed [2], based on a cluster method, which is thus flexible enough to be applied to crystalline solids [3], molecules [4], amorphous solids [5] and organometallic complexes as well as solid surfaces.

The calculation proceeds by dividing the cluster into shells of atoms (each of whose scattering properties are described by a set of phase shifts), and the multiple scattering equations are solved within each shell in turn, with the final step being calculation of multiple scattering between shells and the assembly of the whole cluster. The size of the scattering matrices is reduced by use of any symmetries (mirror or rotational) of the cluster, as we pick out, in the course of the multiple scattering calculation, only those components of the angular momentum expansions which can be coupled by a Hamiltonian possessing these symmetry elements. The calculation can, however, readily deal with systems possessing no symmetry. The reflection matrix thus calculated, when combined with an atomic matrix element linking the core and excited electron states, gives the XANES cross-section in one-electron theory. Many-body decay processes which limit the lifetime of the core hole and excited electron are included in our calculation as a complex (absorptive) potential whose effect is to broaden spectral features on the appropriate scale.

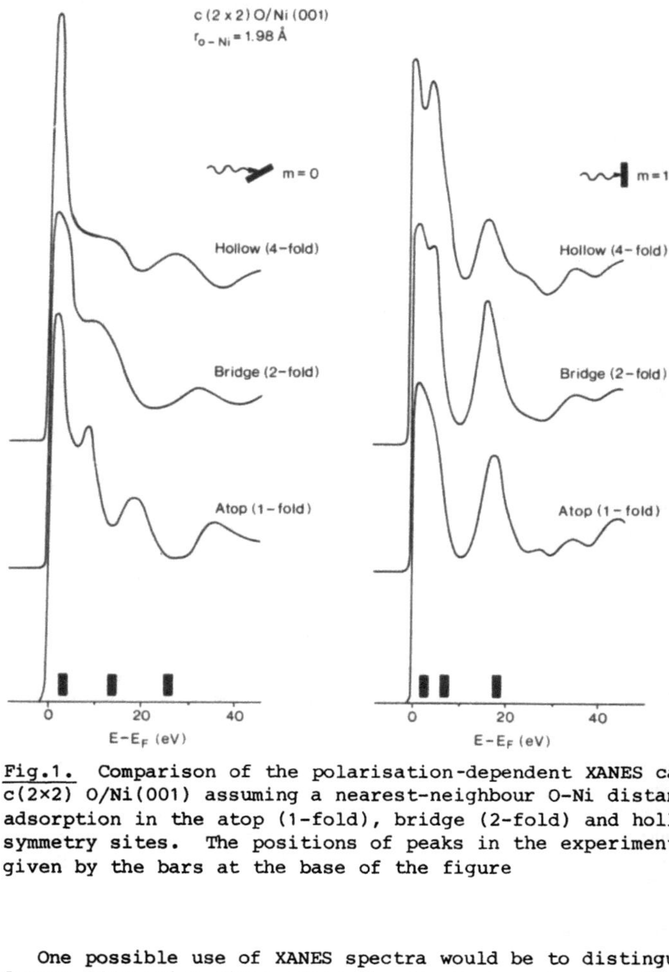

Fig.1. Comparison of the polarisation-dependent XANES calculations for c(2×2) O/Ni(001) assuming a nearest-neighbour O-Ni distance of 1.98 Å and adsorption in the atop (1-fold), bridge (2-fold) and hollow (4-fold) high symmetry sites. The positions of peaks in the experimental spectra are given by the bars at the base of the figure

One possible use of XANES spectra would be to distinguish between different adsorption sites after the adsorbate-substrate distance had been derived from a SEXAFS analysis. Figure 1 shows the results of such a calculation for absorption at the K edge of oxygen on Ni(001), and depicts the XANES spectra for a c(2×2) symmetry with adsorption in the atop (onefold), bridge (twofold) and hollow (fourfold) high symmetry sites, assuming an O-Ni nearest-neighbour distance of 1.98 Å [6].

The phase shifts and matrix elements were calculated from one-electron muffin-tin potentials, constructed according to the Mattheiss prescription [7] for bulk Ni and for the O atoms in their proper surface arrangements. The calculations include a total of around 30 atoms, and have been fully tested for convergence in numbers of scattering atoms and in angular momentum components. In fig.1, only minor changes are visible for the m=1 case, where the polarisation vector is parallel to the surface plane, but major changes are seen with grazing incidence photons. There is thus convincing evidence that this method may be of use in selecting from a range of coordination sites. The bars at the base of fig.1 indicate the positions of peaks in the experimental spectra, and it is seen that there is reasonable agreement with the calculations for the fourfold hollow site.

For this adsorption system of oxygen on Ni(001), there has recently been some controversy over the perpendicular spacing of the c(2×2) overlayer above the surface Ni plane. It had long been accepted that the structure was the same for both the p(2×2) quarter-monolayer and c(2×2) half-monolayer systems, with the adsorbate atoms occupying the fourfold hollow sites at a perpendicular distance ($d_\perp$) of 0.9 Å above the unreconstructed surface Ni layer. However, generalised valence bond calculations [8] indicated that the chemisorption of atomic oxygen on Ni(001) leads to two distinct states. These were identified, by examination of the wave functions and associated charge distributions, as a low-coverage radical state with $d_\perp$ = 0.88 Å and higher-coverage precursor oxide state with $d_\perp$ = 0.26 Å. Comparison of calculated vibrational frequencies [9] for the two states with observed values from electron energy loss spectroscopy leads to identification of the $d_\perp$ = 0.88 Å state with the p(2×2) structure and the $d_\perp$ = 0.26 Å state with the c(2×2) structure.

The XANES calculations for the two possible c(2×2) structures ($d_\perp$ = 0.9 Å and 0.2 Å) are shown in Fig.2. The m=1 polarisation (not shown) predict-

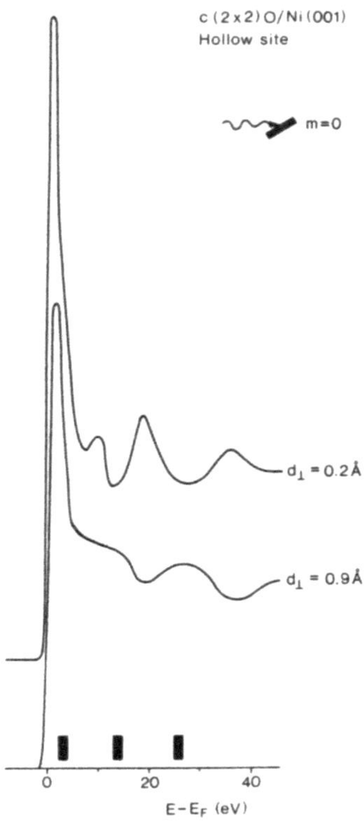

Fig.2. XANES calculations for grazing incidence photons for c(2×2) O/Ni(001) at perpendicular O-Ni layer spacings ($d_\perp$) of 0.9 Å and 0.2 Å. Peak positions from the experimental spectrum are given by the bars at the base of the figure

ably shows little sensitivity to the perpendicular spacing, but the m=0 curves are distinctly different. The experimental spectrum displays peaks at ca.14 eV and 26 eV above the edge, as calculated for the $d_\perp$ = 0.9 Å case, whereas the curve for $d_\perp$ = 0.2 Å has valleys at these energies. This good agreement therefore gives further confirmation of the $d_\perp$ = 0.9 Å assignment, as was found in a recent surface EXAFS study [6].

References

1. For a recent review see D. Norman, Proceedings of X-82 Conference on X-ray Physics and Atomic Inner Shell Processes, AIP Conference Proceedings Series, (to be published).

2. P.J. Durham, J.B. Pendry and C.H. Hodges, Comput. Phys. Commun. 25, 193 (1982).

3. G.N. Greaves, P.J. Durham, G. Diakun and P. Quinn, Nature 294, 139 (1981).

4. A. Bianconi, M. Del'Ariccia, P.J. Durham and J.B. Pendry, Phys. Rev. B, to be published.

5. P.H. Gaskell, D.M. Glover, A.K. Livesey, P.J. Durham and G.N. Greaves, J. Phys. C: Solid State Phys. 15, L597 (1982).

6. J. Stöhr, R. Jaeger and T.Kendelewicz, Phys. Rev. Lett. 49, 142 (1982).

7. L.F. Mattheiss, Phys. Rev. 134, A970 (1964).

8. T.H. Upton and W.A. Goddard III, Phys. Rev. Lett. 46, 1635 (1981).

9. T.S. Rahman, J.E. Black and D.L. Mills, Phys. Rev. Lett. 46, 1469 (1981).

Near Edge Structure in Cerium and Cerium Compounds

B. Lengeler and J.E. Müller

Institut für Festkörperforschung, Kernforschungsanlage,
D-5170 Jülich, Fed. Rep. of Germany

G. Materlik

Hamburger Synchrotronstrahlungslabor, DESY,
D-2000 Hamburg, Fed. Rep. of Germany

The L absorption spectra of γ- and α-Ce and of some Ce alloys have been measured. The L_2 and L_3 spectra show white lines. Those systems in which Ce is compressed exhibit an additional peak (α peak) above the white line. We deduce that α-Ce is not valence fluctuating and that the α peak in α-Ce is due to a core hole mediated excitation of the 4f electron into empty 5d states.

The near-edge structure of the L edges of γ- and α-Ce, of dilute Ce alloys and of some Ce compounds has been measured in order to gain additional insight in the isostructural γ-α phase transition in metallic cerium. The different models proposed to explain this puzzling phase transition have been reviewed in |1|. Based on systematics of the lattice parameter in the rare earths it was argued that Ce changes valence from 3 to 3.7 when changing phase from γ to α. Cohesive energy calculations led Johansson to explain the phase transition as a Mott transition |2|. It has also been argued that α-Ce is a mixed valent system |3|. Based on Compton scattering we came to the conclusion that the 4f electron is not promoted (partially or fully) into the conduction band and that α-Ce is not valence fluctuating (by more than 0.05 electron/atom) |4|. The present paper on the near-edge structure corroborates this statement. Fig. 1 shows the $L_{2,3}$ edges of γ and α-Ce. Both γ- and α-Ce have a white line due to the excitation of 2p electrons into the empty 5d states above the Fermi level. That of α-Ce is smaller and broader by 13 % than that of γ-Ce. α-Ce shows an additional peak (α peak) between 9 and about 25 eV above the edge. Note that there is no shift between the edges of γ- and α-Ce. In figure 2 are shown the edges of some dilute Ce alloys and some Ce compounds. In the systems of fig. 2a the Ce atom is not compressed. The $L_{2,3}$ edges show only a white line similar to γ-Ce. In the systems of fig. 2b, on the other hand, the

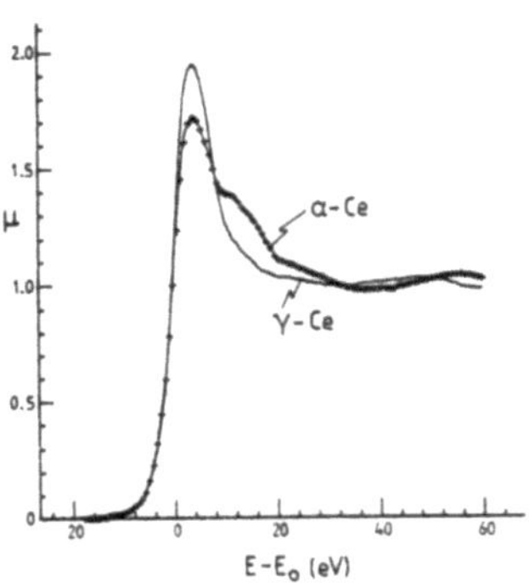

Fig. 1. $L_{2,3}$ edges of γ- and α-Ce

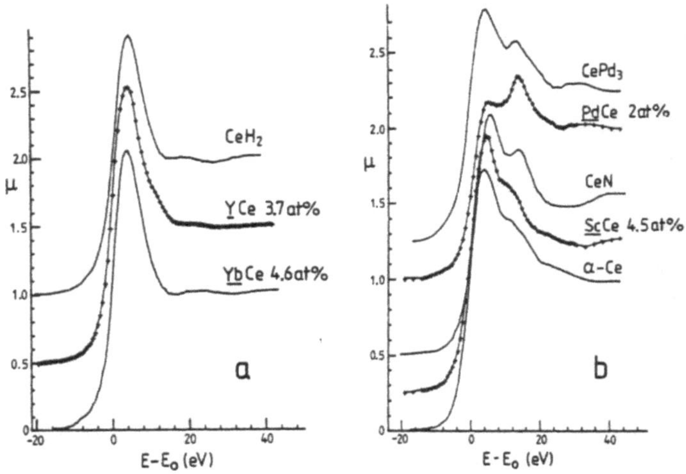

Fig. 2. $L_{2,3}$ edges of Ce in different systems in which Ce is a) not compressed or b) compressed by the surrounding lattice. All measurements were done at 77 K

Ce atoms are compressed and a double structure similar to that in α-Ce is shown up. In PdCe with 2 at% Ce the chemical pressure is the largest for all systems shown. It also shows the most pronounced α peak. We conclude from fig. 2 that the α peak appears whenever the Ce is compressed. On the other hand, an overlap of the 4f orbitals of neighboring Ce atoms is not necessary to produce the α peak, as the dilute alloys show. Fig. 3 shows a calculation of the near-edge structure for Ce with the lattice parameters of γ- and α-Ce. In both cases the same frozen $4f^1$ configuration was assumed. Details are given in |5|. The calculation explains the decrease of the height of the white line and the increase of its width by 13 % when going from γ- to α-Ce by an increase of the overlap of the 5d states in the collapsed phase. The changes observed in the white line have nothing to do with changes in 4f count. This is in contrast to valence fluctuating sys-

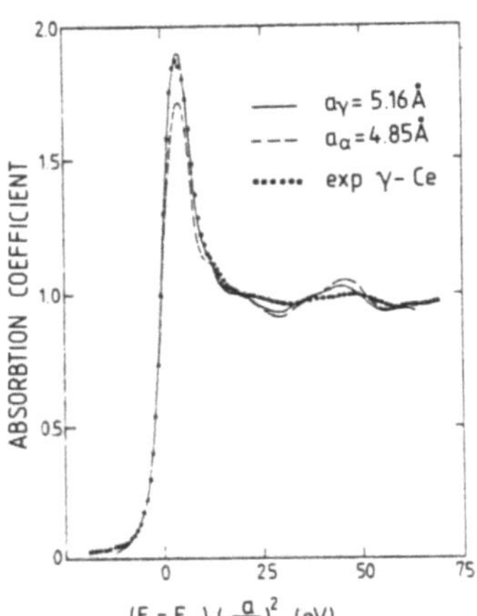

Fig. 3. Calculated $L_{2,3}$ edge of Ce with $a_\gamma = 5.16$ Å and $a_\alpha = 4.85$ Å and the experimental γ-Ce data (dots)

tems containing Sm or Tm where one peak of the double structure increases at the expense of the other |6|. Note also that our calculation which is based on ground-state properties of the system does not reproduce the α peak. We conclude that the α peak is a final state effect and that α-Ce is not valence fluctuating. We propose that it arises from the presence of the core hole by which the 4f electron is excited into the empty 5d states. The binding energy of the 2p electrons differing by 9 eV between a singly occupied and an unoccupied 4f level, and the width of the empty 5d states being 8 eV |5|, we expect the α peak between 9 and 25 eV above the edge. This agrees well with the range where it is observed in fig. 1. This proposition does not need an overlap of 4f orbitals in the compressed state in agreement with the result on dilute alloys quoted above. Finally, we expect only one value of E_0 in the EXAFS of α-Ce according to the present explanation since the α peak is limited to the small energy range quoted above. This prediction has been confirmed in an analysis of the EXAFS of α-Ce. Details are given in |5|.

1. D.C. Koskenmaki, K.A. Gschneider: Handbook on the Physics and Chemistry of Rare Earths, ed. K.A. Gschneider, L. Eyring (1978) p. 337
2. B. Johansson, Phil. Mag. 30, 469 (1974)
3. L.M. Falicov, W. Hanke, M.B. Maple: Valence fluctuations in solids, North Holland Co. (1981)
4. U. Kornstädt, R. Lässer, B. Lengeler: Phys. Rev. B21, 1898 (1980)
5. B. Lengeler, G. Materlik, J.E. Müller: to be published in Phys. Rev B (1982)
6. in ref. 3, pp. 423 and 427

XANES at High-Pressure Phase Transitions

R. Ingalls, J.M. Tranquada, and J.E. Whitmore

Department of Physics, University of Washington,
Seattle, WA 98195, USA

E.D. Crozier

Department of Physics, Simon Fraser University,
Burnaby, B.C. V5A 1S6, Canada

Introduction

For the past several years our group has been observing the effects of high pressure on x-ray absorption spectra. The EXAFS has been shown to be quite sensitive to pressure due to the changing bond lengths and atom-pair, mean-square displacements[1-3]. We have also reported some of the effects of pressure on the XANES. However, these reports have been mainly directed to the high pressure community[4,5]. Our purpose here is to summarize such effects for those interested in XANES, *per se*.

Samarium Chalcogenides

Both SmS and SmSe undergo structureless, insulator-to-metal transitions under pressure. In SmS the transition occurs at about 6.5 kbar. The Sm L_3-edge XANES appear to exhibit a shift of approximately 7 eV, reflecting the "valence change" of +2 toward +3, or rather a configuration change of $4f^6$ toward $4f^5 5d$. Near the transition the XANES resembles a superposition of the two patterns in varying proportions[5].

We have also previously reported the XANES of SmSe[3]. The change to the mixed valence state in SmSe has been found by electrical measurements, etc., to be continuous with pressure, in contrast to SmS. However the Sm L-edge XANES does not show any change until a pressure over 25 kbar. More experiments are planned to find how abrupt the XANES change actually is. The Se K-edge XANES also exhibits a change with pressure.

Zincblende Type Semiconductors

In general, semiconductors of the zincblende structure undergo transitions to tetragonal structure under pressure. The white line of the Ge K-edge loses a shoulder above this transition[6]. The same has been found for that of Ga in GaAs[7]. At about 45 kbar, CuBr undergoes such a transition, and above $\sim$ 62 kbar transforms to the NaCl structure[8]. The XANES at both edges changes remarkably[4]. In the zincblende phase, and probably the tetragonal, the nearest-neighbor coordination is fourfold. That of the NaCl structure is sixfold.

In Fig. 1 is shown the edge positions (first inflection point) precisely determined by comparison with a second edge held at atmospheric pressure (Ni K-edge and Bi L_3-edge, for Cu and Br, respectively). There appear discontinuities at the Zb-to-tetragonal phase change, and then smaller blue shifts through the tetragonal-to-NaCl transition. There is less of an effect in the case of the Cu edge position. It may be argued that the blue shift reflects some charge delocalization as is observed in the mixed-valence systems.

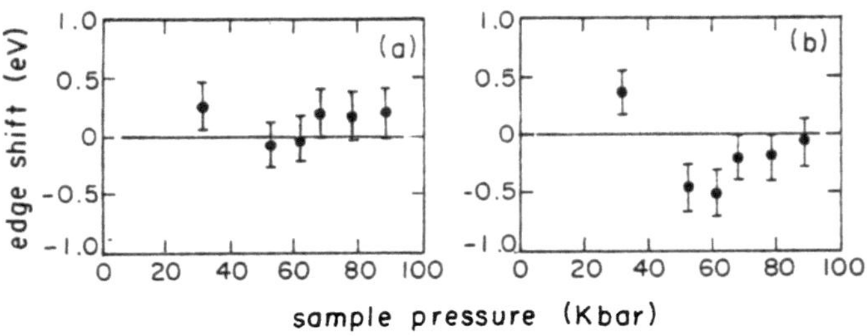

Fig.1. First inflection point shifts in CuBr: a) Cu K-edge, b) Br K-edge

Alkali Halides

Many of the alkali halides undergo a structural phase change from the NaCl structure to the CsCl structure (B1 to B2) under pressure. We have studied [3-5] RbCl ($P_t \sim 5.8$ kbar) and KBr (18 kbar) and RbBr (4.8 kbar). In each case the XANES exhibits a marked change. Recently, using a second absorption (the Tl L_2-edge) we have carefully documented the Rb K-edge in RbCl (Fig. 2). Here we also report new data on both K-edges in RbBr (Fig. 3). It is noted that these edges resemble their counterparts in RbBr and KBr.

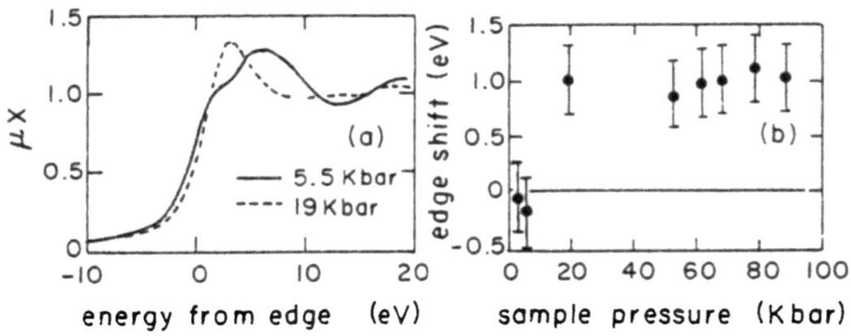

Fig.2. Rb K-edge: a) XANES, b) Shift of the first inflection point

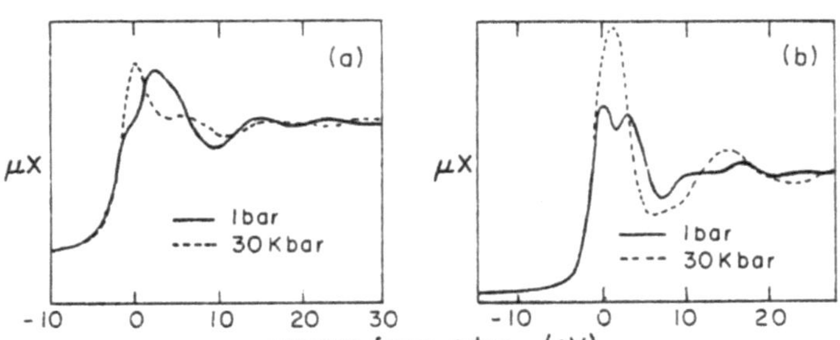

Fig.3. XANES in RbBr a) Rb K-edge, b) Br K-edge

Summary

In general the structural phase transitions brought upon by high pressure cause changes in the neighboring configuration of the absorbing atom, and the position and character of the XANES. It is anticipated that in the case of the alkali halides where the change is from six nearest neighbors to eight such XANES changes as we observe can be ultimately calculated with some accuracy. Several approaches appear in the literature. One of the most straightforward calculates the excited states of the absorbing atom-plus-K-hole, and then treats the effect of the neighboring atoms using crystal field theory[9]. Multiple-scattering approaches, etc., have not yet been carried out for the present systems. It is expected that once such changes in XANES become better understood, they can be applied to unknown structures, of course including the tetragonal phase of CuBr discussed above.

References

This work was supported by the U. S. National Science Foundation under Grants No. DMR 78-24995 and 77-27489 (in cooperation with the U. S. D.O.E.) and by the National Sciences and Engineering Research Council of Canada.

1. R. Ingalls, G. A. Garcia and E. A. Stern, Phys. Rev. Lett. 334 (1978).
2. R. Ingalls, E. D. Crozier, J. E. Whitmore, A. J. Seary and J. M. Tranquada, J. Appl. Phys. 3158 (1980).
3. R. Ingalls, J. M. Tranquada, J. E. Whitmore, E. D. Crozier and A. J. Seary, in EXAFS Spectroscopy: Techniques and Applications, B. K. Teo and D. C. Joy, Eds. (Plenum, 1980), p. 127.
4. R. Ingalls, J. E. Whitmore, J. M. Tranquada and E. D. Crozier, in High Pressure Science and Technology, B. Vodar and P. Marteau, Eds., (Pergamon, 1980), p. 528.
5. R. Ingalls, J. M. Tranquada, J. E. Whitmore and E. D. Crozier in Physics of Solids under High Pressures, J. S. Schilling and R. N. Shelton, Eds., (North Holland, 1981), p. 67.
6. R. Ingalls, E. A. Stern and D. E. Sayers, Stanford Synchrotron Radiation Project Report No. 77/02 (1977), p. 109.
7. O. Shimomura, T. Kawamura, T. Fukamachi, S. Hosoya, S. Hunter, and A. Bienenstock, Ibid Ref. 4, p. 534.
8. V. Meisalo and M. Kalliomäki, High Temp. Press 5, 663 (1973).
9. C. M. Dutta and H. W. Huang, Phys. Rev. Lett. 44, 643 (1980).

Near Edge and EXAFS Studies of Amorphous Pd-Ge and Ni-Zr Alloys

P.H. Gaskell, D.M. Glover, and A.K. Livesey
Cavendish Laboratory, Madingley Road, Cambridge CB3 OHE, UK

P.J. Durham and G.N. Greaves
Science and Engineering Research Council, Daresbury Laboratory, Daresbury, Warrington WA4 4AD, UK

1. Introduction

For amorphous metals, great benefit could accrue if, firstly, pair correlation functions were to be obtained to second or third neighbours - as the recent neutron scattering data of Steeb and co-workers have shown [1]. Secondly, techniques such as NMR and Mössbauer spectroscopy have demonstrated the value of information on the symmetry of the first-neighbour shell in glasses [2]. The recent work of Durham, Pendry and Hodges [3] and Greaves, Durham, Diakun and Quinn [4] has shown that, in principle, information on both second- and higher-neighbour shells and site symmetries are contained in the near-edge X-ray absorption edge structure (XANES). In this paper we describe attempts to ascertain the value of XANES data for two typical metallic glasses $Pd_{80}Ge_{20}$ and $NiZr_2$. Specifically, we have measured XANES and EXAFS data above the Ge and Ni K-edges for amorphous and crystalline alloys using the Daresbury SRS. These data are then analysed by calculating the XANES spectra for candidate structures using the full multiple scattering theory of Durham et al.[5] and conventional curved and plane wave treatments for the EXAFS spectra. In this way it is possible, in principle, to access all the experimental reciprocal space information and, therefore, to derive more complete real-space data. Details of the work on $Pd_{80}Ge_{20}$ have been published earlier [5] and related work on the calcium K edge in silicate and aluminate glasses and crystals is reported elsewhere in these proceedings [6].

2. Summary of results for $Pd_{80}Ge_{20}$

EXAFS spectra of amorphous $Pd_{80}Ge_{20}$ are similar to those from the crystallised material, Fig. 1a, reflecting similarities in nearest-neighbour organisation - particularly the first peak of the Ge-Pd pair distribution function. Although analysis of the EXAFS data is still incomplete, it seems unlikely that there will be any serious disagreement with the results of Hayes et al. on the same alloy [7]. They found a shell of 8.6 (± 0.5) Pd atoms at a mean distance of 2.49 ± 0.01 Å with a further shell of 4 ± 1 Ge atoms at a distance of 3.38 Å.

There are significant differences in the near-edge structure of amorphous and crystalline specimens however (Fig. 1b) which may reflect either differences in site symmetry or in the second-neighbour shell. Calculations of the XANES spectra for several candidate structures have been performed, in which each of these possibilities is examined.

2.1 Symetry of the First-Neighbour Shell

A number of single shell clusters were considered ranging from a tetrahedral shell of Pd atoms around Ge to capped and uncapped trigonal prisms, Fig.2

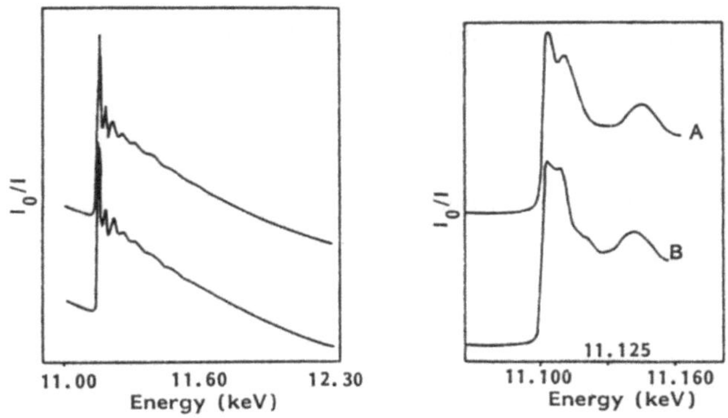

Fig. 1. (a) EXAFS spectra for amorphous $Pd_{80}Ge_{20}$. The lower spectrum is of the same material after crystallisation. (b) Near-edge region of the X-ray absorption spectrum for amorphous (A) and crystalline (B) $Pd_{80}Ge_{20}$

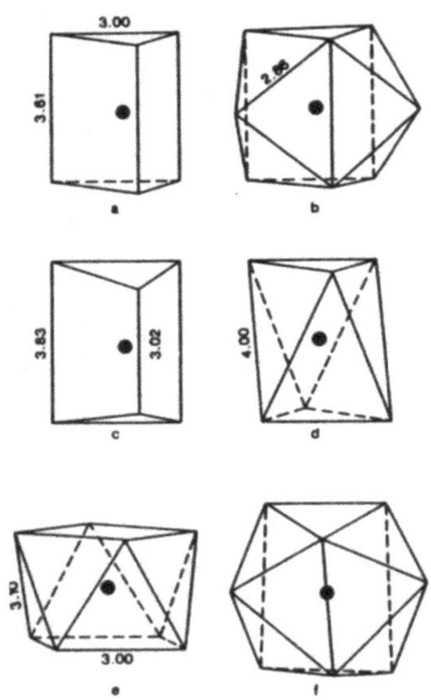

Fig. 2. Polyhedra used as a basis for calculations described in the text. The trigonal prism (a) is capped with two Pd atoms (b). Diagram (c) represents a trigonal prism distorted to simulate the polyhedron found in crystalline Pd_3As (f); (d) and (e) are a trigonal antiprism and Archimedean antiprism respectively

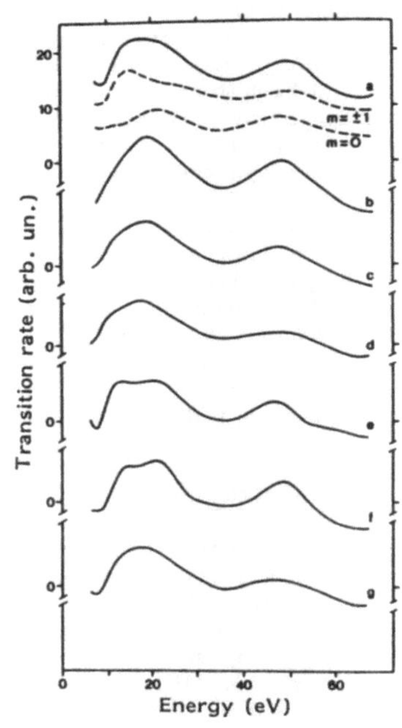

Fig. 3. Near-edge absorption spectra calculated from single coordination shells of Pd atoms: A, uncapped trigonal prism; B, tri-capped trigonal prism; D, regular tetrahedron- E, cube; F, octahedron; G, distorted trigonal prism (Fig. 2c). In A, component spectra (broken) correspond to outgoing p waves along and at right angles to the threefold axis of the trigonal prism

(which are observed in crystalline Pd-Ge alloys). In all cases the Ge-Pd bond length was fixed at 2.5 Å so that EXAFS oscillations from all structures have the same period. Results are shown in fig.3. The integrated intensity of the peak at 50 eV is qualitatively related to the coordination number and its position to the Ge-Pd distance. Changes in site symmetry clearly affect the shape of the XANES spectrum and such changes should be within the limits of detection by current spectrometers.

2.2 Second- and Third-Neighbour Shells

Addition of second and third shells produces more pronounced changes in the XANES spectrum. Figure 4 shows a series of calculations for a second shell of 6 Ge atoms at 3.96 Å (Fig.4b), compared with the previous calculation for a tri-capped trigonal prism. In each case, perfect trigonal symmetry is assumed. The effect is to produce a splitting of the first 'white line' peak at 10 eV and this is observed experimentally. The effect of changes in the radius of the second shell is shown in Fig.5 - values of the Ge-Ge distance being shown on the figure. It is clear that the splitting and relative heights of the two components of the first peak are strongly dependent on the Ge-Ge separation. While the analysis is still incomplete, a Ge-Ge distance of 3.3 - 3.4 Å is indicated, which is somewhat smaller than the value observed in crystals and agrees with the estimates given by Hayes et al. from EXAFS. Secondly, the fact that the first peak is, experimentally, found to be split suggests that the Ge-Ge distribution is relatively sharp (a rough estimate would suggest $\sigma \simeq 0.2$ Å).

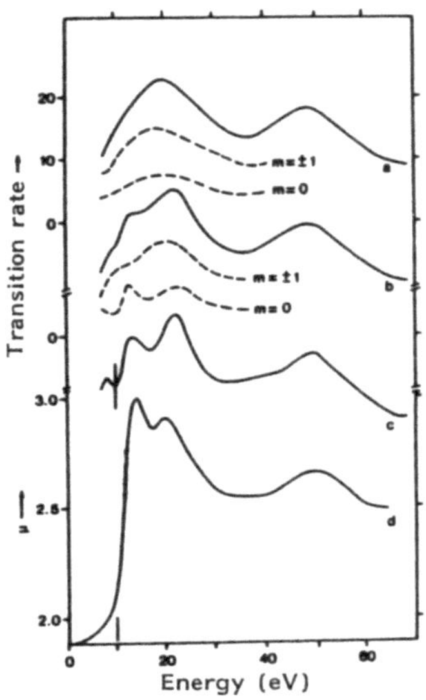

Fig.4. Near-edge absorption spectra for multi-shell clusters: A, trigonal prism with three capping atoms; B, as A, with a second shell of six Ge atoms at a distance 3.96 Å; C, with the second shell and third shell of nine Pd atoms at 4.16 Å; D, experimental near-edge spectrum plotted to the same scale. The vertical bar in C and D represents an estimate of the Fermi energy

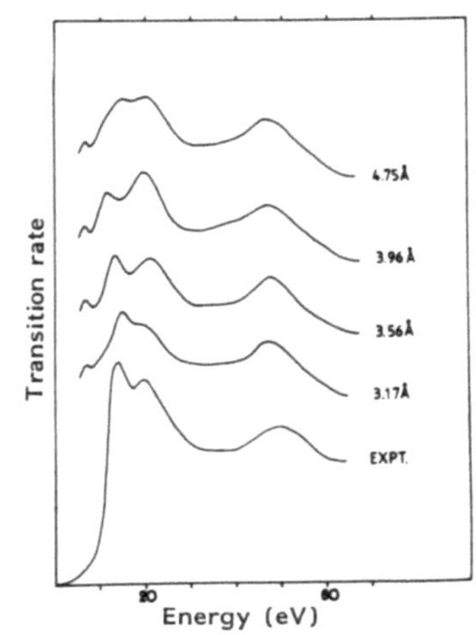

Fig.5. Near-edge absorption spectra for the multishell cluster described in Fig.4c, as a function of the Ge-Ge distance

3. Nickel-Zirconium Alloys

Preliminary results of an analysis of the near edge structure above the Ni edge in amorphous and crystalline $NiZr_2$ are shown in Fig.6. As expected, the amorphous material shows weaker XANES and EXAFS modulations than c-$NiZr_2$ with slightly larger periodicity above 50 eV implying tighter binding or a less symmetric coordination shell. The second peak at 65 eV is split for the crystal, suggesting changes in the contribution of second- and higher-neighbour shells.

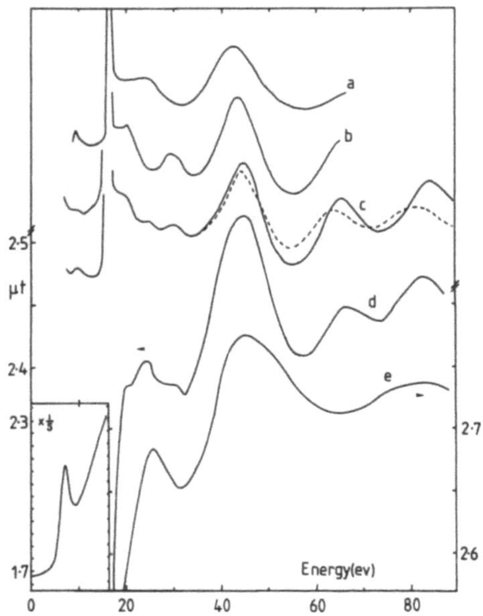

Fig.6. Calculated and experimental Ni-edge spectra. Calculated spectrum for a) 8 Zr atoms at 5.19 A and 2 Ni at 4.95 A from a central Ni atom; b) with a further shell of 4 Ni at 8.66 A, 8 Zr at 8.72 A and 8 Zr at 8.77 A; c) with third and fourth shells of 2 Ni at 9.90 A, 8 Ni at 9.97 A, 8 Zr at 11.22 A, 8 Zr at 11.25 A and 4 Ni at 12.24 A. The maximum angular momentum is 3 except for the dashed line where it is 4. Curves shown in d) and e) correspond to experimental data for the crystalline and amorphous alloys respectively. The energy zero of the upper three computed curves is arbitrary

In c-$NiZr_2$, Ni atoms are equivalent and have high point symmetry (4 mm), so that the multiple scattering programmes can handle large clusters of atoms. Nearest neighbours of Ni are 8 Zr atoms forming an Archimedean anti-prism with Ni atoms capping the square faces. Calculated XANES spectra for this cluster are shown in Fig.6a together with calculations for second and higher shells, Fig.6b,c. The calculation only converges after 4 shells - showing the sensitivity of this region to the presence of higher order shells.

Acknowledgments

Generous support from SERC and Pilkington PLC is gratefully acknowledged.

References

1. P. Lamparter, W. Sperl, E. Nold, G. Rainer-Harbach and S. Steeb, Proc. 4th Int. Conf. Rapidly Quenched Metals, ed. T. Masumoto and K. Suzuki, Jap. Inst. Metals, Sendai (1981) p.343
2. P. Pannisod, D. Aliaga Guerna, A. Amanov, J. Durand, W. L. Johnson, W. L. Carter and S. J. Poon, Phys. Rev. Lett. 44, 1465 (1980)
3. P. J. Durham, J. B. Pendry and C. H. Hodges, Solid State Commun. 38, 159 (1981)

4. G. N. Greaves, P. J. Durham, G. Diakun and P. Quinn, Nature 294, 139 (1981)
5. P. H. Gaskell, D. M. Glover, A. K. Livesey, P. J. Durham and G. N. Greaves, J. Phys.C: Solid State Phys., 15, L597 (1982)
6. R. G. Geere, P. H. Gaskell, G. N. Greaves, J. Greengrass and N. Binstead, this volume
7. T. M. Hayes, J. W. Allen, J. Tauc, B. C. Geissen and J. J. Hauser, Phys. Rev. Lett. 40, 1282 (1978)

XANES Determination of V-Mixed Valence State in $(V_2O_5)_x(P_2O_5)_{100-x}$ Binary Oxide Glasses

I. Davoli, S. Stizza, and M. Benfatto
Istituto di Fisica, Università di Camerino, I-62032 Camerino, Italy

O. Gzowski and L. Murawski
Institute of Physics, Technical University, Gdansk, Poland

A. Bianconi
Dipartimento di Fisica, Università di Roma, La Sapienza, P. Aldo Moro 2, I-00185 Roma, Italy

1. Introduction

The binary oxide glasses exibit interesting electronic properties such as the electron hopping conductivity. In $(V_2O_5)_x(P_2O_5)_{100-x}$ binary amorphous oxides the electrical conductivity is realized by the hopping of the electron between transiton metal ions in the two valency state,e.g. $V^{+4}\rightarrow V^{+5}$ [1].

The primary aim of this investigation was to determine in the same experiment both the effective change of vanadium ions and the local structure around the V ions. We have recently shown that such information can be extracted from the K XANES (X-ray Absorption Near Edge Structure [2,3]) spectra of $(P_2O_5)_{50}(FeO)_{50-x}(V_2O_5)_x$ oxide glasses [4] .

2. Experimental

Absorption measurements were performed at Frascati Laboratory using a channel cut Si(220) single crystal monocromator. Our samples,consisting of a fine powder, could easly be made into homogeneously thin films. The transmission through two homogeneous films was measured to avoid effects of pin holes. A vertical (0.7 mm) entrance slit of the same size as the electron beam was used to have the best resolution.

3. Result and discussion

Fig. 1 shows the XANES of the semiconducting glasses at various V_2O_5 concentrations. The XANES in the continuum part of the spectrum above the Threshold of $1s\rightarrow\varepsilon p$ allowed transitions at nearly 5478 eV which are very smooth,as expected for amorphous materials. The multiple scattering resonance in amorphous materials is expected to be determined only by the first coordinations shell [3] . Difference in relative intensity and energy positions of resonances A and B in the two cases of extreme concentrations x=40 and x=80 have been observed. The"edge region" is shown in the fig. 2. The edge is dominated by a "white line" which indicates that final excited state is p-like although it is in the energy region of 3d vanadium-derived orbitals. In V tetrahedral coordination,i.e. in VO_4 microscopic cluster, a t_2 molecular orbital final state produces a similar white line [5] . A similar white line is expected in vanadium distorted sites with one "double V-O bond " like in V_2O_5 or with two"double V-O bonds" [6]. Such a peak is expected to be due to a localized bound excited state because of the large ($\simeq$ 5eV) electron-hole Coulomb interaction in these systems.

The edge spectra in fig. 2 show clearly that the "white line " is split in two components separated by 1 eV.

The relative intensity of the low energy component is large at low (x=40) V_2O_5 concentration and the high energy component appears as a shoulder. The intensity of this high energy component increases, increas-

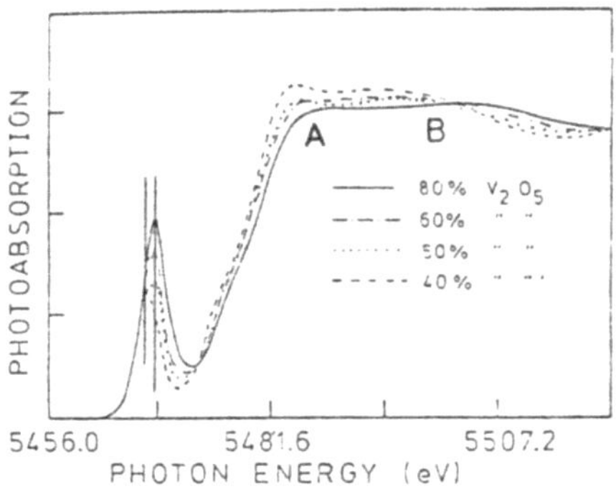

Fig.1. Normalized data of K XANES for V atom in glasses type $(V_2O_5)_x(P_2O_5)_{100-x}$

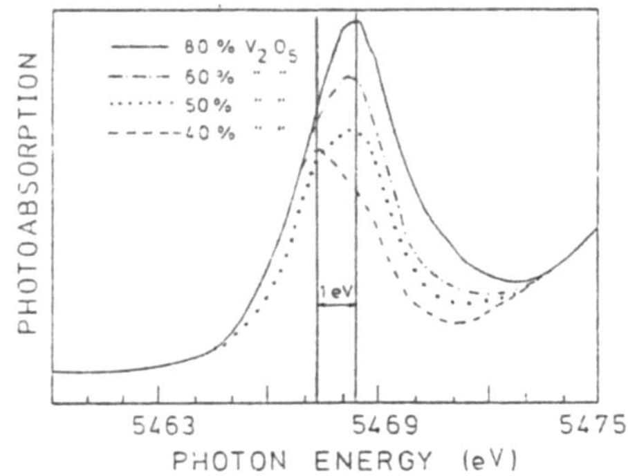

Fig.2. Edge region showing the chemical shift for different V_2O_5 concentrations

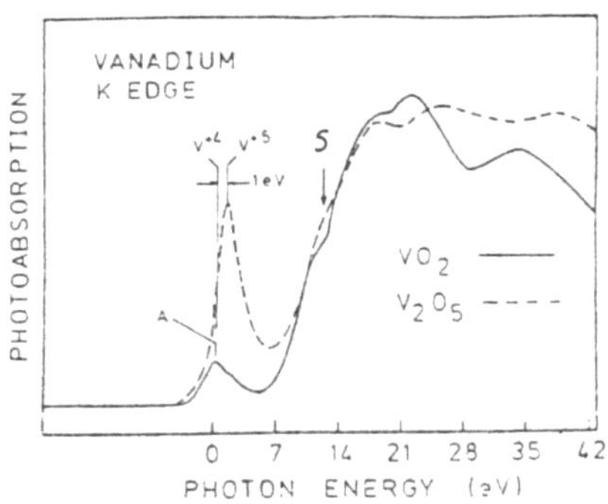

Fig.3. Vanadium K XANES of cristalline samples showing the chemical shift due to two different coordination number

ing the V_2O_5 concentration,and it is pratically the only dominant feature at $(V_2O_5)_{80}(P_2O_5)_{20}$ sample. We have recently demonstrated experimentally that in spite of the difficulty to determine the effective atomic charge in XANES spectra, it is possible to determine its variations in similar systems following the shift of the first excited bound state at the K edge [7,4] .

Fig. 3 shows that this rule is valid for vanadium oxides . In fact the first excited state shifts by 1 eV , nearly the same shift as the core level binding energy of core level measured by XPS, going from the V^{+4} charge state of VO_2 crystal to the V^{+5} charge state of V_2O_5 crystal. We would like to remember that the energy shift of the $1s \rightarrow \varepsilon p$ rising edge *S* does <u>not</u> follow the effective charge state [3].

Therefore we assign the two components of the "white line" in our glasses to excitations in different V ions with two different charge states. The static mixed valence state with V^{+4} and V^{+5} ions is clear for the x=40, 50, 60, samples.

The determinations of the structure of the two ion sites requires the joint analysis of EXAFS and XANES, however at this stage, the large intensity of the white line in both sites indicates the presence of V-O double bounds.

1- N.F. MOTT
J. Non Crystalline Solids 1 , 1 (1968)
2- A. BIANCONI
Appl. Surface Science , 6 ,392 (1980)
3- M. BELLI, A. SCAFATI, A. BIANCONI, S. MOBILIO, L. PALLADINO, A. REALE and E. BURATTINI
Solid State Commun. 35 , 355 (1980)
4- A. BIANCONI, A. GIOVANNELLI, I. DAVOLI, S. STIZZA, L.PALLADINO, O. GZOWSKI, L. MURAWSKI
Solid State Commun. 42 ,547 (1982)
5- F. W. KUTZLER, C.R. NATOLI, D.K. MISEMER, S. DONIACH, K.O. HODGSON
J. Chem. Phys. 73 , 3274 (1980)
6- R. KOZLOWSKI , R.F. PETTIFER , J.M. THOMAS.
in "EXAFS and Near Edge Structure ". This book
7- A. BIANCONI in " EXAFS for inorganic Systems"
C.D. Garner and S. Hasnain,eds. , Daresbury Publications, p. 13 (1981)

L Edge Adsorption Systematics of Some Transition-Metal (Pd, Ag), Main-Group-Metal (Sn, In) and Ionic (CsI) Systems

T.K. Sham

Department of Chemistry, Brookhaven National Laboratory,
Upton, NY 11973, USA

1. Introduction

The title system was selected with two objectives in mind. First, it is desirable to substantiate that near edge absorption in these elements is a density of states effect with a system of which the filling of the valence d, s and p states across the periodic table is well understood. Second, the sensitivity of these L edge white lines to the presence of empty states is of great concern with the hope that this technique would be valuable in studying the behavior of these states upon compound formation and subtle changes in the bonding of metal ions and complexes upon dissolution.

2. Results and Discussion

Measurements were made at SSRL (Stanford Synchrotron Radiation Laboratory) using a focus EXAFS beam line (II-3). Metal samples were either thin foils or evaporated films. Powder was used in the case of CsI. Transmission mode was employed in obtaining all the spectra. The L edges for Pd and Ag metal are shown in Fig.1.

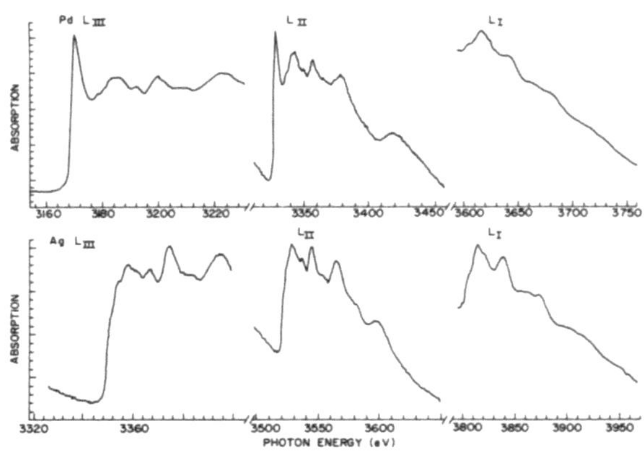

Fig.1. L edge spectra of Pd and Ag metals

It is immediately apparent that while the EXAFS wiggles for both elements are nearly identical, the Pd $L_{II,III}$ edges exhibit an intense resonance absorption (white line) that is absent in the case of Ag. The EXAFS is expected to line up because both Pd and Ag have fcc structure. The Pd white lines therefore must arise from dipole-allowed transitions. Consequently, there must exist high density of unoccupied d states just above the Fermi energy in Pd and not in Ag, (p→s transition is relatively unimportant here). This picture clearly emerges from the calculations by MORUZZI et al. [4] shown in Fig.2. Considering that there is only ~0.3 of a 4d hole in the unoccupied band in Pd at the Fermi level and practically nothing in the case of Ag, this technique is indeed very sensitive to the presence of low-lying, unoccupied d states. The observed intensity ratio of the white lines (L_{III}/L_{II}) is ~2.6/1, somewhat greater than the statistical 2/1. This suggests that there is more $d_{5/2}$ than $d_{3/2}$ character in the 4d hole. However, this is not like Pt metal where the 5d hole is almost entirely $d_{5/2}$ in character[1].

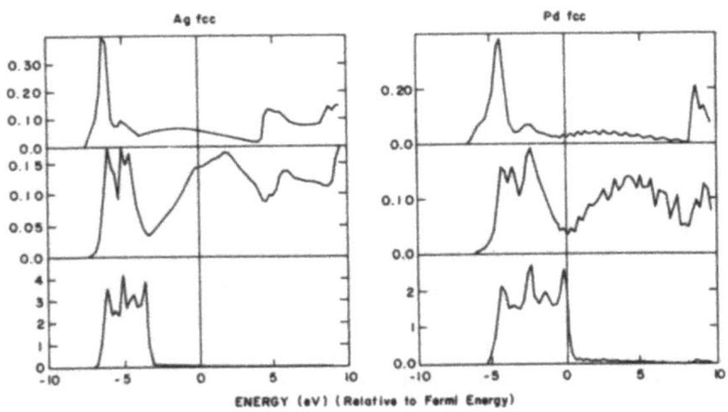

Fig.2. Density of states of Pd and Ag from [4]. From top to bottom is the density of states of s, p and d respectively

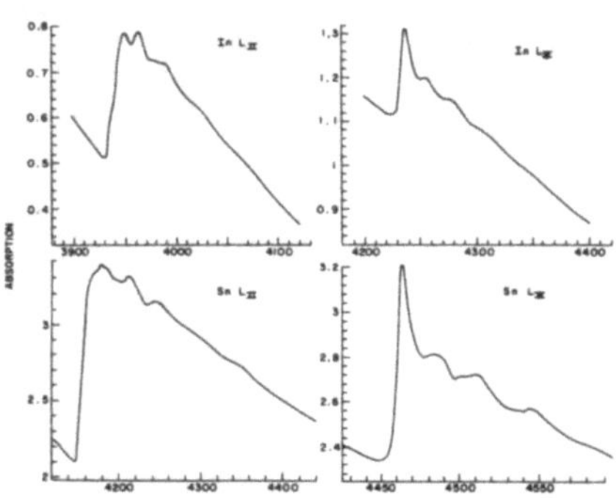

Fig.3. L edge spectra of In and Sn metals. The intense peak at the L_I edge is a white line arising from s→p transition

As we move across the period, we run into the main group elements of which the 5s and 5p states are partially occupied. The L_{II} (L_{III} is almost identical) and L_I edges of In and Sn are shown in Fig.3. The situation obviously turns around in these systems in which only L_I edge shows white line. This observation indicates that there is high density of p state just above the Fermi level and these states are more localized than that of the noble metals. In contrast to the 5p states, the 5s states are nearly occupied and the 5d states are still far away from the Fermi level.

The L edge spectra of CsI are given in Fig.4. Both Cs^+ and I^- have the same local symmetry in solid CsI. Their L edge structures however exhibit a noticeable difference. Cs^+ shows intense white line at the $L_{II,III}$ edges whereas I^- has a considerably weaker absorption at these edges. This observation can be understood in terms of atomic effects in the condense phase by taking into account how the 5d wave function is "collapsed" [5] going from I^- to Cs^+. The radial overlap integral, $A = \langle P_{2,p}|r|P_{5,d}\rangle^2$ for these ions are evaluated. The ratio, $A(Cs^+)/A(I^-)$ turns out to be > 2 qualitatively consistent with the observation.

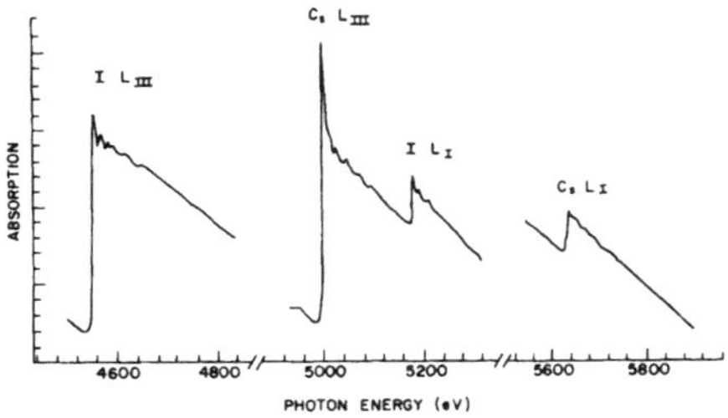

Fig.4. L edge spectra of Cs^+ and I^- in solid CsI

Acknowledgment

I thank the authors of [4] for communicating their results. EXAFS spectra were made at SSRL which is supported by the U.S. NSF through the Division of Materials Research, the U.S. NIH through the Biotechnology Resource program in the Division of Research Resources, and the U.S. DOE. This work was performed at Brookhaven National Laboratory under contract with the U.S. Dept. of Energy and supported by its Office of Basic Energy Sciences.

References

1. M. Brown, R. E. Peierls, and E. A. Stern, Phys. Rev. B 15, 738 (1977).
2. F. C. Brown in "Synchrotron Radiation Research", Eds. H. Winick and S. Doniach, Plenum (1980).
3. P. S. P. Wei and F. W. Lytle, Phys. Rev. B 19, 679 (1979).
4. V. L. Moruzzi, C. D. Gelatt, Jr., and A. R. Williams, to be published.
5. The word "collapsed" is loosely used here, see for example J. P. Connerade Contemp. Phys. 19, 415 (1978).

X-Ray Absorption Spectra of Ruthenium L Edges in $Ru(NH_3)_6Cl_3$

T.K. Sham and B.S. Brunschwig

Department of Chemistry, Brookhaven National Laboratory,
Upton, NY 11973, USA

1. Introduction

In the near edge region of an x-ray absorption spectrum, one sometimes sees a very sharp peak in the vicinity of the edge jump[1]. These absorption peaks arise primarily from atomic-like electric dipole transitions which are final state selective. In connection with the structural and electronic aspects of electron exchange reactions [2,3], we have obtained the L edge spectra of Ru in $Ru(NH_3)_6Cl_3$ at Stanford Synchrotron Radiation Laboratory (SSRL). We should like to show that the three L edges of Ru in $Ru(NH_3)_6Cl_3$ yield information on the distribution of the density of states of the unoccupied molecular ortibals.

2. Results and Discussion

All the Ru L edge spectra were recorded in a transmission mode with the focused EXAFS spectrometer. Si(III) crystals were used. Several spectra were taken for each edge to ensure reproducibility. The specimen was prepared in the form of homogeneous mixture of finely ground powder of $Ru(NH_3)_6Cl_3$ and grease vaseline. The L edge spectra of $Ru(NH_3)_6Cl_3$ are shown in Fig.1. The L_1 spectrum at the edge jump shows a broad structure. The L_{II} and L_{III} spectra exhibit relatively narrow and intense peaks in the near edge region in contrast to the broad L_I edge. The small double-peak structure at ~2820 eV is the Cl K edge. What is perhaps the most striking feature is that while L_2 has only one sharp resonance, B', L_3 has two (A and B). Detail features together with qualitative assignments based on O symmetry are shown in Fig.2. The ionization threshold E_0 of

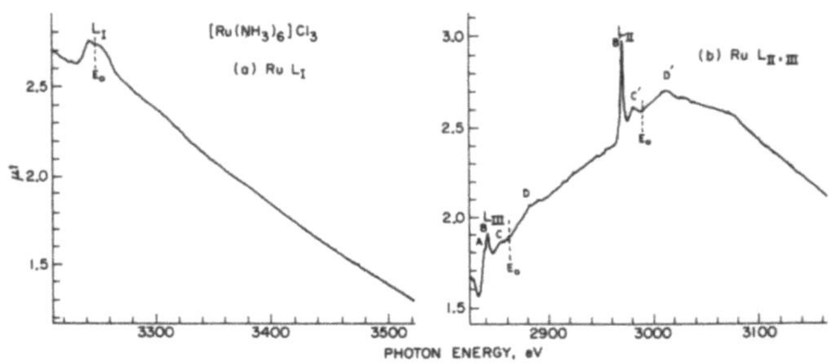

Fig.1. (a) and (b). X-ray absorption spectra of the Ruthenium L_I and $L_{II,III}$ edges respectively; E_0's are obtained by a procedure described in the text

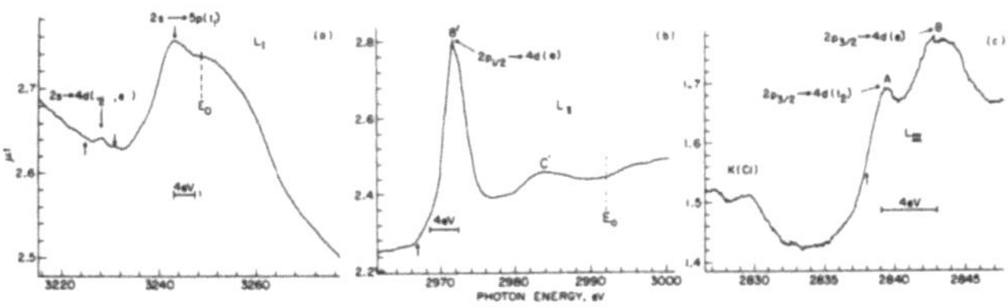

Fig.2. Edge region of $Ru(NH_3)_6Cl_3$: (a) Ru L_I, (b) Ru L_{II} and (c) Ru L_{III}; A, B, and B' are transitions to the 4d final states (see text)

the Ru L_I edge shown in Fig.1 and 2 were obtained from EXAFS analysis. The E_0's of $L_{II,III}$ were located by aligning the theoretical edge positions of McMASTER et al. [4] with E_0 of the L_1 edge.

The main concern here is to understand the anomaly at the $L_{II,III}$ edges. Since B and B' are of the same origin (p→d) based on E_0 considerations, peak A must arise from a transition involving spin consideration which is allowed for the L_{III} edge and forbidden for the L_{II} edge. To understand this observation more quantitatively, we have calculated the dipole transition intensities $\langle\psi_{initial}|r|\psi_{final}\rangle^2$ between $2p_{j,m_j}$ and the unoccupied valence levels of concern by taking into account the electronic configuration, the local symmetry of the complex and spin-orbit effects (Fig.3) in constructing the final state one-electron wave functions (ϕ_1-ϕ_6, u and v in Fig.3b). Since both initial and final state wave functions are in terms of the spherical harmonics (Y_{ℓ,m_ℓ}), the relative intensity of the p→d transition is based on the results of the following integral

$$\langle Y_{\ell m_\ell}|\cos\theta|Y_{\ell' m_{\ell'}}\rangle = (-1)^{m_\ell}\left[\frac{3(\ell+1)(2\ell'+1)}{4\pi}\right]^{1/2}\begin{pmatrix}\ell & 1 & \ell'\\ -m_\ell & 0 & m_{\ell'}\end{pmatrix}\begin{pmatrix}\ell & 1 & \ell'\\ 0 & 0 & 0\end{pmatrix} \qquad [1]$$

Fig.3. (a) Simplified molecular orbital diagram of $Ru(NH_3)_6^{3+}$; according to O symmetry. (b) Energy diagram of the d orbitals of $Ru(NH_3)_6^{3+}$ (d^5 low spin configuration in a strong crystal field); spin-orbit and tetragonal distortion are included. The spin-orbit wave-functions ϕ_1 to ϕ_6 and u and v are in terms of the octahedral basis set and spin-functions. The three allowed 2p→4d transitions are shown with arrows.

It turns out that there are two allowed p→d transitions for the L_{III} edge and only one for the L_{II} edge (Fig.3b) entirely consistent with our observation. Similar procedure enables us to assign the rest of the near-edge region below E_0 as shown in Fig.2 and Fig.3.

The splitting of the p→d transition at the L_{III} edge can be used to extract the crystal field 10 Dq parameter of $Ru(NH_3)_6Cl_3$. Considering a negligible Jahn Teller effect and the final state relaxation to be the same for the split 4d levels, we can write $\Delta E(ev) = 10\ Dq - \zeta_d$. Using tabulated ζ_d of 0.13 ev and the observed ΔE of 3.8 ev (33600 cm^{-1}) we calculate a 10 Dq of 3.93 eV which is in agreement with the usual 10 Dq value for Ru^{III} and Rh^{II} compounds. Finally, it must be noted that since the initial state is exclusively atomic-like, ligand to metal charge transfer which tends to obscure d→d transition in these complexes would be eliminated at the near edge and the technique should be complementary to UV-visible spectroscopy.

Acknowledgment

EXAFS spectra were recorded at SSRL which is supported by the U.S. NSF through the Division of Materials Research, the U.S. NIH through the Biotechnology Resource program in the Division of Research Resources, and the U.S. DOE. This work was performed at Brookhaven National Laboratory under the auspices of the U.S. Dept. of Energy and supported by its Office of Basic Energy Sciences.

References

1. M. Brown, R. E. Peierls, and E. A. Stern, Phys. Rev. B 15, 738 (1977).
2. T. K. Sham and B. S. Brunschwig, J. Amer. Chem. Soc. 103, 1590 (1981).
3. B. S. Brunschwig, C. Creutz, D. H. Macartney, T. K. Sham and N. Sutin, submitted to Disc. Faraday Soc.
4. W. H. McMaster, N. Kerr Del Grande, J. H. Mallett, and J. H. Hubbell, Compilation of X-ray Cross Sections (NTI Service, Springfield, VA. 1969).

Study of XANES at the K Absorption Edge in Some 3*d* Compounds

S.G. Saxena, H.S. Chauhan, Satish Chandra, and K.B. Garg
Department of Physics, University of Rajasthan,
Jaipur-302 004, India

Introduction

Unlike EXAFS, the x-ray absorption near edge structure (XANES) is not even qualitatively well understood at present. Need for studies to extend our understanding of the XANES could not, therefore, be overemphasized. Present efforts comprise an experimental study of the XANES in the vicinity of the K absorption edge in Fe, Co and Cu metals and some of their compounds. The large number of reports published on these materials, for example [1,2,3], do not discuss the XANES as such except in a scattered way.

Results and Discussion

Table 1 represents the results of some earlier as well as our own measurements on chemical shift and occurrence of white lines in the K absorption spectra of Fe and $FeSO_4 \cdot 7H_2O$, Co and $CoCO_3$, and Cu, Cu_2O and CuO. Reported crystallographic data is also incorporated in it to facilitate discussion. As is evident from it, appreciable chemical shifts are observed in all the compounds except Cu_2O. Same is true of the white line.

Table 1. Data on the chemical shift of the K edge, the white line and the crystal structure of the samples studied

System	Shift [ev] Earlier	Present	White Line	Lattice	Lattice Vectors a	b	c	
Fe	-	-	No	fcc	3.64	-	-	[5]
$FeSO_4 \cdot 7H_2O$	9.0[4]	9.0	Yes	mclinic	14.07	6.50	11.04	[6]
Co	-	-	Nc	hcp	2.50	-	4.06	[5]
$CoCO_3$	-	6.1	Yes	h(rh)	4.66	-	14.96	[6]
Cu	-	-	No	fcc	3.61	-	-	[5]
Cu_2O	-0.67[7]	0.0	No	fcc	4.25	-	-	[5]
CuO	3.87[7]	3.6	Yes	mclinic	4.65	3.41	5.11	[5]

Figure 1 shows the K edge of copper and the region immediately beyond it in case of Cu and CuO. A white line is unmistakably present at A in the spectrum of CuO. Same is true of CoCO3 and $FeSO_4 \cdot 7H_2O$ also - see Fig.2.

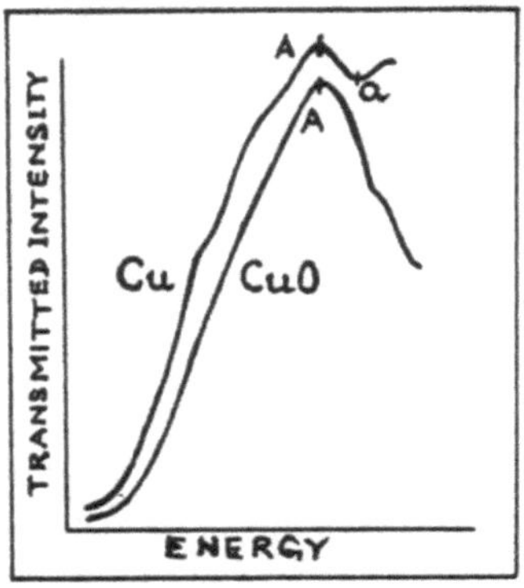

Fig.1. The structure near the K edge of Cu in copper metal and CuO

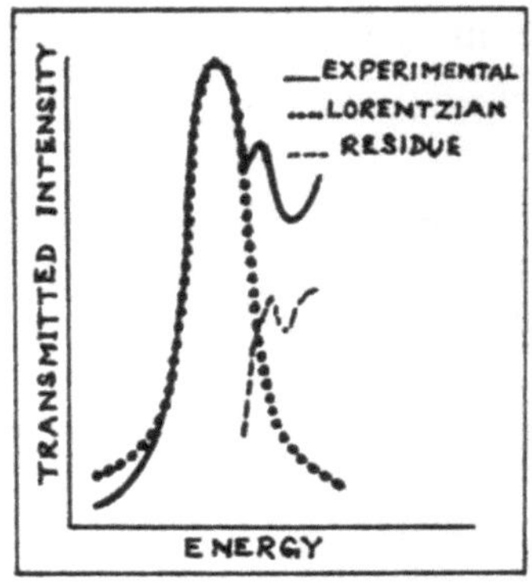

Fig.2. Unsuccessful resolution of the $FeSO_4$ white line into component Lorentzians

A weak and narrow absorption minimum is found to occur immediately beyond the edge in the spectra of the metals, marked a in Fig.1 for the case of Cu. As can be seen from Fig.1 this minimum is engulfed by the white line in CuO and this is true of all three cases wherein a white line makes appearance in the spectrum. To understand the observed shifts and the white lines we now turn our attention to the changes brought about in bonding and interatomic distances by chemical combination. The bond, as is well known, changes from predominantly metallic to substantially covalent whereas the interatomic distances tend to increase, see Table 1. The latter, in turn, leads to dehybridization of the states due to narrowing of the bands. The narrowed 4p band in these 3d compounds would thus result in a high density of unoccupied p-symmetric states near the Fermi level. Transitions from the 1s to these narrowed 4p bands are thus only likely to yield a shift as well as a white line. This white line, it is quite clear, cannot be due to transitions to the narrow d-symmetric states for then it should have been present in the spectra of the metals too. White lines due to d-symmetric states are well known at the L_{II} and the L_{III} edges of 3d metals as well as their compounds. The possibility of the white line being excitonic in origin [8] can be overruled on at least two counts. If it is assumed to originate in formation of excitons-since the materials in question are semi-conducting/insulating, the edge would be situated on the high energy side of the white line. There is no evidence of this on our films. Even if the undulation beyond the white line is to be designated for the edge, the chemical shift, which would now shoot up to about 30 eV, would rule out the possibility. Another conclusive argument against this is the shape of the white line. If it were to be excitonic in origin it should be possible to resolve it in its component Lorentzians [8] . Figure 2 shows the result of such an attempt. The white line observed in case of $FeSO_4 . 7H_2O$ (solid line), on subtraction of a Lorentzian (dotted) from it, yields a residual curve (dashed) to which it is impossible to fit Lorentzians.

Conclusion

To conclude it may be stated that the observed chemical shift and the white line in the K absorption spectra of 3d compounds perhaps have a common origin in the dehybridization and narrowing of the 4p band on chemical combination. The absence of both in case of Cu_2O may perhaps be due to smaller changes in the bonding and interatomic distances and hence lack of appreciable narrowing of the 4p band. Further attempts at identifying the origin of the first absorption minimum in the K spectra of the 3d metals are important in view of the gross changes undergone by it on chemical combination. Its appearance in the case of the metals and Cu_2O and disappearance in the three compounds that show a white line point to its relationship to bonding ?

References

1. W.W. Beeman and H. Friedman, Phys. Rev. 56, 392, (1939)
2. H.P. Hanson and W.W. Beeman, Phys. Rev. 76, 118, (1949)
3. S.I. Salem, C.N. Chang, P.L. Lee and V. Severson, J. Phys. C 11, 4085, (1978)
4. S. Kawata and K. Maeda, J. Phys. F 3, 167 , (1973)
5. W.B. Pearson, Handbook of Lattice Spacings and Structures of Metals and Alloys, Pergamon Press, (1964)
6. S.D. Hodgman, Handbook of Chemistry & Physics, Chemical Rubber Publishing Co., Ohio (1940)
7. L.P. Verma and B.K. Agarwal, J. Phys. C 1 , 1658 (1968)
8. M. Brown, R.E. Peierls and E.A. Stern, Phys. Rev. B 15, 738 (1977)

Generalized Oscillator Strength for K-Shell Near Edge Structures in N_2

R. Camilloni, E. Fainelli, G. Petrocelli[1], and G. Stefani
Istituto Metodologie Avanzate Inorganiche C.N.R., Area Ricerca di Roma, C.P. 10, I-00016 Monterotondo, Italy

Near-edge structures have recently attracted good amount of interest as a novel source of information on local structural properties of the matter. Their study in simple molecules is undertaken with the aim of fully clarifying the fundamental excitation mechanism and the nature of the structural parameters to which they are sensitive. The N_2 molecule in particular has been extensively investigated by using either photons and electrons as excitation sources, the main difference being in the possibility, for the electrons, to transfer a continuum of linear momentum K and of energy ε . Within the framework of the first Born approximation, the target response to the electron impact excitation is described in terms of a bidimensional generalized oscillator strength $f(K,\varepsilon)$, whose equivalent in photoabsorption is the optical oscillator strenght $f_0(\omega)$. In the limit of $K\to 0$, or equivalently high incident electron energy and zero scattering angle, the G.O.S. converges to the O.O.S. and on this basis optically related quantities can be extracted from electron energy loss experiments and compared with photoabsorption experiments. In the region of $K\neq 0$, conversely, where also optically forbidden transitions are excited, unique information can be derived from the G.O.S. dependence on K.

This work is meant to investigate in the region of $K\neq 0$, by means of angle-resolved electron impact energy loss spectroscopy, the inner shell excitation of N_2 which exhibits a sharp peak in the discrete region of the K edge, about 9eV below the threshold for the ionization continuum. A controversy exists on whether the transition ends at a single broken symmetry state $(1s^{-1}\,2\pi)^1\Pi$, or at two almost degenerate states: $(1\sigma_u^{-1}\,1\pi_g)^1\Pi_u$ and $(1\sigma_g^{-1}\,1\pi_g)$ $^1\Pi_g$, which are respectively an optically allowed state and a parity forbidden one. Molecular symmetry considerations are in favour of the latter thesis, while to account correctly for the experimental value of the transition energy, calculations must include relaxation effects and localization of the core-hole, leading to an equivalent core model and to a single optically allowed transition. The experiments have not given up to now a definite answer. The equivalent core model is supported by a high resolution study of the vibrational structure of the electron energy loss peak [1]. On the other hand, the values of the O.O.S. as obtained from photoabsorption studies [2,3] or inferred by extrapolating the G.O.S. measured at small K's [4], span from f =.12±.05 to f =.23±.05 and do not exclude the contribution of a forbidden transition in the electron impact experiments.

---
[1]Guest Fellow

In order to clarify the matter, the absolute value of $f(K,\varepsilon)$ has been measured up to $K=80a_o^{-1}$ at 1400eV and 3400eV incident energies, with a newly constructed (e,2e) type apparatus. Crossed electron-gas beams are used and two electrostatic electron spectrometers rotate in the scattering plane from 0° to 130° with .1° precision and .5° resolution. The angular calibration has been checked by measuring the angular dependence of the elastic cross section, whose value is well known from the literature [5], and of the KLL Auger spectrum, which is expected to be isotropic.

Relative measurements were brought on absolute scale through the intensity ratio with the elastic cross section measured in the same experimental conditions. The energy resolution was .7eV at 1400eV and 1.2eV at 3400eV. The energy loss spectra have been analyzed by fitting the background with a least-squares fitting programme and by assuming for the peak profile the vibrational structure and the line width as measured in [1], folded in the instrumental energy response. The G.O.S. is then calculated from the energy loss cross section following the relation

$$f(K,\varepsilon) = \frac{\varepsilon}{2} K^2 \frac{K_o}{K_1} \frac{d^2\sigma}{d\omega d\varepsilon} \quad , \tag{1}$$

where $K^2 = K_o^2+K_1^2-2K_oK_1\cos\vartheta$, K_o and K_1 are the initial and final momentum of the electron and atomic units are used. Results are reported in Fig.1 together with the other published values which are known to us [2-4,6-9]. Our statistical accuracy was better than 2% and the main contribution to the error, quoted at the 68% confidence level, comes from the accuracy with which the elastic cross section, used for normalization, is known. The dispersion of the data has to be ascribed to the uncertainty in the measure of the e^- current and to the channeltron quantum efficiency.

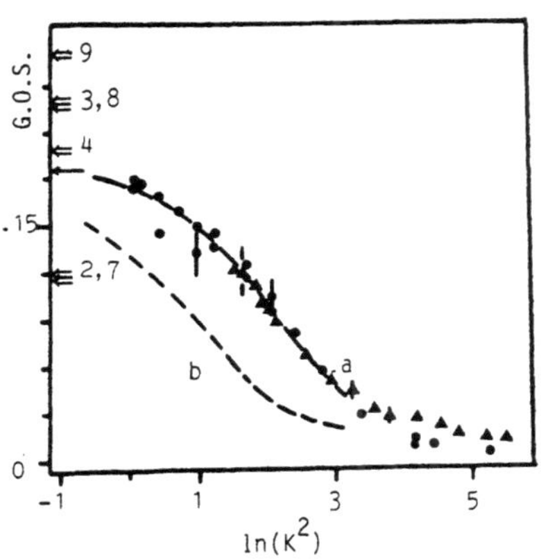

Fig.1. G.O.S. for $^1\Pi_{(u,g)}$ core excitation of N_2. This experiment: E_o=3.4KeV (●), E_o=1.4KeV (▲); theory [7]: (---) $X^1\Sigma_g \to {}^1\Pi_u$, (——) $X^1\Sigma_g \to {}^1\Pi_u + {}^1\Pi_g$. Optical oscillator strength: theory [7,8,9], experiment [2,3,4]

The following remarks can be made:

i) There is a substantial agreement among the present data and data by ODA et al. [6], indistinguishable in the figure from $\ln(K^2)=1.2$ on. The independence of the results from the incident energy fulfills the necessity condition for they can be treated in first Born approximation and interpreted in terms of generalized oscillator strength.

ii) Comparison is done with the only calculation of $f(K,\varepsilon)$ known to us [7]. The full curve (a) is calculated in a configuration interaction scheme, in order to retain the overall symmetry properties for the excited states together with the localization of the atomic K hole, and predicts two transitions to the $^1\Pi_u$ and the $^1\Pi_g$ states almost degenerate in energy (.06eV apart). In order to fit the experimental data the curve has been multiplied by a factor 1.64 and then the agreement is very good. Data can not be fitted anyway by the curve (b) (dashed line), which is relative to the optical transition alone, thus indicating the presence of quadrupole contributions.
iii) If the mentioned disagreement in the absolute value is disregarded and only the calculated shape of the G.O.S. is retained, a value is inferred for the optical oscillator strenght f =.194. That value can be compared with the previous determinations of the O.O.S. also reported in Fig.1 and with f =.20±.02 obtained by applying a least-squares method in fitting our data with a polinomial expansion of the type suggested in [10].

Conclusion

The present study of the $^1\Pi_{(u,g)}$ feature in the N_2 K-edge furnishes measured absolute values of the G.O.S. over the largest range in the K dominium investigated up to now. That allows for a critical matching of the existing e^--impact determinations with optical determinations, those obtained from e^- energy loss appearing consistent among themselves. Further, it is shown how the K dependence of the G.O.S. can reveal forbidden transitions, which contribute only at momenta $K\neq 0$. At the present stage, however, with only one calculation available, to assess the presence of such a contribution would be a model-dependent result and more theoretical work is needed.

1. G.C. King, F.H. Read, M. Tronc: Chem. Phys. Lett. 52, 50 (1977)
2. F. Wuilleumier, M.D. Krause: Int. Conf. on Inner Shell Ionization Phenomena, Georgia Institute of Technology, Atlanta, 1972, p. 773
3. A. Bianconi, H. Petersen, F.C. Brown, R.Z. Bachrach: Phys. Rev. A17, 1907 (1978)
4. R.B. Kay, Ph.E. VanderLeeuw, M.J. VanderWiel: J. Phys. B10, 2513 (1977)
5. R.H.J. Jansen, F.J. deHeer, H.J. Luyken, B. vanWingerden, H.J. Blaauw: J. Phys. B9, 185 (1976)
6. N. Oda, N. Nishimura, T. Osawa: Abs. XI Int. Conf. on Phys. of Elect. and Atomic Coll., Kyoto, 1979, p.408
7. T.N. Rescigno, A.E. Orel: J. Chem. Phys. 70, 3390 (1979)
8. J.L. Dehmer, D. Dill: J. Chem. Phys. 65, 5327 (1976)
9. T.N. Rescigno, P.W. Langhoff: Chem. Phys. Lett. 51, 70 (1977)
10. K.N. Klump, E.N. Lassettre: Chem. Phys. Lett. 51,99 (1977)

High-Resolution Analysis of the $L_{2,3}$ White Lines of Pd

L. Incoccia and S. Mobilio
PULS-CNS Laboratori Nazionali di Frascati, I-00044 Frascati, Italy

M. Benfatto, I. Davoli, S. Stizza
Istituto di Fisica, Università di Camerino, I-62032 Camerino, Italy

A. Bianconi
Dipartimento di Fisica, Università di Roma, La Sapienza, I-00185 Roma, Italy

"White lines" appear at the L_3 and L_2 absorption edge in the XANES (x-ray absorption near edge structure) of transition metals [1] and rare earth compounds.

Recently it has been pointed out that these "white lines" provide a novel method for probing:

a) The change in the number of empty d holes in transition metal catalysts [2].
b) The total angular momentum character $j = 1 \pm 1/2$ of empty conduction band states [3].
c) The "intermediate valence" in mixed valent rare earth compounds [4].

The recent success of one electron-theories [5,6] to explain XANES and their sensitivity to the structure around the absorbing atoms [7,8] have stimulated an increasing interest in their use as local structure probes. However the presence of many-body final state effects, due to the rearrangment of the passive electrons in the atom and in the surrounding environment, after the core hole excitation, can induce changes in the lineshape and give satellite features which complicate XANES interpretation. Therefore it deserves accurate experimental investigation to determine the relevance of final state effects in XANES.

Different final state effects are expected in different systems. In the XANES of atoms [9] and molecules [10] "shake up" satellites above the continuum threshold have been clearly identified.

Recently the presence of a peak [11] in the XANES of insulators has been interpreted as a "shake-up" process by analogy with the "shake-up" peaks in the XPS spectra. However, since the final states in XPS and XANES are different the presence of a "shake-up" peak in XPS is not enough to predict its presence in the XANES. For the XANES of metals there will be a different final state regime depending upon the band structure of the system (bandwidth W and band filling) and the Coulomb interaction Q between the core hole and the valence electrons [12].

In the free electron limit, $Q/W \to 0$, there will be no satellites and the many-body effects induce only a broadening and asymmetry in the threshold lineshape.

In the high correlated electron limit on the other hand $Q/W \to \infty$ and partially filled band, the electron-hole interaction Q splits the final state configurations, so multiplet splitting appears and interference effects between final state configurations plays an important role.

In this work we focused our interest on the "white line" which appears in the L edges of metallic Pd. Pd, like Ni and Pt, belongs to the intermediate case Q W with a nearly filled 4d valence band. Its groundstate configuration $2p^6\ 4d^{9.64}\ 5sp^{0.36}$ can be described by the overlap of two virtually degenerate atomic configurations,

$$\text{A)} \quad 2p^6\ 4d^9\ 5sp^1 \qquad \text{and} \qquad \text{B)} \quad 2p^6\ 4d^{10}\ 5sp^0 \ .$$

At the 2p threshold only the excitation of the configuration A give a $2p \to 4d$ atomic resonance

$$\hbar\omega + A \rightarrow 2p^5\, 4d^{10}\, 5sp^1 \, .$$

In this case no multiplet splitting appears because of the filled $4d^{10}$ configuration and the "infrared singularity" effect at the edge can be disregarded.

Above the edge due to excitation

$$\hbar\omega + B \rightarrow 2p^5\, 4d^9\, 5sp^2$$

there could give a satellite in the XANES of Pd as the one observed in XPS [13].

The experiment has been carried out, on a vacuum beam line, at Frascati Laboratories using the storage ring ADONE operated at low energy 1.2 GeV to suppress the third harmonic content of the beam at 3 keV photon energy. We have used a Si(111) channel cut monochromator with resolution of $\Delta E/E=1.3 \times 10^{-4}$ determined by the rocking curve of the crystal.

Fig. 1 shows the L_3 and L_2 white lines of metallic Pd taken at room temperature.

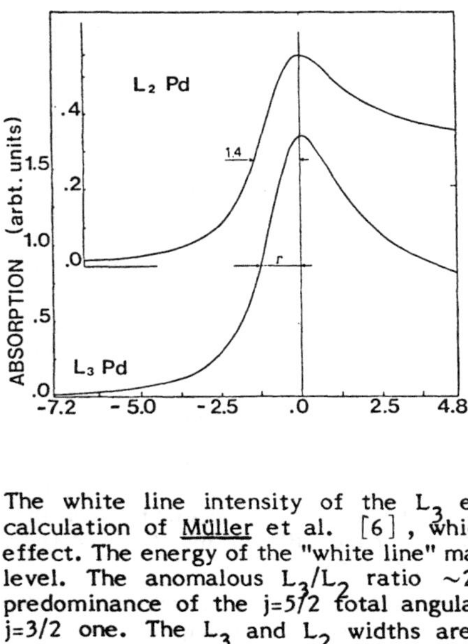

FIG. 1.
High resolution spectra of the white lines at the L_2 and L_3 absorption edge of Pd

The white line intensity of the L_3 edge is well predicted by the one-electron calculation of Müller et al. [6], which does not take into account any core hole effect. The energy of the "white line" maximum corresponds to excitation to the Fermi level. The anomalous L_3/L_2 ratio $\sim$2.75 at the white line maximum is due to predominance of the j=5/2 total angular momentum of the empty 4d hole over the j=3/2 one. The L_3 and L_2 widths are determined by the core hole lifetime and instrumental resolution.

The only evidence of a possible weak many-body effect has been found in the difference between L_2 and L_3 spectra. Fig. 2 shows the difference

$$\alpha_{L_2}(h\nu - 156.7) - \alpha_{L_3}(h\nu)/2.1$$

where 156.7 eV is the spin-orbit splitting and 2.1 the experimental L_3/L_2 statistical weight, deduced from the EXAFS tail. All the high energy XANES features are washed out; only the smooth EXAFS background of the L_3 level at energy of L_2 level remains. At energy corresponding to the white line a sharp negative peak appears because of the high L_3 peak already discussed. The shoulder extending up to 13 eV

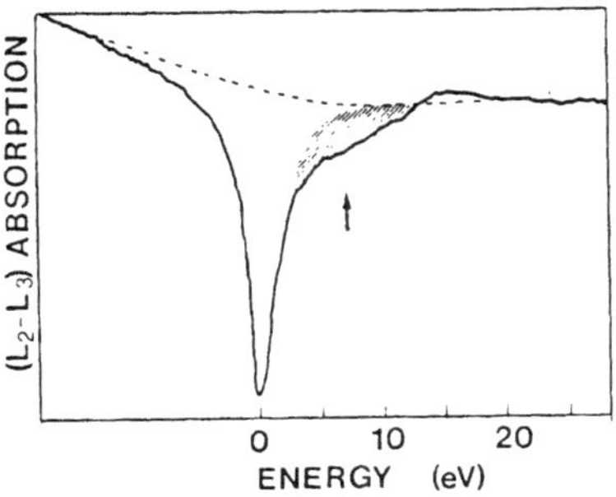

FIG. 2. Difference spectrum between L_2 and L_3 XANES.

$\alpha_{L_2}(h\nu-156.7)-\alpha_{L_3}(h\nu)/2.1$

The shoulder S has been tentatively assigned to a "two holes" two-electrons excited bound state

could be due to the "Two-hole" bound state already seen only in XPS spectroscopy, which is stronger at the L_3 edge than at the L_2 edge.

However this broad weak structure is so weak that it does not appear in the XANES as a characteristic feature and all the main XANES features are due to one-electron effects.

References

1. M. Belli, A. Scafati, A. Bianconi, S. Mobilio, L. Palladino, A. Reale and E. Burattini, Solid State Commun. 35, 355 (1980); A. Bianconi, Proc. of Meeting on EXAFS for Inorganic Systems Daresbury Report DL/SCIR17 (1981) p. 13 and A. Bianconi, this book.
2. M. Brown, R.E. Peirels and A. Stern, Phys. Rev. B15, 758 (1977).
3. F.W. Lyttle, P.S.P. Wei, R.G. Greegor, G.H. Via and J.H. Sinfelt, J. Chem. Phys. 70, 4849 (1979).
4. A. Bianconi, S. Modesti, M. Campagna, K. Fischer and S. Stizza, J. Phys. C 14, 4737 (1981) and A; Bianconi, M. Campagna, S. Stizza and I. Davoli, Phys. Rev. B 24, 6139 (1981).
5. P.S. Durham, J. Pendry, C.H. Hodges, Sol. State Commun. 38, 159 (1981).
6. J.E. Müller, O. Jepsen and J.W. Wilkins, Sol. State Commun. 42, 365 (1982).
7. F.W. Kutzler, C.R. Natoli, D.K. Misemer, S. Doniach and K.O. Hodgson, J. Chem. Phys. 37, 3274 (1980); A. Balzarotti, F. Comin, L. Incoccia, M. Piacentini, S. Mobilio, A. Savoia, Solid State Commun. 35, 145 (1890).
8. A. Bianconi, Dell'Ariccia, P. Durham, J.B. Pendry, Phys. Rev. B, to be published.
9. T.A. Carlson, M.O. Krause and W.E. Moddeman, J. Phys. Collog 32, C4-76 (1971); and F. Wuileumier ibidem C4,-88 (1971).
10. A. Bianconi, H. Petersen, F.C. Brown and R.E. Bachrach, Phys. Rev. A 17, 1907 (1982).
11. E.A. Stern, Phys. Rev. Letters 49, 1353 (1982).
12. A. Bianconi, Phys. Rev. B26, 2741 (1982).
13. B. Chandesris, G. Krill, G. Maire, J. Leconte, Y. Petroff, Solid State Commun. 37, 187 (1981) and references cited therein.

Part IV

Special Crystalline Systems

Structural Studies of Superionic Conduction

T.M. Hayes and J.B. Boyce

Xerox Palo Alto Research Center, Palo Alto, CA 94304, USA

A solid electrolyte with ionic conductivity in excess of 10^{-2} (ohm cm)$^{-1}$ is frequently referred to as a "fast ion" or "superionic" conductor. These materials have been studied extensively in recent years because of their remarkably high ionic conductivity and because of their technological promise as electrolytes in solid state batteries with higher energy and current densities at lower cost than conventional lead-acid batteries. Extended x-ray absorption fine structure (EXAFS) spectroscopy has been a very useful probe of superionic conductors. Analysis of EXAFS spectra from the normal and superionic phases of several cationic conductors has revealed a great deal about correlations between the highly mobile cations and the fixed anion lattice. The nearest neighbor peak in the cation-anion correlation function $p_{c-a}(r)$ was found to be broad and asymmetric at high temperatures, necessitating the development of EXAFS data analysis techniques which could accommodate such peak shapes. It is becoming increasingly apparent that broad and asymmetric peaks are to be found in many materials, for which these EXAFS data analysis techniques should prove equally useful. The detailed information about $p_{c-a}(r)$ revealed by the EXAFS studies has led to the elimination of many models which had been proposed for the structure, and for the conduction process as well. Furthermore, enough insight was gained into the underlying cation-anion interaction potentials to enable a calculation of the DC ionic conductivity as a function of temperature. The results of these EXAFS studies of AgI and the cuprous halides have been published in detail elsewhere [1,2], and are summarized below for the case of AgI.

AgI attains superionic conductivity abruptly with a structural phase transition at 420K. Below 420K, AgI is found in a β phase in which the structure is wurtzite and the ionic conductivity is comparable to that of a normal salt: 10^{-4} (ohm cm)$^{-1}$ at 410K with an activation energy of approximately 0.6 eV. Above 420K, AgI is found in its α phase in which the structure is nominally body-centered cubic and the conductivity is typically superionic: 1 (ohm cm)$^{-1}$ at 430K with an activation energy of approximately 0.05 eV. This is quite close to the theoretical conductivity calculated by assuming that *all* of the cations are *always* moving at thermal velocity. This has led to the notion that essentially all of the cations must participate in the superionic conduction process, and that they cannot linger at their "rest" sites. It is also interesting to note that the entropy change ΔS_t at the $\beta \rightarrow \alpha$ transition is slightly larger than that on melting ΔS_f. This implies that the increase in "disorder" is comparable in these two phase transitions. Furthermore, $\Delta S_t + \Delta S_f$ is approximately equal to the entropy change on melting for a normal salt. This behavior is also typical of superionic conductors and has given rise to the concept of "sublattice melting," in which the attainment of superionic conductivity is associated with the "melting" of the sublattice of the conducting ion species. It will be seen in the following discussion that these concepts feature prominently in the model which we have developed for the structure and for the conductivity of AgI and of the cuprous halides.

The EXAFS on the Ag K edge in AgI, χ_{Ag}, was measured at a variety of temperatures from 77K in the β phase to 575K in the superionic α phase. $\chi_{Ag}(k)$ is shown in Fig. 1 for two representative temperatures. The EXAFS data were Fourier transformed in the complementary variables k and r, and analyzed in r space using the following expression for $\varphi_{Ag}(r)$, the complex Fourier transform of $k\chi_{Ag}(k)$:

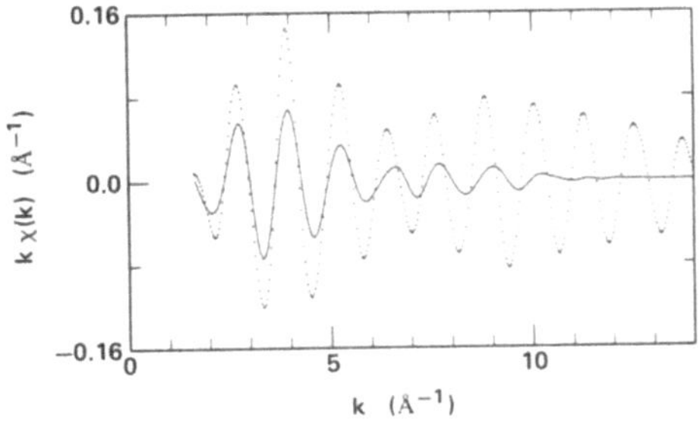

Fig. 1. The EXAFS $k\chi_{Ag}(k)$ on the Ag K-shell absorption in AgI at 77K (dotted line), well below the normal→superionic transition temperature of 420K, and at 471K (solid line), well above that transition

$$\varphi_{Ag}(r) = \sum_{\gamma=Ag,I} \int_0^{\infty} dr'/r'^2 \; p_{Ag-\gamma}(r') \; \xi_{Ag-\gamma}(r-r'). \tag{1}$$

The EXAFS is divided into two contributions. The structural information is in $p_{Ag-\gamma}(r')$, while the details of the electron scattering processes are contained in $\xi_{Ag-\gamma}(r'')$. $p_{Ag-\gamma}(r)$ is the pair correlation function of γ atoms about the excited Ag atom, normalized so that $\int dr\, p_{Ag-\gamma}(r)$ is the total number of γ atoms in the sample. ξ_{Ag-Ag} and ξ_{Ag-I} were extracted from EXAFS data on Ag metal and on β AgI at 77K, respectively. No evidence was found in the spectra for contributions from Ag–Ag nearest neighbor pairs. Data analysis consisted therefore of determining the $p_{Ag-I}(r)$ which best fit each data set. The procedures for accomplishing this are described in detail in [2,3].

A large number of existing models were tried in the analysis of the data on these superionic conductors, of which none were successful in fitting the α-phase data. Each of these models is based on the body-centered cubic lattice of the iodine ions in the α phase, in which the largest prospective sites for Ag ions are sites of distorted tetrahedral symmetry at (1/2,1/4,0). These models included (a) the STROCK model [4] in which Ag ions are distributed over 42 prospective crystallographic sites; (b) the displaced site model of BÜHRER and HÄLG [5] in which the Ag ions occupy sites displaced along (100) directions from the (1/2,1/4,0) sites (towards an edge of the tetrahedron); (c) a single Gaussian peak in $p_{Ag-I}(r)$, corresponding to a harmonic oscillator model; (d) a model in which the Ag ions are displaced in (110) directions from the (1/2,1/4,0) sites (towards a face of the tetrahedron) [6]. While none of these models yielded a very close simulation of the α-phase data, a clear pattern emerged from the values of the parameters corresponding to the best fits for each model, and from the relative degree by which each model fit or failed to fit. Models (a) and (b) were inconsistent with the data. Model (c) yielded the result that the mean nearest neighbor separation and coordination number *appear* to decline with increasing temperature. The apparent decline is less and the simulation closer to the data using model (d), which is distinguished from the other models by the nearest neighbor peak in $p_{Ag-I}(r)$ being asymmetric, skewed toward lower r.

These results led to the formulation of an excluded volume model for the conducting ion distribution in these superionic conductors [2,7]. This model approximates the net effect of the forces acting on the cations with a hard-sphere cation–anion pair potential. [The resulting charge density is convoluted radially with a Gaussian of adjustable width so as to accommodate the slight softness of the actual potentials as well as the thermal motion of the anions. The model then has two adjustable parameters, the sum of the hard-sphere radii and the width of that Gaussian.] Within the excluded volume model, the Ag ions move freely over large interstitial spaces, constrained only by the requirement that they cannot approach closer to an iodine ion than a distance $r_{excluded}$, equal to the sum of the two hard-sphere radii. The resulting Ag ion density is indicated schematically in Fig. 2. Inspection of Fig. 2 reveals that this Ag density will yield a nearest neighbor peak in $p_{Ag-I}(r)$ which is asymmetric in the manner desired, reflecting the shape of the allowed Ag ion locations. Furthermore, it is capable of yielding the sort of behavior expected for a superionic conductor. Consider the position of a Ag ion when it is passing from one enclosed volume into another. Denote by r_{face} the

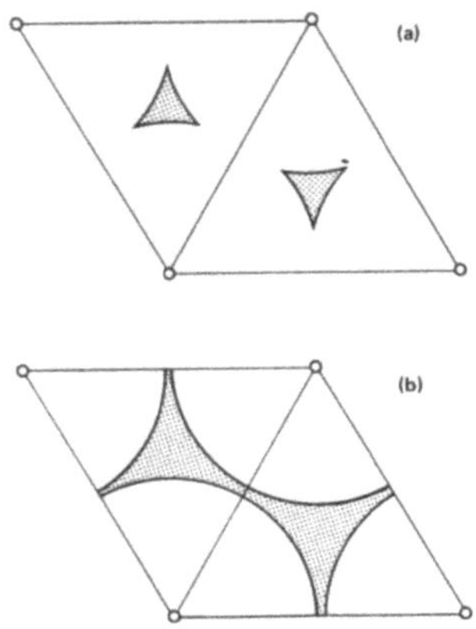

Fig. 2. A schematic two-dimensional representation of the allowed (shaded) and excluded regions for cation centers in the excluded volume model in the (a) insulating and (b) conducting phases. The anion centers are indicated by open circles. In the insulating phase the allowed regions are isolated, whereas in the superionic phase they are connected, allowing for high conductivity

maximum distance which a Ag ion can be from all nearest neighbor iodine ions at that point (e.g., one-half the edge of the triangle in Fig. 2). For a value of $r_{excluded}$ slightly in excess of r_{face}, each Ag ion is confined to an isolated volume such as in shown in Fig. 2a. This corresponds to the case of a normal salt in which conduction requires the ion to surmount a barrier (typically $\approx$ 1 eV), leading to low ionic conductivities. If $r_{excluded}$ is slightly less than r_{face}, on the other hand, then Ag ions can "flow" continuously through the material as shown in Fig. 2b. In the absence of an activation energy, the Ag ion flow is limited by its thermal velocity and the size of the "openings" between adjacent cells. This incorporates the notions about superionic conductivity mentioned earlier: all of the conducting ion species participate more or less continuously in the conduction process, flowing in a liquid-like manner through the lattice of immobile ions.

The $p_{Ag-I}(r)$ which result from this model yield simulations of φ_{Ag} which are extremely close to the EXAFS data. The $p_{Ag-I}(r)$ deduced at four temperatures are shown in Fig. 3. Note that the values of $r_{excluded}$ corresponding to these p_{Ag-I} yield the qualitative behavior which had been hoped for: the Ag ion density in the β phase data corresponds to Fig. 2a (but, of course, in the proper three-dimensional space), while that in the superionic α phase corresponds to Fig. 2b. It is possible to compare the Ag–I pair distributions resulting from molecular dynamics (MD) simulations of the superionic conductors directly with these p_{Ag-I}, or to compare them indirectly with the φ_{Ag} using (1) [8]. In either case, the nearest neighbor peaks in the MD p_{Ag-I} are seen to be too broad and to become appreciable at too low a value of r, although the disagreement with the EXAFS data has been reduced somewhat for more recent MD calculations [9].

It is worthwhile to draw attention at this point to those features of the data which lead directly to the adoption of an asymmetric pair distribution function such as is shown in Fig. 3. The wavelength of the nearest neighbor oscillations in $k\chi(k)$ reflects the mean nearest neighbor separation as $k \rightarrow 0$, but corresponds to the sharp features in $p_{Ag-I}(r)$ as k becomes large. Thus distributions such as those in Fig. 3 will yield EXAFS spectra with a variable wavelength, increasing with increasing k. Such variable wavelengths do indeed characterize the EXAFS spectra measured on superionic conductors at high temperatures. This is quite evident in Fig. 1, wherein spectra at two temperatures are plotted on the same axes. Secondly, the envelope of the oscillations in $k\chi(k)$ will not incorporate the simple $\exp(-2k^2\sigma_r^2)$ behavior expected for a Gaussian peak in $p_{Ag-I}(r)$. The envelope will fall off quickly at small k (as would be expected for a broad Gaussian in r space) but will decrease relatively slowly with increasing k for larger values of k, behaving as though σ_r were decreasing with increasing k. This is also characteristic of measurements on these materials at high temperatures, and is a somewhat less obvious feature of Fig. 1. Thus, deviations in the peak shape from a simple

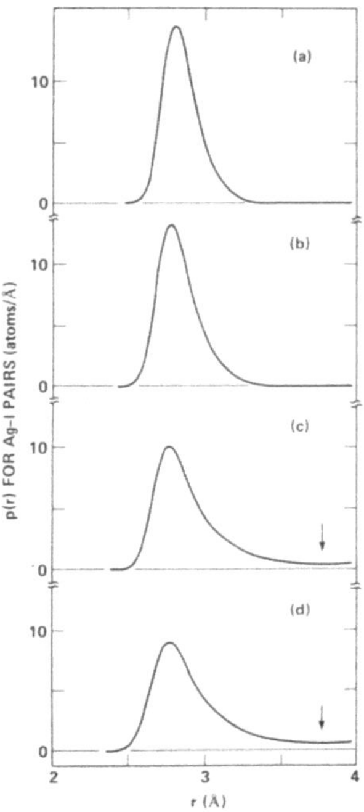

Fig. 3. The radial distribution function of iodine ions about each Ag in AgI in the β phase at (a) 293K and (b) 371K, and in the α phase at (c) 471K and (d) 575K. Each distribution function is normalized to contain four iodine ions in the first peak. The arrows in (c) and (d) locate the contribution from a Ag ion occupying the center of a face between adjacent tetrahedra. Non-vanishing ion density at that location corresponds to a Ag ion distribution similar to that in Fig. 2b

Gaussian will be manifest quite clearly in both the envelope and phase of data which extends over a sufficient range of k space. These effects are discussed in detail elsewhere [3,10].

From the $p_{Ag-I}(r)$ which fits the EXAFS data at each temperature, it is possible to infer both the Ag ion charge density and the effective one-ion potential, $V(\mathbf{r})$, in which the ions move. The charge density is found to compare favorably with those derived from neutron diffraction studies (see discussion in [2]). It is also possible to calculate directly the DC ionic conductivity resulting from the Ag ion flow within this three-dimensional potential. This has been accomplished by modeling the Ag ions as a Boltzmann gas [1]. Input to this model includes the temperature, the masses of the Ag and iodine ions, the Ag ion density, a scattering length (taken to be the distance between adjacent tetrahedral sites), and the $V(\mathbf{r})$ deduced from fitting the EXAFS data at each temperature. The result predicts the measured conductivity within a factor of 2 in magnitude and the correct temperature dependence within a few percent, as shown in Fig. 4. [The structure of β AgI is such that conduction unavoidably involves passage through intervening octahedral sites, the energy of which cannot be predicted from the EXAFS data. It can be seen from Fig. 4, however, that the assumption of an octahedral site energy between 0.5 and 0.6 eV yields agreement in both magnitude and slope with the measured conductivity in this low temperature phase (see discussion in [1]).] Similar results have been obtained for the cuprous halides [1].

In summary, EXAFS studies of these superionic conductors have significantly increased our understanding of the structure of these materials, and also of the unusually high ionic conductivity which they exhibit. In addition, the superionic conductors were among the first materials for which it was demonstrated that the nearest neighbor peak in the pair correlation function is unavoidably asymmetric. The EXAFS data analysis procedures developed to accommodate these peak shapes should prove equally useful in treating other members of this growing class of materials.

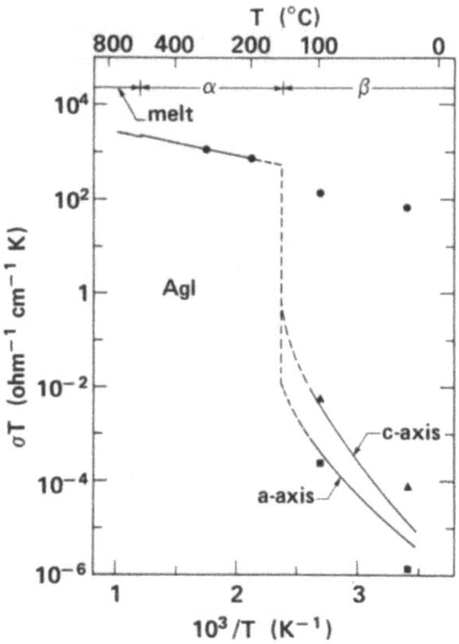

Fig. 4. Predicted conductivity for AgI is shown in the α phase, and in the β phase for octahedral site energies of 0, 0.5, and 0.6 eV (solid circles, triangles, and squares, respectively). The conductivity measured in the β and α phases is also shown (solid lines)

Acknowledgments We gratefully acknowledge the contributions of J. L. Beeby, J. C. Mikkelsen Jr., and W. Stutius to the development of the ideas presented herein. Some of the materials incorporated in this work were developed at the Stanford Synchrotron Radiation Laboratory which is supported by the National Science Foundation (under Contract DMR77-27489), in cooperation with SLAC and the Department of Energy.

References

1. T. M. Hayes and J. B. Boyce, Phys. Rev. B **21**, 2513 (1980).
2. J. B. Boyce, T. M. Hayes, and J. C. Mikkelsen, Jr., Phys. Rev. B **23**, 2876 (1981).
3. T. M. Hayes and J. B. Boyce, "EXAFS Spectroscopy", in Solid State Phys. **37**, ed. by H. Ehrenreich, F. Seitz, and D. Turnbull (Academic Press, New York, 1982), p. 173.
4. L. W. Strock, Z. Phys. Chem., Abt. B **25**, 411 (1934); **31**, 132 (1936).
5. W. Bührer and W. Hälg, Helv. Phys. Acta **47**, 27 (1974).
6. J. B. Boyce, T. M. Hayes, W. Stutius, and J. C. Mikkelsen, Jr., Phys. Rev. Lett. **38**, 1362 (1977).
7. T. M. Hayes, J. B. Boyce, and J. L. Beeby, J. Phys. C **11**, 2931 (1978).
8. T. M. Hayes and J. B. Boyce, J. Phys. C **13**, L731 (1980).
9. A. Rahman and P. Vashishta, in Proc. 1981 NATO Adv. Study Inst. "Superionic Conductors" (Plenum Press), to be published.
10. T. M. Hayes and J. B. Boyce, in "EXAFS Spectroscopy: Techniques and Applications," ed. by B. K. Teo and D. C. Joy (Plenum Press, New York, 1981), p. 81.

EXAFS and Near Edge Structures in Mixed Valent $SM_{1-x}Y_xS$ and the Energy Threshold Puzzle

J.B. Boyce, R.M. Martin, and J.W. Allen
Xerox Palo Alto Research Centers, Palo Alto, CA 94304, USA

1. Introduction

X-ray absorption can provide valuable information on mixed valence materials since it probes two important aspects of the intermediate valence problem. First, the white lines on the L_{II} and L_{III} edges yield a simple and direct measurement of the valence in these materials. Second, the EXAFS can be used to determine possible lattice distortions about the mixed valent ion, thereby providing information on the nature of the electron-phonon coupling. An important step in analyzing the EXAFS data is the determination of the threshold energy, E_0, i.e., that photon energy for which the photo-ejected electron has zero kinetic energy and thus zero wavevector ($k = 0$). This threshold is expected to vary with the valence of the central atom since the atomic charge exerts a substantial effect on the binding energy of a core electron. Before addressing the threshold problem in the mixed valence materials, a summary of our results on the Sm L_{III} edge and EXAFS in $Sm_{1-x}Y_xS$ (x = 0, 0.2, 0.25) and SmP is presented.

2. Sm L_{III} Edge and EXAFS Results

The position in energy of the white line on the Sm L_{III} absorption edge depends on the 4f occupation. For Sm^{2+} ($4f^6$) in SmS the 2p core binding energy is decreased relative to Sm^{3+} ($4f^5$) in SmP because of the additional 2p-4f repulsive Coulomb interaction. This shift is approximately 7 eV, as shown in Fig.1. In the intermediate valence (IV) materials the ground state is a quantum mechanical mixture of the 2+ and 3+ valence states. Thus two white lines are observed, one corresponding to each of the valence states present. This is shown in Fig.1 for $Sm_{0.75}Y_{0.25}S$ at 77K. From the relative intensity of the two peaks a valence of 2.44 ($4f^{5.56}$) at 77K and 2.53 at 300K is determined. The valence at 300K agrees with that from Mossbauer measurements but is some-

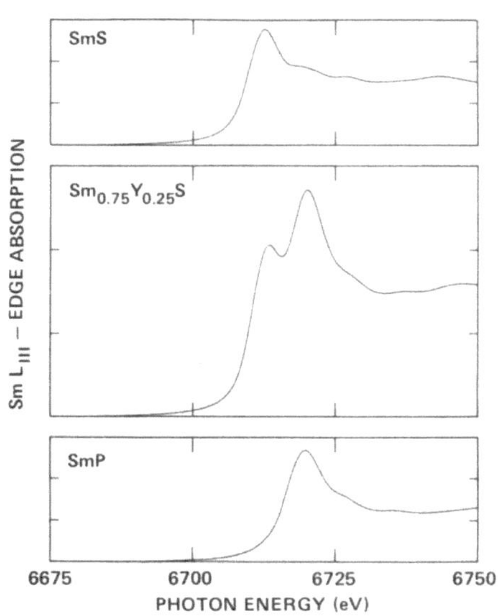

Fig.1. The Sm L_{III} edge absorption of Sm^{2+} in SmS, Sm^{3+} in SmP and both in intermediate-valent (v = 2.44) $Sm_{0.75}Y_{0.25}S$, all at 77K

what lower than the valence of 2.7 from the lattice constant [1].

The EXAFS was also measured in these rocksalt-structured materials in order to determine possible distortions in the near neighbor distribution due to the coupling of the electronic system to the lattice. Distortions for the IV materials might be expected on two accounts. First, there is a large difference, Δr, in ionic radii: $Sm^{3+}-S$ = 2.80 Å and $Sm^{2+}-S$ = 2.98 Å. Second, estimates of the valence fluctuation rate from f^n to f^{n-1} place it in the vicinity of phonon frequencies. As a result, there have been theoretical predictions of distorted Sm-S near neighbor distributions, either a double peaked distribution at 2.80 and 2.98 Å or a broadened distribution with a width comparable to Δr = 0.18 Å [2]. Fig.2 shows the EXAFS on the Sm L_{III} edge transformed to real space for the structurally known standard SmS(v = 2) and IV $Sm_{0.75}Y_{0.25}S$ at 300K (v = 2.53). The first peak from 1.8 to 3.0 Å corresponds to the six S near neighbors in both materials. It is obvious by comparison of these two spectra that neither a double peaked nor a substantially broadened distribution exists in $Sm_{0.75}Y_{0.25}S$. A detailed least squares analysis shows that the near neighbor distribution in the IV material is a single, narrow peak. It is shifted inward by 0.15 Å relative to that in SmS and has a small additional thermal broadening of $\Delta\sigma$ = 0.04 Å, approximately that which is expected for the thermal difference from 77K to 300K. Similar results were obtained in $Sm_{0.8}Y_{0.2}S$ and for both materials at other temperatures from 77 to 300K. Thus despite theoretical predictions, the near neighbors of the IV rare earth ions adopt an average distance and not two different distances corresponding to the two valence states. If proper account is taken of the high degeneracy of the $4f^5$ configurations, then the theoretical estimates of the distortions are much reduced and not inconsistent with these EXAFS results [3].

3. The Threshold Puzzle

In the analysis of the EXAFS data on the IV compounds, allowance was made for the possibility of two different threshold energies corresponding to the two different valence states. From the edge spectra of Fig.1, these two thresholds are expected to differ by about 7 eV. Nonetheless, fits to the data indicate that IV $Sm_{1-x}Y_xS$ has only one threshold energy which is equal to that in divalent SmS. This conclusion is also obvious from inspection of Fig. 2 where the thresholds for SmS and $Sm_{1-x}Y_xS$ were set equal to one another. If $Sm_{0.75}Y_{0.25}S$ actually had two E_0's, then the phase of the real part of the transform for SmS would differ from that for $Sm_{1-x}Y_xS$. The fact that the phases are the same indicates that $Sm_{1-x}Y_xS$ has a single threshold equal to that of SmS. To investigate the origin of this puzzling result, we

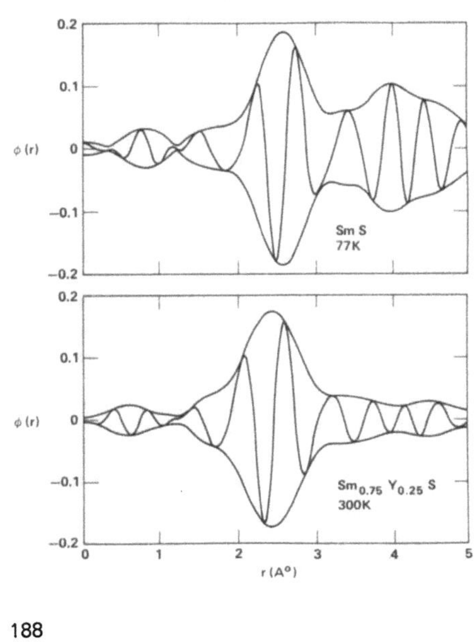

Fig.2. The magnitude (envelope) and real part of the EXAFS on the Sm L_{III} absorption transformed to real space for single-valent SmS (v = 2) at 77K and mixed valent $Sm_{0.75}Y_{0.25}S$ (v = 2.53) at 300K

compared the EXAFS of trivalent SmP with that of divalent SmS. Fig.3 shows the filtered Sm L_{III} EXAFS due to the first neighbor peak in SmS and SmP. The SmS spectra has been adjusted to account for (i) the known difference in the near neighbor spacings, (ii) the differences in backscattering amplitude and phase using the theoretical calculations [4] and (iii) a larger SmS width of $\Delta\sigma^2 \approx 2.5 \times 10^{-3}$ Å^2. With these adjustments the SmS and SmP EXAFS should be equal provided that the E_0 is properly determined. Fig.3a shows the spectra when the E_0's are determined as the half height of the edge, i.e., the E_0 of the SmP is 7 eV higher than that of SmS. For comparison, Fig.3b shows the spectra when the E_0's are set equal to the threshold of SmP. The fact that the spectra agree quite well for thresholds differing by 7 eV and disagree for equal thresholds implies that the E_0's for the differing valence states of Sm are thus expected to differ by 7 eV. Intermediate valence $Sm_{1-x}Y_xS$, nonetheless, does exhibit a single E_0 equal to that of SmS. This single threshold could be understood if there is a rapid relaxation of $4f^5$ to $4f^6$ in the presence of the 2p core hole. (It has been proposed that such a rapid ($\tau < 10^{-16}$ sec) relaxation does occur in $YbAl_3(4f^{13} \rightarrow 4f^{14})$ in the presence of the 4d or 5p hole [5].) However, then the threshold energies for the EXAFS of SmP would be expected to equal that of SmS. An adequate explanation of the threshold puzzle awaits a detailed description of the mixed valence ground state and the effect of the 2p core hole on this state.

We thank J. Schweitzer for the SmP sample. Some of the materials incorporated in this work were developed at the Stanford Synchrotron Radiation Laboratory which is suported by the National Science Foundation (under Contract No. DMR77-27489), in cooperation with SLAC and the Department of Energy.

[1] J. B. Boyce, R. M. Martin, J. W. Allen and F. Holtzberg, in *Valence Fluctuations in Solids*, ed. by L. M. Falicov, W. Hanke and M. B. Maple (North-Holland, Amsterdam, 1981), p. 427; and references contained therein.

[2] See R. M. Martin, J. B. Boyce, J. W. Allen and F. Holtzberg, Phys. Rev. Lett. **44**, 1275 (1980) and references contained therein.

[3] W. Kohn, T. K. Lee and Y. R. Lin-Liu, Phys. Rev. B**25**, 3557 (1982).

[4] B.-K. Teo and P. A. Lee, J. Am. Chem. Soc. **101**, 2815 (1979).

[5] W. F. Egelhoff, Jr., and G. G. Tibbetts, Phys. Rev. Lett. **44**, 482 (1980).

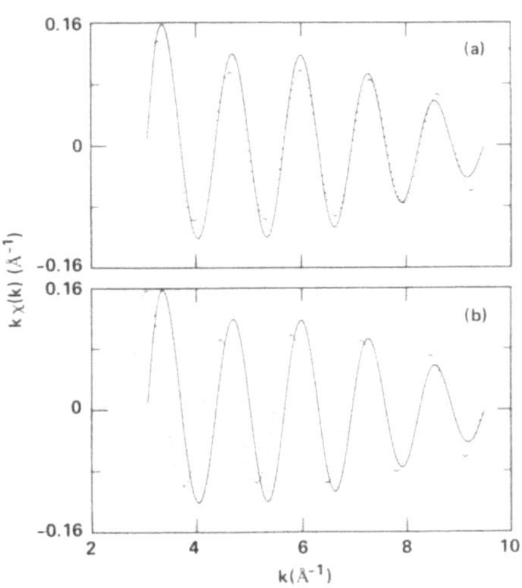

Fig.3. The Fourier filtered EXAFS at 77K due to the first neighbor peak in SmP (solid line) and adjusted SmS (dotted line) for two different SmS threshold energies: (a) different from that of SmP by the known 7 eV shift and (b) equal to that of SmP. The SmS EXAFS has been adjusted to that of SmP by taking into account the known difference in near neighbor spacing, the difference in backscatter using calculated amplitudes and phase shifts and a larger damping in SmS

High-Pressure EXAFS Study on Mixed Valent SmS

G. Krill[1], J.P. Kappler[1], and J. Röhler[2]

LURE, CNRS et Univ. Paris Sud, Bât. 109 C, F-91405 Orsay Cedex, France

1. Introduction

Rare-earth (RE) mixed valent or intermediate valent (IV) compounds present strong phonon anomalies due to the large differences in ionic radii (∿ 0.2 Å) associated with the two valence states of the RE ions [1]. Usually the rate of the valence transition, which depends from the f-d hybridization energy (E_{hy}), is supposed to be small enough (of the same order as the phonon frequencies (∿ 10^{-2} eV)) to allow dynamic e - ph interaction. As EXAFS takes place on a shorter time scale (∿ 10^{-15} sec) than this fluctuation rate (∿10^{-13} sec), such experiments should be useful for the study of the instantaneous atomic configuration surrounding a RE ion in a given valence state [2,3]. The real situation is however not so simple and, as discussed by KOHN et al. [4], the competition between the hybridization and elastic (E_{el}) energies must be considered. If $E_{hy} \gg E_{el}$ the position of the ions surrounding each RE is strongly reduced and even smaller than the EXAFS sensitivity (∿ 0.05 Å). This could explain the fact that no effect has been detected in earlier experiments [2,3]. The estimation of E_{hy} or E_{el} is not easy, moreover these energies are valence dependent ; it is thus difficult to predict what should be the better conditions (type of compounds, values of the thermodynamical parameters, etc.) to observe these dynamic effects in EXAFS experiments.

Recently we published results on IV $SmS_{1-x}O_x$ compounds [5] where the occurrence of a strong damping on the EXAFS Sm-Sm nearest neighbours contribution was discussed. The aim of this communication is to investigate the existence of a similar effect in the high pressure phase of samarium sulphide (SmS) which is the best known example of IV material.

2. Experimental

Experiments have been performed at LURE using the X-ray beam delivered by the DCI storage ring which was operated at 200 mA and 1.72 GeV. All the experimental details concerning our high pressure device have been presented in a previous paper [6]. The EXAFS spectra were taken at T = 300 K for 0, 10, 24, 40 and 72 kbar. Due to the presence of iron impurities in the beryllium high pressure cell, the EXAFS data are limited in energy to the K edge of iron, i.e. ∿ 350 eV above the L_{III} absorption edge of samarium (k ∿ 10 $Å^{-1}$). In order to obtain a good statistic the time of data acquisition was at least 1 hour for each EXAFS spectrum.

---

[1]LMSES (LA 306) ULP, 4, rue Blaise Pascal, F-67070 Strasbourg Cedex, France

[2]Physikalisches Institut, Zülpicherstr. 77, D-5000 Köln 41, Fed. Rep. of Germany

3. Experimental results

The valence of samarium in SmS can be deduced from the L_{III} absorption edges of Sm (see the paper by RÖHLER et al. in this volume). Near $P \simeq 8$ kbars the well-known first-order transition of SmS, where the valence changes from V = 2.1 to V = 2.6, is observed. The crystal structure remains NaCl-type in the whole pressure range.

The Fourier filtered EXAFS spectrum associated to the first S shell (each samarium is surrounded by a 6-sulphur-atoms octahedron) for SmS at P = 10 kbar (i.e. in the IV state) is shown on Fig.1a. The least square fit to the data, using the usual EXAFS formula (1),

$$\chi(k) = (k)^{-1}.N/R^2 \, |f(\pi,k)|.e^{-2R/\lambda(k)}.e^{-2\sigma k^2} \times \sin\left[2k\,R+2\delta_2(k) + \arg(f(\pi,k))\right] \qquad (1)$$

with $k = \hbar^{-1}\sqrt{2m(E-E_o)}$ where E_o is the threshold energy,

and the calculated phase shifts by LEE et al. [7], yield results which are consistent with those given by earlier crystallographical results [8]. On Fig.1b, the pressure behaviour of the lattice constant deduced from our EXAFS results on these first Sm-S shells ($a = 2*R_{Sm-S}$) is compared with that given by X-ray diffraction study [8]. We have no evidence for the existence of two Sm-S distances which should be, assuming that the sulphur atoms follow the change of the ionic radius of the RE, 0.15 Å different. These results are similar to those we report for IV $SmS_{1-x}O_x$ compounds [5] and the earlier ones obtained on IV $Sm_{0.75}Y_{0.25}S$ [3].

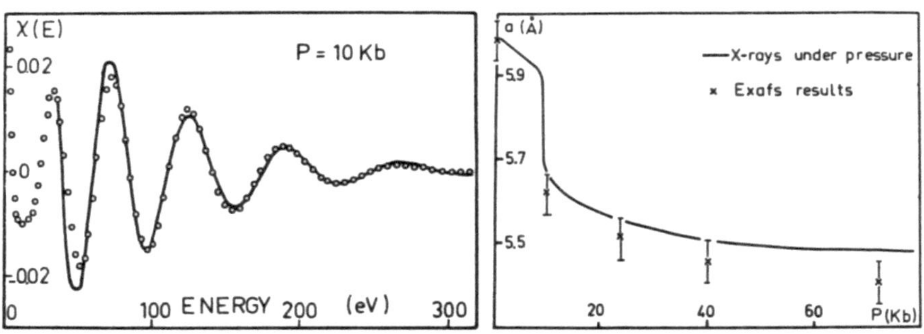

Fig.1 . (a) Fourier filtered EXAFS data for the S neighbours to the Sm ions in SmS at P = 10 kbar (b) Pressure dependence of the lattice constant of SmS as deduced from the EXAFS results

The situation for the twelve samarium ions surrounding the central Sm atom is rather more complicated , as indicated by the Fourier filtered EXAFS data for the second neighbours contribution (Fig.2) particularly in the IV regime (P = 10 kbar) ; similar results hold for P = 24 and 40 kbar. In the figure the full lines correspond to a simulation using (1) with one mean Sm-Sm distance and 12 neighbours. For P = 0 and 72 kbar this hypothesis is rather good, however it cannot be applied in the IV state (e.g. P = 10 kbar). The best fit (dashed line) in this case, is obtained taking into account two mean Sm-Sm distances ($R_1 \simeq 3.98$ Å and $R_2 = 4.25$ Å) and two types of Sm atoms ($N_1 = 8$ at and $N_2 = 4$ at). We can notice that these two Sm-Sm distances are roughly those we can expect for Sm^{2+}-Sm^{2+} and Sm^{3+}-Sm^{3+} pairs.

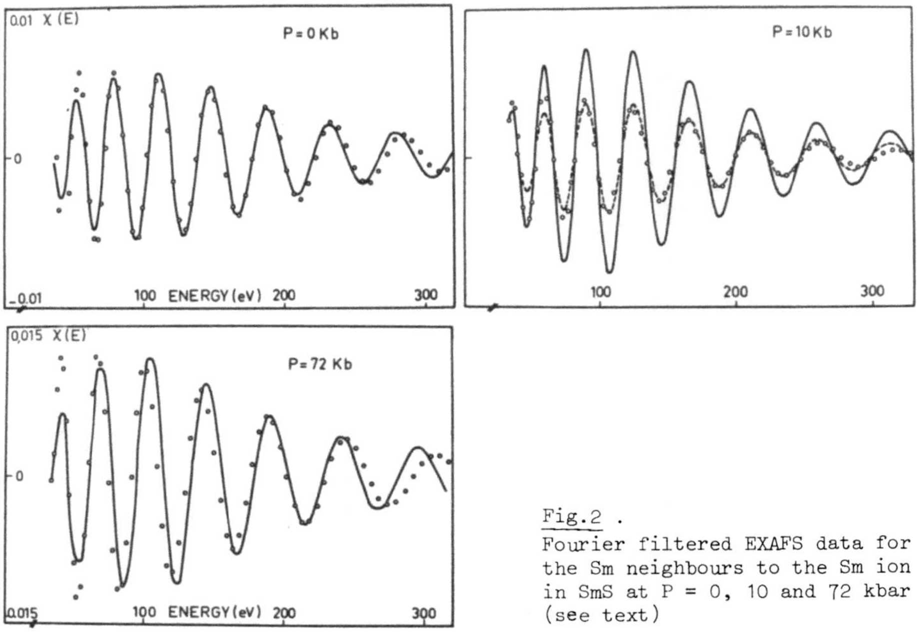

Fig.2 .
Fourier filtered EXAFS data for the Sm neighbours to the Sm ion in SmS at P = 0, 10 and 72 kbar (see text)

4. Conclusion

The results we obtain on the high pressure phase of SmS are identical to those we reported previously on the IV $SmS_{1-x}O_x$ [5]. When the valence approaches an integral value, no effects are observed on the EXAFS spectra. In the IV regime, and for these SmS type compounds, the effect seems to be only observable for the RE shells. The energy balance between E_{hy} and E_{el} must be favourable in this last case. We believe that for other systems an effect should be observed for all shells ; experiments on other IV materials like $EuPd_2Si_2$, $YbAl_2$ etc... are now in progress.

1. Valence Fluctuations in Solids, Int. Conf. of Santa Barbara, North Holland Pub. Comp. Amsterdam (1981).
2. H. Launois, M. Raviso, E. Holland-Moritz, R. Pott and D. Wohlleben, Phys. Rev. Lett. 44, 1271 (1980).
3. R.M. Martin, J.B. Boyce, J.W. Allen and F. Holtzberg, Phys.Rev.Lett. 44, 1275 (1980).
4. W. Kohn, T.K. Lee and Y.R. Lin-Lin, Phys. Rev. B 25, 3557 (1982).
5. G. Krill, J.P. Kappler, J. Röhler, M.F. Ravet, J.M. Léger and F. Gautier, in : Proc. of the Int. Conf. on Valences Instabilites (P. Wachter, ed), Norht Holland (1982).
6. J. Röhler, J.P. Kappler and G. Krill, in : Proc. of the Int. X-Ray and VUV-Synchrotron Radiation Instrumentation Conf. (E. Koch, ed), to be published in Nuclear Instruments and Methods in Physics Research.
7. P.A. Lee, P.U. Cibrin, P. Eisenberger and B.M. Kinward, Rev. Mod. Phys. 53, 769 (1981).
8. A. Chatterjee, A.K. Singh and A. Jayaraman, Phys. Rev. B 6, 2285 (1972).

Peculiarities of the EXAFS Spectra of III-VI Layered Compounds

Mario Piacentini
Gruppo PULS, Laboratori Nazionali dell'INFN,
I-00044 Frascati, Italy

1. Introduction

The III-VI layer compounds GaS, GaSe, and InSe crystallize into stacked hexagonal layers, each consisting of four close-packed monatomic sheets in the sequence anion-cation-cation-anion. The crystallographic structure of a single layer is the same for all compounds but, due to the weakness of interlayer bonds, several polytypes have been observed, corresponding to different stacking sequences of the layers (1).

The electronic properties of these materials have been studied extensively. Recent band structure calculations reproduce accurately their valence bands, that have been mapped experimentally with angular resolved photoemission measurements (2). However, the knowledge of their conduction bands, in particular of the atomic-like symmetry of the conduction states, is not as much advanced.

The aim of this research was twofold:

i) to check if EXAFS measurements and analyses are simpler tools than x-ray diffraction for polytype determination.

ii) to probe the conduction band projected densities of states measuring accurately the near-edge structures at the Ga K edge, Se K edge and In L edges in GaS, GaSe, and InSe respectively.
The highly polarized x ray synchrotron radiation was essential for both purposes. The interpretation of our results - some of which are still preliminary - has revealed some peculiarities in both the EXAFS and the near-edge structures, as we shall see below.

2. The Experiment

The measurements were performed at the x-ray beam line of the PULS laboratory using the synchrotron radiation emitted by the electrons orbiting in the 1.5 GeV storage ring ADONE (3). GaS, GaSe, and InSe single crystals, supported on adhesive tape, were cleaved to a few μm thickness along the basal planes, with the crystal $\underline{c}$ axis perpendicular to the surfaces. The samples were mounted on a vertical shaft that could be rotated with respect to the incident beam and the transmission was measured at several angles of incidence i with p polarization. With this geometry, as shown in fig. 1, the direction of the x-ray polarization versor $\underline{\varepsilon}$ could be changed with respect to the crystal $\underline{c}$ axis.

We performed several checks on the quality of our measurements in order

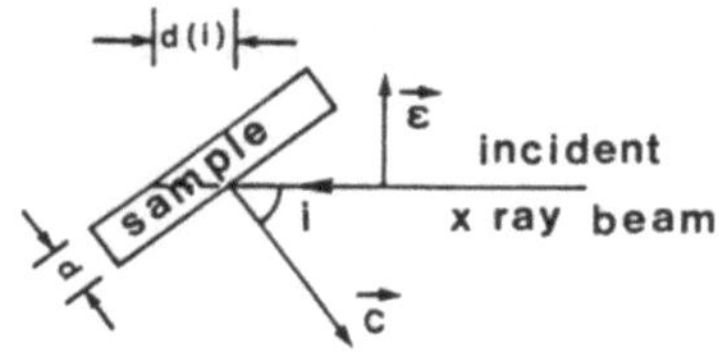

Fig.1. Experimental geometry

to be sure to observe true polarization effects. First, the smooth atomic-like background of all spectra have been normalized to the normal incidence spectrum near the threshold and the normalization constant gave the nominal value of i within ± 1°. This check is based on the isotropy of the atomic absorption coefficient μo, so that in the x-ray region the optical density measured at the angle i is given by

$$\mu o\ d(i)=\mu o\ d(0°)/\cos i,$$

as verified easily from fig. 1. Second, after normalization, the atomic background is the same for all spectra over the whole experimental energy range. Finally, the absorption spectrum of GaS, measured at about 50° s polarization with a special sample holder, agreed perfectly well with the spectra measured at normal incidence. In both cases $\varepsilon \perp c$ and no polarization effects should be present. In fig. 2 the Ga K absorption spectra measured in GaS at 0° and 75° p polarization, are shown to be in very good agreement with each other, except for the noticeable difference between the EXAFS.

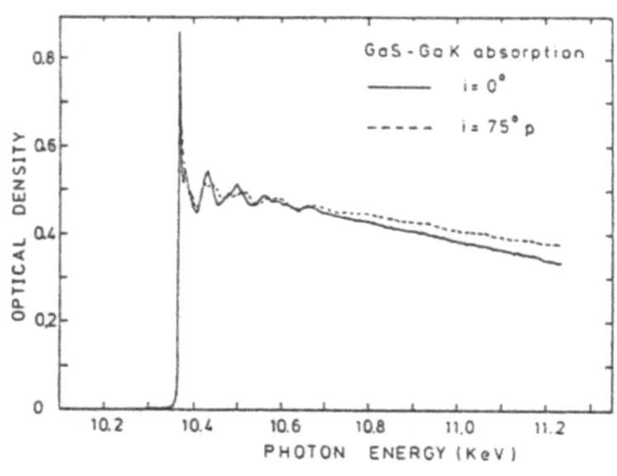

Fig.2. Normalized optical densities of the Ga K absorption in GaS measured at normal incidence and 75° p polarization

3. The Near-Edge Structures

A remarkable dichroism was found at the Ga K absorption edges in both GaS and GaSe, as shown in fig. 3, but not at the In $L_{2,3}$ edges shown in fig.4 (4). This different behavior is easily understood by recalling that taking the crystal c axis as the quantization z axis, dipole transitions are allowed between s and $p_x p_y$ states for $\varepsilon \perp c$ and p_z states for $\varepsilon // c$. The crystal anisotropy splits by a significant amount the empty p-like states into $p_x p_y$ and p_z levels, while the In 2p core states remain practically degenerate. Thus, only the excitation of the core s electrons (Ga 1s and In 2s in the present experiments) must be polarization dependent.

The first structure labelled A in fig. 3 at the Ga K threshold disappears when the polarization is changed from $\varepsilon \perp c$ to $\varepsilon // c$, indicating that the associated final states have $p_x p_y$ symmetry. According to the projected densities of states calculated by Robertson (5) for GaSe with a tight-binding scheme, Ga 4s states contribute to the lowest conduction

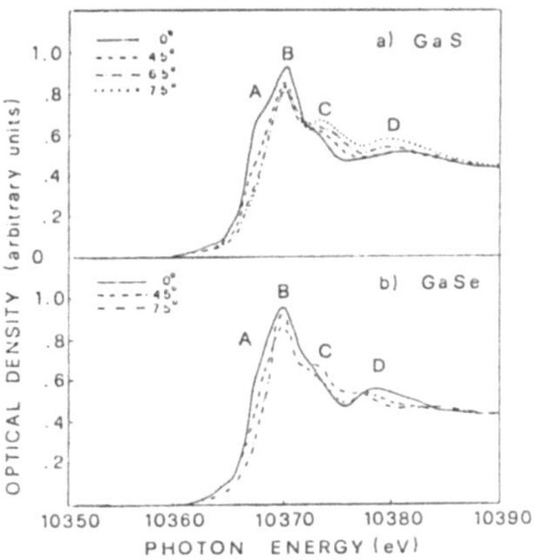

Fig.3. Optical densities at the Ga K absorption edges in GaS (a) and GaSe (b) measured at several angles of incidence, p polarization. The 1.2 eV half-width at half maximum found for all the structures is caused by both the intrinsic lifetime broadening of the Ga 1s (0.8 eV) and the instrumental resolution (0.8 eV at about 10 KeV)

bands and the Ga empty $4p_xp_y$ states form a group of bonding-antibonding bands 3-4 eV above. Due to the strong similarity between the energy bands of these III-VI compounds (2), we assign the structure A of fig. 3 to the latter final states and the threshold peak at 3654 eV of the In L_3 spectrum to the lowest conduction band. The second, weak structure of the In L_3 spectrum, 2.5 eV above, may be assigned to the In $5p_xp_y$ bands and may be the counterpart of the structure A of the Ga K edge. It has a finite transition probability, probably because some hybridization of the In $5p_xp_y$ states with high-lying s states (neglected in Robertson's calculation (5)) occurs at general points of the Brillouin zone. The smooth rise, with possibly minor features superimposed, beginning approximately 5 eV above the In L_3 threshold, may be due to the onset of final d states.

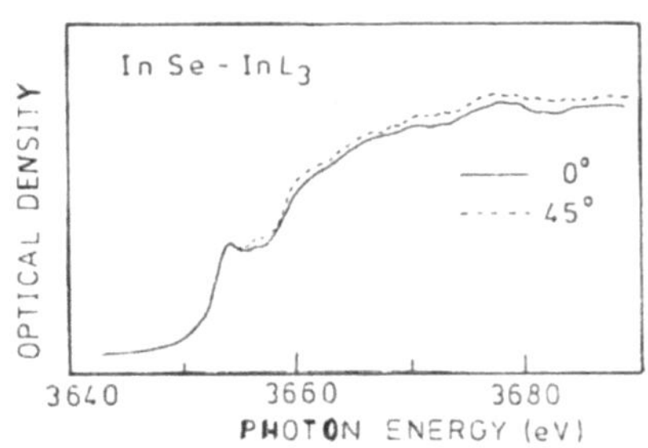

Fig.4. Optical densities at the In L_3 edge in InSe measured at 0° and 45° p polarization. The In L_2 spectrum, displaced at higher energies by the L_2-L_3 spin orbit splitting of 211.3 eV and 0.5 times weaker, is the same as the In L_3 spectrum (4)

4. The EXAFS

The general EXAFS expression for an absorbing K electron, polarized radiation and a single shell of backscatterers (6) becomes

$$k\,\chi_{\perp}(k) = -\left[\frac{3N_j \sin^2\theta_j}{r_j^2}\right]\left[e^{-\frac{2r_j}{\lambda}}\; e^{-2\sigma_j^2k^2}\right]\left[|f_j(\pi)|\sin(2kr_j+\Phi_j(k))\right] \qquad 1$$

$$k\,\chi_{/\!/}(k) = -\left[\frac{3N_j \cos^2\theta_j}{r_j^2}\right]\left[e^{-\frac{2r_j}{\lambda}}\; e^{-2\sigma_j^2k^2}\right]\left[|f_j(\pi)|\sin(2kr_j+\Phi_j(k))\right] \qquad 2$$

in the two limit cases of $\underline{\varepsilon} \perp \underline{c}$ and $\underline{\varepsilon} // \underline{c}$, respectively. θ_j is the polar angle between the $\underline{c}$ axis and the vectors $\underline{r}_j$ connecting the absorbing atom with the atoms in the neighbor shell. The other symbols have the usual meaning (6). The different factors of the EXAFS formula have been grouped in eqs. 1 and 2 within square brackets into three major terms: i) the geometrical factor; ii) the damping factor; and iii) the phase factor

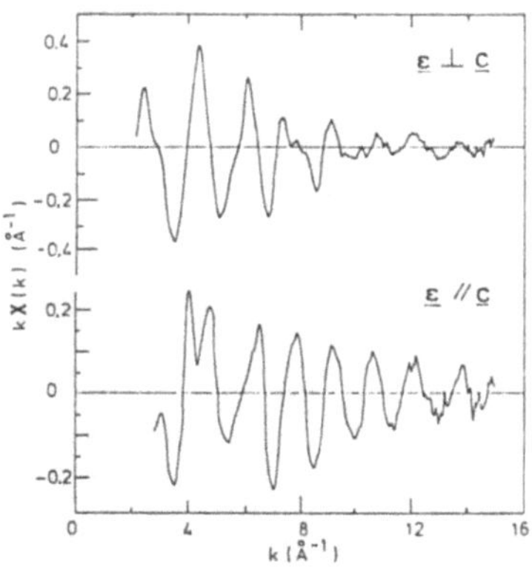

Fig.5. EXAFS spectra associated with the Ga K absorption in GaS for $\underline{\varepsilon} \perp \underline{c}$ and $\underline{\varepsilon} // \underline{c}$

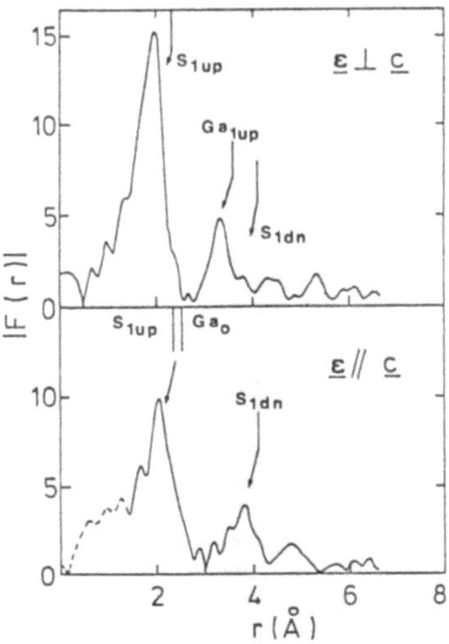

Fig.6. Absolute values of the Fourier transforms of the EXAFS spectra shown in fig.5, calculated without correcting for the phase shifts. The vertical bars indicate the positions of neighbor shells, labelled according to fig. 7, as obtained from the crystallographic parameters

In fig. 5 the EXAFS spectra $k\chi_{\perp}(k)$ and $k\chi_{//}(k)$ of GaS, associated with the Ga K absorption, are presented. The latter spectrum was obtained by subtracting the EXAFS spectrum measured at normal incidence from those measured at the largest angles of incidence (7). The corresponding Fourier transform functions $|F_{\perp}(r)|$ and $|F_{//}(r)|$, shown in figs. 6a and 6b respectively, reflect the differences of the $k\chi(k)$ spectra of fig. 5. In fig. 6 also the positions of neighbor shells as determined from crystallographic data have been indicated with vertical bars near the corresponding peaks of $|F(r)|$. The EXAFS distances are shorter than the diffraction distances, since we did not correct the Fourier transforms for the phase shifts. In fig. 7 the projection on the (110) plane of the crystal structure of β-GaS is shown, and the shells identified in fig. 6 are indicated.

In agreement with eqs. 1 and 2, the shell of 6 Ga atoms lying in the same cation sheet as the absorber, labelled Ga_{1up} in fig. 7, contribute to $F_{\perp}(r)$ only, while the Ga atom below the absorber shows up only in the $F_{//}(r)$ spectrum. The weak structure at 3.80 Å in $|F_{\perp}(r)|$, corresponding to the S_{1dn} shell of fig. 7, is strongly enhanced in the opposite polarization, their intensity ratio being $|F_{\perp}(r)|/|F_{//}(r)|=0.33\pm0.02$, in very good agreement with the theoretical value $tg^2\theta_{3.80\text{Å}}=0.35$.

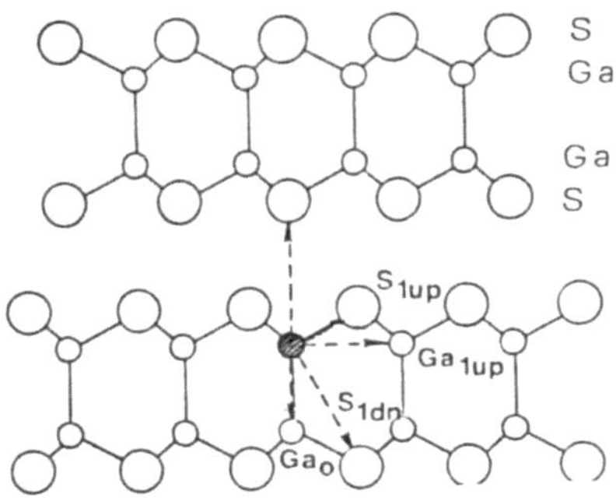

Fig.7. Projection of the crystal structure of β GaS on the (110) plane. The hatched Ga atom is the absorber and the neighbor shells identified in fig. 6 are indicated

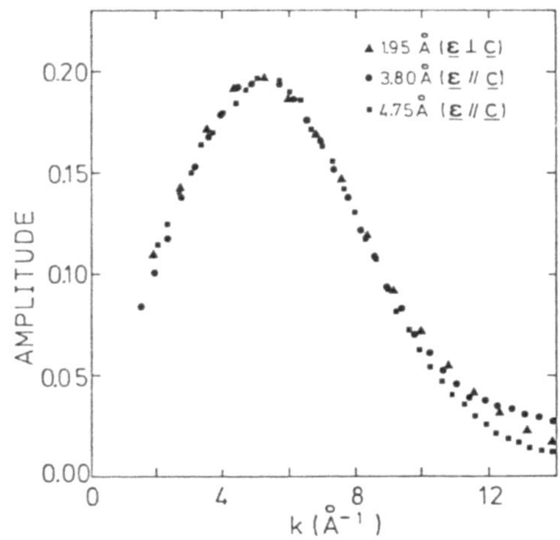

Fig.8. Inverse Fourier transform normalized amplitudes of the peaks in $|F(r)|$ corresponding to the three S shells at 1.95 Å, 3.80 Å, and 4.75 Å

We have performed also the inverse Fourier transforms of some structures of the two F(r) spectra. The amplitude functions A(k) of several well-identified sulfur shells, reported in fig. 8, are in very good agreement with each other after normalization. The peak at 2 Å in $|F_{//}(r)|$ is generated by the two, unresolved shells at 2.33 Å (S_{1up} of fig. 7) and at 2.45 Å (Ga_o in fig. 7) (8). The inverse Fourier transform shows a first beating minimum at k_m=4.78 $Å^{-1}$, from which the separation between the two shells was calculated to be ΔR_{Ga-S}=0.12 Å, in perfect agreement with the crystallographic value, using the relation (9)

$$\Delta R_{ij}=(\pi-\Phi_i-\Phi_j)/2\ k_m \qquad 3$$

where Φ_i and Φ_j are the phase shifts at k_m of the two beating shells. The phase shifts for S and Ge (instead of Ga) tabulated by Teo and Lee (10) have been employed in evaluating eq. 3.

5. Discussion and conclusions

In the previous sections we have discussed mainly well-understood aspects of the dichroism of the x-ray absorption of the III-VI layer semiconductors. Several near-edge structures of the Ga K thresholds in GaS and of the In L_3 threshold in InSe have been assigned to peaks of the conduction band projected densities of states (4). Peaks of $|F_\perp(r)|$ and $|F_{//}(r)|$ follow closely eqs. 1 and 2, as summarized in Tab. 1. In this section we present some points that require a deeper analysis and that are the subject of further investigations.

i) Near the Ga K thresholds in GaS and GaSe also the structures labelled C and D in fig. 3 depend on the polarization. At present we are not able to say if the projected density of states model is sufficient for interpreting them, or if it is necessary to use more sophisticated calculations,

Table I. EXAFS results for the Ga K absorption in GaS. The first column gives the atomic neighbor shells, labelled according to fig. 7. N_j, θ_j and R_j are the coordination numbers, the polar angles and the crystallographic (EXAFS) distances of each shell respectively (distances in Å). The next columns give the results for $|F_{\perp}(r)|$ and $|F_{//}(r)|$. I_T gives the geometrical factors calculated according to eqs. 1 and 2. I_{EXAFS} are the intensities of the experimental peaks, normalized to the † value

					$F_{\perp}$		$F_{//}$		
Shell	N_j	θ_j	R_j	R_j(EXAFS)	I_T	I_{EXAFS}	R_j(EXAFS)	I_T	I_{EXAFS}
S_{1up}	3	62.7	2.33	1.95	0.44	0.49	ΔR_j=0.12	0.12	
Ga_0	1	0	2.448	--	0	0		0.17	
Ga_{1up}	6	90	3.587	3.32	0.47	(0.12)	---	0	
S_{1dn}	3	30.5	4.081	3.80	0.046	0.047	3.80	0.134†	0.134†
S_{2up}	3	75.6	4.278		0.154			0.010	
S Next layer	1	0	4.229	--	0		4.75	0.056	0.05

based on the multiple scattering formalism (11). In addition, the white line at 10370 eV shows a rather peculiar behavior, since it is isotropic within our sensitivity and it seems to be a feature common to other Ga compounds, such as GaAs (12).

ii) If we consider the trigonal prismatic cluster S_3-Ga†-Ga-S_3 around the excited Ga† atom as the fundamental cluster for GaS, it appears from tab. 1 that the intensities of the peaks of $|F_{\perp}(r)|$ and $|F_{//}(r)|$, associated with backscatterers inside the fundamental cluster, scale with respect to each other only for the geometrical factors of eqs. 1 and 2. The variation with the shells of the damping factors, in particular of the mean free path term, seems unimportant. Instead, all the shells outside the fundamental cluster are strongly damped. This is already evident for the two Ga shells Ga_o and Ga_{1up}, whose experimental intensities are reversed with respect to the respective geometrical factors. It is striking that the external sulfur shells do not contribute to the F(r) spectra shown in fig. 6. Similar effects have been observed also in the EXAFS spectra associated with atoms adsorbed on clean surfaces (13). Either an asymmetric pair distribution function due to an anharmonic potential (14), or complete treatment of the Debye Waller factor, that includes the anisotropy as well as the existence of not coupled xy and z lattice vibrations, may be responsible for these peculiar effects.

iii) The $F_{//}(r)$ spectrum shows a strong feature at 4.75 Å, missing in the $F_{\perp}(r)$ spectrum. Such an angular dependence of this peak, together with the shape of the inverse Fourier transform amplitude and the intensity ratio with the other shells, suggests that the peak at 4.75 Å of $|F_{//}(r)|$ is due to a single S backscatterer lying along the c axis from the Ga absorber. GaS crystallizes only in the β modification, and only the S atom in the neighbor layer above the absorber (cfr. fig. 7) satisfies all three requirements. But the crystallographic distance (4.28 Å) is much shorter than the EXAFS determination, unlike all the other shells.

The major difference with respect to the other shells is that in this case the excited electron does not move across a region filled with the valence electrons, but across the region of almost free space existing between the layers.

Acknowledgments

The author is strongly indebted to Prof. F. Bassani for his interest in this research. He also acknowledges the collaboration of all the staff of P.U.L.S. for their collaboration at different stages of this research. Thanks are due also to Dr. D.W. Lynch, Iowa State University, for his hospitality, during which the author performed the analysis of the EXAFS spectra. P.U.L.S. is a program run jointly by the Italian National Research Council (CNR-GNSM) and the Nuclear Physics National Institution (INFN).

References

1. F. Hulliger, Structural Chemistry of Layer-Type Phases, Physics and Chemistry of Materials with Layered Structures, vol. 5 (Reidel, Dordrecht 1976).
2. E. Doni, R. Girlanda, V. Grasso, A. Balzarotti and M. Piacentini; Nuovo Cimento 51B, 154 (1979).
3. A. Balzarotti, F. Comin, L. Incoccia, S. Mobilio; LNF report (1980) unpublished.
4. F. Antonangeli, M. L. Apicella, A. Balzarotti, L. Incoccia and M. Piacentini, Physica 105B, 25 (1981).
5. J. Robertson, J. Phys. C: Solid State Phys. 12, 4777 (1979).
6. P.A. Lee, Phys. Rev. B 13, 5261 (1976).
7. F. Antonangeli, A. Balzarotti, L. Incoccia and M. Piacentini; to be published.
8. The two shells are not resolved in $|F_{//}(r)|$ but they are in both the real and the imaginary parts of $F_{//}(r)$ (Ref. 7).
9. A. Balzarotti, F. Comin, L. Incoccia, S. Mobilio, M. Piacentini and A. Savoia:"Inner Shell and X Ray Phys. of Atoms and Solids",D.J.Fabian, M.Kleinpoppen and L.M.Watson editors, Plenum Press (1981) p. 723.
10. B.K. Teo and P.A. Lee; J. Am. Chem. Soc. 101, 2815 (1979).
11. P.J. Durham, J.B. Pendry and C.M. Hodges; Solid State Commun. 38, 159 (1981) and Comp. Phys. Commun. 25, 193 (1982)
12. M. Piacentini; (unpublished).
13. F. Comin, these proceedings.
14. P. Eisenberger and G.S. Brown, Solid State Commun. 29, 481 (1979).

EXAFS Studies of Disorder in Crystalline Solids

C.R.A. Catlow and L.M. Moroney

Chemistry Department, University College London, 20 Gordon Street, London, UK

A.V. Chadwick

Chemical Laboratory, University of Kent at Canterbury, Kent CT2 7NH

G.N. Greaves

S.E.R.C., Daresbury Laboratory, Daresbury, Warrington WA4 4AD, UK

1. Introduction

This paper describes results of EXAFS experiments to determine the nature of the local structural environment adopted by the trivalent ions Er(III) and La(III) in doped CaF_2. Some structural information about these systems has been obtained from neutron diffraction data [1-3] and on the basis of this and the results of computer simulations of the systems [1,4,6,7], different models have been proposed for the local dopant environment.

However, neutron diffraction measurements represent a sum over all the atomic species and the structural data obtained is of the average unit cell contents. EXAFS is unique in its ability to provide short range structural information and it was decided to exploit this facility to determine if EXAFS experiments could confirm the structural models proposed for the Er and La doped CaF_2.

2. Models for doped CaF_2

CaF_2 has the fluorite structure and can be readily doped with up to 30 mole % of trivalent rare earth ions which considerably enhance the conductivity of these compounds. Spectroscopic studies have shown that the dopant ion, when present in low concentration (i.e. < 0.5 mole %) occupies a lattice cation site with the extra charge compensated for by an additional fluoride ion in a neighbouring interstitial cube-centre site [4].

For higher dopant concentrations, it is thought that aggregations of these simple pairs occurs, which is accompanied by relaxations of lattice ions in the vicinity of the dopant, leading to the formation of more complex clusters. The locations of the ions in these clusters have been determined by HADES calculations of the energetically most stable configurations [1,3,5,7]. These parameters provide a basis for the calculation of theoretical EXAFS structure with which to compare the observed data, if accurate estimates of the Debye-Waller factors can be obtained.

3. Calculation of Debye-Waller factors for CaF_2

The dynamics of an ionic solid can be modelled by the computer programme FUNGUS [8] and estimates of the Debye-Waller factors for each atom calculated. The programme takes a set of structural co-ordinates; in this case, 324 atoms in 27 unit cells of CaF_2 were used, and in the initial step, each atom is given a random velocity. The force on each ion is calculated, assuming that the interaction energy between each pair of ions can be described by an analytic Born-Mayer potential [9,10]. The trajectories of each ion are calculated and new ionic positions determined after a set time interval.

This process is repeated for a number of time steps until thermal equilibrium is reached and the velocity and position of each ion is randomised. From this point, the variances in distance from a central calcium ion to all other ions in the 27 unit cell set were calculated over a period of 10^{-11} seconds. They were averaged to give estimates of the Debye-Waller factors for the first 15 shells (Fig.1)

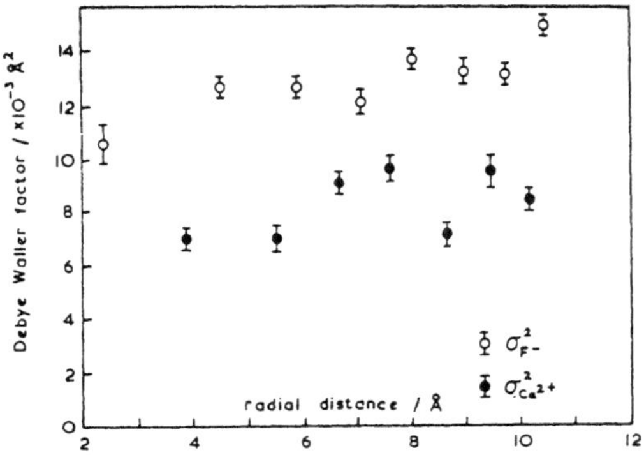

Fig.1. The variation in Debye-Waller factors for Ca^{2+} and F^- ions in CaF_2 with radial distance as calculated from the molecular dynamics programme

4. Comparison of predicted EXAFS with experimental data

The EXAFS of the L_{III} edges of La and Er in the samples of 5 and 10 mole % La(III)-doped CaF_2 and 10, 17, and 30 mole % Er(III) doped CaF_2, showed that the structural environments of the two dopant ions are different.

Comparing the theoretical EXAFS, calculated using the plane wave approximation and programmes in the Daresbury SRS programme library for the various postulated models, with the experimental EXAFS of the Er L_{III} edge, it was seen that the cluster consisting of six dopant ions in an octahedral configuration provided the closest match to the fine structure. Using this model as a starting point for a least squares fit to the data to determine the radial distances precisely, the fit depicted in Fig.2 was obtained. This shows a reasonable degree of similarity between observed and calculated fine structure for the high energy region where the plane wave approximation is valid. The discrepancies between the values of the radial distances from the fit and those calculated from the energetically most favourable configuration, are small.

The comparison of the plane wave EXAFS calculated for the various models with the L_{III} La edge EXAFS showed the best match to the data with the simple dopant-interstitial pair. However, this model requires the occupancy of a cube centre interstitial site by a fluorite ion. The neutron diffraction data for this sample showed no evidence of occupation of these sites. Of the remaining possible models, the structure designated $2/3_1$ [6] which involves a pair of dopants surrounded by three interstitial F^- ions, most closely resembled the experimental EXAFS. It should be noted that this cluster is negatively charged and electroneutrality is probably preserved by the presence of isolated La(III) substitutional ions. Adding the plane wave EXAFS expected for the $2/3_1$ cluster and the EXAFS for such an isolated substitutional La(III) ion in the appropriate ratio to achieve electroneutrality, a substantially closer match to the experimental data is achieved.

Using this in the fitting programme, Fig.3 is obtained. Once again the difference between the fitted and predicted radial distances is small (i.e.

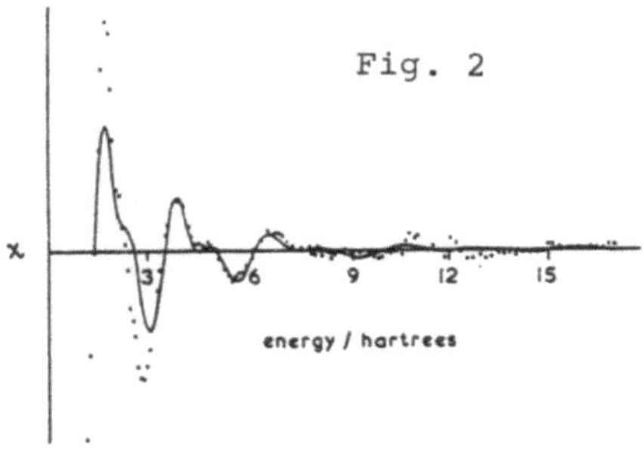

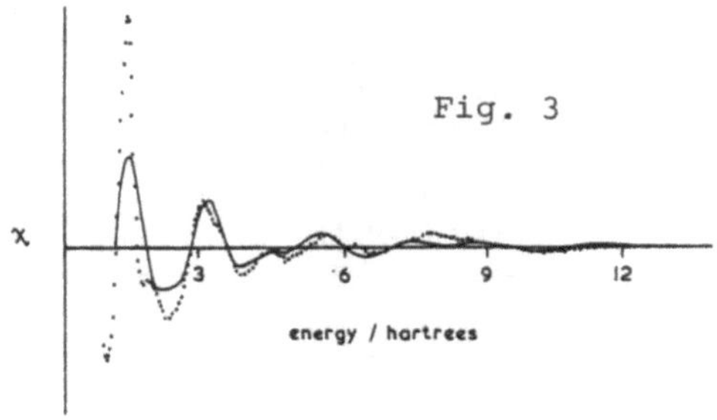

Fig.2. EXAFS of the L_{III} edge of 30 mol % Er-doped CaF_2 and plane wave EXAFS for the octahedral cluster of Er ions

Fig.3. EXAFS of the L_{III} edge of 10 mole % La-doped CaF_2 and plane wave EXAFS for the $2/3_1$ cluster plus charge compensating La substitutional ions

less than 0.03 Å for inner shells). While the fit is not ideal, the major features are reproduced. It is probable that the discrepancies in the fit are due, in part, to shadowing of outer shells in the cubic fluorite structure by inner shells. Furthermore, accurate calculation of the phase shifts used in the plane wave approximation of the EXAFS would require the inclusion of an extra electrostatic term to account for the charge on the cluster.

5. Conclusion

This work has shown that EXAFS is a valuable experimental technique for detecting the different local environments surrounding dopant ions in CaF_2. It has been found that it supports models proposed for the La- and Er-doped CaF_2 based on neutron diffraction data and computer simulation studies.

6. References

1. C.R.A. Catlow, A.V. Chadwick, J. Corish, J.Solid State Chemistry, 1982, in press
2. A.K. Cheetham, B.E.F. Fender, M.J. Cooper, J.Phys. C., 4, (1971), 3107
3. C.R.A. Catlow, J. Corish, A.V. Chadwick, Proceedings of the Fourth Europhysical Topical Conference; Lattice Defects in Ionic Crystals. Radiation Effects (1982)
4. D.S. Moore, J.C. Wright, J. Phys. Chem., 74, 1626, (1981)
5. C.R.A. Catlow, J. Phys. C., 4, L64, (1973)
6. J. Corish, C.R.A. Catlow, P.W.M. Jacobs, S.H. Ong, Phys Rev. B. 25 6425, (1982)
7. P.J. Bendall, D. Phil. Thesis, Univ. of Oxford, 1980
8. J.R. Walker, Computer Simulation in the Physics and Chemistry of Solids ed. C.R.A. Catlow and W.C. Macrodt (Berlin:Springer-Verlag) 1981
9. M. Dixon, M.J. Gillan, J. de Physique, 41, C6-24, (1980)
10. C.R.A. Catlow, M.J. Norgett, J.Phys. C. 6, 1325, (1973)

EXAFS Studies of Krypton Atoms in Matrices of Argon and Nitrogen

W. Malzfeldt, W. Niemann, and P. Rabe
Institut für Experimentalphysik der Universität Kiel,
D-2300 Kiel, Fed. Rep. of Germany

N. Schwentner
Institut für Atom- und Festkörperphysik, Freie Universität Berlin,
D-1000 Berlin, Fed. Rep. of Germany

Abstract

EXAFS measurements at the Kr K-edge (E=14.33 keV) of solid Kr and of Kr in Ar and N_2 at a temperature of 23 K with Kr concentrations between 0.1% and 50% are reported. The measurements have been performed with a double-crystal monochromator at the synchrotron radiation facility HASYLAB in Hamburg. The following results are obtained: a) in solid Kr the known nearest-neighbour distance is confirmed. b) The system 1% Kr/Ar shows a nearest-neighbour distance of 3.81 Å. The Kr atoms take substitutional sites in the Ar lattice. c) In the system Kr/N_2 Kr clusters are observed at all concentrations.

1. Introduction

The spectroscopy of atoms isolated in solid matrices has received considerable interest[1]. Many properties of matrix isolated systems are determined by the local structure around the guests. Conventional techniques for structural analysis fail for these systems: whereas the matrix in general is crystalline, the sites of the guests are statistically distributed in the host. Therefore the geometrical structure is either unknown or has been indirectly concluded from spectroscopic data (e.g. luminescence spectra[2]). EXAFS experiments can provide valuable information about the local structure around the isolated atoms and molecules. This has been shown recently by a study of iron monomers and dimers isolated in solid argon and neon[3,4]. In this paper we report EXAFS measurements of solid Kr and of various concentrations of Kr isolated in matrices of Ar and N_2 at a temperature of 23 K at the K edge of Kr (E=14.33 keV).

In the next section we describe the experimental set-up, and in the last section we present the experimental results.

2. Experiment

The X-ray absorption spectra were measured at the synchrotron radiation facility HASYLAB in Hamburg with a new separated function double-crystal monochromator. With the use of Si 220 crystals an energy range from 6 keV to 17 keV is available. In the vicinity of the Kr K-edge the energy resolution was 3.5 eV. Harmonic rejection was achieved by detuning one crystal with respect to the other. Details about the experimental set-up are reported elsewhere[5]. The samples were prepared in a high-vacuum chamber by freezing the gas mixture on a thin aluminium foil which was cooled to about 23 K by a helium refrigerator. The sample chamber was placed between two ionisation chambers which measure the incident and the transmitted intensities, respectively.

3. Results

The fine structure $\chi(k)$ has been extracted from the absorption background in a conventional way[6]. A Fourier transform of these data shows distinct peaks at distances $R=R_i-\alpha_i$ where R_i is the true interatomic distance and α_i is the linear parameter of the scattering phase shift. No theoretical phase calculations are available for Kr and Ar. Therefore we have used values obtained by interpolation of parameterized phases given by[7]. Utilizing these phases typical errors in the determination of interatomic distances amount to $\pm$ 0.05 Å.

a) Solid Krypton

In solid Kr the EXAFS extends up to 500 eV above the K edge. The structures are damped by a large Debye-Waller factor (Fig.1). The Fourier transform exhibits a large first peak at R=3.69 Å (Fig.2) which gives R_1(Kr-Kr)=4.03 Å after correcting for the phase shift of α(Kr-Kr)=0.34 Å. This nearest neighbour distance agrees in the margins of error with the known distance in solid Kr of 4.00 Å at T=23 K[8].

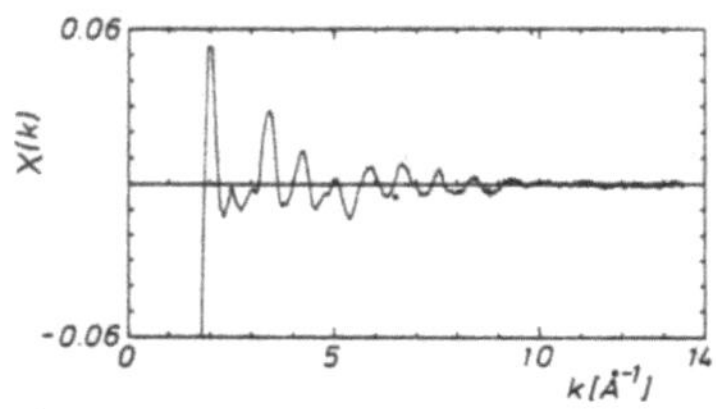

Fig.1. K-edge absorption fine structure of solid Kr at T=23 K

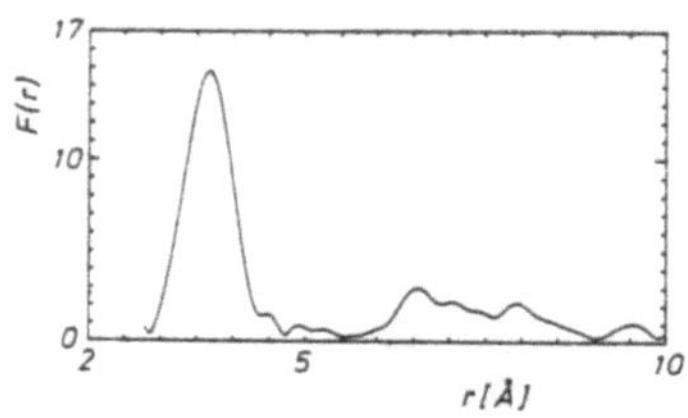

Fig.2. Fourier transform of the spectrum in Fig.1

b) Krypton in Argon

In solid Ar the nearest-neighbour distance at T=23 K is 3.77 Å[8]. Free KrAr molecules have an interatomic distance of 3.82 Å[9]. If Kr atoms are trapped in Ar the fcc lattice structure of the matrix is expected to be maintained, and the Kr-Ar distance should be close to the molecular value at not too high Kr concentrations. The Fourier transform (Fig.4) of the EXAFS spectrum of 1% Kr/Ar (Fig.3) shows a strong nearest-neighbour peak at 3.32 A. Correcting for the phase shift of α(Kr-Ar)=0.49 Å we find R (Kr-Ar)=3.81 Å which is in good agreement with the value for the free molecule. The analysis of the other peaks in the Fourier transform shows that the local structure of the Ar lattice is maintained. We find second and third shells at 5.53 Å and 6.68

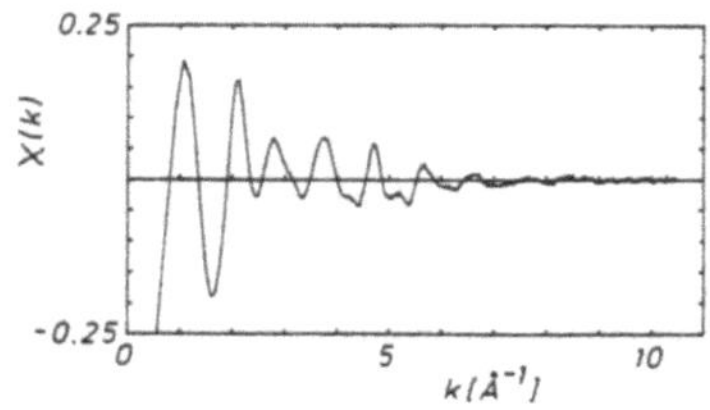

Fig.3. Absorption fine structure of the system 1% Kr/Ar at the Kr K-edge at T=23 K

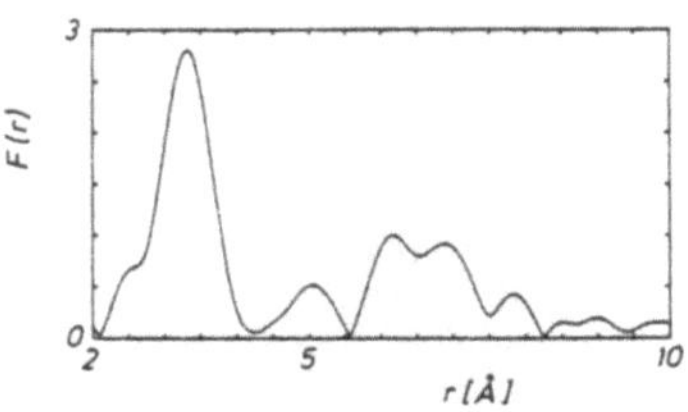

Fig.4. Fourier transform of the spectrum in Fig.3

Å. As expected, these values are between those in pure Ar and pure Kr. The geometrical structure observed at low concentrations is essentially for concentrations up to 10% Kr/Ar. We conclude that the Kr atoms take a substitutional site in the Ar crystal.

c) Krypton in Nitrogen

The spectrum Kr in N_2 shows a damped and noisy EXAFS signal (Fig.5). The Fourier transform exhibits two peaks at 3.0 Å and 3.7 Å which are relatively insensitive to the choice of window function and weighting of the spectrum. The peak at 3.70 Å occurs at all Kr concentrations from 1% to 50%. This peak position is close to the position of the first peak in solid Kr. Therefore we conclude that Kr clustering is observed in solid N_2.

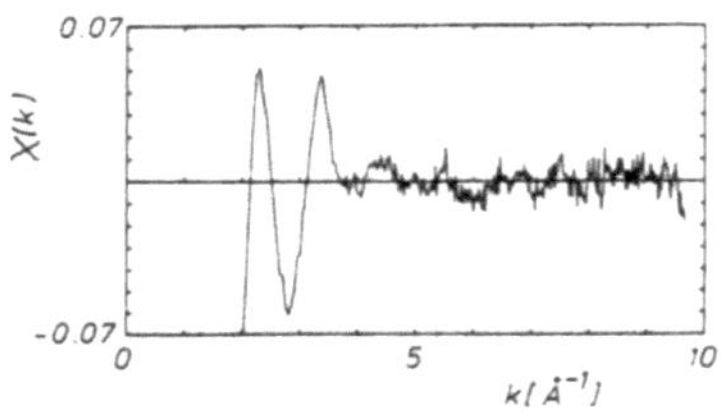

Fig.5. Absorption fine structure of the system 1% Kr/N_2 at the Kr K-edge at T=23 K

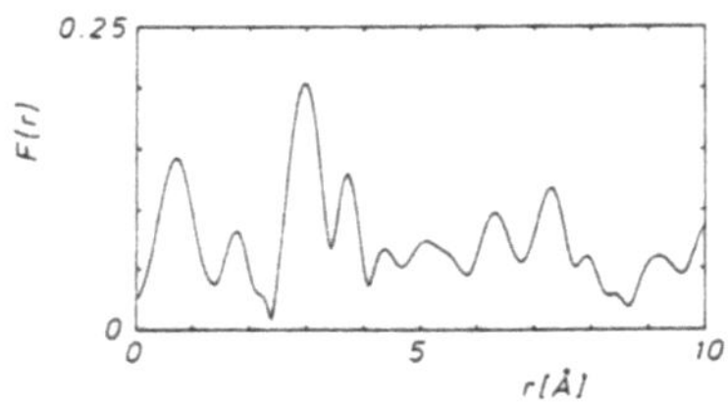

Fig.6. Fourier transform of the spectrum in Fig.5

References

1. Ber. Bunsenges. Phys. Chem. 82 (1978)
2. U. Hahn and N. Schwentner, Chem. Phys. 48, 53 (198o)
3. P.A. Montano and G.K. Shenoy, Solid State Comm. 35, 53 (198o)
4. H. Purdum, P.A. Montano, G.K. Shenoy and T. Morrison, Phys. Rev. B 25, 4412 (1982)
5. W. Malzfeldt, W. Niemann, P. Rabe and R. Haensel, Nucl. Instr. and Meth., in press
6. G. Martens, P. Rabe, N. Schwentner and A. Werner, Phys. Rev. B 17, 1481 (1978)
7. P.A. Lee, B.-K. Teo and A.L. Simons, J. Amer. Chem. Soc. 99, 3856 (1977)
8. J. Hingsammer and E. Lüscher, Helvetia Physica Acta 41, 914 (1968)
9. G. Brual and S.M. Rothstein, J. Chem. Phys. 69, 1177 (1978)

EXAFS and XANES Investigations of Trace V and Ti in Coal

J. Wong, D.H. Maylotte, F.W. Lytle[1], R.B. Greegor[1], and R.L.St. Peters
General Electric Corporate Research and Development,
Schenectady, NY 12301, USA

1. Introduction

Coal contains nearly the entire periodic table as impurity elements. While the common inorganic elements found in coal can be accounted for in the major minerals, the chemical and structural environments of many of the trace elements (i.e., those with a concentration ~1000 ppm) cannot be confidently determined by conventional microscopic, spectroscopic and diffraction techniques. Although an element may occur in only a trace amount it may have significant technological impacts on coal conversion processes through its effects on corrosion, catalyst poisoning, and fugitive emissions. In this paper, we report a combined application of XANES and EXAFS spectroscopy to probe the chemical and structural environment of trace V and Ti in coal.

2. Experimental

The coal used in these experiments came from the top 15cm of the Kentucky #9 seam. It is a high volatile bituminous coal. Parts of this seam are known to be unusually high in vanadium [1]. The vanadium content in our samples ranged up to 1850 ppm. All of the samples studied were in excess of 500 ppm. Ti-containing coal and coal-derived products were obtained from the Pittsburgh Energy Technology Center. XANES and EXAFS spectra were recorded with the EXAFS I-5 spectrometer at SSRL during a dedicated run of SPEAR at an electron energy of 3.0 GeV and a storage ring current of ~100 mA. The synchrotron beam for SPEAR was monochromatized with a channel-cut Si (220) crystal and a 1 mm entrance slit, which yielded a resolution of approximately 0.3 eV at the vanadium K edge at 5465 eV. Fluorescence data were taken with a filter-ion chamber system similar to that described by Stern and Heald [2]. In addition to the measurements on the raw coals, measurements were made on two density-separated fractions ($\leq$1.3g/cc and $\geq$1.4g/cc) and on liquefaction products.

3. Results and Discussion

To aid in the interpretation of the coal spectra a number of spectra were taken of vanadium salts, organometallic vanadium compounds and naturally occurring vanadium minerals. The XANES spectra for a series of relevant V model compounds are shown in Fig. 1. The energy zero in these spectra is taken to be the first inflexion point on the K edge of the vanadium metal. Tetrahedrally coordinated compounds, such as NH_4VO_3 or compounds with a short vanadyl bond in a square pyramid, such as V_2O_4 and vanadyl porphyrin, all exhibit a strong sharp pre-edge peak between 3eV and 6eV above the zero of the energy scale. Octahedrally coordinated compounds such as V_2O_3 show only a weak peak at this location. The main K absorption edge is due to transitions from the vanadium 1s level to the unbound ionization continuum. Peaks that occur prior (i.e., at lower energy) to this edge are due to transitions between the

[1]The Boeing Company, P.O. Box 3999 2T-05, Seattle, WA 98214, USA

Fig. 1

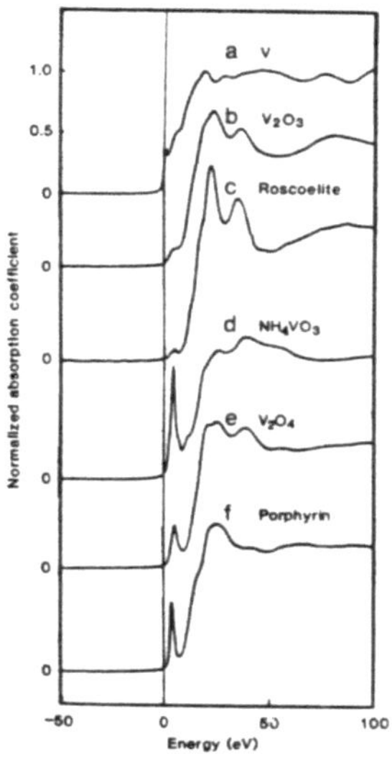

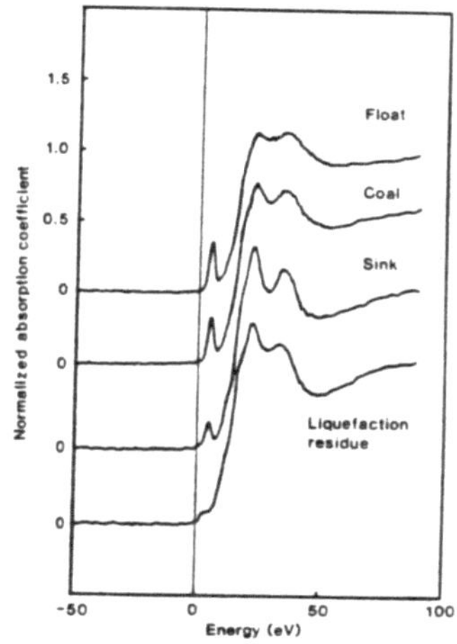

Fig. 2

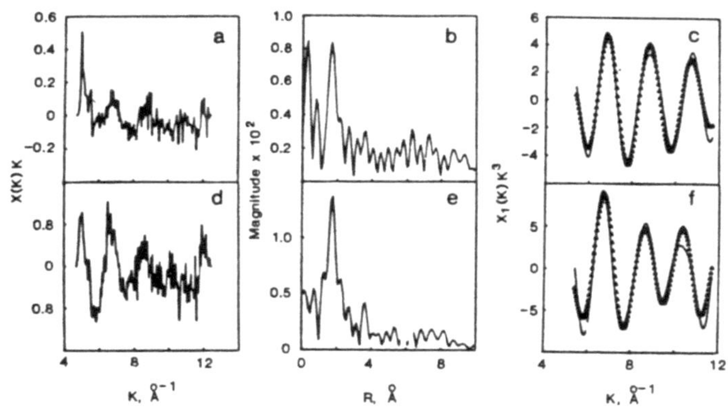

Fig. 3

Fig. 1. Vanadium K edge absorption spectra for (a) the metal, (b and c) octahedral coordination. (d) tetrahedral coordination, and (e and f) square pyramid coordination. Roscoelite is a vanadium mineral with the formula $KAIV_2Si_3O_{10}(OH)_2$.

Fig. 2. Vanadium K edge absorption spectra in coal, its sink and float fractions, and its liquefaction residue.

Fig. 3. X-ray absorption fine structure data for (a to c) float and (d to f) sink fractions of coal samples. (a) Normalized $\chi(k)k$ versus k; (b) Fourier transform of (a): (c) k^3-weighted inverse transform in the k range 4.7 to 11.7 Å^{-1} of (b) in the region 1 to 2.5 Å and (▲) fitted data. A similar sequence is shown in (d) to (f).

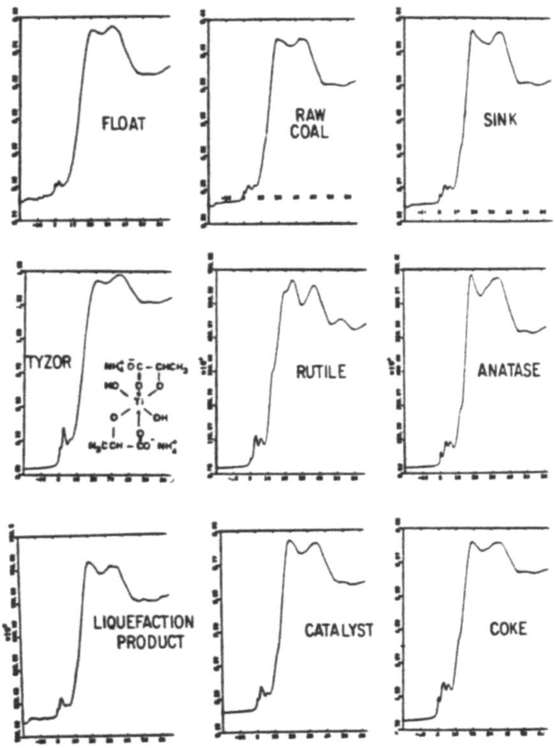

Fig. 4. TI Xanes spectra in coal, model compounds and coal-derived products.

ls levels and empty bound states below the continuum levels. For a transition metal atom in octahedral symmetry the ls to 3d transition is dipole forbidden but is an allowed quadrupole transition. This quadrupole transition gives rise to the low intensity pre-edge peak observed for compounds with octahedral symmetry (e.g., V_2O_3 and Roscoelite). In tetrahedral compounds which have no center of symmetry, the ls to T_2 transition is dipole allowed. The T_2 molecular orbital usually has a large contribution from the atomic 3d orbital of the metal atom. This transition gives rise to the strong peak observed in NH_4VO_3 (Fig. 3) as well as $CrVO_4$ and Vanadinite [3].

In Fig. 2 the K-edge spectra of vanadium in a coal in its sink-float fractions and in its liquefaction residue are shown. All of the samples of coal used in this study showed a pre-edge peak at 4.5 ± 0.2 eV. The intensity of this feature varied over the sink-float samples from a particular raw coal. In all cases the light fraction showed an increase in the peak height relative to the raw coal while the heavy fraction always showed a decrease. This is proof that in this Kentucky #9 coal the vanadium is present in at least two different environments and that they can be separated (at least partially) by a density fractionation. The trend in pre-edge peak intensity suggests that a clean heavy fraction would have a spectrum similar to V_2O_3 or Roscoelite, i.e., V^{+3} in octahedral oxygen coordination.

In the light fraction the energy of the pre-edge peak is at 4.5 eV. This position excludes V^{+5} (e.g., NH_4VO_3, V_2O_5, quinoline) from consideration because their prepeak occurs at 5-6 eV [3]. It also excludes V^{+4} with some nitrogen ligands as in the

the porphyrin and phthalocyanine. For these compounds the pre-edge peak occurs at 3.5-4 eV. The pre-edge peak intensity and position fit very well to V^{+4} in oxygen coordination with one short vanadyl bond. In Fig. 2 it is also possible to compare the edge spectrum before and after liquefaction. It is quite clear that the vanadium environment in the raw coal did not survive the liquefaction procedure, but appeared reduced totally to a trivalent state. This reactivity during liquefaction is probably not restricted to vanadiumm and may be a general path for the release of inorganic elements from the coal minerals into the product stream.

These findings from the pre-edge data can be complemented by an analysis of the EXAFS data from the high energy side of the absorption edge. Using model compounds of known structure a set of backscattering amplitudes and phase shifts may be derived by fitting their EXAFS spectra to the theoretical expressions. These scattering parameters can be used to fit the Fourier-filtered EXAFS data and then used to derive the structural parameters for the unknown material. Details of our Fourier analysis and fitting routines have been published elsewhere [4]. The parameters for oxygen, sulfur and nitrogen were obtained from analysis of the vanadium model compounds.

The EXAFS data for typical "sink" and "float" samples are given in Fig. 3 along with the Fourier transforms and the reverse transforms for the region around the first shell. Fits were tried assuming a short vanadium-oxygen bond and a longer bond to oxygen, sulfur, or nitrogen atoms. The coal EXAFS could not be fitted with sulfur in any combination. Equally good fits could be obtained from nitrogen or oxygen atoms attached to the longer bond. This is not surprising because the atoms are adjacent in the Periodic Table and have similar scattering parameters. However, the near-edge data excluded nitrogen so we conclude that vanadium is oxygen coordinated in this coal. The best fits were obtained with one short bond at ca. 1.6Å and five longer bonds at ca. 20Å. For the light fraction in the "sink-float" separation the results from the pre-edge data and from the EXAFS indicate that the vanadium is tetravalent and in an oxygen environment similar to that in V_2O_4 (this does not exclude metallorganic compounds with this structure for the nearest-neighbor ligands).

The fate of trace Ti in typical coal liquefaction process stream was also elucidated. In the raw coal, titanium exists in at least two chemical and structural environments one of which resemblesthat in a Ti-organo tyzor. Both sites are 6-fold coordinated by oxygens in a distorted octahedral configuration. The organo form was found to persist in both the liquefaction product and the desulfurization catalyst, while the anatase form was found in the coke by-product adjacent to the catalyst (Fig. 4).

Acknowledgment

We are grateful for experimental opportunity at SSRL supported by NSF through the Division of Materials Research and NIH through the Biotechnology Resource Program in cooperation with the Department of Energy.

References

1. P. Zubovic, in Amer. Chem. Soc. Symposium Series 55 221 (1966).

2. E.A. Stern and S. Heald, Rev. Sci. Instru. 50 1579 (1979).

3. J. Wong, F.E. Lytle, R.P. Messmer and D.H. Maylotte, these Proceedings

4. E.A. Stern, D.E. Sayers, F.W. Lytle, Phys. Rev. B11 4836 (1975); G.H. Via, J.H. Sinfelt and F.W. Lytle, J. Chem. Phys. 71, 690 (1979).

Characterization of the Valence States of Iron in Spinels by X-Ray Absorption and Emission Spectroscopy

D. Le Calonnec[1,2], J. Dürr[2], B. Hannoyer[1], M. Lenglet[1], G. Calas[3], and J. Petiau[3]

1. Introduction

In order to use the X-ray absorption spectroscopy for analytical application we have studied the XANES of Fe spinel compounds to follow the effect of the chemical bonding on the structures. Comparisons were made with x-ray emission spectroscopy.

The x-ray absorption spectra (XAS) were performed at the DCI storage ring at LURE [1]. A Si 220 channel-cut crystal was used to monochromatize the white beam. For the K absorption edge of iron the first maximum of the edge is taken as reference value at 7115 ev. The energy resolution at 7 KeV is $\Delta E/E \sim 2.10^{-4}$.

The x-ray emission spectra were recorded on a conventional x-ray fluorescence spectrometer [2]. The tops of the bands $K\beta_{1.3}$ and $K\beta_5$ are recorded by step scanning. A parabolic fit is computed by means of least squares for the intensity values measured. The maximum of the parabola gives the energy of the line with an accuracy of ± 0.10 eV.

2. XAS of trivalent and bivalent iron

Iron III compounds reveal an energy shift of 8 to 9 eV (compared with the iron metal edge). In order to study the effect of the coordinence on the XANES we have selected samples where Fe^{3+} is coordinated by six ($ZnFe_2O_4$) and four oxygen ions ($Li_{0.5}Fe_{0.5}Cr_2O_4$).

Table 1

Compound	Peak positions (ev) *				K-edge shifts/Metal (ev)
	B_1	B_2	C_1	C_2	
$Li_{0.5}Fe_{0.5}Cr_2O_4$	11.4	16.7	22	27	∼ 8.5
$ZnFe_2O_4$		16.1	20.4	27	9.2
$Li_{0.5}Fe_{2.5}O_4$	11.7	16.3	20.4	26.3	8.2

* all peak positions are taken to the first maximum of the derivative spectrum

[1] INSCIR, B.P. N°8, F-76130 Mont Saint Aignan, France

[2] ENSTIMD, F-59508 Douai Cedex, France

[3] Universite Paris VI, 4 Place Jussieu, F-75230 Paris Cedex 05, France

The first derivative of the x-ray absorption spectra (fig. 2) exhibit six main peaks for the tetrahedral coordination and five for the octahedral coordination. These are marked A, B_1, B_2, C_1, C_2 and D [3, 4] in Fig. 1. The relative peak positions are listed in Table 1.

Iron in $Li_{0.5}Fe_{2.5}O_4$ has the two coordination states, its first derivative spectrum exhibits and intermediate structure.

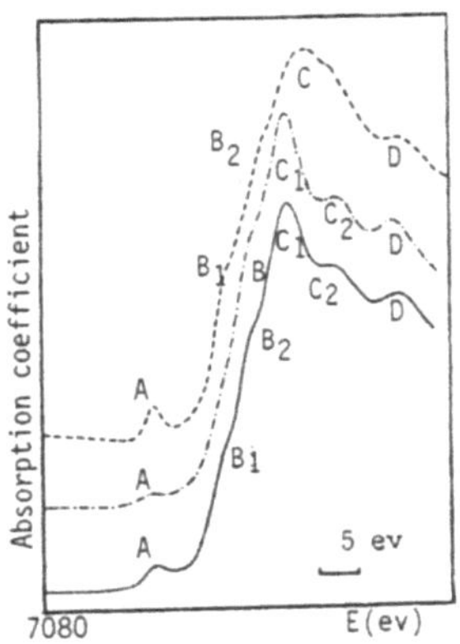

Fig.1. XANES spectra
—— $Li_{0.5}Fe^{3+}_{0.5}[Cr_2]O_4$
—·—· $Zn[Fe^{3+}_2]O_4$
– – – $Fe^{3+}[Li_{0.5}Fe^{3+}_{1.5}]O_4$

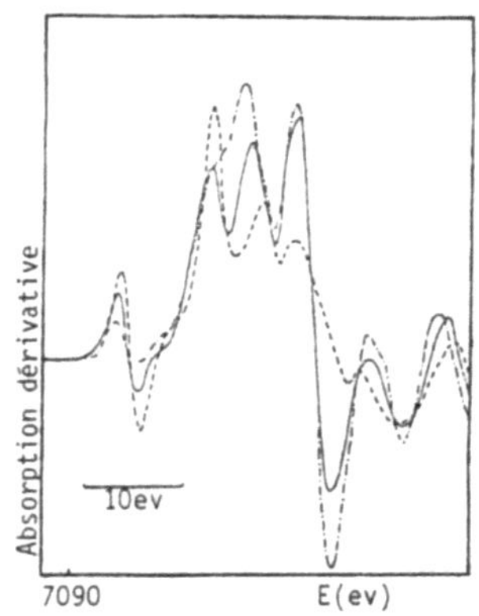

Fig.2. Derivative XANES spectra
—— $Li_{0.5}Fe^{3+}_{0.5}[Cr_2]O_4$
—·— $Zn|Fe^{3+}_2|O_4$
– – $Fe^{3+}[Li_{0.5}Fe^{3+}_{1.5}]O_4$

Iron II compounds such as $FeCr_2O_4$, $FeAl_2O_4$, Fe_2GeO_4 are characterized by an energy shift of 5 to 6 eV. The evolution of the fine structure looks very much like the evolution observed in the trivalent iron spectra.

No simple relation exists between the shift of the edge and the effective charge on the Fe ion, nevertheless these results confirm the fact that the position of the K-absorption discontinuity and the shift of the XANES are a function of the valence state. A more detailed work is going to be published. A similar study concerning copper oxides led to the same conclusion [5, 6].

3. Analytical application

The influence of the chemical bonding on the x-ray emission bands and more precisely on the band $K\beta_5$ is well defined for iron compounds [2, 7, 8]. The shift, relative to the iron metal, of the band $FeK\beta_5$ is located around 2 ev for the ferrous compounds and in the range 1-1.5 eV for the ferric compounds [9].

The chemical shift $\Delta K\beta_5$ of magnetite, 2.5 eV, is remarkably high ; the shift of the K absorption edge, 6.6 eV, is located between the values observed for the two kinds of oxides.

The simultaneous analysis by x-ray absorption and emission spectroscopy of the different terms of solid solutions $(Li_{0.5}Fe_{0.5})_{1-x}Fe_{2+x}O_4$ and $Zn_{1-x}Fe_{2+x}O_4$ [6] shows the interest of these two physical analysis methods for the display of FeII in mixed ferrites (subject to a good knowledge of the environments of the different ions).

One can also mention the charaterization of the chemical bonding in materials such as $Cu_{0.5}Fe_{2.5}O_4$ and $Cu_{1-x}Fe_{2+x}O_4$ $(x < 0.5)$: the observed chemical shifts values, as much for band $K\beta_5$ [9, 10] as for K absorption edge [11], reveal the presence of bivalent iron without ambiguity.

References

[1] D. RAOUX, J. PETIAU, P. BONDOT, A. FONTAINE, P. LAGARDE, P. LEVITZ, G. LOUPIAS, A. SADOC, Rev. Phys. Appl., 1980, 15, 1079.
[2] J. DÜRR, Thèse LILLE, 1980.
[3] G.N. GREAVES, P.J. DURHAM, G. DIAKUN, P. QUINN, Nature, 1981, 294, 139.
[4] M. BELLI, A. SCAFATI, A. BIANCONI, S. MOBILIO, L. PALLADINO, A. REALE, E. BURATTINI, Sol. State Comm., 1980, 35, 355.
[5] B. HANNOYER, J. DÜRR, G. CALAS, J. PETIAU, M. LENGLET, Mater. Res. Bull., 1982, 17, 435.
[6] Unplublished
[7] K.M. KOLOBOVA, S.A. NEMNONOV, E.V. AGAPOVA, Soviet Physics - Solid State, 1968, 10, 1968.
[8] A.S. KOSTER, G.D. RIECK, J. phys. chem. solids, 1970, 31, 2505.
[9] J. DÜRR, A. D'HUYSSER, J.P. BONNELLE, M. LENGLET, Ann. Chimie, 1982, 6, 629.
[10] B. LEREBOURS, J. DÜRR, A. D'HUYSSER, J.P. BONNELLE, M. LENGLET, Phys. stat. Sol. 1980, 62, K175.
[11] B. HANNOYER, J. DÜRR, G. CALAS, J. PETIAU, M. LENGLET, 2nd European Conference on Solid State chemistry, June 1982, EINDHOVEN, (in press).

High-Pressure Study of L_{III} Absorption Edges of Rare-Earth Mixed Valence Systems

J. Röhler[1], G. Krill[2], J.P. Kappler[2], M.F. Ravet[2]

LURE, CNRS et Univ. Paris Sud, Bât. 109 C, F-91405 Orsay Cedex, France

1. INTRODUCTION

The L_{III} X-ray absorption spectra of the rare earth (RE) systems exhibit a characteristic peak at threshold. This peak (a so-called white line) arises from transitions from atomic $2p_{3/2}$ core levels into empty 5d states. Systematic studies of L_{III} spectra of the RE(Sm-Lu) give unambigous experimental evidence that its white line reflects ground-state properties of the absorbing atom. The anomalous features of the Ce edge are discussed below. The white lines of the $4f^n$ RE systems are shifted by 6 - 10 eV towards higher energies relative to those of $4f^{n+1}$ RE systems. This shift is mainly due to the screening of the central potential by an additional 4f electron in the $4f^{n+1}$ configuration. The L_{III} spectra of RE mixed-valence systems exhibit a characteristic double bump structure [1,2], which consists of a superposition of two white lines, separated by about 6 - 10 eV. We derive the valence $v = 2(3)+\nu$ simply from the relative intensities of the two white lines, where ν denotes the fractional 4f occupation. In this paper we report results obtained from high pressure L_{III}-absorption experiments of SmS, $EuPd_2Si_2$ and $CeAl_2$ (see also Ref. 3).

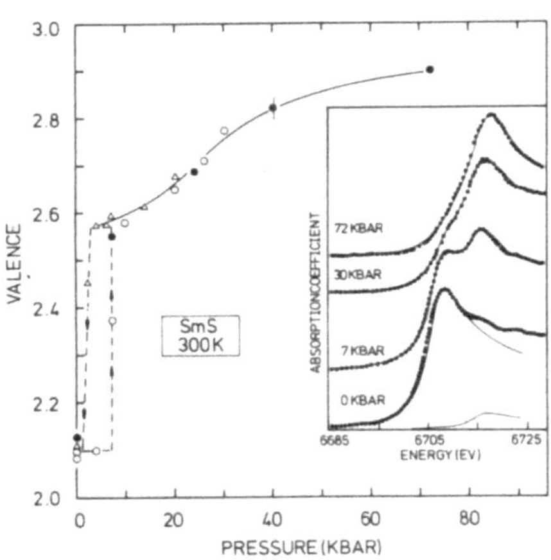

Fig. 1. Pressure dependence of the valence of SmS and typical L_{III} spectra at different pressures. Open and filled circles indicate data taken from different runs. The triangles are data taken at decreasing pressure

[1] II. Physikalisches Institut, Universität zu Köln, Zülpicher Str. 77, D-5000 Köln 41, Fed. Rep. of Germany

[2] LMSES (LA CNRS 306)-4, rue B. Pascal, F-67070 Strasbourg Cedex, France

2. EXPERIMENTAL

The high pressure cell and experimental details have been described elsewhere [4]. Our analysis of the data is restricted to 10 eV around threshold. We followed the simple assumption that $\Delta(E)$ in the edge region may be described by an asymmetric Lorentzian whose width is given mainly by the inverse life time of the core hole and whose asymmetry is given by the existence of the absorption threshold. The fractional superpostion of two asymmetric Lorentzians, fitted to our data, has been foalded with a spectrometer function, assumed to be a Gaussian. Nota bene: the value of the asymmetry parameter, derived from our fits does not have the same theoretical meaning (many-body effects) as, e.g., in the Doniach - Sunjič (DS) theory [5]. Our asymmetry parameter is simply a quantity which describes the amplitude of the absorption threshold. Our visual fits optimize two energy positions, E_2 and E_3, the linewidth Γ, and the intensity ratio I_2/I_3. Γ was found to be within 3 - 5 eV. The width of the spectrometer function is within 1.5 - 2.5 eV. It is noteworthy, that a fit of a superpostion of a DS line shape and an arctan stepfunction to the data forces the DS lineshape to the symmetric case. This result supports an interpretation of the RE white line through atomic ground-state properties.

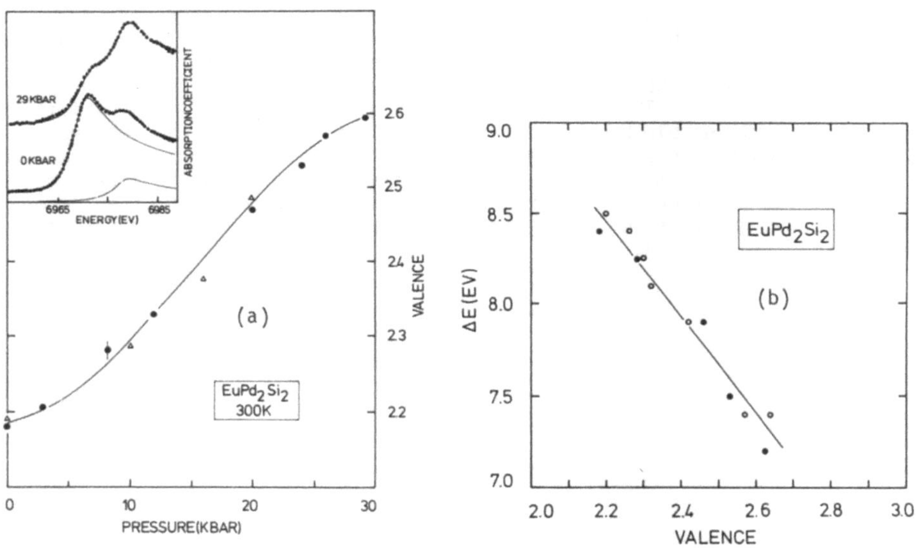

Fig. 2.(a) Pressure dependence of the valence of $EuPd_2Si_2$ and typical L_{III} spectra at different pressures. Triangles indicate data taken at decreasing pressure.
(b) Energy splitting ΔE of the two white lines of $EuPd_2Si_2$ as a function of valence derived from the intensity ratio of the white lines. Filled circles indicate data from the high pressure experiment at 300 K, open circles data from the 0 kbar, low temperature experiment

3. DISCUSSION

The data, represented in Fig. 1, 2a, show a strongly nonlinear pressure-valence relation. In Fig. 2b we show that the splitting of the two white lines, ΔE, scales with the valence as derived from the intensity ratios. This re-

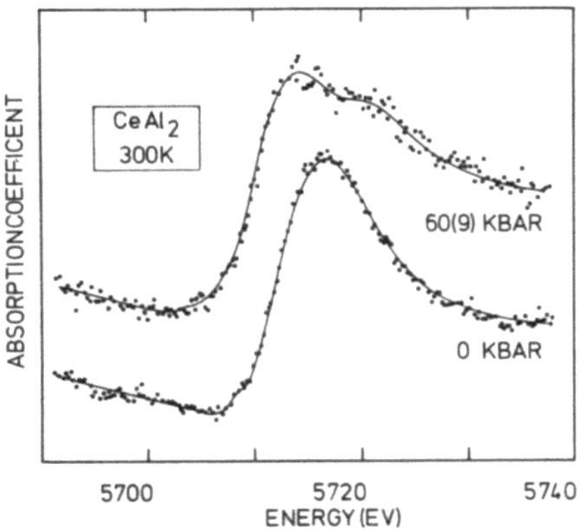

Fig. 3. Zero pressure and 60 kbar spectra of $CeAl_2$. The drawn out line is a guide the eye. The energy calibration is uncertain within ± 2 eV

lation seems to be universal. We have found the same relation from the SmS and TmSe data [3]. We find $\partial(\Delta E)/\partial\nu = -2.7 \pm 0.6$ eV, from isobaric and isothermal data. (Open circles indicate data from a valence shift, induced by cooling the sample down to He-temperature, filled circles are data from the high pressure experiment (Fig. 2a)). The slopes of the individual lines are: $\partial E_3/\partial\nu = -2.2 \pm 0.7$ eV (high energy side); $\partial E_2/\partial\nu = 1.0 \pm 0.5$ eV (low energy side).

The spectra of the Ce L_{III}-edge exhibit some peculiarities. They have a fine structure, not observed at the heavy RE. Furthermore, the spectra of the so-called 4^+ Ce compounds exhibit a double structure and no single white line [6]. As yet there is no unambigous method to derive the valence from the Ce L_{III}-spectra. However, γ-Ce type systems (single peak) and α-Ce type systems (double peak) can be clearly distinguished. The peak on the high energy side is attributed to the $4f^0$ configuration. Fig. 3 represents the zero pressure and 60 kbar L_{III} spectrum of $CeAl_2$. The drawn out line is only a guide to the eye. The high pressure spectrum clearly exhibits the α-Ce double bump feature. This transition to the α phase around 60 kbar is expected from high pressure lattice parameter data [7]. Because our certainty in energy calibration is only within ± 2 eV, we do not draw any conclusion from the energy shift of the 60 kbar spectrum. Further complications of the data analysis arise from statistics, which we hope to improve by use of a smaller Be gasket.

References

[1] E.E. Vainshtein, S.M. Blokhin and Y.B. Paderno, Sov. Phys. Solid State 6, 2318 (1965)
[2] H, Launois, M. Rawiso, E. Holland-Moritz, R. Pott and D. Wohlleben, Phys. Rev. Lett. 44, 1271 (1980)
[3] J. Röhler, G. Krill, J.P. Kappler, M.F. Ravet and D. Wohlleben in Proceedings of the International Conference on Valence Instabilities, North Holland (1982)
[4] J. Röhler, J.P. Kappler and G. Krill, Nucl. Instr. Meth., in press, (1982)
[5] S. Doniach, M. Šunjić, J. Phys. C3, 285 (1970)
[6] K.R. Bauchspieß, W. Boksch, E. Holland-Moritz, H. Launois, R. Pott and D. Wohlleben in Valence Fluctuations in Solids, North Holland (1981), 417
[7] M. Croft, A. Jayaraman, Sol. State Com. 29, 9 (1979)

EXAFS of Antimony L_{III} Edge of Ferroelectric SbSI Above the Phase-Transition Temperature

E. Burattini and N. Cavallo
INFN:PWA Group, Frascati, and Naples University, Italy

G. Cappuccio
CNR, Istituto di Elettronica dello Stato Solido, I-00185 Roma, Italy

G. Dalba and P. Fornasini
Dipartimento di Fisica dell'Università di Trento, I-38059 Povo, Italy

M. Grandolfo and P. Vecchia
Physics Lab., Ist. Superiore di Sanità I-00185 Roma, Italy

Sh.M. Efendiev
Azerbaijan State University, Baku, USSR

1. Introduction

Among the crystals of the type $A^V B^{VI} C^{VII}$, a family of ferroelectric semiconductors characterized by high resistance and relatively narrow forbidden band ($\simeq$ 2 eV), SbSI is the most widely studied for its remarkable properties [1]. Crystals of SbSI display a first-order ferroelectric phase transition of a basically displacive type at a temperature $T_C \simeq 293$ K. Structural investigations have established that the SbSI crystals and their isomorphic analogs are rhombic and belong to the class mmm (space group D_{2h}^{16}) above T_C, and to the class mm (C_{2v}^{9}) below T_C. Additional phase transitions are observed in ferroelectrics of the type under study besides the main ferroelectric-paraelectric phase transition. For example, a phase transition at a temperature of about 233 K is observed in SbSI with a change in symmetry mm → 2 ($C_{2v}^{9} \rightarrow C_{2}^{2}$).

In order to study the possible influence of the paraelectric -ferroelectric phase transition on the onset of the X-ray absorption edges, the temperature dependence of the L_I absorption edge of the Sb ion in SbSI single crystals near T_C has been recently measured [2] . To characterize better the onset of the absorption edges and define the dynamics of the change of crystal structure crossing the transition temperature, in this paper the attention has been focused on the extended X-ray absorption fine structure (EXAFS) of antimony L_{III} edge [3] , consisting of the modulation of the absorption X-ray coefficient μ in the vicinity of the L_{III} absorption edge that characterizes Sb ions in the material. From such modulation, structural information such as the bond distances, the measure of disorder in bond distances due both to thermal and structural disorder, the coordination number, and the chemical identity of the coordination shell are obtainable [4]. Preliminary results of EXAFS spectra of paraelectric SbSI (T=310 K $>T_C$) have been obtained.

2. Experimental

EXAFS measurements have been performed with an absorption spectroscopy apparatus at the Adone Wiggler Facility in Frascati utilizing a Si (111) channel-cut crystal monochromator. The samples were powdered, deposited on filter paper, and enclosed between two sheets of adhesive mylar.

3. Results and discussion

Figure 1 shows the structures above the Sb L_{III} edge in SbSI at $T \simeq 310$ K; the edge energy E_0 was conventionally chosen as the maximum of the first derivative of the signal. The FOURIER transform F(R) of the EXAFS oscillations from $K=3.8$ Å^{-1} to the onset of the L_{II} edge is reported in Fig.2.

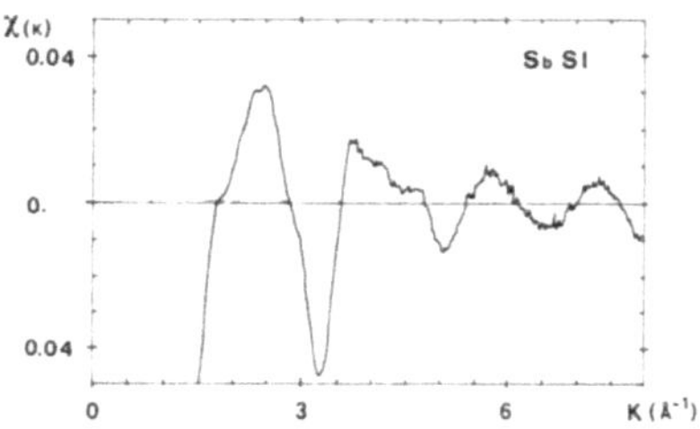

Fig.1. EXAFS oscillations at the Sb L_{III} edge in paraelectric SbSI

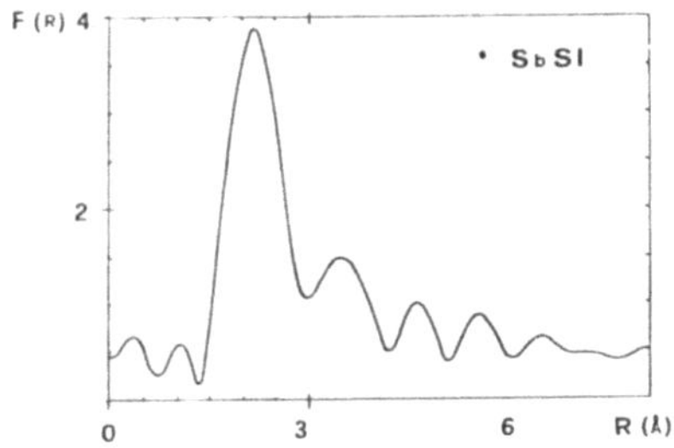

Fig.2. Magnitude of the FOURIER transform

By backtransforming the main peak of the F(R), the contribution χ' (K) of the nearest neighbours has been obtained. The absorber (Sb) and backscatterer (S) phase, interpolated from the values tabulated by TEO and LEE, have been then substracted to the phase of χ' (K), thus obtaining the function $\phi'(K) = 2\ KR + n\pi$.

By adjusting the edge energy so as to have an integer value of the n parameter, the R(K) function shown in Fig.3 (solid line) has been obtained for $\Delta E_0 = +10$ eV. The mean value (dashed line) is $R_E = 2.56$ Å.

The most recent crystallographic measurements [5] give as first neighbours of the Sb atom in SbSI in the paraelectric phase one sulfur atom at 2.492 Å and two sulphur atoms at 2.764 Å.

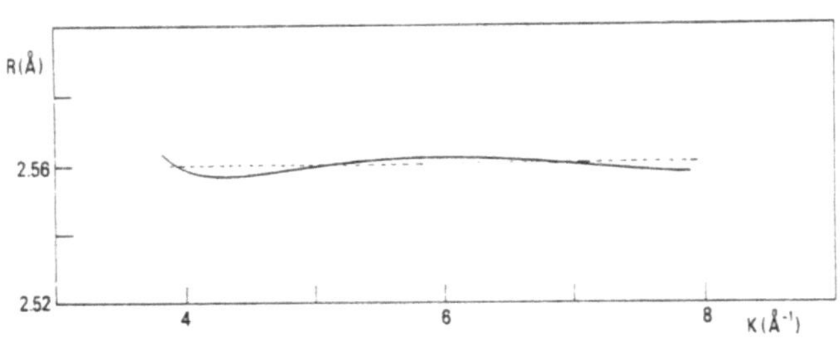

Fig.3. K dependence of the first-neighbour distance Sb-S

The slight discrepancy between crystallographic and EXAFS results could be due to the limited accuracies of the theoretical phases.

Further work is now in progress to study the temperature dependence of EXAFS spectra down to 200 K.

REFERENCES

1. V.M.Fridkin, Ferroelectric Semiconductors, Consultants Bureau, New York and London (1980)
2. E.Burattini, G.Cappuccio, N.Cavallo,Sh.M.Efendiev, M.Grandolfo, and P.Vecchia, Ferroelectrics 43, 211 (1982).
3. E.A.Stern, Phys.Rev. B10, 3027 (1974)
4. E.A.Stern, D.E.Sayers and F.W.Lyttle, Phys.Rev. B11, 4836 (1975)
5. A.Kikuchi, Z.Oka and E.Sawaguchi, J.Phys.Soc.Japan 23, 337 (1967)

Structure of $LaNiO_2$ by EXAFS and X-Ray Diffraction: Local Environment of Ni^{+1}

Pierre Levitz, Michel Crespin, and Lucien Gatineau
C.R.S.O.C.I.-C.N.R.S., 1B, rue de la Férollerie, F-45045 Orleans Cedex, France, and
L.U.R.E.-C.N.R.S., Université Paris-Sud, Bât. 200, F-91405 Orsay Cedex, France

1. Introduction

During the reduction of $LaNiO_3$ perovskite a new phase, $LaNiO_2$, was produced. The oxidation state of Ni in this phase, +1, has been determined by the hydrogen consumption during the reaction and by X.A.N.E.S. and E.X.A.F.S. [1]. The structure of this new compound is tetragonal with :

$$a = b = 3.966 \pm 0.001 \text{ Å, and } c = 3.376 \pm 0.001 \text{ Å} .$$

The $LaNiO_2$ structure has been resolved by (i) X-ray diffraction data of a powder diagram (20 reflections), (ii) the radial distribution of interatomic distances around the nickel calculated by E.X.A.F.S.

2. X-Ray Results

By analogy with the $LaNiO_3$ perovskite structure from which the structure of the reduced compound is derived, nickel is located at the origin, lanthanum at the center of the unit cell and the oxygen atoms at the middle of the square face sides (Fig.1). With these coordinates the residual factor calculated from 20 observed reflections is : $R = \Sigma \, |F_{obs} - F_{cal}| / \Sigma |F_{obs}| = 0.07$.

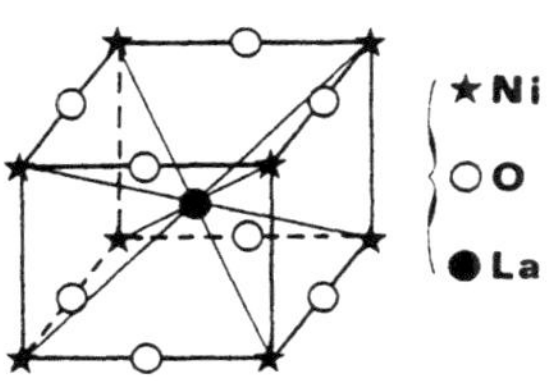

Fig.1. Structure of $LaNiO_2$

3. E.X.A.F.S. Results

3.1 Experimental

The experimental conditions have been described previously [1].

3.2 E.X.A.F.S. Analysis

The radial distribution around the nickel atoms in $LaNiO_3$ and $LaNiO_2$ obtained from the E.X.A.F.S. oscillations are shown in Fig.2. A striking similarity is observed if the profiles of both distributions are compared.

In $LaNiO_3$,
(i) the first peak is due to the nearest neighbours consisting of 6 oxygen atoms at distance Ni-O = 1.918 Å,

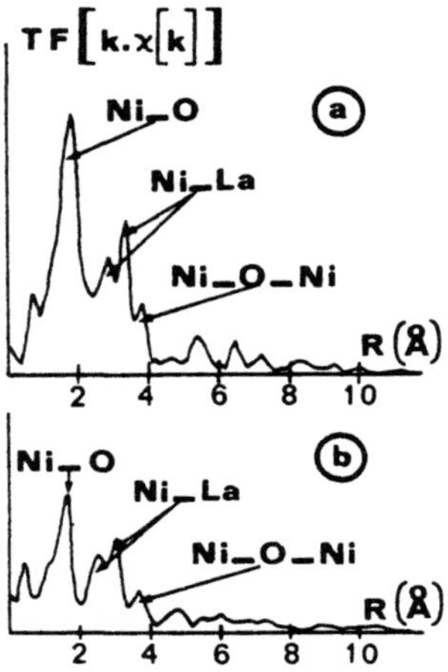

Fig.2. E.X.A.F.S. radial distribution around Ni
a) in $LaNiO_3$
b) in $LaNiO_2$

(ii) the second coordination shell around nickel is composed of 8 Ni-La pairs at an average distance of 3.223 Å. The presence of two peaks corresponding with these La-Ni distances is due the k dependent structure of the backscattering amplitude of La,
(iii) the fourth peak is due to a colinear Ni-Ni pair separated by an oxygen atom. The corresponding coordination shell is composed of 6 atoms at a distance Ni-O-Ni = 3.837 Å.

The radial distributions of $LaNiO_3$ and $LaNiO_2$ being very similar, the peaks of $LaNiO_2$ can be indexed using the data obtained for $LaNiO_3$.

In $LaNiO_2$,
(i) first coordination shell Ni-O. The theoretical relations found by TEO and LEE [2] were used in order to calculate the phase shifts with the conditions shown in Table 1.

Table 1 . Ni-O distances

	$LaNiO_3$	NiO	$LaNiO_2$
Multiplicative factor of $\chi(k)$	k^3	k^3	K^3
Window of T.F. in k space	7-15 $Å^{-1}$	7-15 $Å^{-1}$	7-15 $Å^{-1}$
R from X-Ray data	1.918 Å	2.092 Å	1.983 Å
R from E.X.A.F.S.	1.94 Å	2.08_5 Å	$1,98_5$ Å

Table 2 . Ni-La distances

	$LaNiO_3$	$LaNiO_2$
Multiplicative factor of $\chi(k)$	k	k
Window of T.F. in k space	4-15 $Å^{-1}$	4-15 $Å^{-1}$
R from X-Ray data	6 atoms at R = 3.335 Å 2 atoms at R = 3.284 Å	3.270 Å
R from E.X.A.F.S.	3.35 Å	3.24 Å

(ii) second coordination shell Ni-La. As in the case of the first coordination shell the phase shift calculated by TEO and LEE was used with the conditions exposed in Table 2.

(iii) third coordination shell Ni-O-Ni. A theoretical electronic phase shift is not available to account for this effect. By chemical transferability it is possible to calculate an empirical phase shift from the Ni-O-Ni coordination in NiO oxide (4.183 Å).

The distance in $LaNiO_2$ calculated this way for a window in k space between 7 and 15 $Å^{-1}$ is 3.97_5Å.

This assignment does not take into account the shell corresponding to the Ni-Ni distance along the z axis (3.376 Å). The corresponding peak and the double Ni-La peak are certainly superimposed but the contribution of the Ni-Ni shell is very weak. Indeed there are only 2 nickel atoms in this shell while there are 8 in the Ni-La shell. Moreover the backscattering of heavy atoms like La is much more important than the backscattering of nickel.

4. Conclusion

Table 3 summarises the results obtained with the two techniques used for resolving the structure of $LaNiO_2$.

Table 3 . Comparison between X-ray and E.X.A.F.S. results

Distance	X-Rays	E.X.A.F.S.
Ni-O	1.983 Å	1.98_5 Å
Ni-La	3.270 Å	3.24_0 Å
Ni-O-Ni	3.966 Å	3.97_5 Å

Table 3 demonstrates the good agreement found for the local environment of nickel and it establishes the Ni-O interatomic distance in case of monovalent nickel in coordination 4.

References

1 - Crespin M., Levitz P., Gatineau L. In these Proceedings.

2 - Teo B.K. and Lee P.A. (1979). J. of Am. Chem. Soc., 101, 2815

EXAFS Study of the Disorder Effect Induced in Layered $MnPS_3$ by Intercalation of Cations

A. Michalowicz

Laboratoire de Physicochimie Structurale, Université Paris-Val de Marne, F-94000 Creteil, France and
LURE, CNRS, Bât. 209C, F-91405 Orsay Cedex, France

G. Vlaic

Istituto Guido Donegani S.p.A., I-28100 Novara, Italy and
LURE, CNRS, Bât. 209C, F-01405 Orsay Cedex, France

R. Clement, Y. Mathey

Laboratoire de Physicochimie Minérale and ERA 672, Université Paris-Sud, F-91405 Orsay, France

$MnPS_3$ is a layered compound with a well-known crystal structure [1] and the following radial distribution around the Mn^{2+} ion : 6 sulfur at 2.6 Å, 6 phosphorous at 3.7 Å and 3 manganese at 3.5 Å.

In this work we are interested by the modifications induced in the radial distribution by intercalation of cations between the layers. In a previous work we have shown that an important static disorder effect occurs in the EXAFS spectrum of Mn when $Co(C_5H_5)_2^+$ is intercalated [2]. Qualitatively this effect may be evaluated by the change in the relative amplitudes of the first and second peaks of the radial distribution function (F.T. $(k^3\chi(k))$). It is also possible to evaluate the variation of the Debye-Waller factor $\Delta\sigma^2 = \sigma^2$ (intercalate) - σ^2 ($MnPS_3$) for each shell, by curve fitting using the classical EXAFS formula and TEO-LEE phase shifts and amplitudes. For the second shell of $MnPS_3(Co(C_5H_5)_2^+)_{0.34}$ the results are : $\Delta\sigma^2$ (300 K) = 0.020 Å^2 $\Delta\sigma^2$ (30 K) = 0.008 .

An interpretation of this effect as a distortion of the site is consistent with the dramatic change observed in the magnetic properties of the intercalates [2] [3] . We present here a new result obtained with the intercalated cation CH_3-N-R_3^+ (R = C_8H_{17}). The study of this compound is important since its magnetic properties are intermediate between pure $MnPS_3$ and the $Co(C_5H_5)_2^+$ intercalates [3] . As it is shown on fig. 1 this compound is also intermediate for the structural modifications of the manganese site with $\Delta\sigma^2$ (300 K) = 0.007 Å^2 .

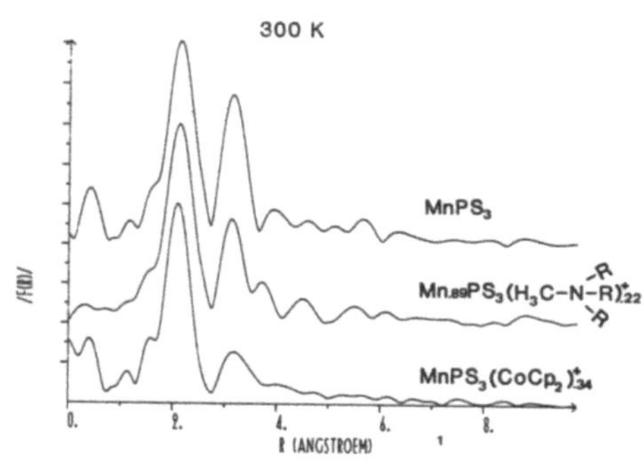

Fig. 1.
EXAFS Fourier transform of $MnPS_3$ and two intercalation compounds

A second interesting case is the study of $Mn_{1-x}PS_3Cu_{2x}$ x=1/6 at both transition metal edges. In spite of the presence of vacancies, the manganese site is only slightly distorted $\Delta\sigma^2$(300 K) = 0.001 Å^2 for the second peak). The hypothesis of an occupation of the manganese vacancies by copper (I) ions is ruled out by the Cu K edge EXAFS spectra (fig. 2)

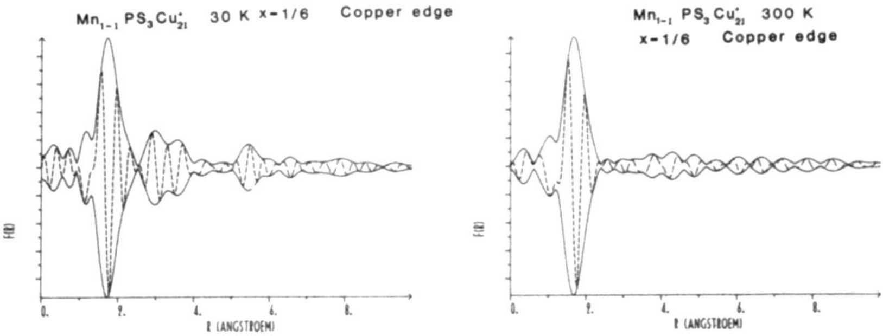

Fig. 2 EXAFS Fourier transforms of $Mn_{1-x}PS_3Cu_{2x}$

A Cu-S distance of 2.20 Å and a strong temperature dependence (the outer shells disapear at room temperature) are consistent with a position of these ions near the sulfur planes and probably outside the layers. This feature could explain the important ionic conductivity of the compound.

1 W. KLINGEN, G. EULENBERGER and H. HAHN, Z. Anorg. All. Chem., 401, 97, (1973).

2 A. MICHALOWICZ and R. CLEMENT, Inorg. Chem., in press (Nov. 1982).

3 R. CLEMENT, J.J. GIRERD and I. MORGENSTERN-BADARAU, Inorg. Chem., 19, 2852, (1980).

Structural Properties of $Cd_{1-x}Mn_xTe$ by EXAFS

F. Antonangeli, A. Balzarotti, N. Motta, and M. Piacentini
PULS, Laboratori Nazionali di Frascati, I-00044 Frascati, Italy
A. Kisiel and M. Zimnal-Starnawska
Institute of Physics, Jagellonian University, Cracow, Poland
W. Giriat
Instituto Venezolano de Investigaciones Cientificas, Centro de Fisica, Caracas 1010A, Apdo. 1827, Venezuela

X-ray diffraction studies on $Cd_{1-x}Mn_xTe$ crystals show a well-defined pattern characteristic for the fcc structure and a linear decrease of the lattice constant with increasing Mn content up to x=0.7 (1). On the other hand, reflectivity spectra measured recently exhibited changes with increasing Mn concentration that could suggest the existence of disorder induced by Mn atoms (2).

To check the local environment around each atomic constituent of these compounds, we have performed EXAFS measurements above the K edge of Mn and the L edges of Te. Data at 300 K and 80 K were collected at the PULS X-ray line utilizing the synchrotron radiation emitted by the ADONE storage ring (3). The samples were powdered $Cd_{1-x}Mn_xTe$ obtained from single crystals grown by a modified Bridgman method and deposited on kapton tape.

The EXAFS modulation $\chi(k)$ above the Te L_{III} edge in pure CdTe, $Cd_{0.5}Mn_{0.5}Te$, $Cd_{0.7}Mn_{0.3}Te$ and the Mn K edge in $Cd_{0.3}Mn_{0.7}Te$ and $Cd_{0.5}Mn_{0.5}Te$ have been extracted from the experimental spectra using the Fourier filtering technique (4). The zincblende CdTe was used as a model compound for the EXAFS analysis.

In Fig. 1 a) we compare the experimental $\chi(k)$ spectrum for the Te L_{III} edge with a model calculation in which crystallographic bond lengths, coordination numbers and tabulated amplitude and phases have been taken (5). Corrections to the usual plane-wave phases for Cd and Te have been considered (6). In addition a constant $S_o^2 = 0.7$ factor which accounts for the relaxation of the passive electrons and chemical effects had to be introduced to match theory and experiment in the EXAFS formula (7):

$$k\chi(k) = \sum_j \frac{N_j}{R_j^2} S_o^2 |f_j(\pi,k)|^2 e^{-2\sigma_j^2 k^2} e^{-R_j/2\lambda} \sin(2kR_j+\phi_j(k)+\Delta\phi_j) .$$

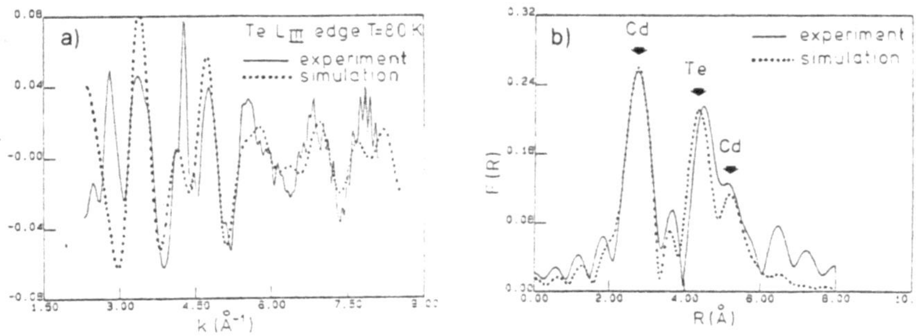

Fig.1. a) Experimental $\chi(k)$ vs k for Te L_{III} edge at 80 K compared with the model calculation (dotted line). b) Radial distribution functions derived from the Fourier transform of the data of Fig.1 a)

Table 1. Parameters used in the simulation for CdTe

Shell	Atom	N_j	R_j(Å)	σ_j^2(Å^2)	$\Delta\phi$(rad)
1	Cd	4	2.80	.0015	-1
2	Te	12	4.58	.003	-1.6
3	Cd	12	5.37	.006	-1
4	Te	6	6.48	.006	-1.6

E_o(eV) = 4338

S_o^2 = 0.7

The parameters utilized in the model calculation are summarized in Table 1. The radial distribution function F(R), obtained from the Fourier transform of both the experimental data and the model calculation are displayed in Fig.1 b). The experimental position of the Te shell disagrees with the model, probably because two different E_o values for Cd and Te shells are required.

In the mixed $Cd_{1-x}Mn_xTe$ (x=0.3, x=0.5) crystals, the first shell is shifted by about 0.6 Å toward lower R values, almost irrespective of Mn concentration. This might be caused by the different backscattering phase shift of Mn and/or by a shortening of the average near-neighbor bond length in the alloy. Our results are consistent with the lattice reduction displayed by X-ray diffraction analysis (1). The presence of the second and third near-neighbors' shells in Fig.2 b) indicates that in the solid solution the zincblende structure around Te is preserved, at least up to x=0.5.

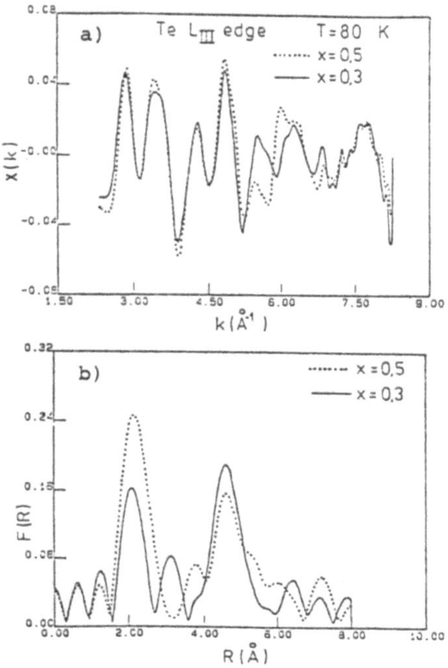

Fig.2.a) Same as in Fig.1 a) for different Mn concentrations. b) Radial distribution functions obtained from χ(k) of Fig.2 a)

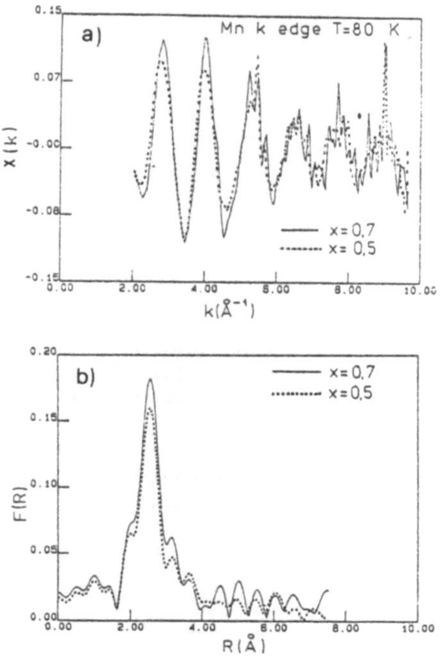

Fig.3.a) χ(k) above Mn K-edge for x=0.5 and x=0.7. b) Radial distribution functions obtained from χ(k) of Fig.3 a)

The Mn K-edge EXAFS of $Cd_{0.3}Mn_{0.7}Te$ and $Cd_{0.5}Mn_{0.5}Te$ measured at 80 K are presented in Fig.3 a). The striking feature of the corresponding F(R) is that only the nearest-neighbor Mn-Te shell is contributing, suggesting that the local environment around Mn is more disordered than around Te. Whether this effect depends on a distortion of the bonds or on disorder in the Mn and Cd positions in the cell cannot be inferred from the present data. According to X-ray diffraction (1), the Mn-Te distance decreases of 0.02 A from x=0.5 to x=0.7. This difference is hardly observable in Fig.3 b). Instead the variation of the Ga-As (In-As) distance was recorded as a function of x from the EXAFS spectra of $Ga_{1-x}In_xAs$ (8).

These problems deserve further studies, which are in progress.

(1)- M. Zimnal-Starnawska, Ph.D. Thesis , Jagellonian University, Cracow (1980).

(2)- M. Zimnal-Starnawska, M. Podgorny, A. Kisiel, W. Giriat, M. Demianiuk, J. Żmija, Submitted to J. Phys. C (1982).

(3)- PULS Activity Report LNF 80/79 (1980)
" " " LNF 82/51 (1982)
(4)- S. Mobilio, F. Comin and L. Incoccia, Internal Report LNF 82/19 (NT)(1982).
(5)- B.K. Teo and P.A. Lee, J. Am. Chem. Soc. 101, 2815 (1979).
(6)- R.F. Pettifer in "Inner-shell and X-ray Physics of Atoms and Solids" Ed. by D.J. Fabian, H.Kleinpoppen and L.M. Watson, Plenum,N.Y. (1980), p. 653.
N. Motta, M. De Crescenzi, A. Balzarotti, submitted for publication to Phys. Rev. B (1982).
(7)- E.A. Stern, B.A. Bunker and S.M. Heald, Phys. Rev. B 21, 5521 (1980).
(8)- J.C. Mikkelsen Jr. and J.B. Boyce (this volume).

Oxidation States of Nickel in Reduced Phases of $LaNiO_3$ by XANES and EXAFS

Michel Crespin, Pierre Levitz, and Lucien Gatineau

C.R.S.O.C.I., C.N.R.S. 1B, rue de la Férollerie, F-45045 Orleans Cedex, France and
L.U.R.E., C.N.R.S. Université Paris-Sud. Bât. 200, F-91405 Orsay Cedex, France

1. Chemical and X-Ray diffraction results

The reduction of $LaNiO_3$ perovskite has been carried out within a recirculation loop by hydrogen. The degree of reduction of the perovskite was measured manometrically by the hydrogen consumption. Moreover the all consummated hydrogen has been transformed into water trapped in a nitrogen getter.

During a reduction at increasing temperature (5°C per hour) the slope of the reduction curve changes several times (Fig.1). The observed plateaus correspond to the formula : $LaNiO_{2.5}$ (A) $LaNiO_2$ (B) $LaNiO_{1.5}$ (C).

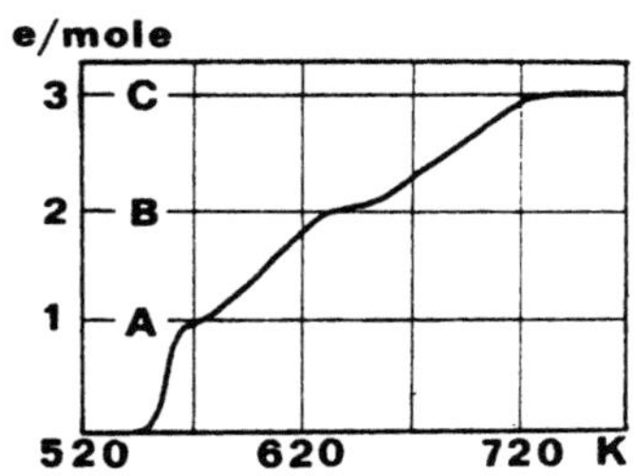

Fig.1. Reduction of $LaNiO_3$ at increasing temperature. The chemical composition of the reduced products are A : $LaNiO_{2.5}$, B : $LaNiO_2$, C : $LaNiO_{1.5}$

For these different stages A, B and C, the following chemical reactions can be written.

(A)	$2\,LaNiO_3 + H_2 \longrightarrow La_2Ni_2O_5 + H_2O$	(Ni^{2+})	(1)
(B)	$LaNiO_3 + H_2 \longrightarrow LaNiO_2 + H_2O$	(Ni^{1+})	(2)
	$2\,LaNiO_3 + 2H_2 \longrightarrow La_2NiO_4 + Ni^{\circ} + 2\,H_2O$	$(Ni^{2+}$ and $Ni^{\circ})$	(3)
(C)	$2\,LaNiO_3 + 3H_2 \longrightarrow La_2O_3 + 2\,Ni^{\circ} + 3\,H_2O$	(Ni°) .	(4)

The X-ray diffraction shows in the A and B stages two new single phases.
A : monoclinic symmetry with parameters : a = 11.068 Å ± 0.002, b = 11.168 Å ± 0.002, c = 7.824 Å ± 0.002, β = 92°21 ± 0.01
B : tetragonal symmetry with parameters : a = b = 3.966 Å ± 0.001, c = 3.376 Å ± 0.001.

Reaction number 3 in which the existence of La_2NiO_4 is assumed (see the structure determined by RABENAU and ECKERLIN [1] [2]) is not compatible with this result.

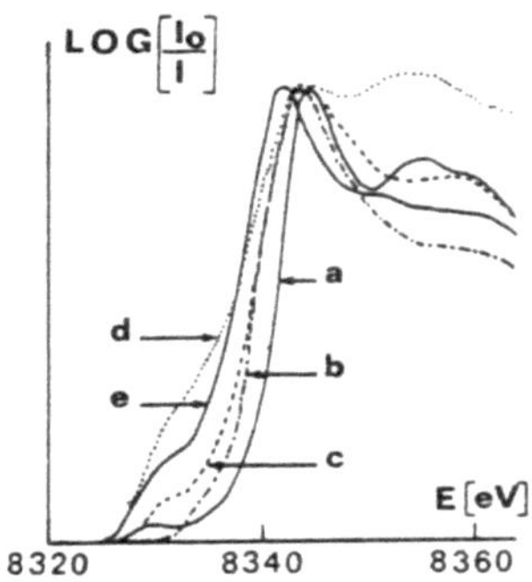

Fig. 2

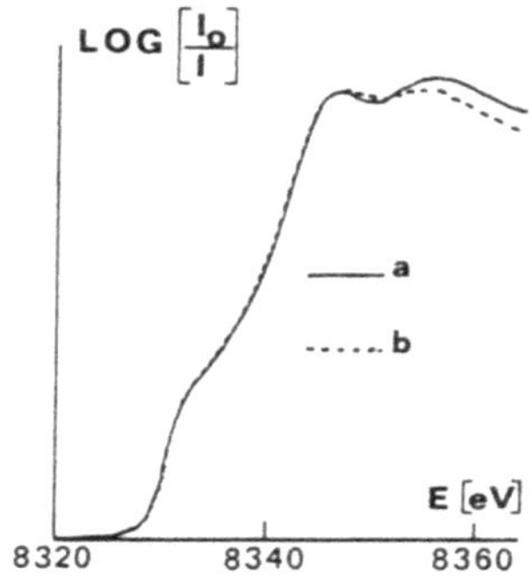

Fig. 3

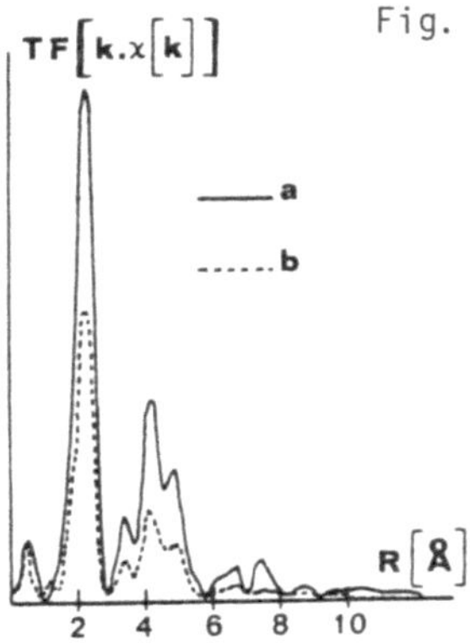

Fig.2. Ni K edge of a : $LaNiO_3$ b : La_2NiO_4, c : $La_2Ni_2O_5$, d : Ni metal, e : $LaNiO_2$

Fig. 3. Ni K edge of a : Ni metal b : Ni in the final stage of the reduction

Fig. 4. EXAFS radial distribution of a : Ni metal, b : Ni in the final stage of the reduction

At the final stage of the reduction (reaction 4) the X-ray diagram shows the La_2O_3 spectrum only. The nickel, in metallic form, appears only after recrystallisation at 1000 K under argon.

These results assume that the nickel, initially in the 3 + stage in $LaNiO_3$, is reduced in 2 + in $La_2Ni_2O_5$ in 1 + in $LaNiO_2$ and is in the metallic stage at the end of the reduction but, in this case, the Ni particles are very small and therefore they cannot be detected by X-ray diffraction.

2. XANES and EXAFS results

The different oxidation states of nickel in the three stages of the reduction has been determined by XANES and EXAFS to confirm the previous results.

2.1 Experimental

The energy selection was done by a channel cut monochromator using the 220 reflection of silicium. The energy increments were 0.25 eV for the measurement of the absorption edges and 1.5 eV for the EXAFS spectrum. The position of the absorption edges can be defined within an accuracy of ± 0.25 eV with respect to the maximum K edge of metallic copper (8991.2 eV). Immediately after the measurement of the nickel absorption edge of each sample the absorption edge of the metallic nickel was determined in order to avoid time-related shifts of the energy calibration.

2.2 Results

Figure 2 shows the K edges of nickel in $LaNiO_3$, $La_2Ni_2O_5$, La_2NiO_4 and $LaNiO_2$ after calibration with metallic nickel. All these curves have the same general shape but they show a shift of 2 eV to lower energies between $LaNiO_3$ and $La_2Ni_2O_5$ and between La_2NiO_5 and $LaNiO_2$. Moreover the absorption edge of $LaNiO_2$ cannot be obtained by a linear combination of the Ni^{2+} and Ni° absorption edges.

Consequently the oxidation states 2+ and 1+ can unambiguously be attributed to nickel in $La_2Ni_2O_5$ and $LaNiO_2$ respectively. The oxidation state of monovalent nickel has scarcely been observed [2,3]. In contrast to the rare examples where nickel was found in the + 1 state, $LaNiO_2$ is stable in air up to 450 K.

In the products of reaction (4) the absortion edge of nickel corresponds exactly to the edge of a standard nickel metal (Fig.3). Moreover the radial distribution function of this nickel as determined by EXAFS has the same shape that the radial distribution of a metallic nickel foil (Fig.4). In the final stage of the reduction the nickel is present as small particles with a structure similar to that of bulk nickel. The dimensions of these particles are between 15 and 25 Å diameters which are the detection limits of EXAFS and X-ray diffraction respectively.

References

1. Rabenau A. and Eckerlin P. (1958), Acta Cryst., 11, 304.
2. Hayes W. and Wilkens J. (1964), Proc. R. Soc., A 281, 340.
3. Gleizes A., Dartiguenave M., Dartiguenave Y., Galy J. and Klein H.F. (1977). J. of Am. Chem. Soc., 99, 5187.

Part V

Liquids and Disordered Systems

Application of EXAFS and XANES to Structural Studies of Metallic Glasses

Denis Raoux, Anne-Marie Flank, and Anne Sadoc
LURE, Bâtiment 209c, F-91400 Orsay Cédex, France

This paper is not an extensive review of the results obtained using EXAFS to investigate the local structure of metallic glasses. They are too numerous to be reviewed in a few pages. Moreover nice comprehensive reviews are available elsewhere (1,2,3). For these reasons, this paper focusses on a few questions. It tries to give some examples of the usefulness of EXAFS as well as of its limitations without going too far in the discussion of the implications of the novel structural insights yielded by EXAFS. These are however already very rich.

Metallic glasses are metastable alloys which exhibit the usual metallic property of high electrical and thermal conductivity. They however have special characteristics which are of great technological interest,such as their mechanical properties (high strength) and large corrosion resistance. However their major interest is the very low magnetic coercivity of the iron-based amorphous alloys, which will hopefully be used for making low losses transformers. This technological interest has recently led to a renewal of structural studies of metallic glasses which were previously considered as scientific curiosities. Metallic glasses cannot be elaborated as pure metals. In order to stabilize the glassy structure, it is necessary to alloy a transition metal with a metalloid such as Si, Ge, P or B, or to combine a late transition metal with an almost full d shell such as Ni or Cu with an early one such as Ti, Y or Zr, or a rare earth atom. Metallic of glasses technological interest are even ternary or even more complex alloys. The properties of the glass depend much on the composition of the alloy. The problem of atom identification in the radial distribution function (RDF) is then of crucial importance. This is a major reason to use EXAFS for structural studies of metallic glasses since it has the ability of probing the environment of each species in a multicomponent system. Suprisingly there are not many examples of EXAFS having been measured on each edge, even in a binary glass.

The major results to be determined are whether or not the glass structure exhibits a chemical ordering and, if it does, how it compares to that found in the crystalline state. This is still controversial for metal-metal (M-M) which can be elaborated over a wide concentration range and for which there is no simple structural model. Metal-metalloid (X-M) alloys have been more extensively investigated by diffraction techniques, so that structural models exist. We will try to demonstrate the ability of EXAFS to yield new structural insights (part II). We also have to discuss how the dramatic effects of disorder may be used to extract structural information which are not available from conventional scattering experiments (part III). Part one is devoted to XANES and edge structures which yield information on the medium range structure, or on electronic properties.

1. Edge Structure and Xanes

Right at the edge, the X-ray absorption measures the local density of empty states of a specified symmetry. A direct comparison of the absorption edges for the glass to those of the elemental components gives some information on the electronic changes experienced by atoms in the amorphous alloy. In the case of metal-metalloid amorphous alloys, DE CRESCENZI et al. (4) have observed a strong reduction of the intensity of the first peak at the Fe K-edge in $Fe_{80}B_{20}$ in comparison with bcc Fe. They claim that this is evidence for a charge transfer from metalloid to metal, causing a decrease of the empty states of the metal and a rather strong iono-covalent bonding of the metalloid to its metal neighbours. Similar effects have been observed later on for other X-M glasses $CO_{80}B_{20}$ (5) or $Co(Ni)_{80}P_{20}$ (6). The direct comparison between the edge for amorphous and crystalline states (5, 6, 7) shows that they have very similar shapes. However for Cu_2Ti, we have observed a shift between the edges of the crystalline and amorphous state, in opposite direction on Ti and Cu edges, which could be interpreted as an indication of a less ionized state in the glass. More systematic studies are needed. Major advances have been performed recently by theoreticians, so that it is now possible to do ab initio calculations of the near edge structure starting at a few eV from the edge jump. We like to mention a first application to the XANES on the Ge K edge in amorphous $Pd_{80}Ge_{20}$ (11) which demonstrates the sensitivity of the calculations to changes in the local symmetry. The authors show that all the principal features of the experimental XANES are reproduced using a three shell cluster. The elementary unit chosen was a capped distorted trigonal prism as in crystalline Pd_3As in agreement with the model of a stereochemically defined structure for metal-metalloid glasses (12). Though the comparison with experiment is only warranted at a qualitative level at this stage, this work is very promising. It makes us very confident than in a near future, XANES interpretation will give useful information on triplet correlation functions, or even higher order correlations. This might become a major bonus for X-ray spectroscopy since all the other scattering techniques (X-rays, neutron diffraction, etc...) are only sensitive to pair correlations functions.

2. Determination of partial radial distribution functions

EXAFS has the ability to probe the local environment of each species in a multicomponent glass while conventional diffraction experiments give information on the average of the atomic arrangements. Of course special techniques such as the anomalous scattering in X-ray diffraction, or the isotopic substitution in neutron scattering may also give partial RDF. However EXAFS has a specific advantage : the energy dependence of the backscattering amplitude $|f(\pi)|$ changes with the atomic number, so that it is possible to identify the backscattering atom with more efficiency than using scattering data. This has been used to investigate chemical ordering in various metal-metal glasses for compositions rich in late transition metals such as Cu_2Ti (9), Ni_2Y and Cu_3Zr_2 (13,14) as well as for $DyFe_2$ and $TbFe_2$ alloys (15). In all cases EXAFS has evidenced a preferential binding between unlike atoms which implies chemical ordering. In an alloy less concentrated in copper, $Cu_{46}Zr_{54}$ which has been studied by two groups (16, 21) there is however no clear indication of a chemical order. We have also measured EXAFS and X-ray scattering for Cu_xY_{1-x} with x ranging from 0.9 to 0.33 (17) and found that the degree of chemical order increases with the copper concentration. Similar conclusions have been derived from a neutron scattering study of Cu_xTi_{1-x} amorphous alloys (10). The mechanism leading to such a composition-dependent chemical ordering is not quite clear ; both covalent and partially ionic bondings may be suggested (3).

For concentrations rich in early transition metals M" such as $M'M''_2$ alloys the structure of the glass has been found to resemble that of the crystalline phases (9, 14, 17) of similar composition. Partial coordination numbers have almost the same values (their ratios are far from what would be expected from the composition). The topologies are also very similar. EXAFS analyses (13, 14, 17) show that each coordination shell in the glass is disordered and may be resolved into two widely apart subshells. For an example, the Cu-Cu RDF is splitted into two subshells at 2.52 and 3.00 Å with partial coordination numbers of respectively 3.75 and 2 (14). This is reminiscent of the crystalline structure of the $Cu_{10}Zr_7$ phase which is very complex and exhibits a lot of internuclear distances spread out over more than 0.5 Å.

It is interesting to notice that the structural order induced by the M'-M" binding dominates the local structure even in a very concentrated alloy such as Cu_9Y (17). This is an almost "pure" amorphous metal. If its structure results from random packing of two different spheres, each yttrium atom would be surrounded by at least 14 copper atoms at a well-defined distance close to the sum of the atomic radii 3.06 Å. Indeed EXAFS on the yttrium K edge shows that yttrium is surrounded by 6 copper atoms at a much shorter distance of 2.90 Å (fig.1). The fit is

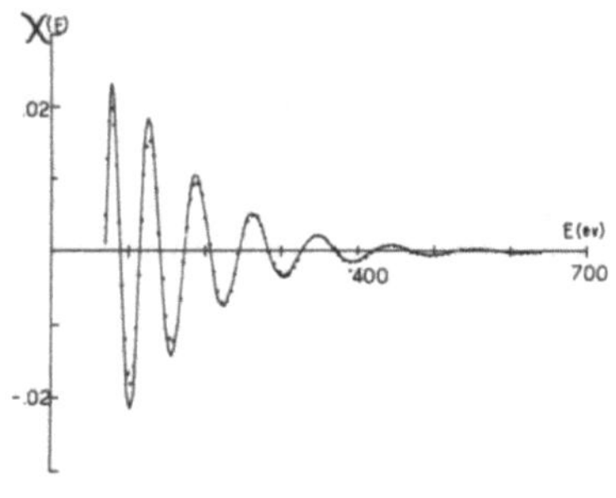

Fig.1. EXAFS for a Cu_9Y amorphous alloy on the yttrium K edge. Data show the Fourier filtered data. The full line is the reconstruction for 6 copper neighbours at 2.90 Å. The amplitude of backscattering by Cu has been extracted from EXAFS on a Cu foil. The phase shift has been determined from EXAFS on a yttrium foil and corrected for the difference between backscattering phase shifts by Cu and Y atoms using TEO and LEE calculations (22)

very good using the Gaussian approximation for a reasonable value of the Debye-Waller parameter $\sigma = .11$ Å. This is exactly the environment of the yttrium atoms in the crystalline phase Cu_5Y : 6 copper neighbours at 2.875 Å (there is no crystalline phase less concentrated in yttrium). A second shell of 12 copper atoms at a much larger distance of 3.23 Å ($d1/d2 \sim \sqrt{3/2}$ as in the bcc structure) does exist in the crystalline state. The EXAFS data do not provide any evidence for it in the glassy state, probably because of a larger disorder. We have also measured the X-ray scattering. The RDF calculated from scattering data show a proeminent slightly asymmetric peak corresponding to a coordination of copper atoms by 9 copper atoms at 2.51 Å in very good agreement with EXAFS on the Cu edge. The weak tail on large distance side can be fitted by Y-Cu pairs at 2.90 Å. There is also a weak and broad peak at about 3.17 Å which can be attributed to the Cu-Y correlations which are washed out in EXAFS because of their larger disorder. There is then a nice agreement between EXAFS data and the less-resolved scattering data. We thus believe that this model is correct. It implies that the structure of the glass is dominated by some local unit reproducing the local structure of the Cu_5Y crystalline phase. Excess Cu atoms in the Cu_9Y glass probably link such units resulting in a larger coordination number for Cu-Cu pairs in the glass than in the Cu_5Y crystalline state (n = 7.2). This yields a model which is very similar to the so-called stereochemically defined unit model which has been proposed by GASKELL for metal-metalloid amorphous alloys (12) with concentration in the range $X_{20}M_{80}$. The local unit is of course different since the dilute species has a larger size that the host atoms in such an alloy, while metalloid atoms are smaller. We point out that scattering data alone could not give such a detailed picture.

These examples do show significantly that EXAFS has already given some new insight in the understanding of the structure of metal-metal glasses.

The determination of the partial RDF's is more difficult for metal-metalloid amorphous alloys because the low Z metalloid atoms have rather weak backscattering amplitudes (X-ray scattering suffers from similar limitations). This has led HAYES et al.(18) to replace Si by Ge in $Pd_{80}Ge_{20}$. Their careful analysis of the spectrum on the Ge K edge gave a very convincing support to the model first proposed by SADOC et al.(20) on the basis of scattering data. HAYES et al. could demonstrate that there is significantly less than one germanium neighbour around a germanium atom. This model implies a very strong chemical ordering by some kind of covalent bonding between metal and metalloid atoms. Other experiments (2, 4, 5, 19, 21) have been made on the metal K edges. To our knowledge LAGARDE et al.(6, 13) have been the only group to measure EXAFS on both the metal and the metalloid K edges for amorphous $Co_{80}P_{20}$ and $Ni_{80}P_{20}$. (the phosphorous edge lies in the soft X-ray range at about 2100 eV). They found no evidence of P-P correlations and a ninefold coordination of phosphorous atoms by cobalt or nickel ones in good agreement with X-ray data (20) and the previous study of $Pd_{80}Ge_{20}$ (18). They however had to introduce asymmetry in the P-Co distribution function in order to get the same mean value of the internuclear distance as that given in (20), and a slighter one for Co-Co distributions. Their analysis is at variance with the one performed by CARGILL (19) for the cobalt K edge of the same glass. CARGILL did not measure the EXAFS on the phosphorous edge so that he had to make some assumptions concerning the relative contributions of Co-Co and Co-P pairs to the EXAFS modulations on Co edge. He assumed that the latter ones were dominant. This implied a greater sharpness of the Co-P pairs distribution which does not seem to be observed from the EXAFS spectrum on the phosphorous edge (6). It also results in a too large distance (2.64 Å) for Co-Co pairs. This controversy gives an example of the difficulties which can been encountered for metal-metalloid glass as specially when only the metal K edge can be measured.

Finally, two points should be made. First, the number of parameters to be determined in order to describe the local structure is large. In a binary alloy, there are at least nine of them, and even more than that if the glass is heavily disordered. In that case it is necessary to consider third-order moments in the RDF, or even higher order ones. In order to determine so many parameters, it is necessary to measure EXAFS on both edges, and even also to make another independent experiment such as X-ray scattering. A single experiment cannot yield more than $n \sim 2\Delta r \Delta k/\pi$ independent values of parameters for a shell isolated in a Δr range and a spectrum extending over Δk. The second remark is related to the disorder effects which will be discussed in part III. They may lead to an underestimation of coordination numbers. This is a serious problem since the question of whether or not chemical ordering occurs depends entirely on the determination of coordination numbers. However, if the degrees of disorder for the partial coordinations are not too different, the relative coordination numbers may be of significance even if the total coordination is underestimated, so that we may be able to characterize the eventual chemical order.

3. Sensitivity of EXAFS to the spread of the internuclear distances

It is well known that EXAFS is much more sensitive to the spread of distances than X-ray or neutron scattering are. This is due to the larger k range spanned in an EXAFS experiment ($2k \sim 30$ Å^{-1}). X-ray or neutron scattering are usually restricted to a 15 Å^{-1} wide domain, so that the RDF derived are less resolved than those calculated by Fourier transforming EXAFS data. This is illustrated by fig.2b and 2c which show the Fourier transforms of EXAFS on zirconium K edge for a $Cu_{46}Zr_{54}$ amorphous alloy in the as quenched state (2b) and after annealing at a temperature higher than the glass transition temperature (2c). In the as quenched sample two peaks are well resolved which correspond respectively mainly to Zr-Cu and Zr-Zr pairs (16, 21). Such peaks are not separated in a scattering experiment, except when using very high energy neutrons from spalliation sources. After thermal relaxation (fig.2c) another peak grows up, which is not evidenced by the EXAFS spectra on the Cu edge. It is due to the ordering of a second shell of about five zirconium atoms at 3.77 Å from the central zirconium atom. A weaker ordering is also found for the low-distance Zr-Zr coordination shell by 5 atoms at 3.15 Å. This is a clear example showing directly the occurence of two sub-shells in the distribution of like atom pairs in a glass. Interestingly it is not found for the

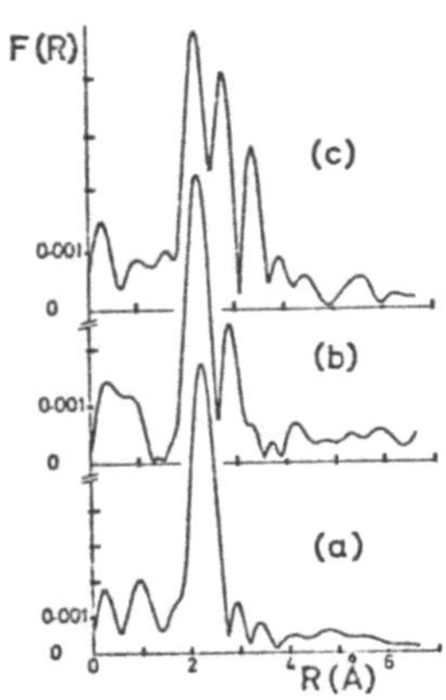

Fig.2. Magnitude of the Fourier transform of EXAFS on the zirconium K edge for an amorphous $Cu_{60}Zr_{40}$ alloy (a), a $Cu_{46}Zr_{54}$ one (b) in the as quenched state, and after a thermal treatment at a temperature higher than the glass transition temperature (c)

crystalline state (16), so that it cannot be attributed to a partial recrystallization of the sample in that crystalline state. It evidences a structural change which is to be related to the irreversible changes in the electronic properties induced by thermal relaxations of metallic glasses. WONG (2) has also reported annealing effects in $Fe_{40}Ni_{40}P_xB_{x-20}$ metglasses which result in a decrease of the σ^2 disorder parameter. These are two promising examples showing that the high sensitivity of EXAFS may be very useful to analyze at an atomic scale the structural relaxation of the metastable glassy state. This is of special importance since the relaxation has very drastic effects on the electronic and mechanical properties.

However EXAFS suffer from an intrinsic limitation because it is usually not analyzed for low values of k(2k < 6 to 8 $Å^{-1}$). This means that information on the broad features in the RDF are lost, so that coordination numbers may be severely underestimated if disorder effects are not carefully examined. EXAFS also yield incorrect values for internuclear distances which correspond to the steepest part of the RDF. This has been demonstrated by EISENBERGER and BROWN for hcp Zn (23) and recognized for metallic glasses by HAENSEL et al. (21) for the first time. It is illustrated by fig.3a which shows the experimental spectrum for Ni_2Y glass on yttrium K edge (dots) and the simulation which has been calculated using the coordination number and distances determined from a joint X-ray and neutron scattering experiment (24).

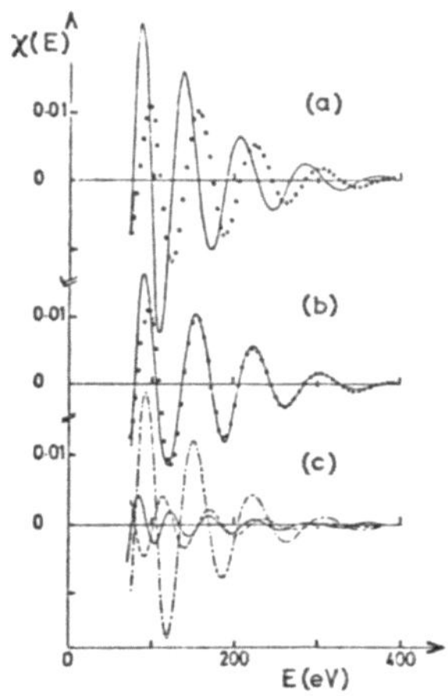

Fig.3. EXAFS of Ni_2Y glass on the yttrium K edge. Experimental data : dots. Solid lines show theoretical reconstructions calculated using TEO and LEE's (22) amplitudes and phase shifts for respectively the structural model given by scattering data (a) and the subshells model discussed in text (b). The curve c shows the partial contributions of -.-.-. 9 Ni atoms at 2.71 Å, ------ 3 Ni at 3.05 Å and 4 Y at 3.40 Å

It is obvious that it does not fit the data. A fit would give a Y-Ni distance of about 2.72 Å, while the scattering experiments yield 2.80 Å, and a coordination number of 9 nickel atoms around the central yttrium atom while the correct value from (24) is 12. Similar difficulties have been encountered for a lot of metal-metal glasses (9, 13, 14, 17) but also for metal-metalloid ones at least if the metalloid atom is smaller than the metallic host, which is the case for boron and phosphorus (5, 6, 8, 21).

Recognizing these difficulties has led to some discouragement in recent years when experimentalists realized that EXAFS was not as straightforward and unambiguous as they have thought. The ambiguity may be passed by making use of scattering data which do not suffer so severely from disorder effects. The mean value of the internuclear distance, and the total coordination number are given by the scattering experiment ; EXAFS then measures departures from the average Gaussian distribution yielded by scattering. This can be done by expanding the RDF into momentums (23), or by a modelling of the structural disorder. This last procedure has been used first by HAENSEL et al.(21) who introduced an asymmetric distribution with a sharp rise followed by an exponential decrease with distance. It allows an easy analytical integration of EXAFS over the RDF and seems to be quite useful to describe structures which are not too heavily disordered, like metal-metalloid alloys (6, 8, 21). For other materials especially for metal-metal glasses $M_xM'_{1-x}$ with x between 0.3 to 0.7, we have found out that this distribution does not simulate well the actual distribution function, and that a more flexible model should be used. A simple way is then to use two more or less overlapping Gaussian distributions (13, 14). Fig.3b shows that such a model fits the experimental curve for the Ni_2Y glass. Around the yttrium atom, it corresponds to an environment of 9 nickel atoms at 2.71 and 3 ones at 3.05 and 4 yttrium at 3.40 Å. This model yields a coordination of 12 nickel atoms at an average distance of 2.80 Å, which is exactly the structural model given by the scattering data (24). Similar results have also been obtained for Ni-Ni pairs (14). Such distortions of the RDF are to be related to very short internuclear distances for unlike atoms. For example in the Ni_2Y case, the Y-Ni mean distance of 2.80 Å is considerably shorter than its value of 2.98 Å in the crystalline state, and than the sum of the atomic radii (3.03 Å). It is then not possible to pack 12 nickel atoms around an yttrium one without a distortion of the RDF. This in turn induces a distortion of like atom pairs. Such a modelling by subshells may be meaningful since it is reminiscent of the structure of the corresponding crystalline intermetallic phases (13, 14, 17). It seems also to describe the structure of metalloid-metal glass for rich metalloid concentration like Ni_2B (7) or $Co_{76}B_{25}$ (5).

4. Conclusions

EXAFS has already given very interesting new insight into the local atomic structure of glasses, especially for metal-metal amorphous alloys.

It has evidenced the occurence of a well-defined chemical ordering at least for alloys which are concentrated in late transition metals, and its dependence on composition. In several circumstances the local structure has been found to be reminiscent of that in the corresponding crystalline state. However disorder effects are severe and they have to be seriously examined. If only EXAFS measurements are available, the data extracted may give a rather wrong picture of the structure if it is too much disordered. EXAFS has to be compared with other diffraction techniques. When used conjunction with them, it gives very detailed information which is not accessible by conventional scattering techniques.

References

1 T. M. Hayes, J. Non-Cryst. Solids, 31, (1978), 57.
2 J. Wong, "Metallic GlassesI", edited by H. J. Guntherodt and H. Beck (Springer-Verlag, Berlin 1981), 71.
3 S. J. Gurman, J. Mat. Science, 17, (1982), 1541.
4 M. De Crescenzi, A. Balzarotti, F. Comin, L. Incoccia, S. Mobilio, D. Bacci, J. Phys. Colloq. C8, (1980), 238.

5 J. M. Dubois, Thèse d'Etat, Université de Nancy, (1981).
6 P. Lagarde, J. Rivory, G. Vlaic, To be published.
7 J. Wong, This book.
8 M. De Crescenzi, A. Balzarotti, F. Comin, L. Incoccia, S. Mobilio, N. Motta, Sol. St. Comm. 37, (1981), 921.
9 D. Raoux, J. F. Sadoc, P. Lagarde, A. Sadoc, A. Fontaine, J. Phys. Colloq. C8, (1980), 207.
10 M. Sakata, N. Cowlam, H. A. Davies, J. Phys. Colloq. C8, (1980), 190.
11 P. H. Gaskell, D. M. Glover, A. K. Livesey, P. J. Durham, G. N. Greaves, J. Phys. C15, (1982), 597.
12 P. H. Gaskell, J. Non-Cryst. Solids, 32, (1979), 207.
13 A. M. Flank, P. Lagarde, D. Raoux, J. Rivory, A. Sadoc, "Proc. 4th Int. Conf. on Rapidly Quenched Metals", (Japan Inst. Metals, Sendai), (1981), 393.
14 A. Sadoc, D. Raoux, A. Fontaine, P. Lagarde, J. Non-Cryst. Solids, 50, (1982), 331.
15 E. A. Stern, S. Rinaldi, E. Callen, S. Heald, B. Bunker, J. Magn. Magn. Mater. 7, (1978), 188.
16 A. Sadoc, A. M. Flank, D. Raoux, P. Lagarde, 5th Int. Conf. The Physics of Non Crystalline Solids, Montpellier 1982, To be published in J. Phys. Colloq.
17 A. M. Flank, Unpublished.
18 T. M. Hayes, J. W. Allen, J. Tauc, B. C. Giessen, J. J. Hauser, Phys. Rev. Lett. 40, (1978), 1282.
19 G. S. Cargill III, in "Proc. 4th Int. Conf. on Rapidly Quenched Metals" (Japan Inst. Metals, Sendai), (1981), 389.
20 J. F. Sadoc, J. Dixmier, in "The Structure of Non Crystalline Materials" edited by P. H. Gaskell (Taylor and Francis, London, 1977), 85.
21 R. Haensel, P. Rabe, G. Tolkiehn, A. Werner, in "Proc. NATO Adv. Study Inst. : Liquid and Amorphous Metals", Edited by E. Luscher (D. Reidel, Dodrecht, 1980).
22 B. K. Teo, P. A. Lee, J. Am. Chem. Soc. 101, (1979), 2815.
23 P. Eisenberger, G. S. Brown, Sol. St. Comm. 29, (1979), 481.
24 D. Givord, A. Lienard, J. P. Rebouillat, J. F. Sadoc, J. Phys. Colloq. 5, (1979), 237.

EXAFS and X-Ray Diffraction in Solutions

Giovanni Licheri and Gabriella Pinna

Istituto di Chimica Fisica e Industriale, Università di Cagliari,
I-09100 Cagliari, Italy

Aqueous solutions of electrolytes have been, among liquid systems, those most widely studied from a structural point of view, mainly because of the interest in the ionic hydration which is a central problem in chemical physics. The most widely used structural technique has been up to now X-ray diffraction which, in spite of its limitations when applied to non-crystalline materials, has provided, in recent years, important information about the structure of solutions [1-3]. The recent development of the EXAFS spectroscopy opens now new perspectives, so that it can be of interest to compare the two techniques in order to understand their capabilities in obtaining a reliable picture of the structural behaviour of aqueous solutions.

Let us recall some basic equations in X-ray diffraction. For a system with randomly oriented local environments and containing more than one atomic species, the "total structure function" i(s), that is the structurally sensitive part of the scattered intensity, is related to the pair correlation functions according to [4,5]:

$$s\, i(s) = 4\pi\rho_o \sum_\alpha \sum_\beta x_\alpha x_\beta f_\alpha(s) f_\beta(s) \int_0^\infty \left[g_{\alpha\beta}(r) - 1\right] r \sin(sr)dr \qquad (1)$$

where f_α is the atomic form factor, x_α is the concentration of the α-type atoms, $s = 4\pi\sin\vartheta/\lambda$ is the momentum transfer for elastic scattering, 2ϑ being the scattering angle and λ the incident wavelength. The sums are taken over the composition unit to which the intensity is normalized. $g_{\alpha\beta}$ is the pair correlation function of β-type atoms around an α atom, defined so that $4\pi\rho_o x_\beta \int r^2 g_{\alpha\beta} dr$ is the total number of β species contained in the sample of mean number density ρ_o. The total structure function is a weighted sum of contributions from each atom pair and obviously, using only one scattering experiment, it is not possible to decompose it uniquely into its independent components, which number $N(N+1)/2$ for a N-component system. In principle, this might be accomplished by taking different independent measurements in which the weight functions are changed, for example using anomalous X-ray scattering techniques as will be discussed by BIENENSTOCK [6] and EISENBERGER [7] elsewhere in this book.

For a single scattering measurement it is useful to visualize the structural information in r space defining a total modified correlation function G(r) through a sine Fourier transform in the complementary variables s and r. This radial function is related to the pair correlation functions $g_{\alpha\beta}$ through a convolution equation [8]:

$$4\pi\rho_o r\, G(r) = 4\pi\rho_o \sum_\alpha \sum_\beta x_\alpha x_\beta \int_0^\infty u\, g_{\alpha\beta}(u)\, \frac{1}{2}\left[T_{\alpha\beta}(r-u) - T_{\alpha\beta}(r+u)\right] du \qquad (2)$$

where $T_{\alpha\beta}$ is a Fourier transform of $f_\alpha f_\beta$ multiplied by a suitable "modification function" [4,5]. The radial function accessible from X-ray diffraction is thus a linear combination of the true $g_{\alpha\beta}$'s, each of which convoluted with the known function $T_{\alpha\beta}$. $T_{\alpha\beta}$ may be visualized as a shape function into which peaks in $g_{\alpha\beta}$ are modified by the effects of electron distributions and data treatment [5].

At this point it is useful to compare the expressions written above with those used to analyse EXAFS data. The EXAFS interference function from the absorption of atom species α can be represented by:

$$k\,\chi_\alpha(k) = 4\pi\rho_0 \sum_\beta x_\beta\, A_\beta(k) \int_0^\infty g_{\alpha\beta}(r)\exp(-2r/\lambda)\sin\left[2kr+\Phi_{\alpha\beta}(k)\right] dr \qquad (3)$$

which shows similarities with (1). In order to interpret the structural information in real space one usually takes a complex Fourier transform in the complementary variables 2k and r. Using an approach introduced by HAYES [9,10] the Fourier transform of the EXAFS may be written by:

$$F(r) = 4\pi\rho_0 \sum_\beta x_\beta \int_0^\infty g_{\alpha\beta}(u)\ \frac{1}{2}\left[\Gamma_{\alpha\beta}(r-u)+\Gamma^{*}_{\alpha\beta}(-r-u)\right] du \quad . \qquad (4)$$

In analogy with X-ray diffraction, (4) expresses the radial function F(r) as a sum of the true pair correlation functions convoluted with the peak shape functions $\Gamma_{\alpha\beta}$ which involve the details of the electron scattering processes and the effects of data treatment.

If it is assumed that the distribution of distances about a mean value R is Gaussian, with root-mean-square deviation σ, (1) and (3) become respectively:

$$s\ i(s) = \sum_\alpha x_\alpha \sum_j f_\alpha f_j \frac{N_{\alpha j}}{R_{\alpha j}} \exp(-\frac{1}{2}\, s^2\, \sigma^2_{\alpha j})\sin(s\, R_{\alpha j}) \qquad (5)$$

$$k\,\chi_\alpha(k) = \sum_j A_j \frac{N_{\alpha j}}{R^2_{\alpha j}} \exp(-2\, R_{\alpha j}/\lambda)\exp(-2k^2\, \sigma^2_{\alpha j})\sin(2k\, R_{\alpha j}+\Phi_{\alpha j}) \qquad (6)$$

where the j sum is over the different shells around the origin atom. In both cases the modulation induced by thermal or static disorder is a Debye-Waller like factor. Note that in the case of disordered systems, the σ's have the same meaning in EXAFS and X-ray diffraction and both differ from the one defined in crystallography which represents the root-mean-square deviation of an atom about its mean position.

The comparison between the diffraction and EXAFS equations clearly shows analogies and differences between the structural information obtainable with the two techniques. The most evident analogy consists of the fact that in both cases the structural information appears as pair correlation functions, the extraction of which can be considered the ultimate aim of both experimental studies. The most evident difference consists of the presence of a double sum over the atoms species in the diffraction equations.

As a consequence some disadvantages in X-ray diffraction are evident: i) the presence of the concentration x_α makes diffraction data strongly dominated by the prevalent species, so that the structural information about low concentration species is suppressed; ii) diffraction data contain all the N(N+1)/2 pair correlation functions. Therefore, in the total radial function it is unlikely to find a peak ascribable to only one $g_{\alpha\beta}$, so that a direct peak analysis is difficult and can lead to wrong evaluation of coordination numbers.

In X-ray diffraction it is thus common to resort to modeling techniques in order to determine the structural parameters. But a model which is to be tested against diffraction data must involve the whole structure and meet certain conditions in order to be realistic and tractable [4,5]. In short, the model must describe the distance spectrum around each species taken as origin atom, fulfilling the requirements that structurally equivalent atoms have the same distance spectra and that the distance spectra of different atoms are geometrically compatible. This means that we have to construct a sort of space lattice and then embody in it the effects due to the progressive loss of positional correlation as distance increases. Obviously, the result of this approach cannot be proved to be unique, and this represents a major complication in interpreting diffraction data.

On the other hand, there are differences which reveal some disadvantages of EXAFS: i) the pair correlation functions enter multiplied by r in X-ray diffraction and by $\exp(-2r/\lambda)$ in EXAFS. As a consequence, in EXAFS the long range structural information is almost completely lost; ii) the typical range of momentum space covered by the two techniques, is different being $s = 0.5 \div 16$ Å^{-1} in X-ray diffraction and $2k = 6 \div 30$ Å^{-1} in EXAFS. This gives rise in EXAFS to a greater resolution in real space, but the absence of low k data makes EXAFS strongly sensitive to the damping in amplitude due to the disorder term. In disordered systems, such as solutions, this restricts the structural information essentially to the nearest-neighbour shell.

As a consequence of this comparison the two techniques appear to be complementary, in that in studying the same system, they are more sensitive to different aspects of the pair correlation function [9-12]. EXAFS gives a better and more direct description of the nearest-neighbour environment, while diffraction detects better the longer range order. Both these aspects are important to understand the structural behaviour of a system, so that by a combined use of the two techniques one can take advantage of the potentialities of both and overcome their limitations.

It would be interesting to compare the results obtained by EXAFS and X-ray diffraction on the same systems when both techniques are able to obtain the same information in quite a direct way. For electrolyte solutions this situation can occur for the first hydration shell of some metal ions; in this case the diffraction G(r) usually presents a well-defined peak [1] which can be analysed directly. Unfortunately, the available data on exactly

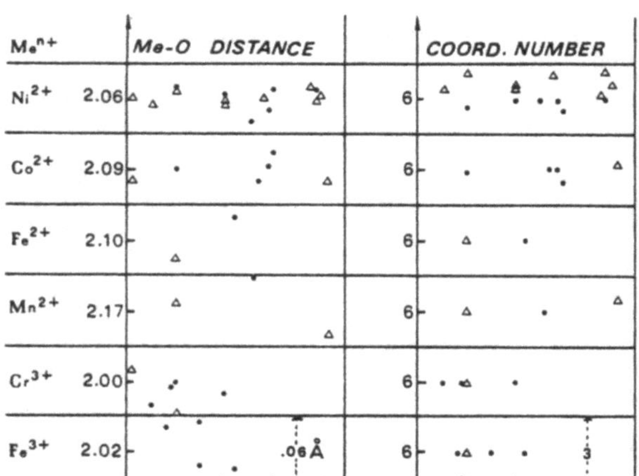

Fig.1. First hydration shell in MeX_n aqueous solutions, with $X = Cl^-$, NO_3^-, ClO_4^- (dots: X-ray diffraction; triangles: EXAFS)

the same solutions are not many for such a comparison to be extensive. Nevertheless Fig.1 reports distances and coordination numbers for the first hydration shell of different metal ions in aqueous solutions with chloride, nitrate or perchlorate as counter ions [13-30]. For these solutions cationic hydration can be considered specific to the cation and not significantly disturbed by the nature of the counter ion, so that an overall comparison is meaningful. It appears to be a good agreement between the results of the two techniques.

Looking at the aqueous solutions studied by EXAFS, some remarks are in order: only in a few cases EXAFS data cover concentrations not accessible to diffraction and studies on coordination of the anions are very few. Yet the great advantages of EXAFS consist in the possibility of studying low concentrations and of choosing independently both ions as origin for the structural information. To our knowledge, the only studies within this outline are those carried out by the ORSAY and MESSINA groups who have studied some divalent metal chloride and bromide solutions [31-33], revealing in the case of Zn and Cu ions the existence of chemical complexes involving the anion and embedded in extended structures.

A problem extensively investigated by EXAFS [13-15,32] and X-ray diffraction [15-21] as well as by neutron diffraction [34,35], incoherent neutron spectroscopy [36], Raman spectroscopy [37], regards the structuring of the Ni^{2+} ion. This has been a point of some discussion in the recent years and a subject of controversy. Here we wish to report some results of our EXAFS and X-ray diffraction investigations on aqueous solutions involving this ion. This can be an example of a complementary use of the two techniques. The main question of this study regards the point of whether inner or outer sphere, or no nickel-chloride coordination is present in nickel chloride concentrated solutions.

We first considered solutions of nickel nitrate [38]. In them, only nickel-oxygen coordination is present, so they can be used as a reference for studying nickel chloride solutions. The EXAFS Fourier transforms of four solutions examined, ranging in concentration from 0.5 to 4 Molar, were very similar one to the other, and similar to the one obtained from the hydrated crystals, suggesting that the same nickel-water coordination, which is known to be in the crystal [39] exists also in solution. This indication was confirmed by interpreting the filtered spectra in terms of one shell of oxygen atoms around the nickel ion. The crystal was used as model compound and amplitude and phase shifts were obtained from it fixing the structural parameters to the known values r = 2.06 Å and N = 6. About the same values, within experimental errors, are found for the solutions in the whole concentration sample.

As a second step, we examined two nickel chloride concentrated solutions about 2 and 4 molar in concentration. Fig.2 displays Fourier transforms and Fourier filtered spectra for the chloride solutions compared with those for nitrate solutions of about the same concentration. This direct comparison indicates, without any further data treatment, that the structure of the nearest-neighbour shell of the nickel ion is the same both in nitrate and chloride solutions, that is, about six water molecules. No first sphere nickel-chloride coordination is revealed. These findings are also confirmed by least-square fitting the filtered spectra. In Table 1, the best fit values of distances and coordination numbers are presented, together with those we obtained by X-ray diffraction [18], those obtained by SANDSTROM [14] and those obtained by NEILSON and ENDERBY in a neutron diffraction study [34]. All data point to the same result for the first coordination shell of the nickel ion.

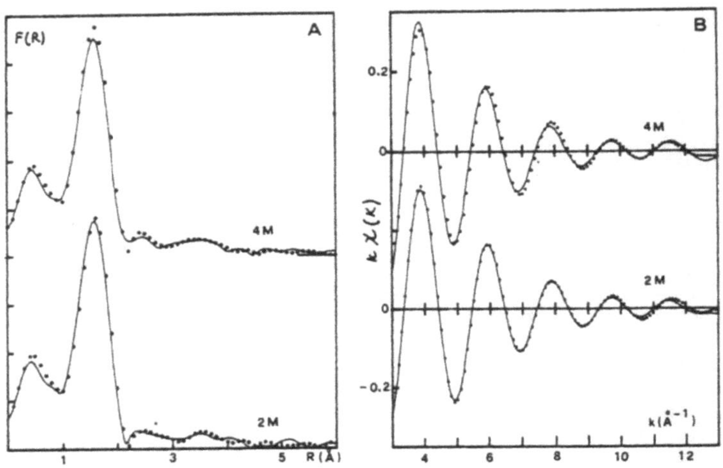

Fig.2. Fourier Transforms (A) and first shell filtered spectra (B) for $Ni(NO_3)_2$ (dots) and $NiCl_2$ (full lines) aqueous solutions

Table 1. Ni^{2+} Hydration in $NiCl_2$ aqueous solutions

	EXAFS		X-ray[18]		EXAFS[14]		Neutron[34]	
Molarity	R,Å	N(±20%)	R,Å	N	R,Å	N(±20%)	R,Å	N
~ 4	2.06(1)	7.2	2.07(1)	6	2.07(1)	6.2	2.07(3)	5.8(2)
~ 3					2.06(1)	7.0	2.07(2)	5.8(2)
~ 2	2.06(1)	6.7	2.06(1)	6				

A second aspect of this structural investigation concerns the possible presence of outer sphere chloride ions around the cation. The radial functions of Fig.2 do not show any evidence of other shells beyond the first as no significant structure emerges among the ripples in the region beyond 2 Å. This is not surprising if only weakly interacting water molecules or chloride ions are present beyond the first hydration shell. However, it must be observed that the Fourier transforms of Fig.2 appear different from the ones reported by SANDSTROM [14] which present a shoulder in the peak extending to about 4.5 Å. SANDSTROM explains this structure as due to the presence of at least one other coordination shell with about 3 outer sphere chloride ions at a distance of 3.1 Å.

Information about the structuring beyond nearest environment can be provided by X-ray diffraction data. We found that a model which involves hexahydrated ions and a second shell of only water molecules around the cation fits very well X-ray diffraction data [17,18]. It is worth noting that this structural picture has been confirmed by neutron diffraction results [34,35]. Anyway, alternative models involving outer sphere chloride ions were tested, but only in the 4 molar solution a weak evidence of nickel-chloride correlation was found at a distance of about 4.3 Å. On the contrary, no acceptable fit could be obtained with models where chloride ions approach more closely the hydrated cations: diffraction data are incompatible with a short nickel-

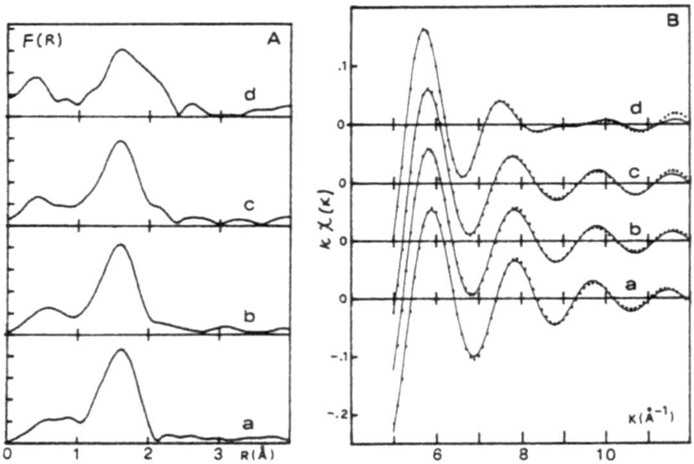

Fig.3. A: Fourier transforms for aqueous solutions with Cl^-/Ni^{2+} ratio equal to 2(a), 4(b), 5(c) and for $NiCl_2$ hydrated crystal (d). B: Experimental (dots) and fitted (full lines) filtered spectra for the same cases as A

-chloride distance and with the unusually short chloride-water distances it would imply. We can therefore conclude that even in concentrated solutions, short distance nickel chloride coordination does not occur.

The situation can change if the ratio chloride to nickel increases beyond the value of two it has in stoichiometric solutions. Here some preliminary EXAFS results are reported on two solutions in which nickel-chloride close coordination is expected to be present. The solutions were obtained by adding lithium chloride to nickel chloride solutions 2 molar in nickel. The ratios chloride to nickel were respectively 4 and 5. In this case diffraction results [40] suggest inner sphere complex formation between nickel and chloride ions.

Fig.3A shows a comparison among the Fourier transforms of four cases: the stoichiometric solution, the two enriched solutions and the nickel-chloride hydrated crystal, in which the first shell around the cation is composed of 4 water molecules (R = 2.05) and 2 chloride ions (R = 2.36) [41]. It is apparent that the solutions display some evolution towards the crystal, suggesting an increasing degree of complex formation as the chloride-nickel ratio increases. For the analysis of the filtered spectra (Fig.3B) the crystal was used as model compound to obtain chloride backscattering amplitude and nickel-chloride phase shift, fixing the oxygen amplitude and nickel-oxygen phase shift to the values obtained from the nickel-nitrate crystal.

Table 2 gives the best fit results for distances and coordination numbers obtained with a two-shell fit. The evidence of nickel chloride coordination is clear, despite the high estimated uncertainties of the chloride parameters. On the other hand, an attempt to reproduce the experimental functions with only oxygen atoms gave rise to an apparent mismatch.

In conclusion, we again wish to emphasize that X-ray diffraction and EXAFS contain complementary information. Indeed, at present, the inherent capabilities of both techniques often suffer of limitations due to experimental constraints; but this is a further reason for making use of both in approaching the structure of aqueous solutions.

Table 2. Distances and coordination numbers for $NiCl_2$ 2M with different $[Ni^{2+}]/[Cl^-]$ ratios

$[Cl^-]/[Ni^{2+}]$	R_{Ni-O}(Å)	N_{Ni-O}	R_{Ni-Cl}(Å)	N_{Ni-Cl}
2	2.06±.01	6.5±1	-	-
4	2.05±.01	6.0±1	2.38±.03	0.7±.3
5	2.05±.01	5.3±1	2.38±.03	1.0±.3

The EXAFS measurements were made at the PULS Facility, Laboratori Nazionali di Frascati. Special thanks are due to Dr. S. MOBILIO and Dr. L. INCOCCIA for helpful discussions.

References

1 R.Caminiti, G.Licheri, G.Piccaluga, G.Pinna and M.Magini, Rev. Inorg. Chem., 1, 333 (1979).
2 J.E.Enderby and G.W.Neilson, in: "Water, a Comprehensive Treatise", ed. by F.Franks, Vol. 6, Plenum Press, N.Y., 1979.
3 G.Palinkas and E.Kalman, in: "Diffraction Studies on Non-Crystalline Substances", ed. by I.Hargittai and W.J.Orville-Thomas (Elsevier, Amsterdam, 1981).
4 A.H.Narten and H.A.Levy, Science, 167, 1521 (1970).
5 A.H.Narten and H.A.Levy, in: "Water, a Comprehensive Treatise", ed. by F.Franks, Vol. 1 (Plenum Press, N.Y., 1972).
6 A.Bienenstock, this book.
7 P.Eisenberger, this book.
8 J.Waser and V.Schomaker, Rev. Mod. Phys., 25, 671 (1953).
9 T.M.Hayes, J. Non-Cryst. Solids, 31, 57 (1978).
10 T.M.Hayes, P.N.Sen and S.H.Hunter, J. Phys. C, 9, 4357 (1976).
11 R.F.Pettifer, Proc. of the Daresbury Study Week-end 28-29 March (1981), DL/SCI/R 17.
12 S.Mobilio and L.Incoccia, LNF-82/44(P) Laboratori Naz. Frascati Internal Report (1982).
13 D.R.Sandstrom and H.W.Dodgen, J. Chem. Phys., 67, 473 (1977).
14 D.R.Sandstrom, J. Chem. Phys., 71, 2381 (1979).
15 G.Licheri, G.Paschina, G.Piccaluga, G.Pinna and G.Vlaic, Chem. Phys. Lett., 83, 384 (1981).
16 W.Bol, G.J.A.Gerrits and C.L.van Panthaleon van Eck, J. Appl. Cryst., 3, 486 (1970).
17 R.Caminiti, G.Licheri, G.Piccaluga and G.Pinna, Far. Disc. N° 64, 62 (1978).
18 R.Caminiti, G.Licheri, G.Paschina, G.Piccaluga and G.Pinna, Z. Naturforsch., 35a, 1361 (1980).
19 H.Ohtaki, T.Yamaguchi and M.Maeda, Bull. Chem. Soc. Japan, 49, 701 (1976).
20 M.Magini, J. Chem. Phys., 74, 2523 (1981).
21 M.Magini and G.Giubileo, Gazz. Chim. Ital., 111, 449 (1981).
22 D.R.Sandstrom and J.M.Fine, SSRL Rep. 78/09, 77 (1978).
23 T.I.Morrison, A.H.Reis, E.Gebert, L.E.Iton, G.D.Stucky and S.L.Suib, J. Chem. Phys., 72, 6276 (1980).
24 T.K.Sham, J.B.Hastings and M.L.Perlman, Chem. Phys. Lett., 83, 391 (1981).
25 T.I.Morrison, L.E.Iton, G.K.Shenoy, G.D.Stucky and S.L.Suib, J. Chem. Phys., 75, 4086 (1981).
26 W.Bol and T.Weltzen, Chem. Phys. Lett., 49, 189 (1977).

27 R.Caminiti, G.Licheri, G.Piccaluga and G.Pinna, Chem. Phys., 19, 371 (1977).
28 R.Caminiti, G.Licheri, G.Piccaluga and G.Pinna, J. Chem. Phys., 65, 3134 (1976).
29 M.Magini, J. Inorg. Nucl. Chem., 40, 43 (1978).
30 R.Caminiti and M.Magini, Chem. Phys. Lett., 61, 40 (1979).
31 A.Fontaine, P.Lagarde, D.Raoux, M.P.Fontana, G.Maisano, P.Migliardo and F.Wanderlingh, Phys. Rev. Lett., 41, 504 (1978).
32 P.Lagarde, A.Fontaine, D.Raoux, A.Sadoc and P.Migliardo, J. Chem. Phys., 72, 3061 (1980).
33 G.Galli, G.Maisano, P.Migliardo, C.Vasi and F.Wanderlingh, Solid State Comm., 42, 213 (1982).
34 G.W.Neilson and J.E.Enderby, J. Phys. C 11, L 625 (1978).
35 J.E.Enderby and G.W.Neilson, Rep. Progr. Phys., 44, 593 (1981).
36 N.A.Hewish and J.E.Enderby, Phys. Rev. Lett., 48, 756 (1982).
37 M.P.Fontana, G.Maisano, P.Migliardo and F.Wanderlingh, J. Chem. Phys., 69, 676 (1978).
38 G.Licheri, G.Pinna and G.Navarra, to be published.
39 F.Bigoli, A.Braibanti, A.Tiripicchio and M.Tiripicchio, Acta Cryst., B27, 1427 (1971).
40 M.Magini, G.Paschina and G.Piccaluga, J. Chem. Phys., 76, 1116 (1982).
41 R.Kleinberg, J. Chem. Phys., 50, 4690 (1969).

Local Order in Oxide Glasses

G.N. Greaves

Science and Engineering Research Council, Daresbury Laboratory, Daresbury, Warrington WA4 4AD, England

Abstract

This paper reviews the local structure revealed by EXAFS that surrounds glass formers, intermediates and modifiers in oxide glasses. The modifier environment is shown to be the one most sensitive to changes in property from one composition to the next.

1. Introduction

Although oxide glasses featured prominently in early models for glass structure [1], in most cases it is the more recently studied amorphous semiconductors that have provided the best examples of these initial ideas [2]. Elemental amorphous semiconductors like germanium, arsenic or selenium have enabled the principal themes of disordered networks - ring statistics, bond angle and dihedral angle distributions - to be explored. Whilst there is little doubt these concepts are of relevance to the structure of oxide glasses, it is also clear the problem is more complicated for these materials. In the first place ring sizes are generally larger than in elemental systems and so ring statistics tend to be out of range for most diffraction techniques. Secondly the strong electronegativity of oxygen means all bonds in oxide glasses are partially and in some cases almost completely ionic. Finally oxide glasses are usually multicomponent, containing modifying additives as well as glass forming or network components. These difficulties are compounded by the experimental fact that oxide glasses are generally composed of light atoms which are weak scatterers of x-rays and, in so far as x-ray spectroscopy is concerned, have K absorption edges falling in the soft x-ray region [3]. Nevertheless the EXAFS technique is bringing something of a renaissance to modelling the amorphous state by its ability to probe comprehensively the local structure of oxide glasses [4].

The central problem in understanding the structure of oxide glasses is to ascertain how the overall topology accommodates both modifying and network components to maintain a stable but disordered structure. Figure 1 illustrates in two dimensions how such a balance might be achieved [4]. Both ionic and covalent bonds are included and all atoms are fully co-ordinated. EXAFS and XANES studies have done much to substantiate this picture. For instance until the recent observation of sodium K edge EXAFS [11] there was no direct evidence that modifiers in oxide glasses might possess distinct local order. The local geometries of glass-forming and glass-modifying cations obtained from EXAFS studies are assembled in Table 1. The first shell radii and co-ordination numbers are indicative of both covalent and ionic species. The relative bond strengths are also reflected in the different Debye-Waller factors, e.g. for Si–O this is 0.001 $Å^{-2}$

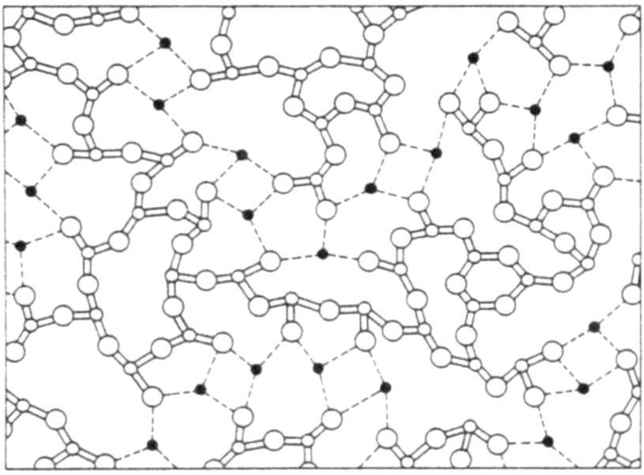

Fig.1. Fully co-ordinated random structure for a modified glass. Modifier cations are shown by small filled circles and glass former cations by small open circles. The large open circles are oxygens; these are 2-fold co-ordinated within the network and 3-fold co-ordinated between network regions. Ionic bonds are shown by dotted lines and covalent bonds by solid lines

Table 1. Local structure of oxide glass formers, intermediates and modifiers obtained from XANES and EXAFS. Modifiers like sodium adopt different co-ordinations in oxide glasses of different composition

Bond	Co-ordination	Bond Length/Å	Debye-Waller Factor $Å^{-2}$	Reference
Si-O	4	1.6(1)	0.001	[11]
Ge-O	4 - (6)	1.7(1) - 1.8(3)	0.003	[7,8,9]
As-O	3	1.8(0)	0.002	[29]
Ti-O	4 - (6)	1.8(1) - (2.1(5))	0.003	[18]
V-O	5 - 6			[20]
Cr-O	4 - (6)			[22]
Mn-O	6	2.1(1)		[16]
Fe-O	4 - 6	1.9(1) - 2.0(7)		[15,16,22]
Co-O	4 - 6	1.9(8) - 2.0(4)		[22]
Ni-O	6	2.0(3)		[16,21]
Zn-O	4			[29]
Na-O	5 - 6	2.2(5) - 2.4(3)	0.005 - 0.03	[11,23]
Ca-O	8 - 9	2.4 - 2.8	0.03	[25]

compared to 0.03 $Å^{-2}$ for Ca-O. It is this mixture of strong and weak bonds that enables the physical properties of oxide glasses, like viscosity and density, to be tailored for specific applications.

2. Glass Formers

The immediate atomic environment of glass forming cations like silicon and germanium has been well studied using x-ray and neutron diffraction [5,6]. EXAFS measurements on germania [7] and germanate glasses [8,9] and silica

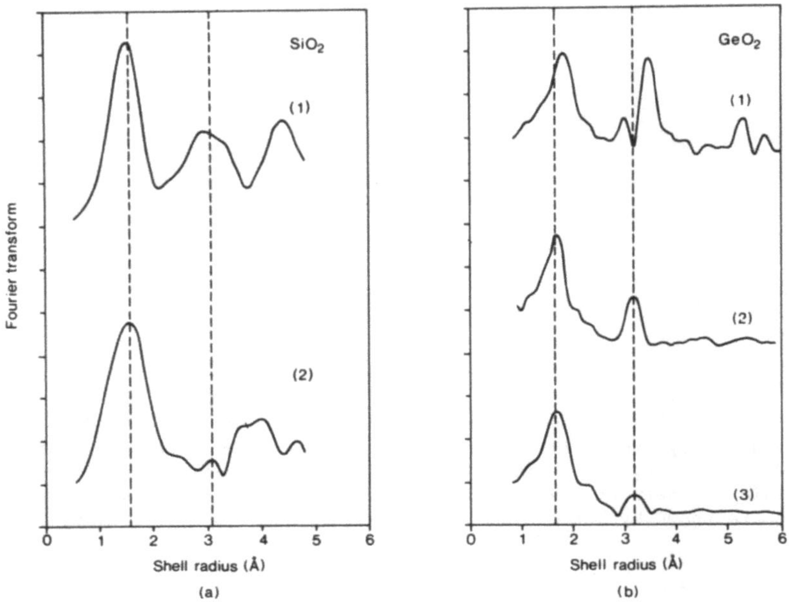

Fig.2. Phase shift corrected Fourier transforms of EXAFS of crystalline and vitreous silica [10] and germania [9]. Dotted lines mark the first and second shells for the glasses

and silicate glasses [10,11] have demonstrated that equivalent and in some cases improved structural information can be obtained from x-ray absorption spectroscopy. For example, the EXAFS Fourier transforms of the K-edge EXAFS of germanium and silicon for vitreous and crystalline GeO_2 and SiO_2 are shown in Fig.2. The first shell radii for both glasses - Ge-O and Si-O - are the same as in the α-quartz structures (1.74 and 1.61 Å respectively). Detailed curve fitting confirms the 4-fold co-ordination of germanium and silicon.

Despite the short wave vector cutoff for EXAFS, the second shell comprising germanium or silicon atoms can also be discerned in both the α-quartz structures and the glasses. This is indicated by dotted lines in Fig.2. The existence of both first and second shells enables the Ge-O-Ge and Si-O-Si bond angles to be deduced. Furthermore, by comparing the Debye-Waller factors for the glasses with those for the α-quartz structures the distortion of these angles in the glasses can be estimated. The average bond angles and their spread reported for vitreous germania and silica are 130° ± 6.5° [7] and 160° ± 20° [11] respectively. These values lie close to those analysed from x-ray diffraction measurements [12,13]. The corresponding discrete oxygen bond angles for α-quartz GeO_2 and α-quartz itself are 130° and 144° respectively. The spread in bond angle found for the glasses is strong evidence for non-crystalline topology. This is particularly true for vitreous silica where the average bond angle is also longer.

A similar treatment of the oxygen K-edge EXAFS of vitreous silica grown on crystalline silicon [14] yields the same bond length as the silicon K-edge EXAFS together with an O-Si-O bond angle of 108°, which lies close to the expected tetrahedral angle.

Figure 2 also includes the atomic distribution for rutile germania obtained from the germanium K edge [9]. In this polymorph the germanium atoms are octahedrally co-ordinated. The difference between the rutile and α-quartz structures are made clear in Fig.2. The first shell radius of rutile is longer than in α-quartz and the second shell comprising germanium atoms is split. Increases in the Ge-O bond length have been observed in the alkali germanate glasses Li_2O-GeO_2 [8] and Na_2O-GeO_2 [9] as the alkali content is increased. This suggests a mixture of tetrahedral and octahedral sites for germanium in these glasses. Of particular interest is the retention of the α-quartz Ge-Ge peak at 3.1 Å in Na_2O-GeO_2 glasses [9], indicating much of the tetrahedral topology of germania persists in the modified glasses. It would clearly be of interest to measure the environment of the alkali in these binary germanate glasses.

3. Transition Metal Oxides

At the impurity level transition metals in oxide glasses occupy discrete sites associated with well-characterised absorption bands at optical wavelengths. It is well-known that by altering the composition of the base glass or the oxidation state of the melt changes in colour in the visible can be achieved. At alloying concentrations transition metal oxides adopt a structural role intermediate between the extremes of glass former and glass modifier. They also act as a medium for small polaron transport: self-localised charge hopping between metal sites of different valency [15].

The ambivalent structural role of transition metal oxides in glasses has recently been studied by several groups using XANES and EXAFS. Figure 3 gives two examples. The near edge structure of iron for a reduced pyroxene glass (1) is compared to crystalline pyroxene (3) and crystalline FeO_{1-x} (2) [16]. In crystalline pyroxene the octahedral ferrous site is distorted whereas in FeO_{1-x} it is more-or-less regular. The qualitative similarity of (1) and (2) suggests that in the glass divalent iron generally occupies fairly regular octahedral sites. The pre-edge peak evident at 7109 eV is characteristic of transition metal oxides where 1s-3d dipole transitions are allowed because of some hybridisation of the 3d states of the metal with the oxygen 2p states. This feature is enhanced for oxidised iron-containing glasses [17], indicating a sizable fraction of ferric ions in tetrahedral sites. Parallel EXAFS studies reveal a decrease in Fe-O distances in going from reduced to oxidised glasses [17].

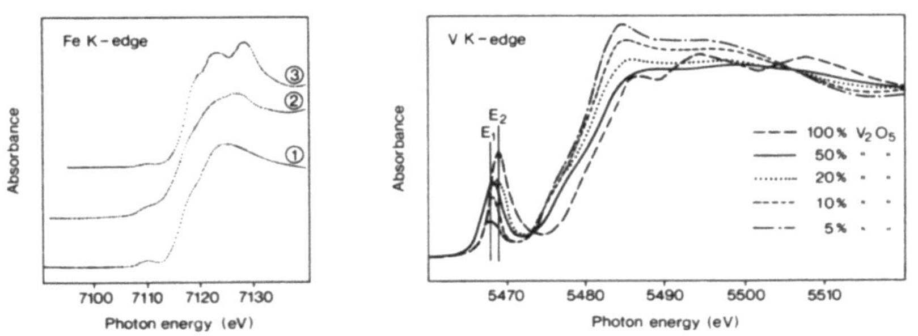

Fig.3. XANES for Fe [22] and V [20] in various oxide glasses. See the text for details

Bivalent sites for other transition metals in oxide glasses have been reported (see Table 1) - notably for titanium [18] and zinc [19] in binary silicates where the transition metal is predominantly 4-fold co-ordinated.

Returning to Fig.3, the near-edge structure of vanadium for a series of vanadium-iron-phosphate glasses measured by Bianconi and co-workers is reproduced [20]. Small polaron conduction in vanadium phosphate glasses has been well studied [21]: the electron hopping between V^{4+} and V^{5+} centres. XANES spectra for the glasses display a clear trend, as the concentration of V_2O_5 is decreased - the structure changing from a broadened version of crystalline V_2O_5, where vanadium is 5-fold co-ordinated, to a profile similar to crystalline VO_2, where the co-ordination is 6-fold. There are complementary changes in the pre-edge peak around 5468 eV. This is strong for hybridised 5-fold co-ordinated vanadium (E_2) and weak for regular 6-fold co-ordinated vanadium (E_1).

4. Modifiers

Modifying oxides like soda and lime have a more decisive effect on the properties of glasses like silica and germania. Huge changes in viscosity result, the optical gap shrinks from around 9 eV to 5 or 6 eV, glasses become ionic conductors, strong broad modes appear around 200-300 cm^{-1} and so on. The most likely explanation seems to be that these changes result from a substantial fraction of the structure being composed of oxygen ionically bonded to modifying cations as indicated in Fig.1. Such modified regions will lower the overall viscosity, alter the distribution of oxygen 2p states at the top of the valence band, provide numerous interstitials for ionic transport, and like ionic crystals will absorb in the far infrared [11]. We will see that, just as in the minerals, it is the local structure in the modified regions rather than in the network which is most sensitive to these changes.

EXAFS is the only local structural probe to date capable of revealing the atomic environment of modifying cations in oxide glasses. The relative weakness of ionic bonds compared to covalent bonds can be seen in the Deybe-Waller factors listed in Table 1. This results in the masking of correlations involving modifier cations by the strong network correlations in the x-ray or neutron scattering profile [3]. In the EXAFS of modifier

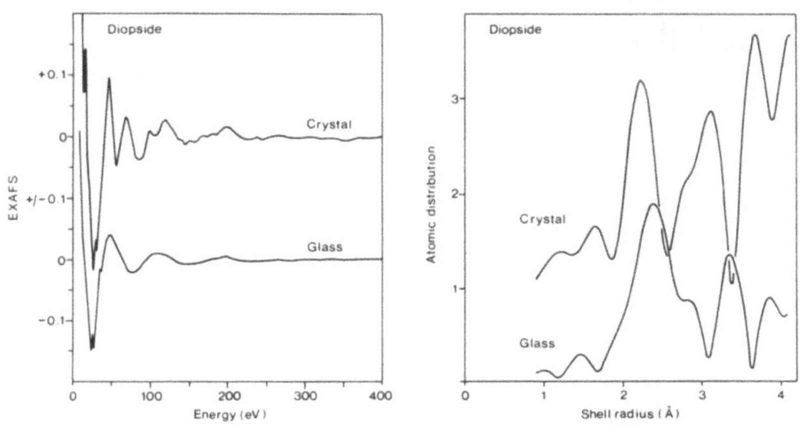

Fig.4. Calcium K-edge EXAFS spectra for crystalline and vitreous diopside ($CaMgSi_2O_6$) together with phase shift corrected Fourier transforms [3,24]

cations in silicates, for instance, the strong Si-O and O-O correlations play a subsidiary role. The K-edge EXAFS of calcium [3,24] and sodium [11,23] in various oxide glasses are shown in Figs.4 and 5, together with the corresponding Fourier transforms. Taking these in turn, there are several important observations.

The calcium EXAFS is from the mineral diopside ($CaMgSi_2O_6$) which is a monoclinic pyroxene where calciums and magnesiums bind chains of SiO_4 tetrahedra together. But for the obvious differences in overall topology, we might expect the same chain-like structure to persist in the glass. The densities however, are very different - the glass being 10% less than the crystal. Because local order in the network remains intact, it is clear from Fig.4 that the longer Ca-O distances in the glass must be largely responsible for its lower density.

In a parallel study of crystalline and amorphous anorthite ($CaAl_2Si_2O_8$) the local structure of calcium was observed to be different from diopside, as expected, but similar for the two polymorphs [3]. The mineral anorthite has a three-dimensional feldspar network. The similarity of the two calcium EXAFS suggests that an equivalent but disordered network persists in the glass.

The sodium EXAFS for the three glasses shown in Fig.5 are all different. This reflects the situation in crystalline silicates and borates. Sodium is 5-fold co-ordinated in sodium meta- and di-silicate [26,27] but 7-fold co-ordinated in sodium metaborate [28]. In ternary silicates the

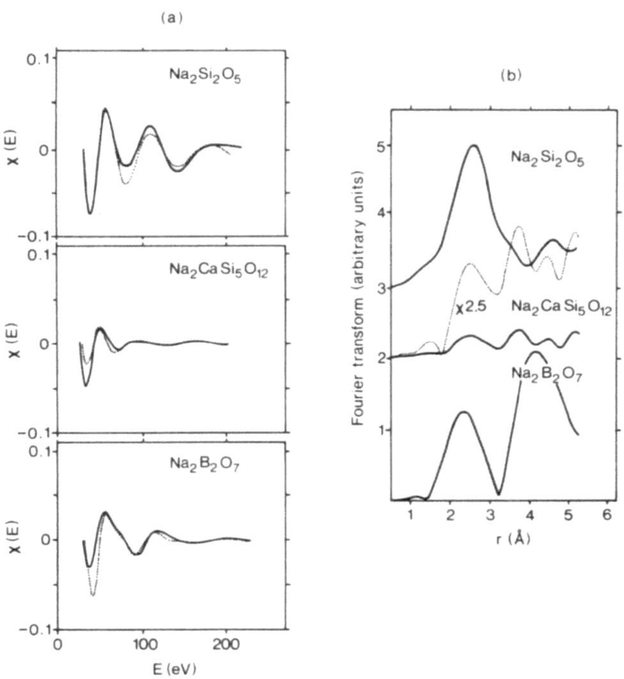

Fig.5. EXAFS and Fourier transforms for the sodium K edge in various oxide glasses [23]. See the text for details

presence of calcium or magnesium typically spreads out the oxygen shell, leaving only two or three oxygens at the bonding distance of ~ 2.3 Å. Curve fitting the EXAFS spectra of the glasses shown in Fig.5 brings out the same structural trends [11,23].

The local structure of modifiers like sodium and calcium, then, clearly changes with glass composition and is even sometimes distinct from the crystalline polymorph. Nonetheless for a given composition all the evidence is that the local environment is well defined. The Debye-Waller factor is in proportion to the bond strength and there is little static disorder [11].

The complement of the modifying cation is the "non-bridging" oxygen. This is better thought of as an "inter-network" oxygen having a similar environment to the modifying cation and a co-ordination higher than the "bridging" oxygen within the network regions. This is the background to the model introduced earlier in Fig.1. Modifying and glass forming cations as well as both varieties of oxygen all have a fully coordinated local environment.

References

[1] Zachariasen, W.H., J. Am. Chem. Soc. 54, 3841 (1932)

[2] E.g. Polk, D.E., and Bondreaux, D.S., Phys. Rev. Lett. 31, 92 (1973); Greaves, G.N., and Davis, E.A., Phil. Mag. 29, 1201 (1974); Long, M., Galison, P., Alben, R., and Connel, G.A.N., Phys. Rev. B13, 1821 (1976).

[3] Greaves, G.N., and Raoux, D., Structure of Non-crystalline Materials II, (Taylor & Francis, London) in press 1982.

[4] Greaves, G.N., La Recherche 137, - in press 1982.

[5] Wright, A.C., and Leadbetter, A.J., Phys. Chem. Glasses, 17, 122 (1976).

[6] Porai-Koshits, E.A., J. Non-cryst. Solids, 25, 87 (1977).

[7] Sayers, D.E., Stern, E.A. and Lytle, F.W., Phys. Rev. Lett. 35, 584 (1975).

[8] Cox, A.D. and McMillan, P.W., J. Non-cryst. Solids, 44, 257 (1981).

[9] Lapeyre, C., Petiau, J. and Colas, G., Structure of Non-Crystalline Materials II, (Taylor & Francis, London) - in press 1982.

[10] Greaves, G.N., Fontaine, A. Lagarde, P., Raoux, D., Gurman, S.J., and Parke, S., Recent Developments in Condensed Matter Physics, p.225 (Plenum, New York and London 1981).

[11] Greaves, G.N., Fontaine, A. Lagarde, P., Raoux, D., and Gurman, S.J., Nature, 293, 611 (1981).

[12] Karle, J., and Konnert, J.H., Phys. Rev. Lett. 36, 823 (1976).

[13] Mozzi, R.L., and Warren, B.E., J. Appl. Cryst. 2, 164 (1969); Da Silva, J.R.G., Pinatti, D.G., Anderson, C.B., and Rudee, M.L., Phil. Mag. 31, 713 (1975).

[14] Stöhr, J., Johansson, L., Lindau, I., and Pianetti, P., Phys. Rev. B20, 664 (1979).

[15] Austin, I.G., and Mott, N.F., Adv. Phys. 18, 41 (1969).

[16] Calas, G. and Petiau, J., Structure of Non-crystalline Materials II, (Taylor & Francis, London) - in press 1982.

[17] Petiau, J., Calas, G., Bondot, P., Lapeyre, C., Levitz, P. and Loupias, G., EXAFS for Inorganic Systems. Daresbury Laboratory Report DL/SCI/R17 p.127 (1981).

[18] Sandstrom, D.R., Lytle, F.W., Wei, P.S.P., Greegor, R.B., Wong, J., and Schultz, P., J. Non-crystalline Solids, 41, 201 (1980).

[19] Pettifer, R.F., EXAFS for Inorganic Systems, Daresbury Laboratory Report DL/SCI/R17 p.57 (1981).

[20] Bianconi, A., Giovannelli, A., Davoli, I., Stizza, S., Palladino, L., Gzowski, O., and Murawski, L., INFN report LNF-2(R) (1982).
[21] Greaves, G.N., J. Non-crystalline Solids, 11, 427 (1973).
[22] Petiau, J., and Calas, G., Proc. of V Int. Conf. on Physics of Non-Crystalline Solids, (Montpellier, 1982).
[23] Greaves, G.N., Fontaine, A., Lagarde, P., and Raoux, D., EXAFS for Inorganic Systems, Daresbury Laboratory Report DL/SCI/R17 p.115 (1981).
[24] Binstead, N., Greaves, G.N., and Henderson, C.M.B., (in preparation, 1982).
[25] Geere, R.G., Gaskell, P.H., Greaves, G.N., Greengrass, J., and Binstead, N., (this conference, 1982).
[26] McDonald, W.S., and Cruickshank, D.W.J., Acta Cryst. 22, 37 (1967).
[27] Pant, A.K., and Cruickshank, D.W.J., Acta Cryst. B24, 13 (1968).
[28] Krogh-Moe, J., Acta Cryst. B30, 578 (1974).
[29] Gurman, S.J., and Pettifer, R.F., Phil. Mag. B40, 345 (1979).

EXAFS and XANES Spectra of Calcium Silicate Glasses

R.G. Geere and P.H. Gaskell

Cavendish Laboratory, Madingley Road, Cambridge CB3 OHE, England

G.N. Greaves

SERC Daresbury Laboratory, Daresbury, Warrington WA4 4AD, England

J. Greengrass

Pilkington Brothers, Lathom, Ormskirk, Lancs.

N. Binstead

Dept. of Geology, University of Manchester, Manchester

1. Introduction

According to conventional views of the structure of oxide glasses, alkali and alkaline earth ions occupy holes of random size and variable symmetry provided by the silicon-oxygen framework. The nearest-neighbour shell of oxygens around Ca would thus be expected to be broad - perhaps so broad as to be undetectable by normal structural techniques. Although some indirect evidence has been accumulated against this view, the most direct and persuasive arguments for an alternative viewpoint derives from the EXAFS studies of GREAVES et al.[1]. The extended fine structure above the Na and Ca K edges corresponds to reasonably well-defined shells of oxygens with mean distances and coordination numbers similar to those observed in crystals. The standard deviation, σ, of the Na-O and Ca-O distribution is in the range 0.07 - 0.17 Å [1,2].

The EXAFS spectrum for Ca in silicate glasses, however, is dominated by the nearest-neighbour oxygen shell. Figure 1 shows the EXAFS spectrum for a glass of composition 0.475 CaO, 0.475 SiO_2, 0.05 $A\ell_2O_3$, together with spectra for two crystalline polymorphs of CaSiO . Whereas the EXAFS spectra

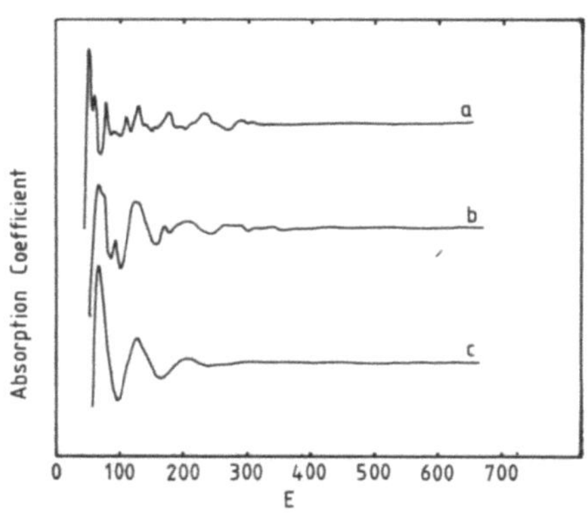

Fig. 1. EXAFS spectra of a) psuedo-wollastonite, b) wollastonite, c) a calcium silicate glass of composition 0.42 CaO, 0.55 Cao, 0.03 $A\ell_2O_3$

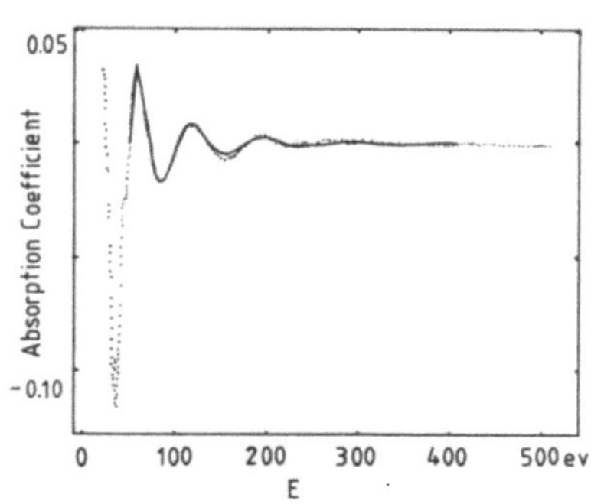

Fig. 2. Computed fit to EXAFS spectrum of the calcium silicate glass (see fig. 1) using a single shell of oxygen at a mean distance of 2.28 Å

for both crystalline specimens clearly show contributions from second- and higher-neighbour shells, the EXAFS spectrum for the glass over this range can be adequately represented by a single shell of oxygens at a mean distance of 2.28 Å with σ = 0.2 Å - fig 2. It is important to analyse the low energy, XANES region where information on second and higher shells is contained. Any information gained on, say, the Ca-Ca distribution could provide vital evidence for or against an alternative model for glasses in which Ca atoms might lie in distorted sheets similar to the planes observed in crystalline silicates.

The purpose of this paper is two-fold. Firstly, we report measurements of the EXAFS and XANES spectra for a number of Ca-containing alumino-silicate glasses and several 'model' crystalline compounds in an attempt to qualitatively identify those structural features which strongly affect the low energy fine structure above the Ca edge. Secondly, we provide a brief statement of preliminary attempts to analyse XANES spectra using the multiple scattering theory of DURHAM, PENDRY and HODGES [3].

2. Near Edge Spectra of Calcium-containing Crystals

Previous work on this subject has been largely due to BIANCONI and co-workers [4] who have measured XANES spectra in a number of organic and inorganic materials. Although many of their reported XANES spectra are qualitatively similar, $CaCO_3$-calcite exhibits additional features - most noticeably a sharp peak at an energy above the edge $E \simeq 20$ eV. This suggests that XANES may be sensitive to details of the structure other than the first-neighbour pair distribution function.

In an attempt to understand qualitatively the features of the XANES spectra of Ca-containing solids and to provide a check on the phase shifts required for XANES calculations on glasses, absorption spectra for several crystalline materials were recorded - fig.3. Two polymorphs of $CaSiO_3$ were studied - wollastonite and pseudo (p)-wollastonite in the nomenclature of TAYLOR [5], together with two forms of $CaCO_3$, calcite and aragonite. The near-neighbour environment around Ca is given - table 1. Calcite and wollastonite have similar first-neighbour shells - 6 oxygens at about 2.4 Å -

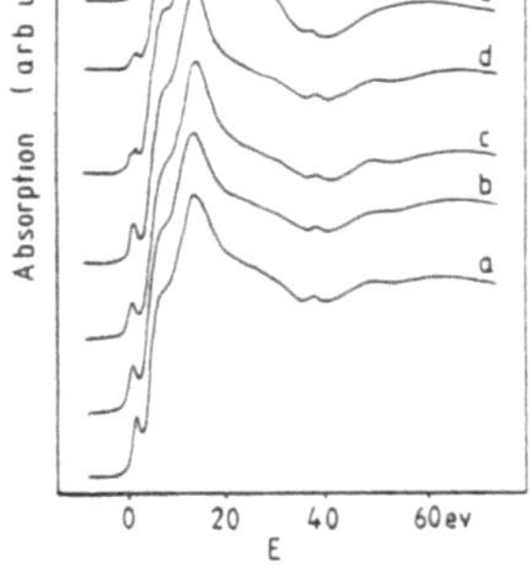

Fig.3. XANES spectra for (a) a calcium aluminate glass of composition 0.53 CaO, 0.08 SiO_2, 0.34 Al_2O_3; calcium aluminosilicates of composition xCaO, (0.97-x)SiO_2, 0.03 Al_2O_3 with (b) x = 0.53, (c) x = 0.42, (d) x = 0.32. Spectra for (e) wollastonite, (f) p-wollastonite, (g) aragonite and (h) calcite, are also shown

Table 1. Near-neighbour shell geometry and coordination around Ca in several crystals

Calcite	Aragonite	Pseudo-wollastonite	Wollastonite
O 2.35 0 6	O 2.52 0.1 9	O 2.54 or 2.30 0.15 0.02 8 2 and	O 2.40 0.09 6.3
C 3.21 0 6	C 3.14 0.25 6	2.62 0.04 6	Si 3.67 0.25 6
O 3.46 0 6	Ca 3.98 0.05 6	Si 3.39 or 3.24 0.25 0.02 6 4 and	Ca 3.63 0.08 4.7
Ca 4.04 0 6	O 4.15 0.13 15	3.69 0.03 2	O 3.6-4.5 broad 13
C 4.26 0 2		Ca 3.96 0.08 6	
O 4.57 0.20 16		O 4.48 0.22 18	

Entries in each column give the type of neighbour (omitting C and Si), mean and standard deviations of the distribution for each shell and the coordination number.

whereas aragonite and p-wollastonite have 9 and 8 neighbours at a larger mean spacing (∿ 2.5 Å). Each pair of materials exhibits broadly similar features from 40 - 60 eV, but details of the spectrum in the XANES region are different, most noticeably in the strong peak at 20 eV which is observed in calcite but not in wollastonite (or in the other crystals). Changes in the symmetry of the first-neighbour shell may be responsible, but experience with calculation of XANES spectra for Pd-Ge alloys [6] suggests that symmetry differences alone are unlikely to produce effects of the magnitude observed experimentally. Differences in second- and higher-neighbour shells appear more likely and it may be significant that calcite alone has a third shell of oxygens: the third shell consists of Ca atoms in the other crystals. The 'pre-edge' feature at about 4 eV is also interesting. This is split only in the spectra for calcite and wollastonite.

3. XANES Spectra for calcium-containing Glasses

In contrast to the weakness of the EXAFS spectra for glassy silicates relative to spectra for the crystalline forms (fig. 1) the XANES spectra of calcium silicate glasses (xCaO,$(0.97-x)SiO_2$, 0.03 Al_2O_3 with $x = 0.32 - 0.53$) is relatively rich (fig.3). Moreover, although features of the XANES spectrum of wollastonite are clearly more prominent than the fine structure observed in the glass, the spectrum of wollastonite is remarkably similar to those for amorphous calcium silicates. It is also remarkable that XANES spectra for a series of silicate glasses containing 30 - 50% CaO show almost identical XANES spectra. This applies also to glasses which are (visibly) phase-separated and, somewhat surprisingly, to a calcium aluminate glass

(0.53 CaO, 0.34 $A\ell_2O_3$, 0.08 SiO_2), which is also almost indistinguishable from the silicates. (In earlier work, differences have, however, been noted in Ca-containing phosphate glasses [2].)

The close similarity between the principal features of the XANES spectrum of wollastonite and the corresponding glasses suggests similar environments for the Ca atom. Wollastonite is a linear chain silicate whereas p-wollastonite contains three-membered Si_3O_9 rings [7]. The former structure might be expected to be more appropriate for the glass, but it would be difficult to detect this by indirect methods - for example the volume per oxygen atom is almost identical in each phase.

4. Calculation of Near-edge Spectra

In order to proceed further, it is necessary to calculate the near-edge spectra for the different environments around Ca. To date, it has only been possible to make pilot calculations for the most symmetrical structure - calcite. Calculations have been performed using the method of DURHAM et al.[3] for an octahedral cluster of oxygens around Ca and for two-, three- and six-shell clusters (Fig.4).

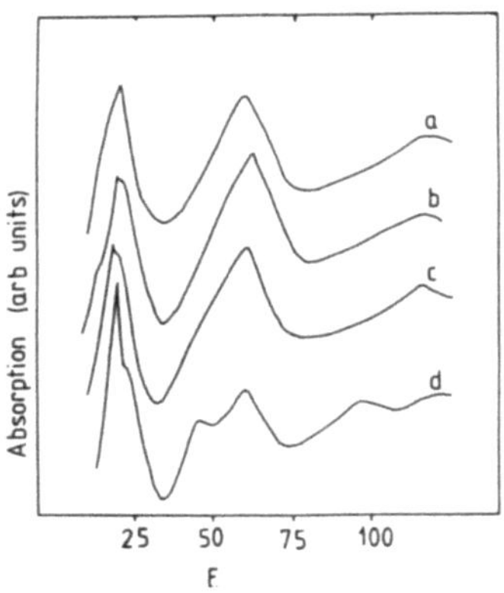

Fig.4. Near-edge spectra calculated for the following clusters surrounding a central Ca atom: a) a single shell of O atoms at 2.35 Å, b) with a further shell of carbon at 3.21 Å, c) with a third shell of oxygen atoms at 3.6 Å, d) with three further shells at 4.04 Å, 4.26 Å, 4.29 Å. Note energy is related to the muffin tin zero and does not necessarily correspond to experiment

Although the calculated spectra reproduce the structure observed in most of the Ca-containing crystals and glasses - namely sharp features at about $E \simeq 12$ eV - there is, surprisingly, no indication whatsoever of intermediate features - particularly the strong peak at 20 eV seen in calcite, and to a lesser degree in wollastonite. We do not yet understand the reason for this discrepancy.

Acknowledgments

Experimental measurements were performed at the Daresbury SRS. Two of us (PHG and RGG) acknowledge with gratitude generous support from Pilkington PLC.

References

1. G. N. Greaves, A. Fontaine, P. Lagarde, D. Raoux, S. J. Gurman, Nature 293, 611-616 (1981)

2. G. N. Greaves, D. Raoux, Structure of Non-Crystalline Materials II (1982) (in press)
3. P. J. Durham, J. B. Pendry, C. H. Hodges, Solid State Comm. 38, 159-62 (1981)
4. A. Bianconi, S. Doniach, D. Lublin, Chem. Phys. Lett. 59, 121-124 (1978)
5. L. Heller , H. F. W. Taylor, Crystallographic Data for the Calcium Silicates, H.M. Stationery Office 1956
6. P. H. Gaskell, D. M. Glover, A. K. Livesey, P. J. Durham, G. N. Greaves, J. Phys. C: Solid State Phys. 15, L597-604 (1982)
7. T. Yamanaka, H. Mori, Acta Cryst. B37, 1010-1017 (1981)

EXAFS and XANES Investigation of the Coordination of Technetium in Borosilicate Glass

M. Antonini[1], C. Caprile[2], A. Merlini, J. Petiau[3], and F.R. Thornley[4]

Physics Division, Joint Research Centre, I-21020 Ispra (Va), Italy

Abstract

Among the various radionuclides contained in glasses storing high level nuclear waste, ^{99}Tc is of particular interest because it is not fixed on geological materials and has the highest initial leaching rate. Since these properties depend on the valence state and the local atomic coordination of Tc in the glass, we decided to investigate them by using EXAFS and XANES spectroscopy at the K edge of Tc. The samples were thin films of TcO_2 and $NaTcO_4$, and a borosilicate glass containing 2% Tc by weight. The analysis of the spectra indicates a major contribution of a single shell of neighbours and seems to exclude the possibility that these neighbours are oxygen atoms in the Tc glass. A good agreement was obtained by assuming that atoms surrounding Tc are also Tc. The first-neighbour distance was in accord with the value obtained for the crystalline structure of metallic technetium.

1. Introduction

Highly radioactive waste products generated in nuclear reactors, after reprocessing, must be collected and stored for long periods of time to protect the environment. A possible way of achieving a very stable nuclear waste form consists of incorporating the different radionuclides in borosilicate glasses. After this treatment the nuclear waste is protected by backfill barriers and stored in stable geological repositories [1]. However, some active species may leach out of the storage matrix and eventually reach the biosphere. In this respect, among the various radionuclides contained in these glasses, ^{99}Tc, a transition metal, is of special concern because of its high leaching rate [2]. The Tc leaching and solubility properties are strongly related to its oxidation states, the most stable ones being Tc(IV) and Tc(VII). High valence compounds, such as $NaTcO_4$, are soluble in ground water, while low valence oxides, such as TcO_2, are only slightly soluble. The atomic environment and chemical bonding of technetium in glass are unknown. Absorption X-ray spectroscopies (XANES and EXAFS) are direct sensitive methods to investigate these problems. Here we report results obtained during the first stage of this investigation.

[1]Università di Modena and GNSM, I-41100 Modena, Italy

[2]Università di Padova, I-35100 Padova, Italy

[3]Université P.M. Curie, 4 Place Jussieu, F-75230 Paris Cedex 05, France

[4]University of Strathclyde, Glasgow G4 0NG, U.K.

2. Experimental

Due to the long-lived radioactivity of ^{99}Tc and to the high volatility of its oxides, the preparation and handling of samples required much care. Two reference compounds, TcO_2 and $NaTcO_4$, where Tc is coordinated (IV) and (VII) respectively, were prepared in addition to a borosilicate glass with 2 wt% of Tc. The TcO_2 film, about 15 μm thick, was electrolytically deposited on graphite. The $NaTcO_4$ film, about 50 μm thick, was thermally deposited on a light glass after mixing NH_4TcO_4 with 0.1 M solution of NaOH and extracting the sodium pertechnetate in insoluble 2-hexanone. Some difficulties were encountered to obtain homogeneous layers of these compounds. Five and 10 mm thick borosilicate glasses, whose composition is given in Table I, were prepared by slowly cooling the melt in a graphite crucible under argon atmosphere. This procedure might have created a reducing environment for the Tc compound used to prepare the glass.

Compound	wt%
SiO_2	53.2
B_2O_3	16.6
Al_2O_3	5.5
Na_2O	21.2
NH_4TcO_4	3.6

Table I. Glass composition; the concentration of NH_4TcO_4 corresponds to 2 wt% of Tc

The X-ray absorption measurements were carried out at the DCI synchrotron radiation source of L.U.R.E., Orsay (France), using the EXAFS-1 beam line spectrometer, equipped with a channel-cut Si(400) crystal in order to achieve a resolution of about 4 eV at the K edge of technetium (E_k = 21044 eV). The energy scale was calibrated by locating the K edges of pure zirconium and molybdenum.

3. Results and Discussion

The near-edge absorption of the two reference compounds are compared with that of the Tc glass and reported in Fig. 1 after background subtraction and height normalization. Shifts of about +6 eV were found in the edge energy between the Tc glass and TcO_2 and between TcO_2 and $NaTcO_4$. These shifts are of the same order as those observed in other 2nd series transition metals in different oxidation states [3]. A pre-edge maximum at 21042 eV is also present in $NaTcO_4$. This maximum is interpreted as arising from the 1s → 4d transition which may take place because of the tetrahedral coordination of Tc in $NaTcO_4$ [4,5]. By contrast, no similar effect was found in TcO_2, in which the metal atom is surrounded by six oxygens in a distorted octahedral coordination [6]. A pre-edge structure is not present in the Tc glass, a circumstance excluding a high valence state of technetium in our glass sample. The EXAFS spectrum of $NaTcO_4$ and of the Tc glass are clearly different in location and amplitude of the EXAFS oscillations (Fig. 2). They were analysed using computer programme elaborated by MOBILIO et al. [7]. Phase shift functions were taken from

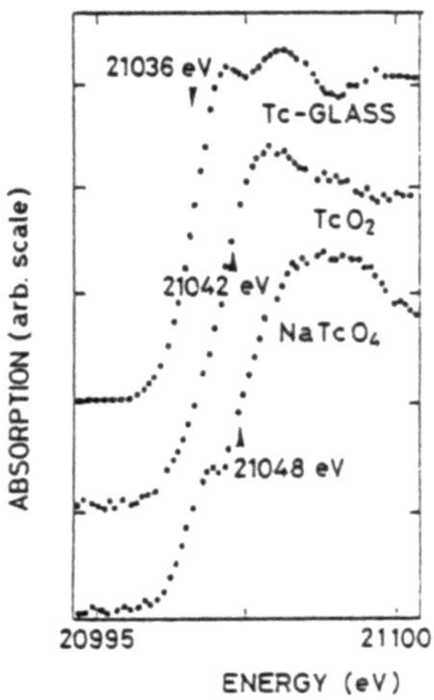

Fig. 1. Near-edge X-rays absorption spectra of $NaTcO_4$, TcO_2 and of the Tc glass. Arrows indicate the position of the edges obtained from derivative curves

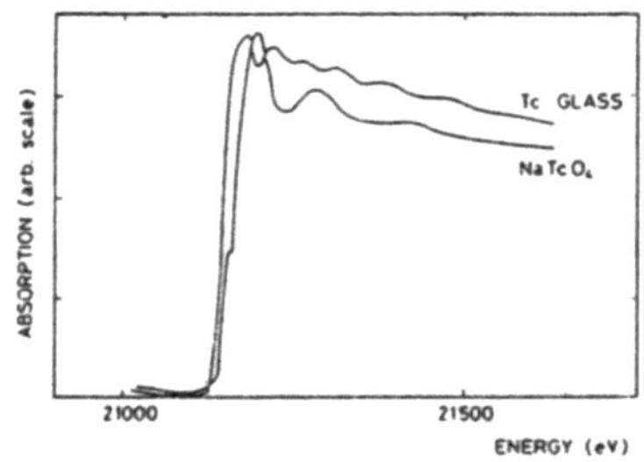

Fig. 2. EXAFS spectra of $NaTcO_4$ and of Tc glass

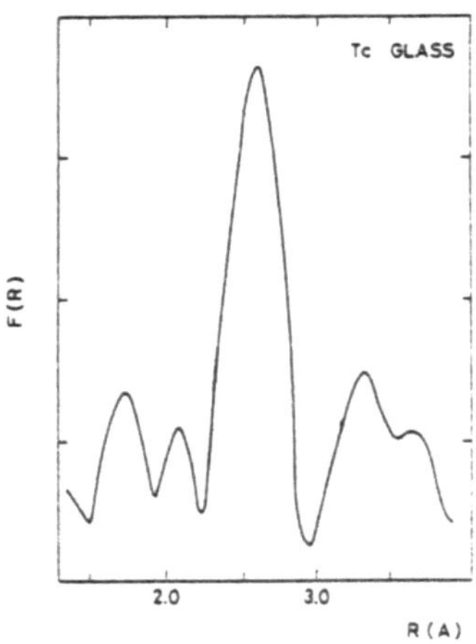

Fig. 3. Atomic radial distribution in the Tc glass

TEO and LEE [8] and interpolated when necessary. The results of the analysis for $NaTcO_4$ agree with the nearest-neighbours distance of 1.71 Å between Tc and its four oxygen neighbours obtained from the crystalline structure of $NaTcO_4$ [4]. In the Tc glass, a typical major contribution from a single shell of atoms is deduced from the analysis of the radial distribution function (Fig. 3). Such a contribution is attributed to direct Tc-Tc bonds, whose length is about 2.7 Å, a value corresponding to the first-neighbour distance in crystalline metallic technetium [9]. This conclusion is also in agreement with the lower edge energy of Tc in glass.

Although the presence of Tc cannot be definitely excluded, the preliminary results reported above indicate that our Tc glass contained mainly metallic technetium. Further measurements are in progress to investigate the effects of an oxidizing atmosphere during sample preparation upon the oxidation state of Tc in a borosilicate glass.

Acknowledgements

We wish to thank the staff of the L.U.R.E. Laboratory for their assistance and hospitality; Dr. G. Bidoglio and Mr. G. Lazzari, JRC Ispra, and Dr. P. Mazzi, CNR, Padova (Italy) for their invaluable advice and help during sample preparation.

References

1. Proceedings of the International Conference on the Scientific Basis for Nuclear Waste Management, Boston (USA), Nov. 16-19, 1981, S.V. Topp ed., Plenum Press (1982).
2. D.J. Bradley, C.O. Harvey and R.P. Turcotte, Techn. Rep. Pacific Northwest Lab., PWL-3152, UC-70 (1979).
3. U.C. Srivastava and H.L. Nigam, Coord. Chem. Rev. 9, 275 (1972-1973).
4. K. Schwochau, Z. Naturf., 17, 630 (1962).
5. F.W. Kutzlehr, C.R. Natoli, D.K. Misemer, S. Doniach and K.O. Hodgson, J. Chem. Phys. 73, 3274 (1980).
6. D.B. Rogers, R.D. Shannon, A.W. Seeight and J.L. Gillson, Inorg. Chem. 8, 841 (1969).
7. S. Mobilio, F. Comin. L. Incoccia, Laboratori Nazionali di Frascati Techn. Rep. LNF-82/19 (NT) (1982).
8. B.K. Teo and P.A. Lee, J. Am. Chem. Soc. 101, 2815 (1979).
9. C.L. Money, C.L. Rose, Phys. Rev. 72, 1269 (1947).

Germanium Environment in Simple and Multi-Component Oxide Glasses

C. Lapeyre[1], J. Petiau[2], and G. Calas

Laboratoire de Minéralogie-Cristallographie, Universités de Paris 6 et 7, LA CNRS 09, F-75230 Paris Cedex 05, France and

LURE, CNRS, Bât. 209C, F-91405 Orsay Cedex, France

1. Introduction

Germanium is an important oxide glass component, classically compared to Si as a network-forming element, but with possible coordination changes as in the alkali germanates. In multicomponent oxide glasses EXAFS is a method of choice for studying Ge environment because of its atomic selectivity. Furthermore Ge-containing glasses are receiving increased interest due to the use of germanium to create a refractive index gradient in optical waveguide fibres. This application gives special importance to the knowledge of Ge local structure in multicomponent glasses.

Up to now EXAFS results have been published on vitreous GeO_2 [1, 2] and Li_2O-GeO_2 glasses [3]. We present here results concerning vitreous GeO_2 and binary glasses based on GeO_2 : GeO_2-SiO_2, GeO_2-B_2O_3 and GeO_2-Na_2O where important structural changes can be suspected. This helps in understanding the Ge behaviour in ternary glasses and more generally in the SiO_2-GeO_2-B_2O_3-Na_2O system.

2. Experimental

The studied glasses were prepared by melting oxide components at 1200°C, apart the GeO_2-SiO_2 system which melts at higher temperatures and was prepared at the solar furnace of the CNRS (Odeillo). EXAFS experiments were performed on powder samples using the DCI storage ring of LURE (Orsay) at the Ge K-edge. The experimental set is the conventional one for transmission measurements. The beam is monochromatized by a "channel-cut" Si crystal using the 220 reflection. The incident and transmitted photon intensities are measured in ionization chambers and stored after voltage-frequency conversion, in a PDP 11/04 computer which also monitors the monochromator rotation [4].

The EXAFS oscillations are analyzed using Fourier transform method. A first transform gives a partial radial distribution relative to the central Ge atom (see below Figure 1). Back-transforming of each peak of this radial distribution gives the contribution of each shell to the EXAFS spectrum. Distances are calculated using theoretical phase-shift calculations. Great care was taken to use the same analysis parameters (background substraction, choice of the energy threshold, Fourier windows) in order to get the best precision on distance variations among the samples. We then make a clear distinction between the precision attained on absolute distance ($|\Delta R| \simeq 0.01$ Å) and that based on distance comparisons ($|\Delta R| \simeq 0.003$ Å).

[1]Thomson CSF, L.C.R. Corbeville

[2]Ecole Normale Supérieure de Montrouge

3. Results

3.1. Vitreous GeO_2

Fourier transform of EXAFS of pure vitreous GeO_2 is reported in Fig. 1a. It shows the presence of two peaks due to Ge-O and Ge-Ge pairs. The determined Ge-O distances are slightly shorter than in the hexagonal form of crystalline GeO_2 (1.73$_4$ Å as compared to 1.73$_9$ Å). This seems to be a general feature of the distances relative to the first atomic shell in glass [5]. The absence of any broadening of the Ge-O peak in the Fourier transform proves that the rigidity of the tetrahedra is maintained in the glass. Ge-Ge distances are the same in the glass and in the crystal (3.14 Å). The position, height and width of the Ge-Ge peak in vitreous in GeO_2 permit to say that about the half of Ge-Ge pairs are connected by a Ge-O-Ge angle of 130° ± 6°. This gives a further confirmation of the angular distribution deduced from RDF [6] or from Raman spectra [7].

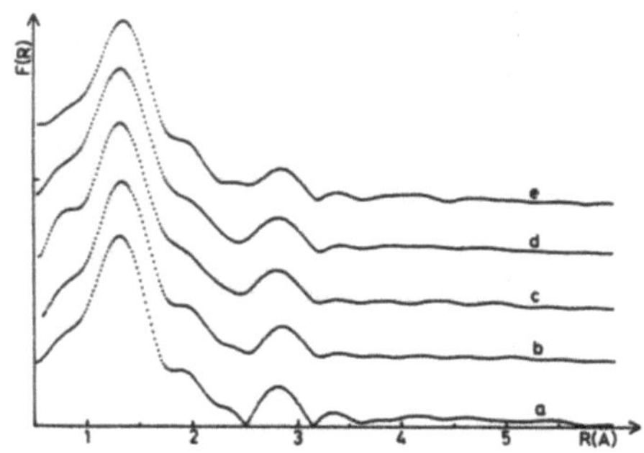

Fig. 1.Partial distribution function in sodium germanate glasses.
a. pure GeO_2
b. 0.9 GeO_2 - 0.1 Na_2O
c. 0.8 GeO_2 - 0.2 Na_2O
d. 0.7 GeO_2 - 0.3 Na_2O
e. 0.65 GeO_2- 0.35 Na_2O

3.2. Na_2O-GeO_2 glasses

Fig. 1 shows the similarity between vitreous GeO_2 (curve a) and the glasses of the Na_2O-GeO_2 system, particularly concerning the presence of a Ge-Ge peak. The Ge-O distances are greater than in vitreous GeO_2 and increase with the Na_2O content, exhibiting an apparent maximum value of 1.766 Å at about 20% molar Na_2O. EXAFS does not allow to show the existence of two distinct subshells, but the mean value obtained can be related to the presence of 6-fold Ge in the glass structure [8]. If we take as extreme values, 1.734 Å and 1.883 Å for typical Ge-O distances for Ge in 4-fold and 6-fold sites, we obtain a maximum of 20% of octahedral sites in the 0.8 GeO_2 - 0.2 Na_2O glass. This value is similar to that determined by Cox and Mc Millan [3] in Li_2O-GeO_2 glasses, and also coherent with the variations of the physical properties of alkali germanates as a function of the alkali content. The measured Ge-Ge distance is identical to that obtained in vitreous GeO_2, but the weaker intensity can be due to the existence of 6-fold Ge which partly hinders the tetrahedron correlation observed in vitreous GeO_2. However the existence of this peak even at the highest Na_2O content (35% molar) confirms the existence of tridimensional network domains not affected by "depolymerization" processes.

3.3. GeO_2-SiO_2 and GeO_2-B_2O_3 glasses

The environment of Ge is different in these two systems. In germano-silicate glasses, the first shell is only observed. The Ge-O distances obtained (1.73$_1$ Å) are slightly shorter than in vitreous GeO_2. These results indicate a glassy framework interconnected with GeO_4 and SiO_4 tetrahedra.

In borate glasses, different results are observed. In the GeO_2-rich glass, we observe similar behaviour as in pure GeO_2 (Ge-O distance of 1.74 Å and presence of a Ge-Ge peak), showing that the GeO_2 framework is maintained. In the B_2O_3-rich glass (0.8 B_2O_3-0.2 GeO_2), only one shell is observed, corresponding to a large Ge-O distance (1.75$_5$ Å) connected with BO_3 units.

3.4. Multicomponent glasses

Ternary glasses of the Na_2O-B_2O_3-GeO_2 system show a distinct behaviour as a function of their Na_2O content. In alkali-poor glasses, the same results are observed as in the corresponding glasses of GeO_2-B_2O_3 system. In alkali-rich glasses, the interaction between Na_2O and GeO_2 is shown by increased Ge-O distances and by the presence of a Ge-Ge peak. When SiO_2 is also present, the interconnection of GeO_4 and SiO_4 tetrahedra is predominent. EXAFS was used to follow the Ge environment during phase separation and leaching in these glasses [9].

References

[1] W.F. NELSON, I. SIEGEL : Phys. Rev. 127, 2025-2027 (1962).

[2] D.E. SAYERS, E.A. STERN : Phys. Rev. Let. 35, 584-587 (1975).

[3] A.D. COX, P.W. Mc MILLAN : Journ. Non Cryst. Solids 44, 257-264 (1981)

[4] D. RAOUX, J. PETIAU, P. BONDOT, G. CALAS, A. FONTAINE, P. LAGARDE, P. LEVITZ, G. LOUPIAS, A. SADOC : Rev. Phys. Appl. 15, 1079-1094 (1980).

[5] J. PETIAU, G. CALAS : in "EXAFS for inorganic systems" (Proceedings of the Daresbury Conf., Mars 1981), 127-129.

[6] A.C. WRIGHT, A.J. LEADBETTER : Phys. Chem. of glasses 17, 122-145 (1976).

[7] F.L. GALEENER : Phys. Rev. B 19, 4292-4297 (1979).

[8] H. VERWEIJ, J.H.H.M. BUSTER : Journ. Non Cryst. Solids 34, 81-89 (1979).

[9] C. LAPEYRE, J. PETIAU, G. CALAS, F. GAUTHIER, J. GOMBERT : Bull. Minéral. (to appear).

Local Structure in Mixed Electrolytes

F. Aliotta[1], G. Galli, G. Maisano, P. Migliardo, and C. Vasi[1]
Istituto di Fisica and GNSM del CNR, Università di Messina,
I-98100 Messina, Italy

Recent studies on the static and the dynamical properties in $ZnCl_2$, $ZnBr_2$, $CuBr_2$, $CuCl_2$ aqueous solutions [1,2,3], allowed one to say unambiguosly that at very high concentrations, these systems locally reproduce the structure of the corresponding hydrated crystals with a middle range order.
The possibility of investigating the local structure in ionic liquids, as strong electrolytes, was greatly enhanced by EXAFS spectroscopy [4,5]. To clarify better the complex nature of the cation-anion and cation-water interactions, EXAFS and Raman measurements have been performed on II-I mixed aqueous electrolytic solutions of the $(MBr_2+NBr_2)\cdot nH_2O$ type ($M,N=Cu^{2+}$, Zn^{2+}, Sr^{2+}). The EXAFS measurements were performed at the PULS facility using the synchrotron radiation from ADONE storage ring. An improved Fourier-filtering curve fitting method has been used to process the data [6].
In Fig. 1 we show, as an example, the experimental EXAFS oscillations and the fitting results for the $ZnBr_2+SrBr_2$ aqueous solutions.

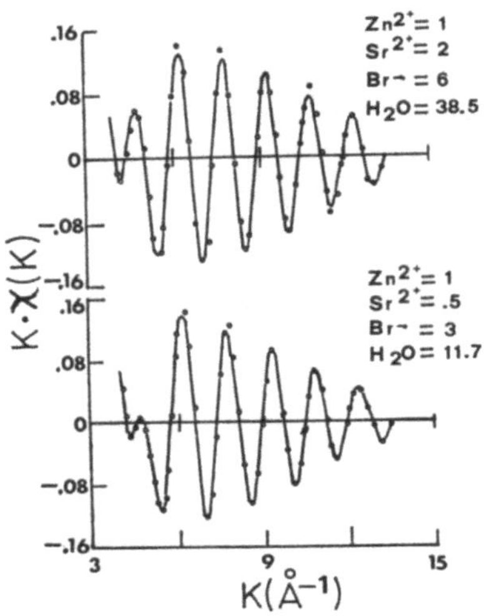

Fig.1. Zn-edge Fourier-filtered EXAFS spectra of $ZnBr_2+SrBr_2$ mixed electrolytes at various concentrations. Dots: filtered experimental data. Continuous line: theoretical fit. Note the variation of the beating phenomenon between the oxygen (low K) and bromine (high K) contributions as the Br^-/Zn^{2+} increase

[1] Istituto di Tecniche Spettroscopiche del CNR, I-98100 Messina, Italy

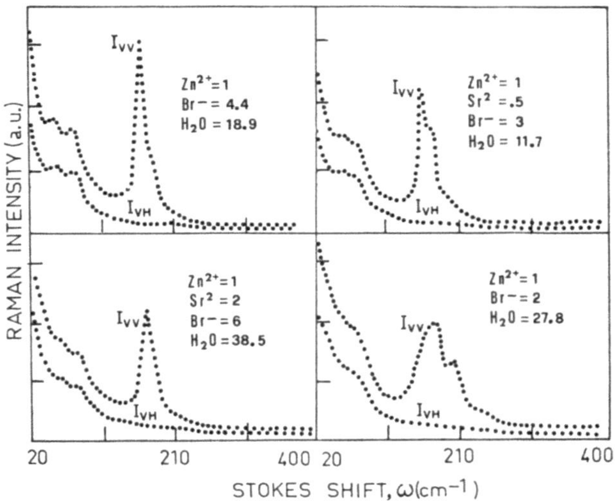

Fig.2. I_{VV} and I_{VH} Raman spectra for $ZnBr_2$+$SrBr_2$ mixed electrolytes. The plot up at left shows the Raman results for $ZnBr_2$+ hydrobromic acid solutions

The whole lot of Raman results is condensed in Fig.2.

With regard to our EXAFS measurements, table 1 summarizes the results [6]. Note that the coordination number $\bar{N}_j$ is to be regarded as "mean value". Hold, in fact, the relation $\bar{N}_j=N_{true}\cdot\alpha$, if α is the fractional part of cations implied in complexes like $(MeBr_n)^{2-n}$. If one considers that the zinc-oxygen distance slowly decreases, adding $CuBr_2$ or $SrBr_2$ in the solutions with $ZnBr_2$, while the zinc-bromine distance remains almost constant; then the opinion may be strongly supported that the ionic pairs originated by zinc-halide bindings are exclusively the tetrahedral $ZnBr_4^{2-}$ units. This hypothesis is moreover supported by Raman results. In fact, following the band evolution in the I_{VV} spectra in ($ZnBr_2$+$SrBr_2$) solutions versus Br^- content, here is evidence that the vibrational model centered at ω_1=175 cm^{-1}, ω_2= 68 cm^{-1}, ω_3=208 cm^{-1}, ω_4=86 cm^{-1} typical of the tetrahedral symmetry, change only their relative strength. For high value of the bromine content in the ($ZnBr_2$+$SrBr_2$) solutions α=1. Because $(1-\alpha)$ gives us the fractional contribution of the oxygen shell, it seems to be an apparent contradiction with the experimental $\bar{N}_{Ox}$=4.3. It may be explained by the existence of an anionic coordination with water and formation of hydrated $ZnBr_4^{2-}\cdot(H_2O)_4$ ionic pairs. As a conclusion, our results show that in the investigated electrolytic mixture, Sr^{2+} and Cu^{2+} behave as bromine donors, like hydrobromic acid; metal-halogen complexes are mainly zinc based; in ($SrBr_2$+$CuBr_2$) mixture, the copper alone gives rise to ionic-pairs formation.

Table 1. Theoretical fitting results of the EXAFS spectra

Systems: $ZnBr_2+CuBr_2$

Zn edge	$Zn^{2+}/Br^-=.5$	$Zn^{2+}/Br^-=.37$	$Zn^{2+}/Br^-=.25$	$Zn^{2+}/Br^-=.13$
$\bar{N}_{Br}$	1.8	2.9	3.5	4
$\bar{N}_{Ox}$	6.2	6.5	5.4	4.3

Cu edge	$Cu^{2+}/Br^-=.25$	$Cu^{2+}/Br^-=.37$	$Cu^{2+}/Br^-=.5$
$\bar{N}_{Br}$	.3	.8	1.6
$\bar{N}_{Ox}$	6.4	6.5	5.5

Systems: $ZnBr_2+SrBr_2$

Zn edge	$Zn^{2+}/Br^-=.33$	$Zn^{2+}/Br^-=.17$
$\bar{N}_{Br}$	3.8	4
$\bar{N}_{Ox}$	4.3	1.8

Systems: $CuBr_2+SrBr_2$

Cu edge	$Cu^{2+}/Br^-=.33$
$\bar{N}_{Br}$	1.5
$\bar{N}_{Ox}$	5.6

References

1. P.Lagarde, A.Fontaine, D.Raoux, A.Sadoc, P.Migliardo, J.Chem.Phys. 72, 3061 (1980).
2. F.Aliotta, G.Maisano, P.Migliardo, C.Vasi, Phys.and Chem. of Liquids 12, 69 (1982)
3. M.Fontana, G.Maisano, P.Migliardo, F.Wanderlingh, J. Chem. Phys. 69, 676(1978)
4. G.Galli, G.Maisano, P.Migliardo, C.Vasi, F.Wanderlingh Solid State Comm. 42, 213(1982)
5. D.R.Sandstrom and F.W.Little, Ann. Rev. Phys. Chem. 30, 215(1979)
6. More detailed discussion on this subject will appear on "Nuovo Cimento D"

Lattice Relaxation Around a Br^- Ion in $KCl_{1-x}Br_x$

T. Murata[1], P. Lagarde[2], A. Fontaine[2], and M. Raoux[2]

LURE-CNRS, Bât. 209C, F-91405 Orsay-Cêdex, France
(Associé à l'Université Paris-Sud)

1. Introduction

Heavy halogen ions substitute easily in alkali halide crystals and even form dimers and trimers... with increasing concentration and finally gather into clusters. This effect has mainly been investigated by the observation of the change in the shape of the absorption bands [1] which appears in the tail of the fundamental absorption of the host crystal in the ultraviolet range. This has been interpreted as a localized exciton. This effect has been confirmed by emission spectra recorded with an excitation at the localized exciton energy.

These optical measurements [2] imply a displacement of the first-neighbour ions around the solute ion. EXAFS is the only adequate technique to probe the local environment around a dilute foreign ion and to measure the magnitude of the relaxation. EXAFS data were collected on the $KCl_{1-x}Br_x$ system for values of x lower than 10 %. Stable solid solutions are expected in the whole concentration range.

2. Experimental procedures and results

Exafs data were collected at 30 K on samples ground in a very fine powder dispersed in paraffin. The experimental set-up installed at LURE has been described elsewhere [3].

In Figures 1 and 2 are shown the radial distribution functions (a) and the filtered back-Fourier transforms of the first shell (b) for the x = .04 and x = .10 samples. Using the scattering amplitudes, the phase shifts calculated by TEO and LEE [4], and the relevant parameters shown in the captions, the fits of the curves have been calculated following the standard one-electron single scattering formula [3]. This theoretical data provide the interionic distances in pure KBr with good accuracy.

The first shell radius around the Br^- ion is found to be 3.18 Å for x = .04 and 3.20 Å for x = 0.10. In order to compare these first shell radii with the average interionic distances in the ionic solutions, X-ray diffraction measurements were also performed at room temperature. Using the value of the thermal expansion coefficient of the host crystal (4.10^{-5}/K) the X-ray data were converted to 30 K values. The Exafs results differ by 0.06 Å and 0.08 Å from the average values given by BRAGG reflections for the two samples.

[1]Department of Physics, Kyoto University of Education, Fukakusa, Fushimi-ku Kyoto 612, Japan

[2]Physique des Solids, Bât. 510, F-91405 Orsay, France

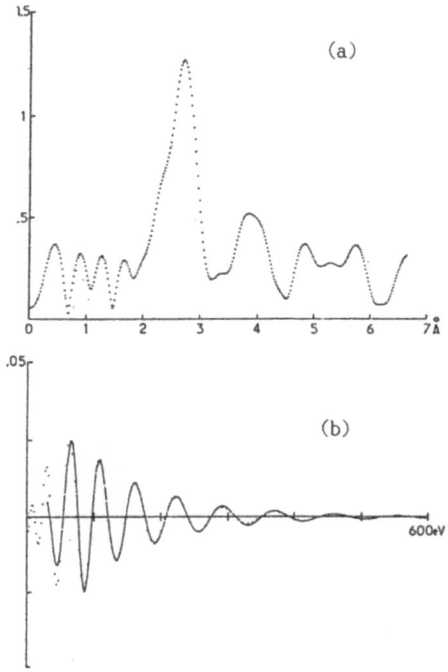

Fig.1a:Radial distribution for x=.04
b:Dots:Filtered back-Fourier transform of the first shell.Solid line:Theoretical fit with ΔE_o=12eV, 6K at R=3.18Å σ=.13 a.u. and γ =.43 a.u.

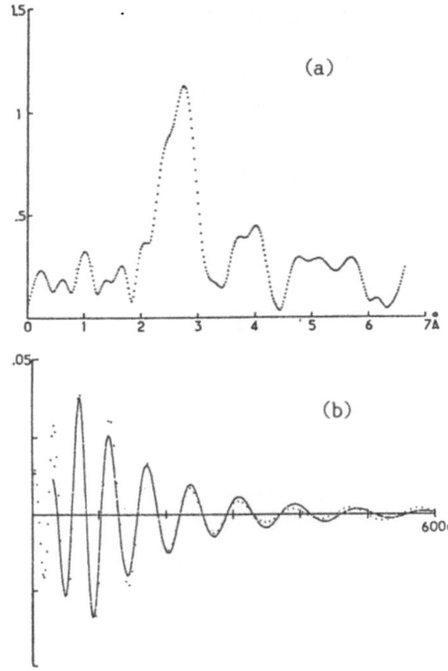

Fig.2a:Radial distribution for x=.10
b:Dots:Filtered back-Fourier transform of the first shell.Solid line:Theoretical fit with ΔE_o=10eV. 6K at R=3.20Å,σ=.14a.u. andγ=.35 a.u.

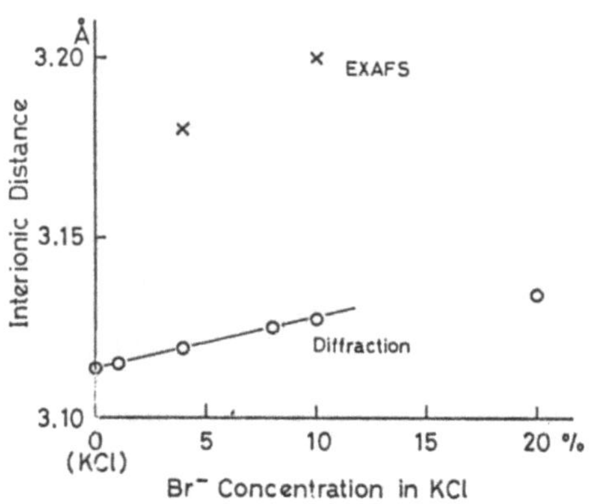

Fig.3 Interionic distances versus concentration of bromine ions in KCl.Circles:Results obtained by diffraction measurements. Crosses: Results from EXAFS

3. Discussion

In the case of Al metallic alloys [5] the first shell distances around an isolated solute atom differ by a factor 2 to 3 from the values calculated via the elastic theory for a continuous medium using the slope of VEGARD's law. Similar treatment cannot be applied to the present system because the host matrix is ionic and the substitution is on second-neighbour sites.

The observed interionic distances around the Br^- ion are larger than the average value as shown in Figure 3, but smaller than the interionic distance in pure KBr (3.29 Å). Moreover a slight increase with concentration is observed. This suggests that dimers or trimers are formed at very low concentration because of the elastic interaction induced by larger solute atoms. This is consistent with optical measurements : NAKAGAWA et al. [2] have observed drastic change in the absorption band at 20 mol. % that they have interpreted as a percolation transition. At lower concentrations, small changes of the band position have been observed. The change of the shape of the 1st and 2nd peaks of the Fourier transforms (Fig.1a, Fig.1b) is more evidence for the cluster birth in these systems. Further analysis of such clues will be presented elsewhere.

Acknowledgements

The authors would like to thank Pr J. CHAPELLE and Mr GODARD for growing crystals. They also thank the Laboratoire de l'Accélérateur linéaire for dedicated shifts of the DCI storage ring.

References

1 T. Murata, J. Phys. Soc. Jpn. (1968) 25. 1632 ; N. Nagasawa, ibid.(1969) 27, 1535.

2 H. Nakagawa, J. Phys. Soc. Jpn. (1973) 34, 410 ; H. Nakagawa, K. Toyoda, Y. Nakai, ibid.(1975) 38, 1067.

3 D. Raoux et al., Rev. Phys. Appl. (1980) 15, 1079.

4 B.K. Teo and P. Lee, J. Am. Chem. Soc. (1979) 101, 2815.

5 D. Raoux, A. Fontaine, P. Lagarde, A. Sadoc, Phys. Rev. (1981) B 24, 5547.

Elastic Core Effect in Isovalent Solid Solutions

Aldo Craievich[1], Elisabeth Dartige[2], Alain Fontaine[2], and Denis Raoux[2]

L.U.R.E., Bât. 209c, F-91405 Orsay-Cedex, France

1. Introduction

The elastic core effect (i.e. the change of the nearest-neighbour distance b around a solute atom) has been measured in three aluminium alloys [1]. This relaxation of the crystal around the substitutionnal site drives the macroscopic effect which is the change of lattice parameter a . According to Vegard's law, at least for low concentrations c , the dependence on c is linear. Using the slope (da/ac) we may calculate the host-solute interatomic distance ($b+\delta b_{CM}$) via an elastic theory which treats the host crystal as an elastic *continuous medium* (C.M.)[2,3].

The C.M. relationship can be written down as $da/ac = (\gamma/y)(\delta b/b)$ (γ is due to the image forces induced by the finite dilution. It is equal to about 1.5 for all isotropic metals. y is the pure geometrical ratio between the volume allowed to one atom (Ω) versus the spherical volume of the first shell $(4/3\pi)b^3$, y = 0.169 for f.c.c.).

The EXAFS results show that although the continuum analysis gets the sign of the change in the first shell distance correct, it seriously underestimates the magnitude of the change (Table 1).

Table 1. Elastic core effect from EXAFS and continuous medium theory

	δb (Å) EXAFS	da/ac	δb (C.M.)	δb(EXAFS)/δb(C.M.)
Al-Mg [1]	+ 0.08	+ 0.109	+ 0.04	2.0
Al-Cu [1]	- 0.125	- 0.120	- 0.041	3.1
Al-Zn [1]	- 0.020	- 0.020	- 0.007	2.9
Cu-Ag	+ 0.068	+ 0.153	+ 0.043	1.6
Mg-Zn	- 0.182	- 0.150	- 0.053	3.4

This may be interpreted as being due to the discrete atomic nature of the host crystal and (or) also to the long range nature of the electronic forces in the case where the solute atom is of different valency to the host atom. This last idea was addressed by the present study of two isovalent solid solutions (Cu-Ag and Mg-Zn).

2. Samples

The Cu-4 at.% Ag was water-quenched from 780°C after 2 hours of annealing at this temperature. The dissolution of silver atoms is easy. Duwez has shown that is even possible to make solid solution in the whole range of concentration using the splat-cooling technique. The second alloy was prepared using this technique.

[1]CBPF, Av. Wenceslau Bras 71, Rio de Janeiro, Brazil
[2]Lab. Phys. Solides, Bât. 510, F-91405 Orsay-Cedex, France

The standards used to check the phase shifts were respectively pure silver samples and the previous studied Al-0.83 at. % Zn system which is very similar to the Mg-Zn in terms of phase shifts involved in the EXAFS analysis. The phase shifts of the atomic pair (solute-host) is only slightly changed to take into account that the Z number of the back scatterer has decreased by one unit [1].

All EXAFS data have been collected with samples being kept at 30 K.

3. EXAFS Analysis

The analysis was achieved with the Fourier-filtering method, using the one-electron single scattering formula. The results were displayed on Fig.1 and Fig.2, with the procedure first suggested by Martens et al. [4].

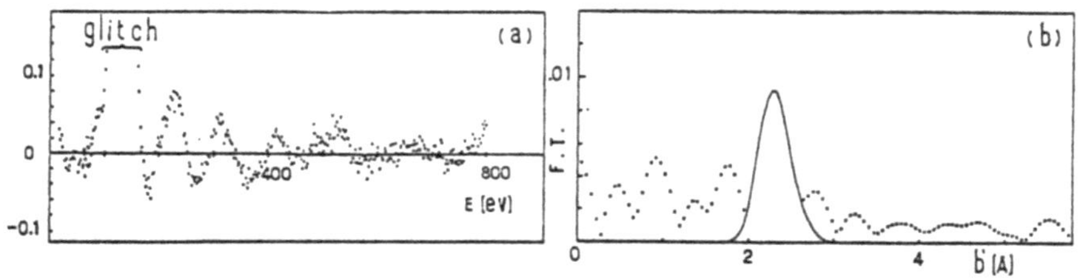

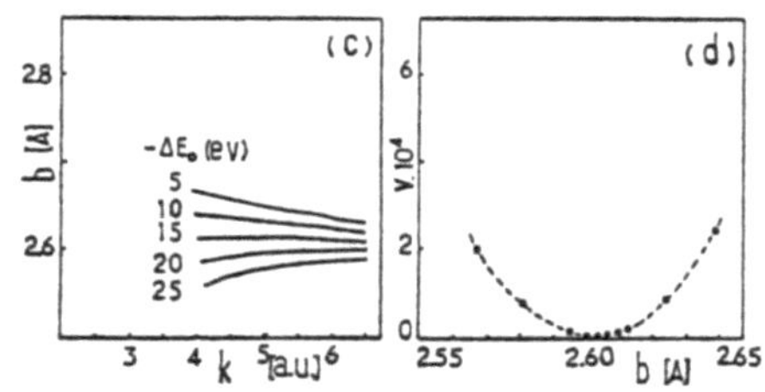

Fig.1. Cu-4%at.Ag (Ag edge)

a) EXAFS spectrum.
b) Fourier transform.
c) Determination of the first shell radius by the Martens and Rabe method: radius b versus photoelectron momentum k for several threshold energies E_o.
d) Variance V

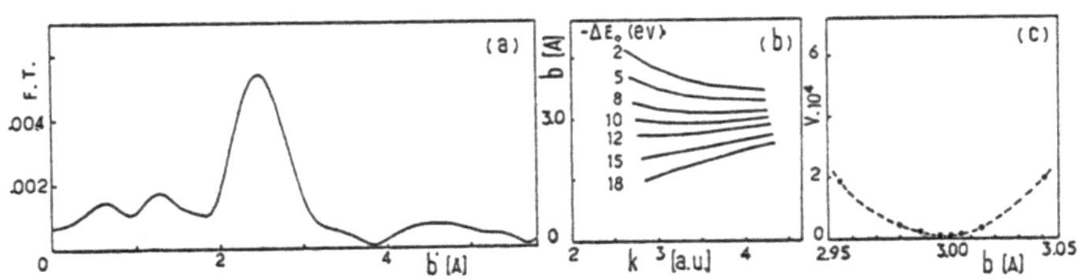

Fig.2. Mg-3%at.Zn (Zn edge).

a) Fourier transform, b) b(k) display and c) Variance

4. Results and Discussion

The results on these two isovalent systems (Table 1) do not provide a definite answer to the addressed issue since if the ratio δb(EXAFS)/δb(C.M.) is only 1.6 for the Cu-Ag solid solution, it rises up to 3.4 for the Mg-Zn one. As a first comment we have to underline that the contribution of electronic forces to build the low value of the interatomic Zn-Mg distance cannot be rejected since the distortion is large: an effective charge of about 0.4 electrons ($2*3\delta b/b$) has to be screened.

Besides that, it may be possible that the hexagonal structure of the magnesium lattice leads to an anisotropic elastic core effect. In such a bimodal first-neighbours distribution it is well known [5] that the EXAFS experiment gives the shortest distance which, in this case, increases the δb(EXAFS) value. In order to check this possibility we shall perform EXAFS experiments on single crystal using the polarization of the synchrotron radiation.

Whatever the correctness of the Mg-Zn distance is, it remains that all ratios δb(EXAFS)/δb(C.M.) are larger than one and thus the C.M. theory is not relevant to describe the connection between the microscopic cause ($\delta b/b$) and the macroscopic effect (da/ac).

5. Soft Sphere Model (S.S.M)

In order to take into account the discrete nature of the crystal we propose a soft sphere model. The main assumption of this calculation is to give one specific volume for each atom which is consistent with the notion of Goldschmidt radius, and to allow changes in the 'shape' of the atom. Let us attribute $\Omega+\delta\Omega$ for the atomic volume of the solute atom in a very dilute medium built by the host atoms of volume Ω.

In that case $3da/a$, which is the total change of volume per average atom, is equal to $c\delta\Omega/\Omega$. If we consider the image forces, the relationship can be written as $3da/a = \gamma c\delta\Omega/\Omega = \gamma c\ 3\ \delta r_s^h/r_s$, where r_s is the radius of the solute atom and δr_s^h is geometrically related to $\delta\Omega$. Thus to accommodate the solute atom the radius of the first shell changes by $\delta b \simeq \delta r_s^h$, while b approaches $2r_s$ if δb remains weak. Thus $da/ac = \gamma\delta b(S.S.M)/(0.5b)$.

Let us notice that the assumption of the atomic softness permits a ratio $(\delta b/b)/(\delta r_s^h/r_s)$ not strictly equal to 0.5.

The δb(S.S.M) value derived from this relationship is roughly three times higher than the δb(C.M) calculated via the continuous medium theory. This last relationship seems more likely since the EXAFS measurements are distributed around δb_{SSM} for the investigated solid solutions (Table 1).

This model allows to have positive or negative effects, and to keep a non-distorted first shell even if the b value decreases. Conversely a hard sphere model leads to an amorphous-like organisation of the first shells which cannot be compatible with the triple periodicity of the virtual crystal which still gives sharp Bragg reflections.

6. Conclusion

This preliminary study of isovalent solid solutions does not bring evidence of the prominent part of electronic forces in the failure of the C.M. theory. It seems more suited to describe the volume expansion of the solid solution in terms of individual atoms acting merely with their own volumes which accommodate each other, mainly by changes of their shape.

References

1 D.Raoux, A.Fontaine, P.Lagarde, A.Sadoc, Phys.Rev.B24, 10, 5547, (1981)
2 J.Friedel, Phil.Mag., 46, 514, (1955)
3 J.D.Eshelby, Sol.State Phys., 3, 79, (1956)
4 G.Martens, P.Rabe, N.Schwentner, A.Werner, Phys.Rev.B17, 1481, (1978)
5 P.Eisenberger, G.S.Brown, Sol.St.Com., 29, 481, (1979)

Transition-Metal and Metalloid Contributions to EXAFS in Amorphous Alloys and Crystalline Compounds

G. S. Cargill III, W. Weber,* and R. F. Boehme
IBM Thomas J. Watson Research Center
Yorktown Heights, New York 10598, U. S. A.

EXAFS measurements were made at CHESS for an amorphous $Fe_{78}P_{22}$ alloy, for crystalline Fe, and for crystalline "Fe_2P", which was actually $Fe_{1.7\pm0.2}P$ and contained a small amount of FeP. $\chi(k)$ functions were calculated from $\mu(E)$ data measured at 8, 80, and 300 K. R-space windows, 1Å to 3Å, were used to extract the near-neighbor components $k^2\chi_{nn}(k)$ by Fourier back-transformation (see Fig. 1). Near-neighbor model $k^2\chi(k)$ functions for crystalline Fe and "Fe_2P" were calculated using

$$\chi(k) = \sum_j \frac{N_j}{kr_j^2} \exp(-2k^2\sigma_j^2) F_j(k) D_j \sin[2kr_j + \phi_j(k)] \tag{1}$$

where the symbols have their usual meanings. Distance distributions, $N_j(r_j)$, for Fe and Fe_2P are shown in Fig. 2. Also shown is the distance distribution appropriate for a mixture of Fe_2P and FeP in which 23% of the Fe is in FeP.

Fits were made at each temperature to the bcc Fe data, using two neighbor shells and varying one scale factor $D_{Fe}=D_1=D_2$, one mean square width $\sigma_{Fe}{}^2=\sigma_1{}^2=\sigma_2{}^2$, and a shift in the zero of the energy scale ΔE_0. Teo-Lee [1] phase shifts and backscattering amplitudes were used. Best-fit values of D_{Fe} were nearly the same for all temperatures, 0.535 ± 0.015, and σ_{Fe} increased with increasing temperature: 0.051Å, 0.052Å, 0.076Å.

Fits for the "Fe_2P" data were made using the distance distribution in which 23% of the Fe atoms are in FeP, slightly more than the 18% expected for $Fe_{1.7}P$. The σ_j's were all set equal to the values determined for bcc Fe, and the best-fit values of D_P and D_{Fe} were determined at each temperature. Results were $D_P=0.48\pm0.04$ and $D_{Fe}=0.25\pm0.02\approx D_P/2$. (Fitting with only the "pure" Fe_2P distance distribution gave $D_{Fe}\approx D_P/3$.) These results suggest that using the Teo-Lee backscattering amplitudes [1] for Fe and P with the same scale factor D would significantly over-estimate the Fe backscattering contribution in crystalline Fe_2P. $|\Phi_2(r)|$ functions calculated for the experimental and best-fit model $k^2\chi_{nn}(r)$ functions are also shown in Fig. 1. The split peak for "Fe_2P", largely a division between Fe-P and Fe-Fe contributions, is well reproduced, suggesting that the calculated phase shifts [1] are "transferable" and reasonably accurate, in spite of the problems with backscattering amplitudes.

Fits were also made for the amorphous $Fe_{78}P_{22}$ alloy using two gaussian shells with $r_{FeFe}=2.61$Å, $N_{Fe-Fe}=10$, $r_{Fe-P}=2.32$Å, and $N_{Fe-P}=3$ [2]. D_j values given above for crystalline Fe_2P were used, and only σ_{Fe-Fe}, σ_{Fe-P}, and ΔE_0 were varied in the fitting process. Values obtained at 8, 80, and 300 K were 0.069, 0.069, and 0.087Å for σ_{Fe-P} and 0.176, 0.177, and

*Present address: Sektion Physik der Ludwig-Maximilians Universität München, W. Germany.

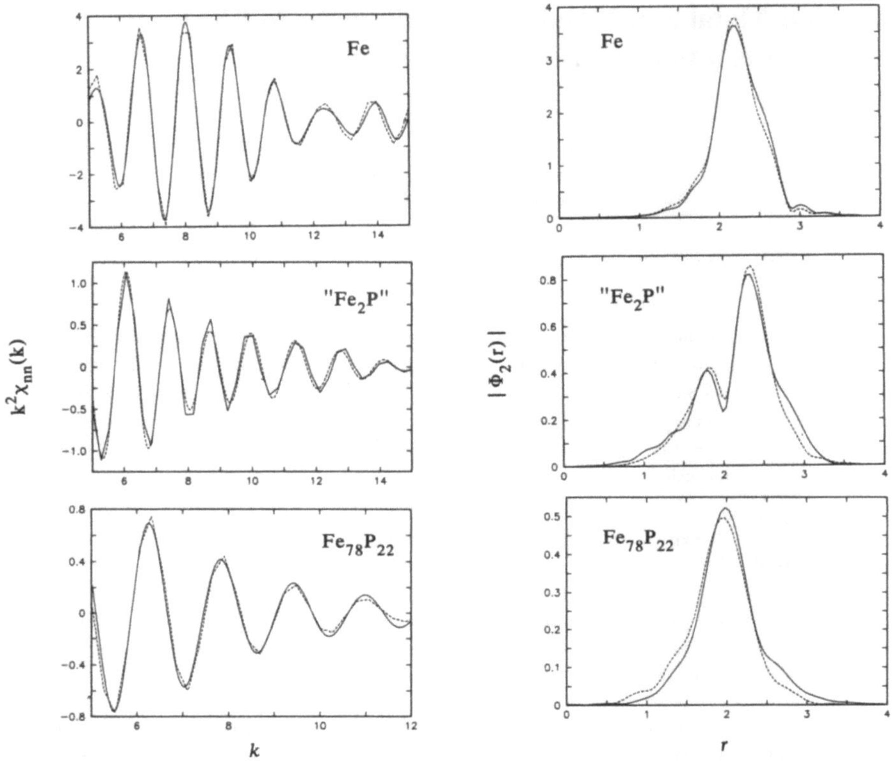

Fig. 1. EXAFS data obtained at 80 K (solid lines) as described in text, and results of best-fit model calculations (dashed lines)

0.221Å for $\sigma_{Fe\text{-}Fe}$. These σ values are expected to include both static and thermal disorder, although for the crystalline materials only thermal disorder is involved. Having $\sigma_{Fe\text{-}Fe} \approx 2\sigma_{Fe\text{-}P}$ is in rough accord with data from x-ray and neutron scattering experiments. Most room temperature scattering data for Fe, Co, and Ni-based amorphous metal-metalloid alloys indicate $\sigma \approx 0.17$Å (0.4Å FWHM) for metal-metal pairs.

We have also examined effects of asymmetrical distance distributions g(r) on EXAFS distribution functions $|\Phi_2(r)|$ and on x-ray scattering distribution functions $4\pi r^2\rho(r)$, since such asymmetries have been proposed as a reason for EXAFS distances being smaller than x-ray scattering metal-metal distances in amorphous alloys [3,4]. For the distribution

$$g(r) = \frac{1}{a} \exp\left(-\frac{r-r_0}{a}\right) \qquad r \geq r_0 \tag{2}$$

shown in Fig. 3(a), the EXAFS $|\Phi_2(r)|$ for k=4-15Å^{-1} looks symmetrical and peaks at the sharp edge r_0 rather than at the average distance $\langle r \rangle = r_0 + a$. However, the radial distribution function $4\pi r^2\rho(r)$ with k_{scat}=0.2-17Å^{-1} (corresponding to k_{EXAFS}=0.1-8.5Å^{-1}) better reproduces g(r) but also peaks much closer to r_0 than to $\langle r \rangle$. For r_0=2.3Å and a=0.6Å, taking peak values in $|\Phi_2(r)|$ and in $4\pi r^2\rho(r)$ gives a shift of only 0.14Å. For a=0.25Å, the shift is 0.10Å.

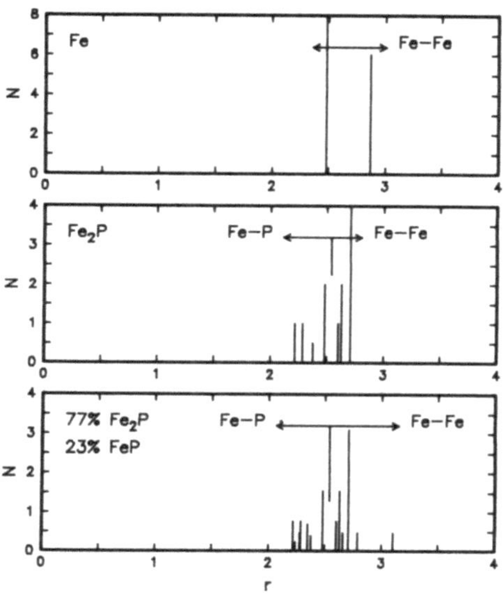

Fig. 2 (above) Crystalline near neighbor distance distributions.

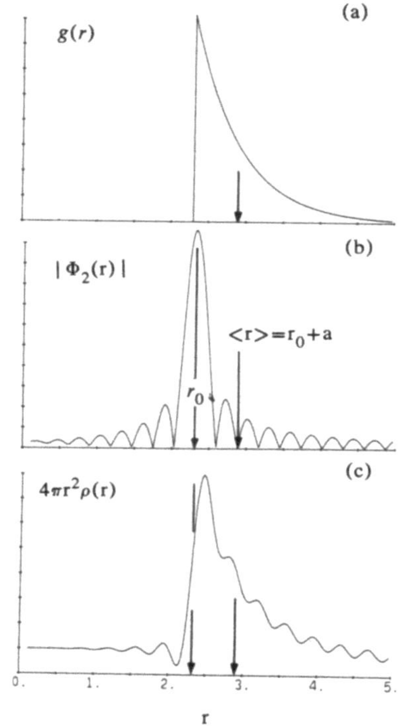

Fig. 3 (right) Asymmetrical peak (a), as seen in EXAFS (b) and in x-ray scattering (c).

An asymmetry of the degree needed to give a shift of 0.14Å, as in Fig. 3, would be very apparent in x-ray or neutron scattering distribution functions for amorphous transition metal-metalloid alloys but is not seen for these materials. Displacements of 0.1Å or less due to peak asymmetries cannot be ruled out by presently available scattering data.

The present results support an earlier suggestion [5] that EXAFS for amorphous transition metal-metalloid alloys is dominated by metal-metalloid nearest-neighbor contributions rather than by metal-metal contributions, because of the greater sharpness of the metal-metalloid distance distribution and an unexplained additional weakness in Fe backscattering, which is also apparent in EXAFS measurements on crystalline Fe_2P.

1. B. K. Teo and P. A. Lee, J. Am. Chem. Soc. *101*, 2815 (1979).
2. Y. Waseda, et al., Sci. Rep. Res. Inst. Tohoku Univ. *A26*, 202 (1977).
3. M. De Crescenzi, et al., Solid State Commun. *37*, 921 (1981).
4. R. Haensel, et al., in **Liquid and Amorphous Metals**, (Sijthoff & Nordhoff, Alphen aan den Rijn, The Netherlands, 1980), p. 459.
5. G. S. Cargill III, in **Proc. 4th Int. Conf. on Rapidly Quenched Metals**, (Japan Inst. Metals, Sendai, Japan, 1982), p. 389.

EXAFS Studies of Glassy $Ni_{66}B_{33}$: A Multi-Shell Modeling of the Ni-Ni Distribution

J. Wong

General Electric Corporate Research and Development,
Schenectady, NY 12301, USA

1. Introduction

The formation of transition metal-metalloid ($TM_{100-x}M_x$) glassy alloys by rapid quenching from the melt is usually confined to a limited composition range of about $15 \leq x \leq 25$ at.%. However, glassy alloys of the type TM-M substantially removed from the 80:20 atomic ratio have been prepared. Takaheshi et al. [1] have found non-crystalline X-ray diffraction patterns in Ni-B samples at 18.5 and at 40 at.% B. Inoue et al. [2] have reported on metallic glasses in the Ni-B and Co-B systems containing 33-43 at.% B and 35-41 at.% B, respectively. More interestingly, amorphous $Ni_{66}B_{33}$ crystallized homogenously to a single Ni_2B phase of known crystal structure (that of Al_2Cu), in which there is only one nonequivalent Ni site [3]. The existence of the $Ni_{66}B_{33}$ composition in both the crystalline and glassy phases makes it an attractive model system for quantitative EXAFS investigation of the local atomic environment in metal-metalloid glasses.

The local structure of a Ni atom in crystalline Ni_2B consists of 4 boron atoms at a distance of 2.14Å, one Ni atom at 2.37Å, and two Ni atoms at 2.42Å and two sets of four Ni atoms at 2.63Å and 2.70Å [3]. This crystal phase can be used to advantage to model the local structure of Ni in the corresponding glassy phase. The crystallographic data (coordination numbers and bond distances) are used as fixed inputs to the Fourier filtered EXAFS of the crystal to simulate a set of self-consistent phase shifts and amplitude functions for the Ni-B and Ni-Ni pairs. These scattering parameters were then transferred as fixed inputs to the Fourier-filtered EXAFS of the glassy phase to determine bond distances and coordination numbers.

2. Experimental

Preparation of master alloys and glassy ribbon by chill block melt-spinning has been described in detail elsewhere [4,5]. EXAFS experiments were performed with the EXAFS I-5 spectrometer at SSRL during a SPEAR run at an electron energy of 2.2 GeV and injection current of ~15mA. Spectra were scanned with a channel-cut Si(220) crystal and a 1 mm entrace slit, which yielded a resolution of ~0.7 eV at the Ni K-edge at 8333 eV.

3. Results and Discussions

The normalized EXAFS spectrum at 77K for crystalline Ni_2B is plotted as $\chi.k$ vs. k in Fig. 1(a). The corresponding Fourier transform is given in Fig. 1(b) which exhibits two resolved radial structure peaks in the region 1-3Å. These two peaks arise from the back-scattering of the innermost boron and immediate four Ni shells in the crystal structure. The corresponding EXAFS at 77K and Fourier transform for the $Ni_{66}B_{33}$ glass at 77K are given in Fig. 2(a) and 2(b), respectively. In the glass the Fourier transform consists basically of a single broad asymmetric radial peak centered at approximately 2.1Å.

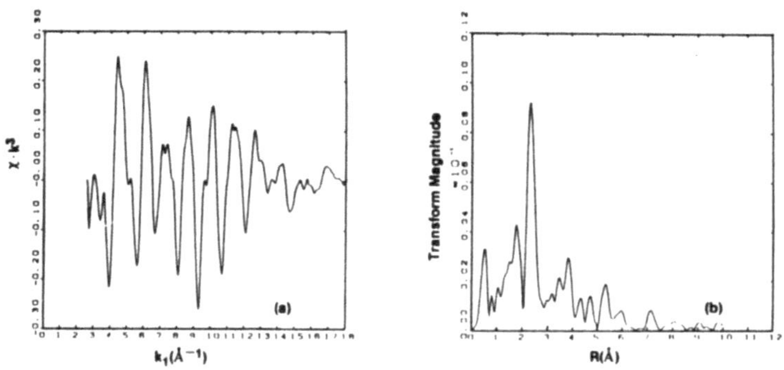

Fig. 1

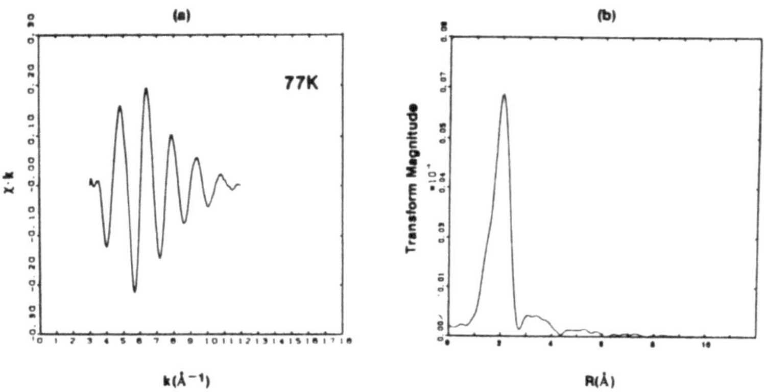

Fig. 2

Fig. 1. (a) Normalized EXAFS Plotted as $\chi \cdot k$ vs k and (b) Fourier Transform for Crystalline Ni_2B at 77K

Fig. 2. (a) Normalized EXAFS and (b) Fourier Transform for Glassy $Ni_{66}B_{33}$ at 77K

To elucidate the local structure giving rise to the observed EXAFS in the glass, we first used the known structural parameters of the crystal and its Fourier-filtered EXAFS in the region of 1-3Å to generate a set of self-consistent phase shifts for the Ni-B and Ni-Ni pairs. The parameterized phase shift values of Lee et al. [6] were used as initial inputs. The boron and Ni envelop functions $f_j(\pi,k)\exp(-2r_j/\lambda)$ were generated from the inverse Fourier-filtered EXAFS spectra of crystalline Ni_2B in the region 1-2Å, and fcc Ni metal in the region 1-3Å, respectively. This then constitutes a 5-shell simulation, i.e., one Ni-B shell and four Ni-Ni shells. The results are shown in Fig. 3 where the curve denotes the Fourier-filtered crystal spectrum and the points denote the simulated spectrum.

The generated Ni-B and Ni-Ni phase shifts are then used as fixed input parameters to simulate the glass spectrum in the same region of 1-3Å to determine the structural parameters. The simulation was done systematically and sequentially with one boron shell; one Ni shell; one B shell + one Ni shell; one B shell + two Ni shells; two B shells + one Ni shell; one B shell + three Ni shells; two B shells + two Ni shells; and one B shell + four Ni shells. The simulation with one B shell + three Ni shells yields the best fit with physically meaningful parameters. The simulation is shown in Fig. 4. The derived structural parameters for the glassy $Ni_{66}B_{34}$ are given in the following table.

Shell	Scatterer	N_j	r_j (Å)
1	B	3.8	2.11±0.03
2	Ni	4.3	2.24±0.09
3	Ni	3.0	2.46±0.06
4	Ni	3.9	2.63±0.12

It is noted that the nearst boron shell about a central Ni in the glass remains very similar to that of the crystal, but there is considerable rearrangement of the outer Ni shells. In the glass, the nearest 11 Ni atoms are now distributed into three shells with bond distances measurably closer to the central Ni atom. This local rearrangement and closer packing of the Ni shells are indeed reflected in the local electronic structure of the glassy phase, and are manifested in spectral changes observed in the near edge absorption features in the vicinity of the Ni K-edge in going from the crystal to the glass (Fig. 5).

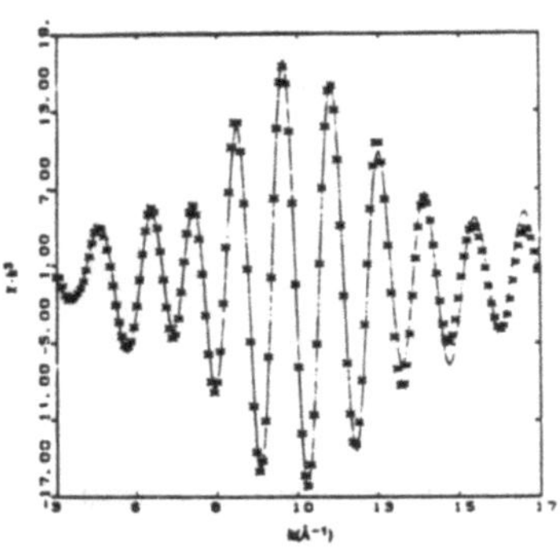

Fig. 3. Experimental (Line) and Simulated (Points) EXAFS for Ni_2B in the Region 1-3 Å about the Central Absorbing Ni Atom

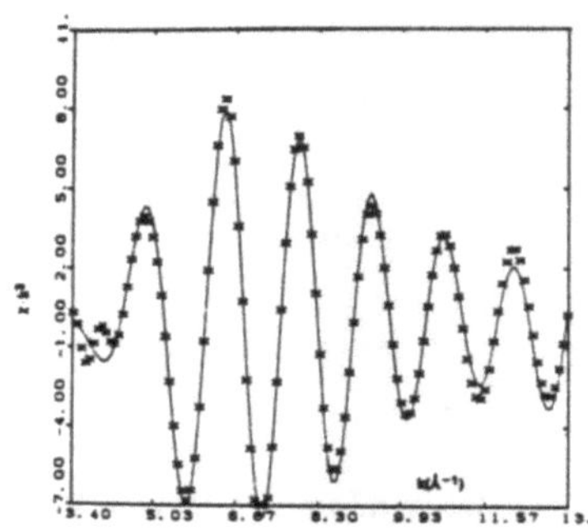

Fig. 4. Experimental (Line) and Simulated (Points) EXAFS for Glassy $Ni_{66}B_{34}$ in the Region 1-3 Å about the Central Absorbing Ni

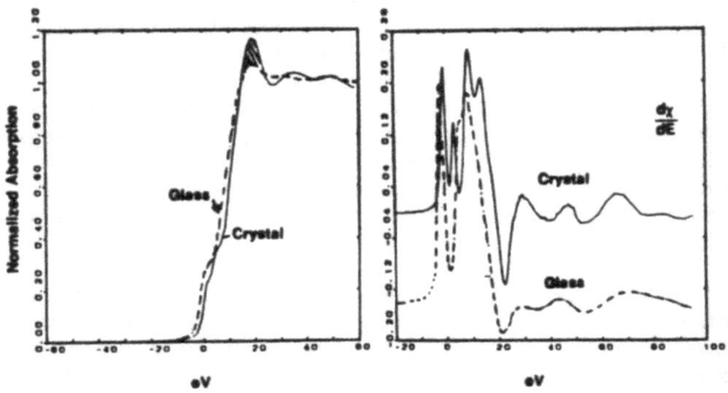

Fig. 5. K-Edge Spectra of Ni in Crystalline and Glassy Ni_2B

4. Concluding Remarks

Highlights of our experimental findings in this novel high metalloid $Ni_{66}B_{33}$ glass as modeled by a multi-shell simulation with the corresponding Ni_2B crystal are (1) around the central Ni atoms there exist both B and Ni "nearest" neighbors; (2) the structure of the B coordination shell remains similar to that in the crystal and (3) as in the crystal, the Ni neighbors exist in sub-shells, but substantially rearranged and measurably closer to the central Ni atom.

The findings we feel lend novel structural insight into local atomic structure in simple binary metal-metalloid glasses, for which conventional diffraction methods yield only an average metal-metal distribution. Determination of sub-shell structure in other (well-chosen) metal-metal glass sytems has also been reported by Sadoc and co-workers [7].

Acknowledgment

The author is thankful for experimental opportunity at SSRL supported by NSL through the Division of Materials Research and the NIH through the Biotechnology Resource Program in the Division of Research Resources in cooperation with the Department of Energy. This work was partially supported by the US Department of Energy under contract DE-AC02-79ER-10382.

References

1. M. Takaheshi, C.O. Kim, M. Koshimura and T. Suzuki, Jap. J. Appl. Physics 17, 741 (1978).

2. A. Inoue, A. Kitamura and T. Masumoto, Trans. JIM 20, 404 (1979).

3. E.E. Havinga, H. Damsma and P. Hokkeling, J. Less Comm. Metals 27, 169 (1972).

4. H.H. Liebermann and J. Wong, J. Non-Crystal. Solids 45, 195 (1981).

5. H.H. Liebermann, Mat. Sci. Eng. 43, 203 (1980).

6. P.A. Lee, B.K. Teo and A.L. Simons, J. Amer. Chem. Soc. 99, 3858 (1977).

7. A. Sadoc, D. Raoux, P. Lagarde and A. Fontaine, J. Non-Crystal Solids (to be published).

EXAFS Study of $Co_{1-x}B_x$ Glasses

J.M. Dubois[1], J. Goulon[2,3], G. Le Caer[3], and P. Lagarde[3]

1. Introduction

The usefulness of EXAFS in studying the local order around the atom probe is now clearly demonstrated by an increasing number of papers. However, in disordered materials, this technique does not reproduce the average pair distances measured by other methods such as X-ray diffraction unless the asymmetry of the pair potential is taken into account. In order to overcome this drawback EISENBERGER and co-workers [1,2] write the modulated part of the X-ray absorption cross section as:

$$\chi(k) = - C(k)[S^2(k) + A^2(k)]^{\frac{1}{2}} \sin[2k\bar{R} + \psi(k) + \Sigma(k)] \quad (1)$$

where k is the photoelectron wave vector, $\bar{R}$ the average pair distance, C and ψ the usual amplitude and phase shift terms respectively and $\Sigma(k) = \tan^{-1}A(k)/S(k)$. S(k) and A(k) account for the symmetric and asymmetric parts of the pair distribution function respectively and are defined in [1,2]. Defining σ^n as the nth reduced moment of the pair distribution function limited to the first shell by the rapid damping of the photoelectron ($\sigma^1 \equiv \bar{R}$), the supplementary phase shift and amplitude terms can be approximated by [1,2].

$$\Sigma^2(k) \simeq - (4\sigma^2/\bar{R})k - 8\,(\sigma^3/15)k^3 \quad (2)$$

$$(S^2(k) + A^2)^{\frac{1}{2}} \simeq [1 + \Sigma^2(k)]^{\frac{1}{2}}\, e^{-2k^2\sigma^2} \quad (3)$$

which are valid in the limits $k\sigma^2/\bar{R} < 1$ and $k^3\sigma^3 \le 1$ (4).

In cobalt borides the backscattering amplitude of the boron can be neglected in comparison with cobalt. Thus measurements of the EXAFS spectrum at the cobalt edge gives the symmetry of the Co-Co nearest-neighbour coordination. This paper reports on the EXAFS spectra of hcp Co and orthorhombic Co_3B (used as model compounds) and $Co_{81.5}B_{18.5}$, $Co_{78}B_{22}$ and $Co_{74}B_{26}$ glasses. The discussion is then devoted to a comparison of the models of POLK [3] and of GASKELL [4] in the light of the experimental results.

[1]Laboratoire de Métallurgie associé au CNRS (LA159), Ecole des Mines, Parc de Saurupt, F-54042 Nancy Cedex, France

[2]Laboratoire de Chimie Théorique, Université de Nancy I, C.O. 140, F-54037 Nancy Cedex, France

[3]LURE, Bâtiment No. 200, F-91405 Orsay Cedex, France

2. Experimental

2.1 Amorphous samples

Approximately 20 μm thick ribbons were cast from master sintered alloys by melt spinning on a copper wheel under a reducing atmosphere of He-5%H_2 to prevent any reaction of the melt with the quartz crucible.

2.2 EXAFS

The measurements were made at room temperature in transmission geometry. The amorphous samples were mounted onto a slide frame with adhesive tape. The crystalline samples were crushed and mixed with finely powdered boron nitride and stuck onto adhesive tape. Lead tapes were used to define the edges of the samples and obstruct the holes. However, some of these holes remained and thus, as is described in [5], it was not possible to obtain the coordination numbers directly.

The spectra were recorded at the K edge of cobalt at the LURE facility using a Si(220) channel-cut monochromator. The data analysis followed the method of EISENBERGER et al.[1,2], using the phase shifts and backscattering amplitudes tabulated for cobalt by LEE et al.[6]. The validity of these parameters was checked by measuring the average pair distances in the model compounds hcp Co and Co_3B ($\bar{R}$ = 2.51 Å and $\bar{R}$ = 2.59 Å respectively) after fixing the origin E_o of the photoelectron in both samples. While this is trivial in metallic cobalt due to the hexagonal symmetry ($\sigma^3 \simeq 0$), the choice of E_o in Co_3B is hindered by the supplementary phase shift (see (1)) corresponding to the highly distorted first Co shell of that compound [7]. Therefore the optimal values $\{E_o,\sigma^3\}$ were successively refined until the maxima of the imaginary part and absolute value of the Fourier transform of $\chi(k)$ were superimposed with the actual value of $\bar{R}$. During this procedure σ^2, which only plays a secondary role in (1), was fixed at a reasonable value, σ^2 = 0.02, and experience has shown that the optimal set $\{E_o,\sigma^3\}$ is sharply defined. Finally, the E_o value found for Co_3B was retained as the energy origin for the glasses.

Figure 1 displays the edge structure of all the samples on an arbitrary scale. At half height the hcp Co edge is shifted from that of Co_3B by + 17 eV.

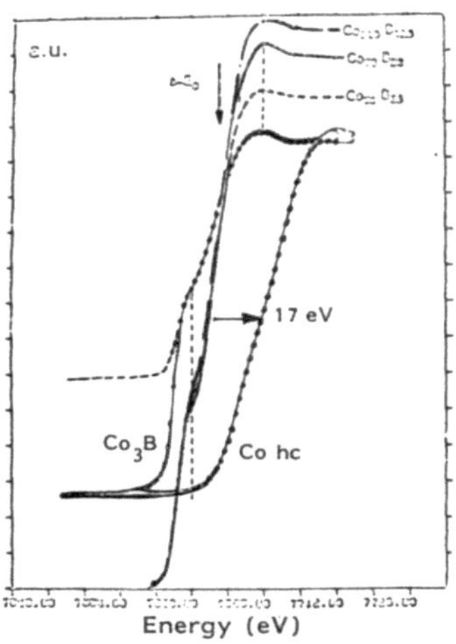

Fig. 1. Edge structure of the different samples

The boron-containing samples exhibit a comparable pre-edge structure which is not shown by the pure metal. It seems likely, therefore, that the densities of unoccupied bonding states are similar in the amorphous and crystalline borides but differ from that of the pure metal. Such a conclusion was already inferred from XPS studies of similar alloys [8].

The inelastic emission and scattering process were taken into account by a k-dependent amplitude term

$$A(k) = A_o(k)\ e^{-2\bar{R}/\lambda(k)} \tag{5}$$

with $\lambda(k) = k/\Gamma$ where λ is the photoelectron coherence length and Γ depends on the backscatterer and with $A_o(k)$ (inelastic emission amplitude) as proposed by STERN et al.[9].

The contribution of the first shell was first filtered in real space with the help of a KAISER window [10] and its contribution back Fourier transformed according to SHULMAN et al.[11]. The moments $\{\bar{R},\sigma^2,\sigma^3\}$ of the radial distribution function in the first cobalt shell were then optimised by using a non-linear optimisation routine developed by GOULON [12].

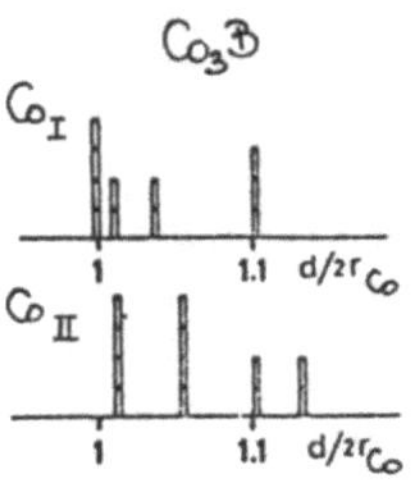

Fig.2. Histograms of the Co-Co pair distances in Co B (I and II denote the two different Co sites after [7] in Co diameter units)

As shown by Fig.2 which presents the number of Co first neighbours of the two crystallographically distinct cobalt sites of the Co_3B structure, the first Co shell of this boride can be split into two subshells corresponding to pair distances respectively in the ranges $2.5 \le \bar{R} < 2.75$ Å and $2.75 \le \bar{R} < 3$ Å. Fits with these two subshells, which fulfil the limit conditions (4) and decrease the residual factor RF, were thus performed. However, this splitting is to be considered as arbitrary and only the moments of the whole shell (calculated from the moments of the subshells) are reported here.

3. Results

The experimental EXAFS modulations $k\chi(k)$ are shown in Fig.3 for the three amorphous samples, together with a superimposition of the experimental spectra of a-$Co_{74}B_{26}$, Co_3B and hcp Co. This clearly shows that the Co_3B boride is the more suitable model compound for the glasses. These spectra are in good agreement with those published by SCHMUKLE et al.[13].

The experimental contributions to $\chi(k)$ of the first shell in the glasses and Co_3B and the corresponding calculated spectra are shown in Fig.4. The results of the fits are given in Table 1.

The average interatomic distances measured in the glasses agree well with the distances reported from X-ray diffraction experiments ($\bar{R}$ = 2.57 Å for x = 0.17 and $\bar{R}$ = 2.61 Å for x = 0.23 [14]). The value of σ^2 for hcp Co is similar to the Debye-Waller factor published by RAMADAND and RAMJI RAO [15] and thus represents the thermal contribution to the pair distance scatter,

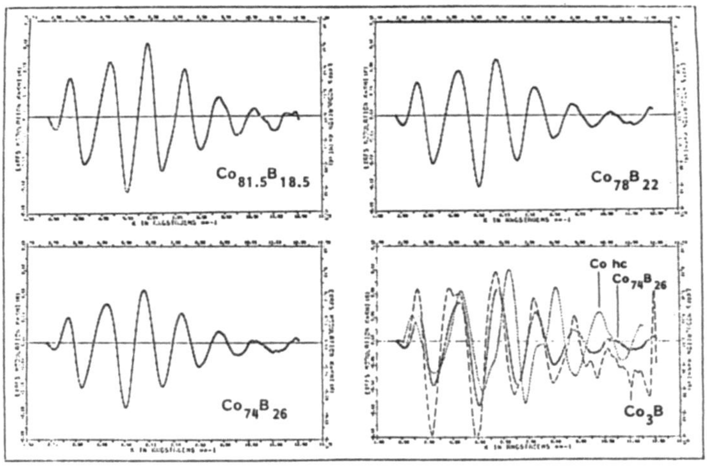

Fig.3. Experimental EXAFS modulations of the glasses, hcp Co and Co_3B. (A Bragg glitch distorts the Co_3B spectrum above $k = 11$ Å^{-1})

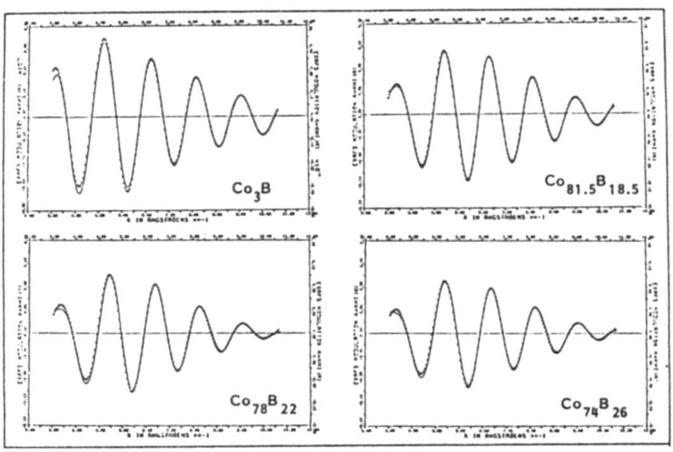

Fig.4. Experimental (dots) and calculated (solid line) first shell contribution to $\chi(k)$ in the glasses and Co_3B

Table 1.

Sample	$\bar{R}$[±0.02 Å]	σ^2[Å^2]	σ^3[Å^3]	RF
hcp Co	2.51	0.007	0.0	0.0049
$Co_{81.5}B_{18.5}$	2.55	0.011	0.0023	0.0024
$Co_{78}B_{22}$	2.57	0.016	0.0032	0.0055
$Co_{74}B_{26}$	2.57	0.016	0.0032	0.0061
Co_3B	2.59	0.025	0.004	0.005

while σ^3 is indeed found to be negligible in the pure metal. In the borides both σ^2 and σ^3 increase, showing that the shell deforms as a result of the introduction of boron into the structure.

4. Discussion

As $\bar{R}$, σ^2 and σ^3 are not independent parameters, the evolution of $\bar{R}$ and σ^3 versus σ^2 (i.e. as a function of the sample being examined) is shown in Fig.5. This figure clearly shows that the average shell symmetry continuously evolves in the glasses between the symmetry of hcp Co and the low symmetry characteristic of Co_3B. This result is not consistent with a rigid metallic skeleton in which the boron atoms progressively fill the largest voids and thus indicates that metallic glasses should not be considered as interstitial solutions with a BERNAL-like structure as was suggested by the model of POLK [3].

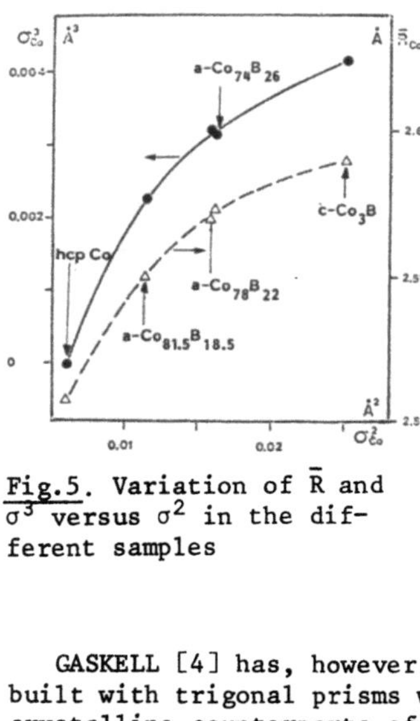

Fig.5. Variation of $\bar{R}$ and σ^3 versus σ^2 in the different samples

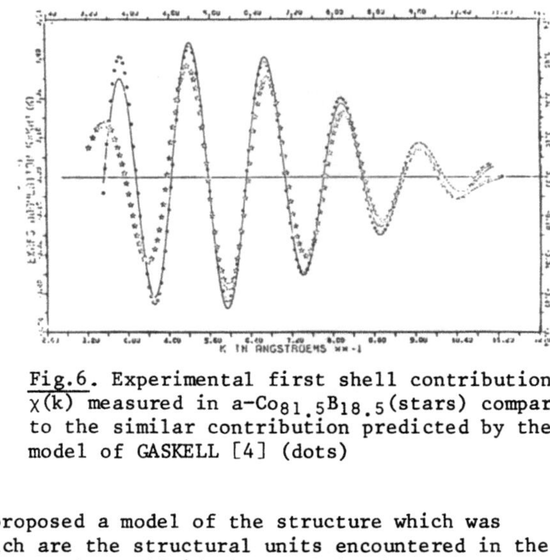

Fig.6. Experimental first shell contribution to $\chi(k)$ measured in a-$Co_{81.5}B_{18.5}$ (stars) compared to the similar contribution predicted by the model of GASKELL [4] (dots)

GASKELL [4] has, however, proposed a model of the structure which was built with trigonal prisms which are the structural units encountered in the crystalline counterparts of the metal-metalloid glasses. As the structure of Co_3B corresponds to the structural building rules adopted by GASKELL in [4], his model can be rescaled in the Co/Pd atomic radii ratio to predict the coordination shell of a-$Co_{80}B_{20}$. A satisfactory agreement is in fact observed between the experimental first shell contribution of a-$Co_{81.5}B_{18.5}$ (Fig.6, stars) and that calculated from the coordinates of the model by broadening the pair distances with the thermal scatter measured in hcp Co (σ^2 = 0.007)(Fig.6 dots). Also superimposed on this figure is the fitted spectrum to the model contribution which yields: $\bar{R}$ = 2.53 Å, σ^2 = 0.014 Å, σ^3 = 0.001. These values compare satisfactorily with the experimental data in table 1.

Therefore, it seems that the composition change of the glasses may not depend on the filling of interstitial vacancies, but more likely on the change of structural units as we proposed recently in [16]. Indeed this change entails a composition-dependent distortion of the cobalt shell as is actually observed and new experiments on samples richer in boron are in progress to clarify this point further.

Acknowledgements We are indebted to the staff of the Laboratoire de l'Accélérateur Linéaire and of the Laboratoire pour l'Utilisation du Rayonnement Electromagnétique (Orsay) for the provision of beam facilities. Special acknowledgement is due to the EXAFS group at LURE for fruitful discussions and to Dr. R. Cortes for his advice during the runs. We are grateful to Dr. C. Tete for help with sample preparation, to Dr. P. H. Gaskell for his model coordinates and valuable encouragement, and to A. K. Livesey for a critical reading of the manuscript.

References

1 P. Eisenberger, G. S. Brown, Sol. State Comm. 29, 481 (1979)
2 P. Eisenberger, B. Lengeler, Phys. Rev.B 22.8, 3552 (1980)
3 D. E. Polk, Acta Met. 20, 485 (1972)
4 P. H. Gaskell, J. Non Cryst. Sol. 32, 207 (1979)
5 J. Goulon, C. Goulon-Ginet, R. Cortes, J. M. Dubois, J. de Physique 43, 539 (1982)
6 P. A. Lee, B. K. Teo, A. L. Simons, J. Am. Chem. Soc. 99, 3156 (1977); B. K. Teo, P. A. Lee, A. L. Simons, P. Eisenberger, B. M. Kincaid, J. Am. Chem. Soc. 101, 2815 (1979)
7 S. Rundqvist, Acta Chem. Scand. 12, 658 (1958)
8 M. Matsuura, T. Nomoto, F. Itoh, K. Suzuki, Sol. St. Comm. 33, 895 (1980)
9 E. A. Stern, S. M. Heald, B. Bunker, Phys. Rev. Lett. 42, 1372 (1979)
10 H. J. Harris, IEEE Proc. 66, 51 (1978)
11 R. G. Shulman, P. Eisenberger, B. M. Kincaid, Ann. Rev. Bioeng. 7, 559 (1978)
12 J. Goulon, EXAFS Program, Institut Universitaire de Calcul Automatique, Université de Nancy I
13 F. Schmuckle, P. Lamparter, S. Steeb, Z. Naturf. 37a, 571 (1982)
14 Y. Waseda, H. S. Chen, Phys. Stat. Sol.a 49, 387 (1978)
15 A. Ramanand, R. Ramji Rao, Can. J. Phys. 58, 384 (1980)
16 J. M. Dubois, G. Le Caer, Vth Int. Conf. Phys. Non Crystalline Solids, Montpellier (1982). To appear in J. de Physique

EXAFS on $(AgI)_x(Ag_2O\ n\ B_2O_3)_{1-x}$ Superionic Glasses

G. Dalba, A. Fontana, O. Fornasini, and G. Mariotto
Dipartimento di Fisica, Università di Trento, I-38059 Povo, Italy
F. Rocca
Istituto per la Ricerca Scientifica e Tecnologica, Trento, I-38059 Povo, Italy
E. Bernieri, M.R. Masullo, and A. Morone
Istituto di Fisica dell'Università di Napoli, Via Tari 3, I-80138 Napoli, Italy

1. Introduction

In the last years the superionic materials have raised a great interest in physicists and chemists for their characteristic property of high ionic condutivity [1]. Most of the new materials which have been used for application as solid electrolytes [2] were crystalline materials in which the structure allows high ionic conductivity. Among superionic conductors, silver iodide is one of the most studied for the motion of conducting ions through the sites: a low potential barrier implies a weak bond between the mobile ion and the reticular structure. Now our interest for ionic compounds is turned to glassy materials. They have the advantage of easiness of preparation in a large range of dimensions. Moreover their composition can be almost continuously changed. For these considerations we have been stimulated to study borate glasses containing silver iodide: in particular $(AgI)_x\ (Ag_2O\ n\ B_2O_3)_{1-x}$ glasses at various x and n concentrations.

In this paper we report preliminary EXAFS measurements: our aim is to compare our findings with existing models for diffusion and structure in vitreous electrolytes, borate glasses in particular.

2. Experimental

The samples were made by heating weighed quantities of reagents: silver nitrate, Silver Iodide and Boron Oxide. The melt samples were poured in stainless steel molds kept at 300 K. EXAFS and XANES measurements are under way at the Adone Wiggler Facility, utilizing a Si(111) channel-cut monochromator. Due to temporary machine difficulties, it was not possible to reach the Ag K-edge energy. Preliminary results are shown on the Ag and I L_{III} edges in samples at room temperature.

3. First results

In Fig.1 we report the absorption L_{III} edge of silver in AgI, $Ag_2O\ 4\ B_2O_3$ and $(AgI)_{0.3}\ (Ag_2O\ 4\ B_2O_3)_{0.7}$: we can easily note the presence of a white peak in the two glassy compounds.

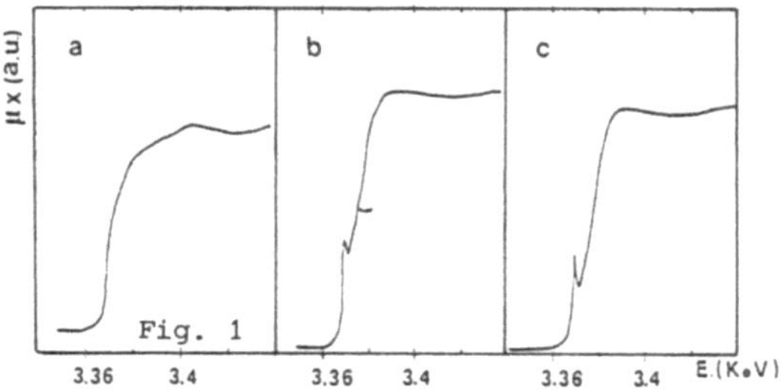

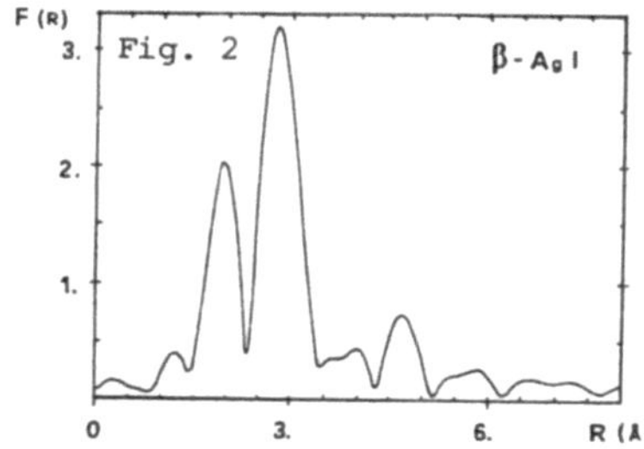

Fig.1. AgL_{III} edge a) AgI; b) glass x=0.0 n=4 c) x=0.3, n=4

Fig.2. Fourier Transform of EXAFS signal on I L_{III} edge in AgI

Our preliminary analysis seems to indicate that varying the AgI concentration in the glasses, the coordination of Ag atoms changes and it tends to that of pure AgI, where the white peak is not present.

We have used the AgI as a standard sample for EXAFS measurements. In Fig.2 we report the Fourier transform of the EXAFS signal on the L_{III} edge of iodine in AgI. The peaks at 2.8 Å and at 4.7 Å can be assigned respectively to the 4 Ag nearest neighbours and to the 12 I atoms of the second shell. By utilizing the RABE method and the theoretical phases of TEO and LEE we have obtained a value of the Ag-I distance equal to 2.836 Å in good agreement with the values obtained for the β-AgI both by crystallographic methods and by previous EXAFS measurements on the Ag K-edge [3].
Fig. 3 shows the phase shift Φ_{I-Ag} for the atomic pair I(absorber), Ag(backscatterer) as calculated from the experimental data, using the crystallographic value of the distance I-Ag (2.81 Å); the dots represent the values calculated by TEO and LEE.

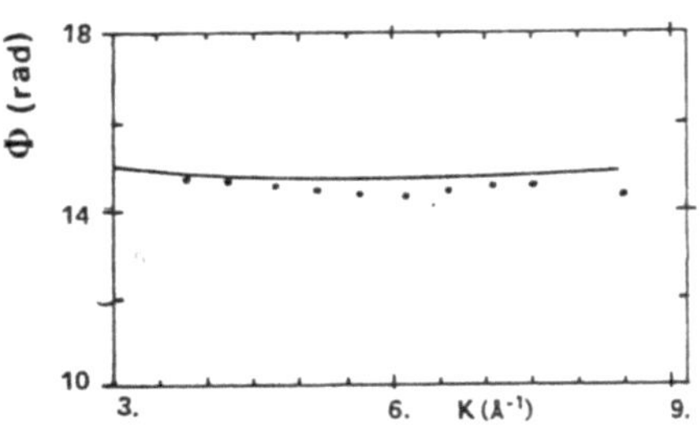

Fig.3. Phase shift in Ag-I

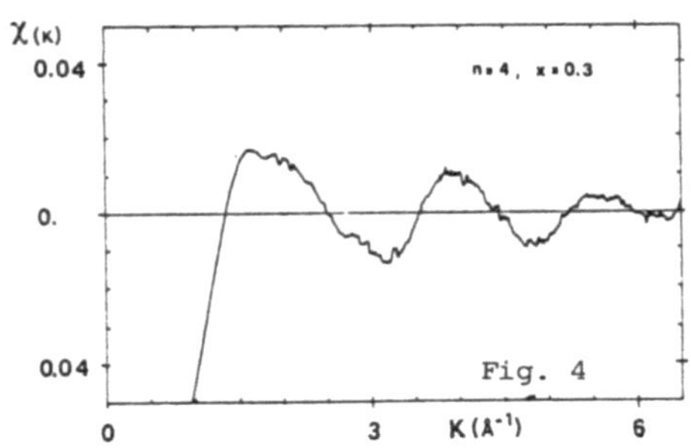

In fig.4 the EXAFS of the AgL_{III} edge in $(AgI)_{0.3}(Ag_2O\ 4\ B_2O_3)_{0.7}$ is shown.

1. M.B.Salamon (ed): Physics of Superionic Conductors, Topics Curr.Phys. 15 (Springer, Berlin, Heidelberg, New York 1979)
2. S.Geller (ed): Solid Electrolytes, Topics Appl. Phys. 21 (Springer, Berlin, Heidelberg, New York (1977)
3. J.B.Boyce, T.M.Hayes, J.C.Mikkelsen jr.: Phys.Rev.B 23, 2876 (1981)

Short-Range Order in $Fe_{78-x}Ni_xB_{18}Mo_4$ (x = 38, 20, 10, 0) Metallic Glasses as Deduced by EXAFS Data

F. Comin[1], L. Incoccia, S. Mobilio, and N. Motta[2]
PULS, Laboratori Nazionali di Frascati dell'INFN,
I-00044 Frascati, Italy

Structural information on metallic glasses (MG) is still the central problem to understand their unique characteristics [1].

Many efforts centered on binary metal-metalloid MG clarified some common features of their radial distribution function (RDF), namely the presence of an asymmetry in the first peak, whose amount depends on the type of metalloid [2], the splitting of the second peak [3], and the avoidance of the metalloid atoms as first neighbours of another metalloid [4].

Some work has been done also in more complicated glasses [5] where the standard X-ray diffraction is unable to provide the different atomic contributions to the total RDF. The EXAFS technique with its inherent atomic selectivity seems then to be the most suitable tool to cast light on the short range order around each specific atom environment in complex MG.

In this paper we report the EXAFS study of some quaternary glasses of the type $Fe_{78-x}Ni_xB_{18}Mo_4$ ($x = 38, 20, 10, 0$).

EXAFS and near edge data above the Fe and Ni K edges, both at LNT and RT, were collected at the X-ray beam line [6] of the PULS facility in Frascati on small strips of samples prepared by splat quenching. Particular care was taken in order to avoid glitches [7] or other non-linear effects.

The data analysis was done using the EXAFS program package developed at PULS [8] and using an R-space approach [9].

The main features that the spectra exhibit following the sequence from the two standard samples Metglas 2603 and Metglas 2826 ($Fe_{80}B_{20}$ and $Fe_{40}Ni_{40}B_6P_{14}$) to the Mo containing samples at progressively decreasing Ni content can be resuméd in the following way.

<u>Peak position</u>

a) No particular trend is observed in the FT of the Fe spectra, the peak position being $R = 2.02 \pm 0.02$ A. This indicates [2] the presence of a sharp and asymmetrical RDF around the Fe atoms in all the sample investigated.

---

[1]Present address: Bell Laboratories, Murray Hill, N.J. 07974, USA
[2]Scuola Normale Superiore, I-56100 Pisa, Italy

b) On the other hand, the peak position of the Ni FT moves slightly away (from 2.04 ± 0.02 to 2.09 ± 0.02 A) upon introduction of Mo.

FT magnitudes

The relevant results are shown in Fig. 1 where we note:

a) In the Mo containing glasses the absolute magnitude of the FT for the Fe atoms (solid lines) is strongly decreasing upon decreasing the Ni concentration.

b) The thermal sensitivity of the Fe FT magnitudes is greatly increasing in going in the same direction. The broken line represents the value of $\Delta\sigma^2/\sigma_{77K}$ calculated following the suggestion of ref. [5].

c) The FT magnitude of the Ni spectra is nearly constant in all the samples, even though the Mo containing samples display an enhanced thermal sensitivity.

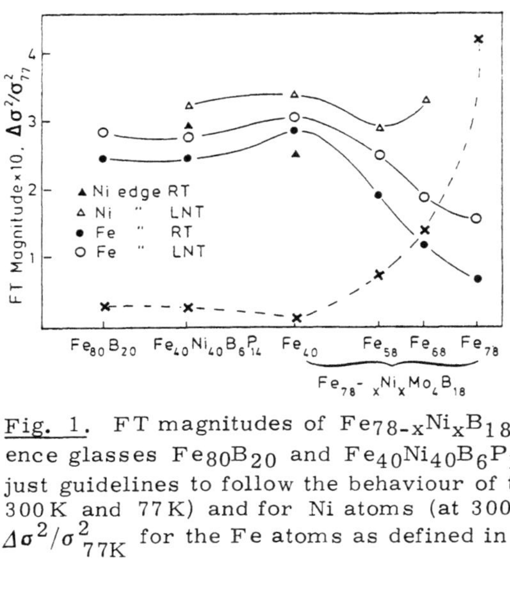

Fig. 1. FT magnitudes of $Fe_{78-x}Ni_xB_{18}Mo_4$ samples and of reference glasses $Fe_{80}B_{20}$ and $Fe_{40}Ni_{40}B_6P_{14}$. The solid lines are just guidelines to follow the behaviour of the FT's for Fe atoms (at 300 K and 77 K) and for Ni atoms (at 300 K). The broken line is $\Delta\sigma^2/\sigma^2_{77K}$ for the Fe atoms as defined in ref. [5]

All these observation can be interpreted in the following way: in the MG $Fe_{40}Ni_{40}B_6P_{14}$ both the Fe and Ni spectra show the poor thermal sensitivity typical of a MG related to the stiff interacting local potential [2]. The introduction of a small number of Mo atoms ($Fe_{40}Ni_{38}B_{18}Mo_4$) changes only the thermal behaviour of the Ni spectrum. This is due to a softening of the potential well around the Ni atoms only, suggesting that the local environment of Ni is different with respect to the preceding sample.

It has been reported recently [10], in fact, that during the crystallization process there is the enucleation of two different phases: a Fe-Ni-Mo solid solution and a Fe-B glass.

It seems reasonable, then, to assume a similar two-phases situation also in the glass, at least for the coordination of B (mainly with Fe) and of Mo (mainly with Ni). The increased thermal sensitivity of the Ni spectrum shows that the typical local stiffness of a glass is lost in the Mo-Ni rich phase. The progressive decrease of the Ni content in the sample with x = = 20, 10, 0 causes an increase of the number of Fe atoms present in the solid solution coordinated with Mo. This causes a decrease of the magnitude of the Fe FT since, as it is known [10], the Fe atoms in the solid solution are in a fcc or a bcc environment depending on the Ni content of the sample.

Furthermore, the enhanced thermal sensitivity of the Fe spectra is related to the increased number of Fe atoms in a broader potential well (solid solution) with respect to the ones in a hard potential environment typical of an homogeneous MG.

In summary we have shown that EXAFS data on complex metallic glasses strongly support the existence of a compositional short range order. In particular we found that in Mo containing transition metal glasses, there exists a preferential coordination between Mo and Ni atoms that can lead to a structure of Mo–Ni rich islands (solid solution) embedded in a Fe-metalloid rich glassy structure.

1. See for example: "Glassy Metals I", ed. by H. J. Guntherodt and H. Beck (Springer Verlag, 1981).
2. S. Mobilio and L. Incoccia, Proceedings of the "Workshop on EXAFS Data Analysis in Disordered Systems", Parma 1981 (in press on Riv. Nuovo Cimento); D. Raoux, this Proceedings; R. Haensel, P. Rabe, G. Tolkiehn and A. Werner, Proceedings NATO Adv. Study Inst. : "Liquid and Amorphous Metals", ed. by E. Lusher (Reidel, 1980); M. De Crescenzi, A. Balzarotti, F. Comin, L. Incoccia, S. Mobilio and N. Motta, Sol. St. Comm. 37, 921 (1981).
3. G. S. Cargill III, in "Solid State Physics", ed. by H. Erenreich, F. Seitz and D. Turnbull (Academic Press, 1975), Vol. 30.
4. T. M. Hayes, J. W. Allen, J. Tauc, B. C. Giessen and J. J. Hauser, Phys. Rev. Letters 40, 1282 (1978).
5. J. Wong, in "Glassy Metals", ed. by H. J. Guntherodt and H. Beck (Springer Verlag, 1981).
6. F. Comin, L. Incoccia, S. Mobilio and A. Balzarotti, Frascati Report, to appear.
7. F. Comin, L. Incoccia and S. Mobilio, J. Phys. E 15 (1982), in press.
8. S. Mobilio, F. Comin and L. Incoccia, Frascati report LNF-82/19 (1982).
9. L. Incoccia and S. Mobilio, these Proceedings.
10. G. Cocco, L. Schiffini, L. Battezzati and A. Lucci, Journ. of non Cryst. Solids, to appear.

Part VI

Catalysts

EXAFS on Catalysis

Pierre Lagarde

LURE, Bât. 209C, Université Paris-Sud, F-91405 Orsay, France

The power of EXAFS spectroscopy for catalysts studies is now well established, and it seems that this technique is particularly well suited since the major qualities of EXAFS find their full use there.

In order to understand the role of the important element, generally speaking a noble metal, involved in the catalytic activity, the main point to solve first is a structural one : we have to determine which type of bonding develops between a metallic atom and the other ones, or with the support, or eventually with the reactants. Since the metal is in a very dilute and dispersed form - typically about 1 % in weight for particles of some 20 Å of extent - the problem is the determination of the local structure of very small clusters, without long range order, built up from heavy elements on a light support. These conditions are typical of the problems that EXAFS is able to solve, and we find here a strong similarly to the study of metalloproteins.

Beside these structural problems, another important point in understanding the catalytic activity is the electronic structure of the metal : this will give ideas about the amount of charge from the metal during the catalytic reaction. The study of the edge structure, by comparing with the bulk metal, can give useful answers to this problem.

Because of the simplicity of an EXAFS experiment, without any special sample preparation, these types of studies can be done in different physico-chemical conditions, like the temperature of preparation, and even during the catalytic process itself. This last point is going to develop strongly when the new technique of EXAFS, in a fast data collection mode, will become available .

After a description of the results which can be obtained by XANES on electronic properties of catalysts, we will discuss what has been done on supported metal catalysts, which are may be the most usual in industry. Fewer works have been devoted to homogeneous catalysis, and we just will discuss them shortly. Finally, we will give some considerations on the future of EXAFS for catalysis, specially in connexion with Exafs in dispersive mode.

Because of the high X-ray absorption cross section of heavy metals, compared to that of the matrix on the correct energy domain, the thickness of the samples is of the order of few millimeters, which is quite easy to achieve. Nevertheless, the experimental set-up has to preserve the physico-chemical integrity of the sample, since it is well known that small amounts of reactive gases, like oxygen, can disturb drastically the properties of the system. Therefore, most of the work is done on vacuum tight cells, with possible heating, cooling and gas flowing for treatments, where the EXAFS data are taken "in situ", in conditions as close as possible from the actual ones. An example of these sample holders has been described by Lytle and co-workers [1].

I. Near-edge studies

Most common metals involved in heterogeneous catalysis show, like Pt or Ir, in the metal state, a strong white line just above the jump at the L edges. A measure of the shift in energy of this edge, and of the change of the area under the peak, of coming from the bulk metal to the catalyst, give an evaluation of the change of the number of unfilled d states in the sample since the L_{II} and L_{III} transitions end at d states of the element. As the following studies will show, the valence structure of the metal is dependant on various physico-chemical conditions, like the nature of the support, the temperature of preparation or the chemisorption of a gas on the catalyst.

In an earlier work, Lytle et al. [1] looking at the L_{III} edge of metal, compounds and catalysts based on Au, Pt and Ir, concluded that an increase of the number of vacant d states versus the metal foil occurs, this tendency being reinforced by adsorption of oxygen. Comparing these values with those obtained from the corresponding oxides, the less ionic character of the metal-oxygen bond in the catalyst versus the definite compound becomes evident.

Recently, A.N. Mansour et al. [2] have done a careful study of Pt/SiO_2 and Pt/Al_2O_3 as a function of support condition and reduction temperature. They have looked at both the L_{II} and L_{III} edges, and give a quantitative evaluation of the electron deficiency. While the effect of SiO_2 on Pt is less important, and the reduction is complete at 473 K, a temperature of 723 K must be reached in the case of Pt/Al_2O_3. Their results are shown in Table I, together with those on Pt/TiO_2 from the same group [3] which allow some conclusions aboutthe change of the chemisorption of H_2 or CO as a function of the reduction temperature.

II. Structural studies

The major part of catalysts consists of finely dispersed metal (Pt, Ir, Au) on a non-reactive matrix with a very large surface to volume ratio , like SiO_2, Al_2O_3 or TiO_2. In that case, the metal appears as small clusters (average size less than 15 Å), of may be different shapes, with different interactions with the support ; all these properties are important for the activity of the catalyst, and are very difficult to approach with the conventional structural techniques like X-ray diffraction or electron microscopy.

A) Monometallic catalysts

Most of the work has been done on reduced catalysts, and by looking at the first shell, although useful information can be extracted from further distances, as we will see later. The classical way to analyse the data, by Fourier filtering and fitting, making use of the bulk metal as a reference material and calculated phase shifts [4], is now widely used.

While sligtly different results are reported in literature, showing the dependance on the support and on the physico-chemical constraints on the sample, some general tendencies seems to show up.

The first shell can be analyzed by using only metal atoms, with distances, in the case of catalysts under hydrogen, not very far from the bulk value, as shown in Table II [5,6,7]. This and other measurements on outer shells, as we will see, suggest that the cubic symmetry is preserved in these systems, while some theoretical works [8,9] would infer that for small particles, the icosahedral structure would be energetically preferred. The interaction with the support, while weak since no definite bonding but the metal-metal one can be detected, has may be a consequence in a systematic shortening of the first-neighbor distance by about 0.02 Å ; but this decrease can also be due to the small size of the clusters.

The presence of the icosahedral symmetry has been found in Pt/zeolite under vacuum by Moraweck et al. [7], at the same time that a contraction of the nearest-neighbor distance of about 3 % occurs.

EXAFS allows also a determination of the average occupancy of the different shells of neighbors ; an interesting feature of the structure of these small clusters found by EXAFS analysis of the first shell is that the mean fluctuations of the distances are not very large compared to the bulk value, despite the high number of surface atoms ; then the coordination number determination is accurate with the classical formulation of EXAFS. As shown also in Table II, all these coordination numbers are much lower than the value of 12 of the fcc structure, which is in line with an average size of 15 Å in most of the samples.

With this information , hypothesis on the actual shape of the clusters can be done : one way is to compare, as Greegor and Lytle did [6], coordinations numbers from EXAFS and chemisorption results, but sometimes, a direct observation of the relative behavior of the more distant shells is a strong indication of the shape of the crystallite. As shown by Lagarde et al. [11], this shape is strongly dependent on the nature of the support, and the EXAFS spectra are definitely different (Figure 1 and Figure 2).

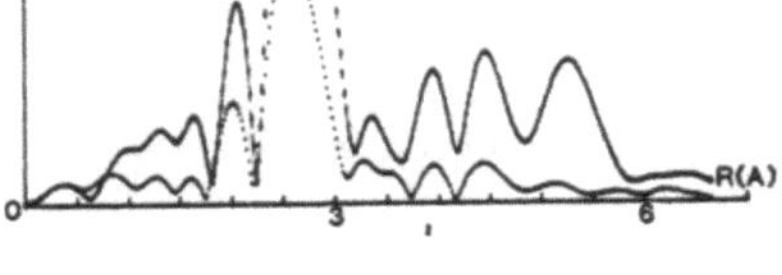

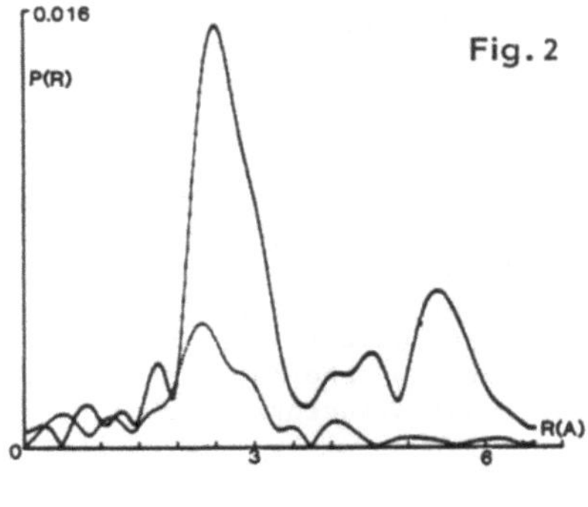

Fig. 1. Magnitudes of the Fourier transforms of Pt metal (solid line) and 1% Pt/ γAl_2O_3 reduced in H_2 (dotted line).
The ratios between the first, second, and third shells in both cases are in good agreement with a model of cluster of 13 atoms of Pt in the fcc structure

Fig. 2. Magnitudes of the Fourier transforms of Pt metal (solid line) and 1% Pt/ ηAl_2O_3 reduced in H_2 (dotted line).
In the catalyst, a Pt atom has about 4 first neighbors, 4 second neighbours and no third one: the model could be a raft parallel to the <100> plane

Another structural information which can be extracted from EXAFS analysis is the degree of disorder, around the average structure, that the metal atoms suffer; the surface atoms in the clusters are expected to be more asymmetrically bonded to their neighbors than a bulk atom, and EXAFS is very sensitive to the overall shape of the radial distribution function. Very little work has been devoted up to now specifically to this problem which has been nevertheless studied recently, on Pt/SiO_2, by Marques et al. [10]. These authors use a Morse potential to model the structural disorder experimentally obtained on the catalyst compared to the Pt foil. While the need of such a treatment of the asymmetry for a Pt foil is in line with classical studies of the behavior of metals, it seems that the catalysts exhibit a slightly larger temperature dependance associated with a static disorder which could be the consequence of constraints due to the support.

While the reduced system is the ultimate stage of the usual catalysts, and has therefore been extensively studied, the different stages of preparation, starting from the precursor to the calcined compound, have also been looked at by EXAFS as well as the effect of adsorption of gases, like oxygen.

Changes in platinum structure in coming from the impregnated support, up to the reduced sample under H_2 have been followed by Lagarde et al. [11] and the presence of PtO_2 in the calcined sample is evident by comparing with the reference compound (Figure 3). Adsorption of oxygen starts by chemisorption on Ru/SiO_2, and the formation of oxidized particles appears at 400°C, giving an EXAFS spectra very close to that of RuO_2, as shown by Lytle et al. [12] ; these results are in agreement with the work of Joyner [13] on Pt/SiO_2 exposed to air where he finds only the Pt-O distance at 1.94 Å, without any evidence for Pt-Pt bonding.

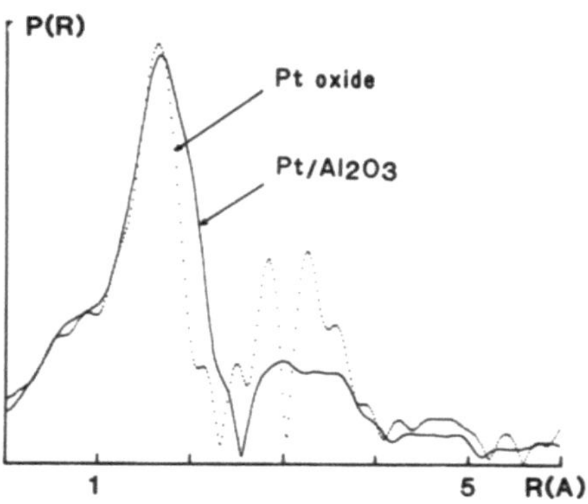

Fig. 3. F.T. of Pt oxide and 1% Pt/Al_2O_3 calcined at 530°C. The first peak in the catalyst is due to oxygen and chlorine bonding while the peak around 3 Å is an indication for the existence of Pt oxide

Among various techniques Friedman et al. [14] made use of EXAFS to look at various Cu/Al_2O_3 calcined catalysts, with different metal loading and treatment temperatures ; they have been able to distinguish between several environments of the cupric ion, which enters tetragonal or octohedral spinels sites, or has a local order resembling aluminate or copper oxides phases.

Table I

Catalyst	Reduction temperature (K)	Nb of Unfilled d states per Pt atom	Reference
0.7 % Pt/SiO_2	723	0.36	2
1 % Pt/Al_2O_3	473	0.38	2
1 % Pt/Al_2O_3	723	0.35	2
Pt foil	-	0.30	2
3.1 % Pt/TiO_2	698	0.33	3
1.7 % Pt/TiO_2	698	0.33	3

Table II

Catalyst	R (Å)	R (bulk)	Coord. Nb.	Reference
Os/SiO_2	2.702	2.705	8.3	5
Ir/SiO_2	2.712	2.714	9.9	5
Ir/Al_2O_3	2.704	2.714	9.9	5
Pt/SiO_2	2.774	2.775	8.0	5
Pt/Al_2O_3	2.758	2.775	7.2	5
$Pt/\alpha.Al_2O_3$	2.75	2.775	6	11
$Pt/\eta.Al_2O_3$	2.72	2.775	4	11
Pt/zeolite	2.76	2.775	-	7

B) Bimetallic catalysts

Industrial systems are often made by bimetallic, instead of monometallic catalysts, because of their striking properties. The selectivity of EXAFS versus the element sampled makes it very suitable for these studies. As in monometallic systems, and using a slightly more sophisticated method of analysis because the presence of two scatterers, EXAFS gives distances and coordination numbers.

Moreover, the most interesting feature of these analyses by EXAFS is that the two metals, which are immiscible in the bulk, give bimetallic clusters, but segregate, one of them forming a shell around a spherical core made by the other metal. This has been shown by Bassi et al. [15] on Ru-Au/MgO, with the help of other physico-chemical methods, and also by EXAFS only by Sinfelt et al. [16,17] on Ru-Cu/SiO_2, Pt-Ir/Al_2O_3 and Pt-Ir/SiO_2. The main evidence for this comes from the coordination numbers, which are almost monometallic for one of the metals, while the other one is almost equally distributed among the two elements.

C) Other studies in catalysis

Various studies on catalysts under different conditions have been published. For instance, Joyner et al. [18] have studied cobalt porphyrins catalysts supported on carbon, either the bare system, where carbon and nitrogen shells are present at respectively 1.95 and 3.08 Å, or the heated sample where just the single shell of nitrogen remains. Clausen et al. [19] have studied sulfided Co-Mo/Al_2O_3 catalysts, where they found the presence of a MoS_2-like structure.

Another important kind are the Ziegler-Natta catalysts, based on Ti halides ; as a part of characterization of these systems, Vlaic et al. [20,21] have undertaken the study of various Ti compounds as α-$TiCl_3$, $TiCl_2$; their results compare well with the crystallographic values, when they are reliable (α-$TiCl_3$) while the discrepancy of 0.03 Å in $TiCl_2$ is may be due to an inaccuracy of the cristallographic work. Finally, Reed et al. [22] have studied Wilkinson's catalysts on both the Br and the Rh edges of ($Rh(PPh_3)_3Br$) : depending on the amount of cross-linking, the EXAFS results suggest dimeric or monomeric species for the polymer.

III. Conclusions and some future possibilities

Because of the large industrial interest in catalysts, and the power of EXAFS to understand the structure of these dilute and very small clusters, as well as the amount of information extracted from near-edge studies, X-ray spectroscopy on catalysts is going to show a large development in the future. Most of the work has been done up to now on "as prepared" systems, or catalysts at different stages of their preparation, and a large amount of work has to be done on catalysts in duty, to understand the role of the metal when the chemical reaction takes place. This would need a faster data collection, of the order of a minute, which could be obtained with the new developed "EXAFS in dispersive mode" [22]. Beside their technical interest, supported metal catalysts present striking features for fundamental physicists, about the stability, the symmetry, the electronic structure of very small clusters of metal atoms for instance. EXAFS does not show a strong bonding between these clusters and the support, nevertheless their actual shape is strongly dependant on the support, even on the crystallographic face of this support, as shown in different aluminas. All these points are interesting to investigate, independently of the economical consequences of the results.

References

1 - F.W. Lytle, P.S.P. Wei, R.B. Greegor, G.H. Via and J.H. Sinfelt, J. Chem. Phys. 70, 4849 (1979).

2 - A.N. Mansour, J.W. Cook Jr., D.E. Sayers, R.J. Emrich and J.R. Katzer, to be published in J. of Catalysis.

3 - D.R. Short, A.N. Mansour, J.W. Cook Jr., D.E. Sayers and J.R. Katzer, to be published in J. of Catalysis.

4 - B.K. Teo and P.A. Lee, J. Am. Chem. Soc. 101, 2815 (1979).

5 - G.H. Via, J.H. Sinfelt and F.W. Lytle, J. Chem. Phys. 71, 690 (1979).

6 - R.B. Greegor and F.W. Lytle, J. of Catalysis 63, 476 (1980).

7 - B. Moraweck and A.J. Renouprez, Surf. Science 106, 35 (1981).

8 - J.J. Burton, Catal. Rev. 9, 209 (1974).

9 - M.B. Gordon, F. Cyrot-Lackmann and M.C. Desjonqueres, Surf. Science 80, 159 (1979).

10 - E.C. Marques, D.R. Sandstrom, F.W. Lytle and R.B. Greegor, J. Chem. Phys. 77, 1027 (1982).

11 - P. Lagarde, G. Vlaic, T. Murata, E. Freund, H. Dexpert and J.P. Bournonville, Proceedings of this Conference.

12 - F.W. Lytle, G.H. Via and J.H. Sinfelt, J. Chem. Phys. 67, 3831 (1977).

13 - R.W. Joyner, J. Chem. Soc. Faraday I76, 357 (1980).

14 - R.M. Friedman, J.J. Freeman and F.W. Lytle, J. of Catalysis 55, 10 (1978).

15 - I.W. Bassi, F. Garbassi, G. Vlaic, A. Marzi, G.R. Tauszik, G. Cocco, S. Galvagno and G. Parravano, J. of Catalysis 64, 405 (1980).

16 - J.H. Sinfelt, G.H. Via and F.W. Lytle, J. Chem. Phys. 72, 4832 (1980).

17 - J.H. Sinfelt, G.H. Via and F.W. Lytle, J. Chem. Phys. 76, 2779 (1982).

18 - R.W. Joyner, J.A. Robert Van Veen and W.M.H. Sachtler, J. Chem. Soc. Faraday Trans. I78, 1021 (1982).

19 - B.S. Clausen, H. Topsoe, R. Candia, J. Villadsen, B. Lengeler, J. Als-Nielsen and F. Christensen, J. Phys. Chem. 85, 3868 (1981).

20 - G. Vlaic, J.C.J. Bart, W. Cavigiolo , S. Mobilio and G. Navarra, Z. Natur. 36a, 1192 (1981).

21 - G. Vlaic, J.C.J. Bart, W. Cavigiolo , S. Mobilio and G. Navarra, Chem. Phys. 64, 115 (1982).

22 - J. Reed, P. Eisenberger, B.K. Teo and B.M. Kincaid, J. Am. Chem. Soc. 100, 2375 (1978).

23 - A.M. Flank, A. Fontaine, A. Jucha, M. Lemonnier and C. Williams, J. Phys. Lettres 43, L315 (1982).

EXAFS Studies of Metal-Ion Exchanged Layer Structures

Lucilla Alagna, Tommaso Prosperi, and A.G. Tomlinson

I.T.S.E., Area della Ricerca di Roma del C.N.R., C.P. 10
Monterotondo Staz., I-00016 Roma, Italy

The zirconium phosphates are the most intensively studied members of a vast class of layer-structured inorganic ion-exchangers with potential end-uses ranging from catalysis to ion transport [I]. Our interest lies in monitoring changes in site availability to transition metal ions after exchange, dehydration, and other treatments, with reference to catalytic applications. An extensive series of materials with first-row transition metal ions is under investigation. For high-loaded (20-100%) materials containing Co(2+) or Ni(2+), there is electronic spectral evidence for coordination number changes as the materials are dehydrated (c.n. 6→4) and the layers subsequently destroyed (c.n. 4→5) [2].

All the materials investigated to date were prepared and chemically and physically characterised as described previously [2]. EXAFS spectra were registered on the PULS Project spectrometer, Adone Synchrotron, Frascati, in the absorption mode at room temperature. $CoRh_2O_4$ (a spinel with only T_d CoO_4 moieties) and NiO were chosen as model compounds for extracting phase parameters. All materials chosen for study were those providing the clearest evidence for the presence of a single environment about the metal ion [2].

Identical data-handling procedures were used in all cases. Extracted, filtered exafs spectra were Fourier transformed, phase shifts applied, and distances corrected for k dependence via R vs k (RABE) plots [3]. Typical plots (of some of the worst cases) are shown in Fig.1. Average M-O distances obtained are listed in the Table. Application of the standard equation :

$$\ln A_u/A_m = 2\Delta\sigma^2k^2 + \ln\left[N_m/N_u(R_m/R_u)^2\right] \qquad (1)$$

(symbols have their usual meanings) gave coordination numbers to no better than $\pm 20\%$. Consequently, coordination numbers are as inferred in ref.2.

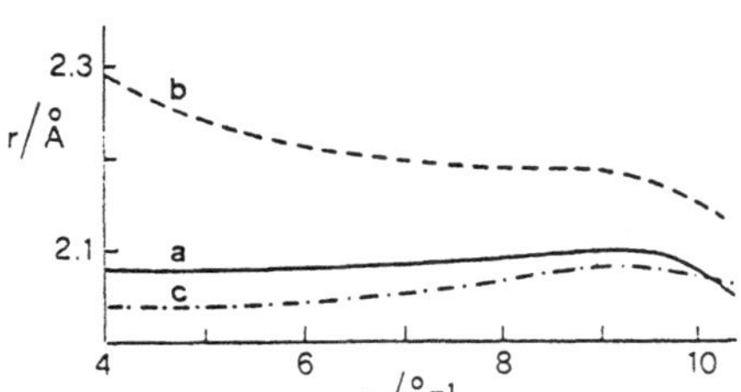

Fig.1. Typical R vs k plots

a. α-$ZrNi(PO_4)_2.4H_2O$

b. α-$ZrNi(PO_4)_2$

c. α-$Ni_{0.33}Na_{0.44}Zr(PO_4)_2$

Table Average M-O bond lengths

Material[a]	Inferred site geometry	M-O (Å)[b]
α-$ZrCo(PO_4)_2.4H_2O$	ca O_h	2.12(3)
α-$ZrCo(PO_4)_2$	ca T_d	2.02(2)
$CoZr(PO_4)_2$	ca D_{3h}	2.02(2)
$Co_{0.33}Na_{0.44}Zr(PO_4)_2$	ca C_{4v}	2.05(2)
α-$ZrNi(PO_4)_2.4H_2O$	ca O_h	2.08(1)
$\alpha ZrNi(PO_4)_2.H_2O$	ca O_h	2.06(3)
α-$ZrNi(PO_4)_2$	ca T_d	2.20(4)
$NiZr(PO_4)_2$	ca T_d	2.06(2)
$Ni_{0.33}Na_{0.44}Zr(PO_4)_2$	ca D_{3h}	2.08(2)
γ-$ZrNi_{0.20}H_{1.80}Zr(PO_4)_2$	ca C_{4v}	2.07(1)
γ-$ZrNi_{0.45}H_{1.55}(PO_4)_2$	ca C_{4v}	2.07(1)

[a] $MZr(PO_4)_2$ = material after destruction of layer structure (900°C); [b] errors from l.m.s. fit of R vs k curve.

Correlations with literature data on known M-O bond lengths are, in general, good [3]. However, two important points emerge. First, bond length ranges for "ideal" geometries overlap, so the simple rule of thumb "shorter bond length = tendency to tetrahedral geometry" is inapplicable. Second, there is a synergism between metal-ion requirements and site rigidity (see especially Ni-O in α-$ZrNi(PO_4)_2$, for which a distorted tetrahedral geometry appears to be present [3]).

Clearer results as to the nature of the site are obtained from the EXAFS spectra of the Cu(2+) analogues (for which electronic spectra provide less information on site geometry [2]). To ensure that long, axial Cu...O lengths were not lost, the model chosen was Cu(pyridine N-oxide)$_4$$(ClO_4)_2$ [4]) Well-separated Cu-O, second-shell Cu...N, and long (3.38 Å) third-shell Cu...O peaks were clearly present in the filtered Fourier transforms, as required by the known crystal structure.

Fourier transforms of the filtered EXAFS spectra show clear evidence for coordination of further PO_4(3-) groups on dehydration (Fig.2). Coordination number determination using (1) gave plots showing that the first shell in both tetrahydrate and anhydrous layer form corresponds to four oxygen atoms (Fig. 3). Inverse transforms of the first shell with extended k ranges brought to light Cu...O distances at $R + \varphi$ = 2.36 Å (tetrahydrate) and ca 2.5 Å (anhydrous layer form). These values are in line with extensive literature plots of Cu-O vs Cu...O for tetragonal octahedral complexes [5].

As for the Co(2+) and Ni(+) analogues, the large inter-layer distance present means that there is little difficulty in accommodating a $Cu(OH_2)_4$+2O (from PO_4(3-) between the layers. More interesting is the situation after total dehydration, when the inter-layer distance decreases [2]. Were the layers to merely collapse after dehydration, the only site available (given the short inter-layer distance and reasonable Cu-O distances) would be a trigonal site

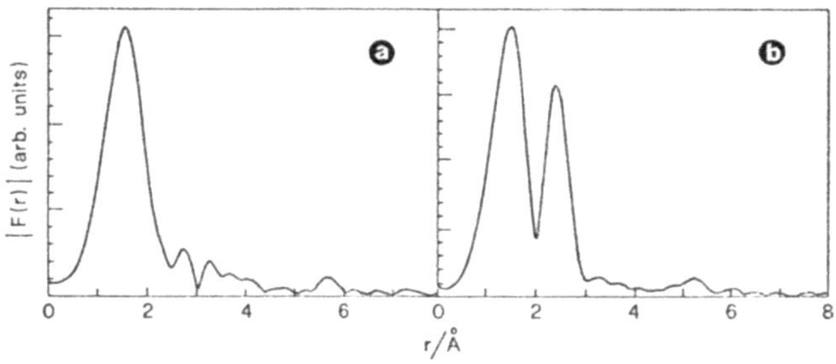

Fig.2. Fourier transforms
a. α-$ZrCu(PO_4)_2 \cdot 4H_2O$
b. α-$ZrCu(PO_4)_2$

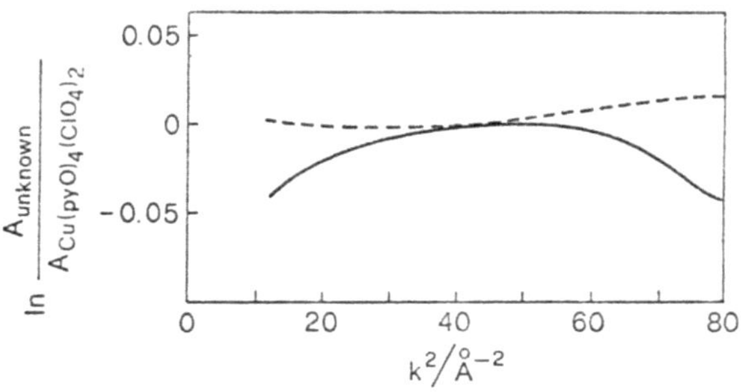

Fig.3. Coordination No. determination
(——) α-$ZrCu(PO_4)_2 \cdot 4H_2O$
(- - -) α-$ZrCu(PO_4)_2$

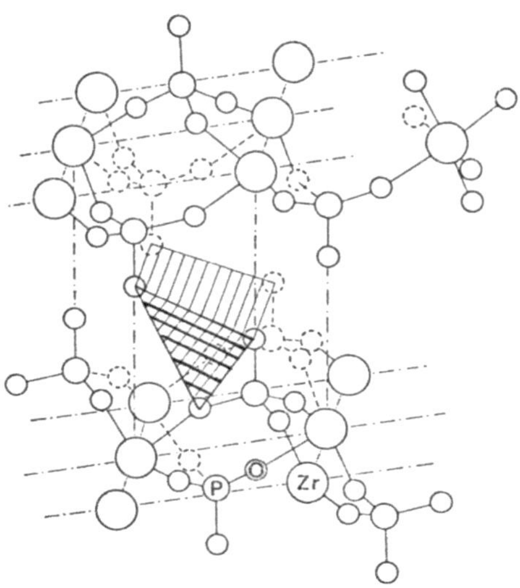

Fig.4. Structure of parent unexchanged α-$Zr(HPO_4)_2 \cdot H_2O$ [7] (schematic).

(cross-hatched) Trigonal site assuming a rigid cavity

(////) Site suggested by EXAFS spectroscopy for anhydrous 100% exchanged Cu(2+) material; layers slide to accommodate ca D_{4h} CuO_4+2O unit

formed by two ionic oxygen atoms above and below the Zr-containing planes, together with a lattice oxygen (Fig.4). The EXAFS results show this is not the case, the Cu(2+) is in a tetragonal octahedral site. The layers must then also slide (and, perhaps, twist) after dehydration in order to accommodate the metal ion. This site plasticity contrasts with the rigidity of zeolites, where trigonal sites have also been suggested [6].

R- φ distances for second shell (i.e. P, from back-scattering curves of model compounds [8]) vary widely, between 2.5 and 2.9 Å, showing that large changes in M-O-P angles occur between materials. This provides further evidence for a relatively plastic site.

In conclusion, even first-shell analysis is capable, in favourable cases, of identifying sites in layer-structured materials (with inputs from other spectroscopic techniques). However, both extraction of φ(M...P) from model compounds and quantitation of possible second-shell focussing effects should allow the $M(PO_4)_n$ units to be completely defined in future[8].

Acknowledgments

We thank Drs S.Mobilio and L.Incoccia (PULS Project) for many helpful suggestions, and Prof.A.La Ginestra and Dr C.Ferragina for continuing collaboration.

References

1. A.Clearfield,"Industrial Uses of Zirconium Phosphate",Magnesium Elektron Ltd.(U.K.),1978
2. L.Alagna,C.Ferragina,A.La Ginestra,and A.A.G.Tomlinson, J.Chem.Soc.Dalton Trans.,1981, 2376
3. L.Alagna,T.Prosperi,A.A.G.Tomlinson,C.Ferragina,and A.La Ginestra, J.Chem. Soc.Faraday Trans.I,1982, in press
4. L.Alagna and A.A.G.Tomlinson, J.Chem.Soc.Faraday Trans.I,1982, in press
5. J.Gazo,I.B.Bersuker et al., Coordination Chem.Rev.,1976,19, 253
6. K.Klier,J.Hutta,and R.Kellerman, ACS Symposium Ser.,1977,40, 108
7. A.Clearfield and G.D.Smith, Inorg.Chem.,1969,8, 431.
8. L.Alagna,T.Prosperi,and A.A.G.Tomlinson, unpublished work

Fluorescence EXAFS of $MgCl_2$-Supported Titanium Trichloride

Gilberto Vlaic, Jan C.J. Bart, Walter Cavigiolo

Istituto G. Donegani, I-28100 Novara, Italy

Alain Michalowicz

LURE Bât. 209C, Université Paris-Sud, F-91405 Orsay, France

1. Introduction

Ziegler-Natta catalyst components, prepared by co-milling of $MgCl_2$ with $TiCl_3$ (ARA, HRA) in dry inert atmosphere, together with the model compounds δ-$TiCl_3$ (a form of $TiCl_3$ with disorder in the stacking of the Cl-Ti-Cl sheets) and $TiCl_4$, were studied by means of fluorescent EXAFS /1/ in order to improve the signal-to-noise ratio with respect to conventional transmission measurements. In fact, in spite of the relatively high concentration of titanium in the samples, the presence of over 75% strongly absorbing element (Cl) in the region of interest and the dishomogeneity of the catalyst samples make collection of reliable data in transmission mode impossible.

2. Data Collection and Analysis

The samples δ-$TiCl_3$ (I), $TiCl_4$ impregnated on dry BN (II), 4.8 wt% $TiCl_3$ ARA. $MgCl_2$ (III) and 8.0 wt% $TiCl_3$ HRA.$MgCl_2$ (IV) were loaded in 10 μm thick polythene bags (5x1 cm), kept in dry N_2 until the measurements and were mounted in dry atmosphere in a plexiglass sample holder provided with an aluminated kapton window. The measurements of the K edge of titanium were performed on the EXAFS-1 x-ray beam line DCI at LURE, Orsay (1.56 GeV, 110 mA) using a Ge(111) channel-cut crystal monochromator and proportional multiwire detector. In these operating conditions the intensities of the higher harmonics were negligible. Metallic titanium was used as a reference to the energy scale (E_o(K)Ti = 4964.5 eV at the maximum of the first derivative). EXAFS spectra were accumulated about 20 times for each sample over 750 eV with ΔE = 2 eV, and signal-averaged. The EXAFS signal was extracted as usual by removing the

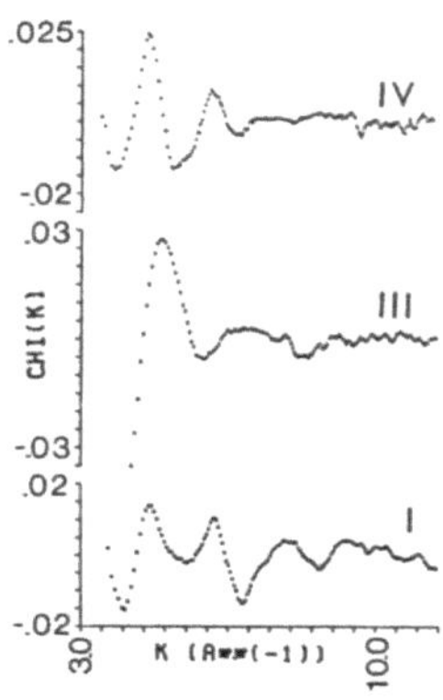

Fig.1. EXAFS of I, III, IV

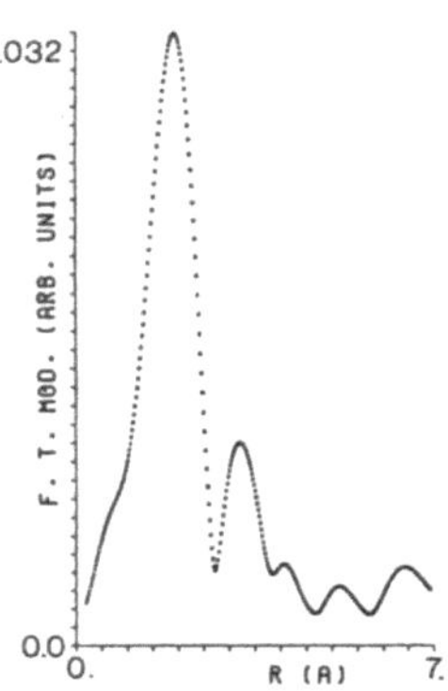

Fig.2. F.T. of $TiCl_3$

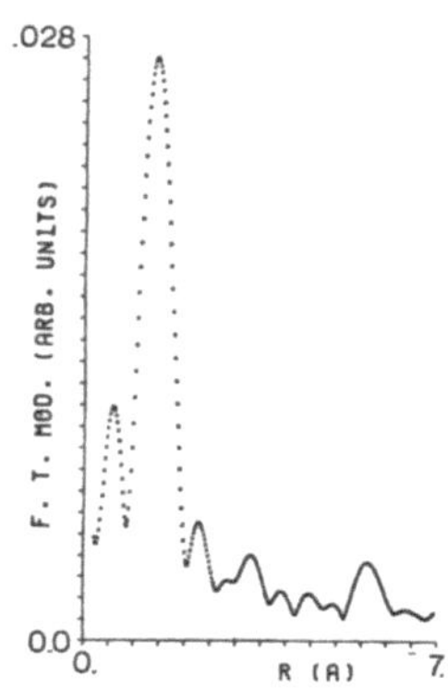

Fig.3. F.T. of $TiCl_4$

pre-edge background and by fitting the monotonous atomic background at the high-energy side with a fourth-order polynomial, followed by normalization. Spurious remainders of the background were eliminated by neglecting the first three or four out of 1024 coefficients in the Fourier analysis of the spectra before back-transformation. Figure 1 reports the EXAFS spectra for I, III and IV. The X(k).k data were Fourier transformed between $4.2 < k < 11$ Å^{-1} ($k = \sqrt{2m(E-E_o)}/\hbar$ with E_o being the threshold energy) following standard procedures and a window as in ref. /2/; the moduli of FT for samples I-IV are reported in Figs.2-5. The back-Fourier transform was applied to the first main peak of each sample except for IV, where strong overlap between the first and second shell called for antitransformation of both simultaneously. Using the concept of chemical transferability of the phase shifts /3/ and the parametrized formulation due to LEE et al. /4/ we have extracted the phase parameters from the sinusoidal term of the EXAFS signal obtained by backtransformation of the first peak of $TiCl_3$ (octahedron; Ti-Cl = 2.46 Å) /5/ followed by dividing of X(k) vs. k by the amplitude envelope. The parameters were then transferred to samples II and III and the Ti-Cl bond distance and energy shift ΔE_o were treated as variables. The fits for I-III are shown in Fig.6. $TiCl_4$ was used to check the reliability of the parameters and the experimentally derived Ti-Cl distance of 2.14(3) Å is only slightly shorter than that of 2.160(1) Å according to neutron diffraction work /6/. The Ti-Cl bond length of III is 2.06(3) Å.

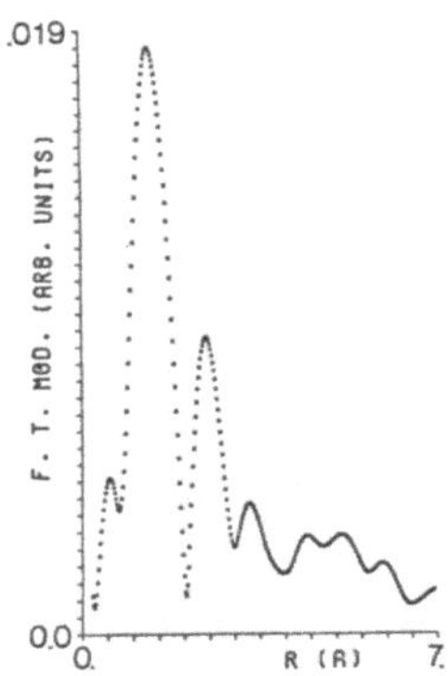

Fig.4. F.T. of III

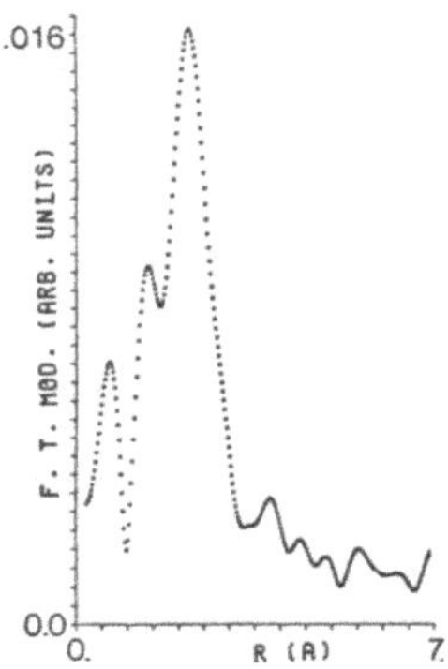

Fig.5. F.T. of IV

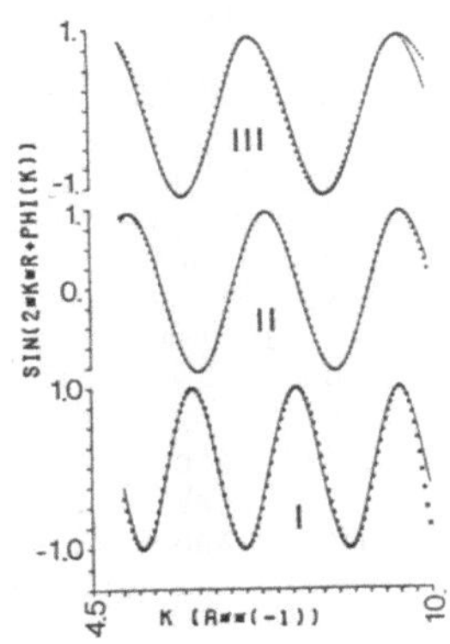

Fig.6. Fit of sine term

Coordination numbers were obtained by means of a first-order polynomial fit of the relation

$$\ln A_r(k)/A_s(k) = \ln N_r/N_s + 2 \ln R_s/R_r + 2(\sigma_s^2 - \sigma_r^2)k^2 \quad (1)$$

for the moduli (A) of the signals of the model compound $TiCl_3$ (r) and the unknown samples (s). Calculated values are N=3.8 for $TiCl_4$ (theor. 4.0) and N=3.2 for catalyst III. The fits for samples II and III with respect to I are shown in Fig. 7.

In case of sample IV a two-shell fit was carried out with the experimental values of $TiCl_3$ and the parametrization of TEO et al. /7/. Calculated Ti-Cl bond lengths of 2.05(3) and 2.50(3) Å are close to those of III and $TiCl_3$, respectively. The ratio of the coordination numbers in octahedral and trigonal sites is about 2.0. The fits of δ-$TiCl_3$ and IV are shown in Fig.8.

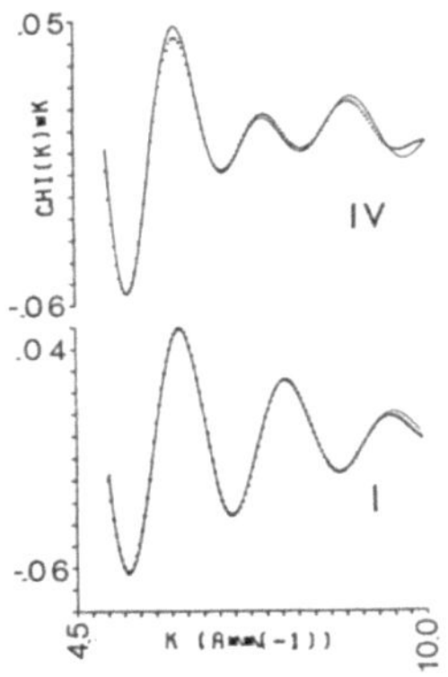

Fig.8. Fit of X(k).k vs k

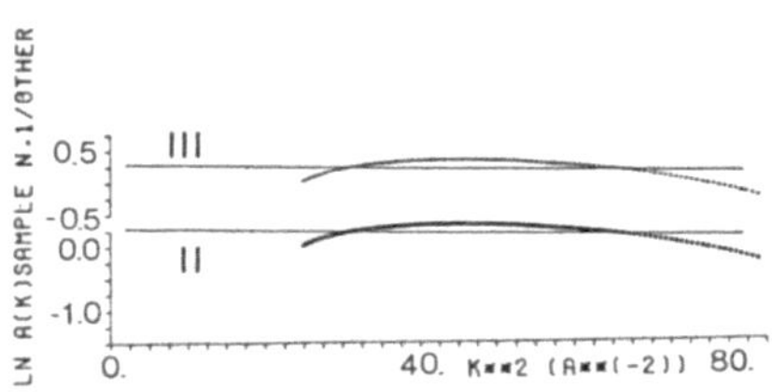

Fig.7. Fit of eq. (1) for II and III

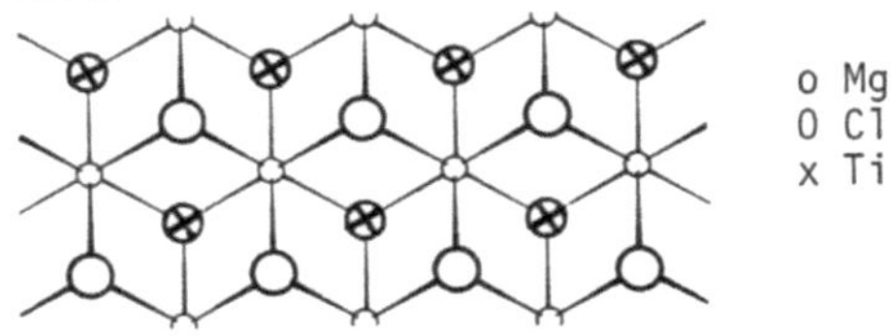

Fig.9. $MgCl_2$ sheet and Ti^{3+} positions

3. Discussion

Co-milling of solid $TiCl_3$ and $MgCl_2$ (layer structures with the latter being thermodynamically more stable) induces a surface reaction. Whereas δ-$TiCl_3$ is octahedrally coordinated, 4.8 wt% $TiCl_3.MgCl_2$ has a very short Ti-Cl bond (2.06 Å) with CN of 3.2, indicating a trigonal planar coordination. Three-fold hollow cation sites (r = 0.27 Å) are available at the (003) surface plane of $MgCl_2$ (Fig.9). The bonding is not unlike that of $Ti(N(SiMe_3)_2)_3$ /8/ with Ti-N = 1.929(4) Å and $r^{III}Ti(III)$ = 0.26 Å. For 8.0 wt% $TiCl_3.MgCl_2$ our data indicate that a maximum of about 4 wt% $TiCl_3$ has interacted with $MgCl_2$, as above, leaving a residue of δ-$TiCl_3$. This is in accordance with independent chemical evidence, indicating about 5 wt%.

Although EXAFS does not provide a clue about the location of the chlorine atoms (of $TiCl_3$), these are likely to interact with the coordinatively unsaturated Mg atoms at the fractured side-planes of the $MgCl_2$ particles (δ-$MgCl_2$ is the result of shear exerted in the direction of the (003) plane and leads to sliding of the Cl-Mg-Cl double layers). For charge neutrality reasons, the Ti ions are then supposed to be in proximity to these anion positions.

4. References

1. J. Jaklevic, J. A. Kirby, M. P. Klein, A. S. Robertson, G. S. Brown and P. Eisenberger, Solid State Comm. 23, 679 (1977).
2. G. Vlaic, J. C. J. Bart, W. Cavigiolo, S. Mobilio and G. Navarra, Chem. Phys. 64, 115 (1982).
3. B. M. Kincaid, P. Eisenberger and P. H. Citrin, Phys. Rev. Lett. 36, 1346 (1976).
4. P. A. Lee, B. K. Teo and L. Simons, J. Am. Chem. Soc. 99, 3856 (1977).
5. G. Natta, P. Corradini and G. Allegra, J. Polym. Sc. 51, 399 (1961).
6. J. B. van Tricht, J. Chem. Phys. 66, 85 (1977).
7. B. K. Teo, A. L. Simons, P. Eisenberger and B. M. Kincaid, J. Am. Chem. Soc. 99, 3854 (1977).
8. P. G. Eller, D. C. Bradley, M. B. Hursthouse and D. W. Meek, Coord. Chem. Rev. 24, 1 (1977).

An EXAFS Study of a Highly Dispersed Rh/Al_2O_3 Catalyst

D.C. Koningsberger, T. Huizinga, H.F.J. van't Blik,
J.B.A.D. van Zon, R. Prins, and D.E. Sayers[1]

Laboratory for Inorganic Chemistry and Catalysis, Eindhoven University of Technology, NL-5600 MB Eindhoven, The Netherlands

Abstract

The X-ray absorption spectrum of the Rh K-edge of a highly dispersed reduced Rh/Al_2O_3 catalyst shows EXAFS oscillations due to Rhodium-Rhodium nearest neighbors. Adsorption of CO at room temperature on the reduced catalyst significantly decreases the amplitude of the EXAFS oscillations. This implies that CO adsorption on very small metal crystallites leads to a disruption of a significant number of the Rhodium-Rhodium metallic bonds.

These results show that the Rh in this system is not monatomically dispersed but·is in the form of very small crystallites whose structure is disrupted by the adsorption of CO.

I Introduction

Several conflicting proposals have been published for the structure of highly dispersed Rhodium metal on Al_2O_3. Electron microscopy studies (1) have indicated that very small two-dimensional metal rafts are present. On the other hand infrared results obtained from CO adsorbed on the metal catalyst (2) (3) have been interpreted at low Rhodium loadings, (0.5 wt%), as being due to isolated Rhodium atoms.

Uncertainties also exist about the oxidation state of Rhodium on Al_2O_3. Electron microscopy shows that metallic Rhodium clusters are present. However, the Rhodium dicarbonyl species formed after CO adsorption exhibit infrared bands, which are similar to those of $Rh_2(CO)_4Cl_2$, suggesting that Rhodium atoms on Al_2O_3 are in the +1 oxidation state.

To investigate the structure of these highly dispersed Rhodium catalysts X-ray absorption studies were carried out before and after CO adsorption. Preliminary results of these measurements will be presented.

II Experimental

EXAFS experiments were performed at the Stanford Synchrotron Radiation Laboratory (SSRL) on station I-5 in February 1982

[1]Department of Physics, North Carolina State University Raleigh, North Carolina, 27650, USA

(Ring Energy 3 GeV; Ring Current 40-80 mA). The EXAFS was measured at 140 K in a controlled environment cell (4).

The Rh/Al_2O_3 catalyst was prepared by pore volume impregnation of the support with an aqueous solution of $RhCl_3$ leading to a metal loading of 0.5 wt%. The catalyst was pressed into a thin (∿ 0.2 mm) self-supporting water and mounted in the sample cell. The sample was reduced by heating at a constant rate (5 K/min) under flowing hydrogen to 593 K. The sample was held at that temperature for about one hour and cooled down to room temperature. Reduction studies in our laboratory verified that after this treatment reduction of the catalyst is complete.

After in situ EXAFS experiments (under 100 kPa H_2) the catalyst was evacuated at 573 K for one hour. After cooling in vacuum (10^{-2} Pa) to room temperature the sample was exposed to 100 kPa CO and the EXAFS spectrum was again measured in situ.

To further characterize the sample H_2 and CO chemisorption measurements were also performed in our laboratory. The temperature treatments of the catalyst during these experiments were the same as those described above. Hydrogen chemisorption resulted in an adsorbed H/Rh value of 2.1 and chemisorption of carbonmonoxide resulted in a CO/Rh value of 3.0 indicating a very highly dispersed material.

III Results and Discussion

The X-ray absorption spectrum of the K edge of Rhodium shows EXAFS oscillations due to a Rhodium coordination around the Rhodium atoms (see fig. 1a). This contradicts the interpretation that the system consists solely of isolated Rh atoms, which was derived from the infrared results. However, based on a preliminary analysis the Rh-Rh coordination number is very low (∿ 5), indicating a highly dispersed system.

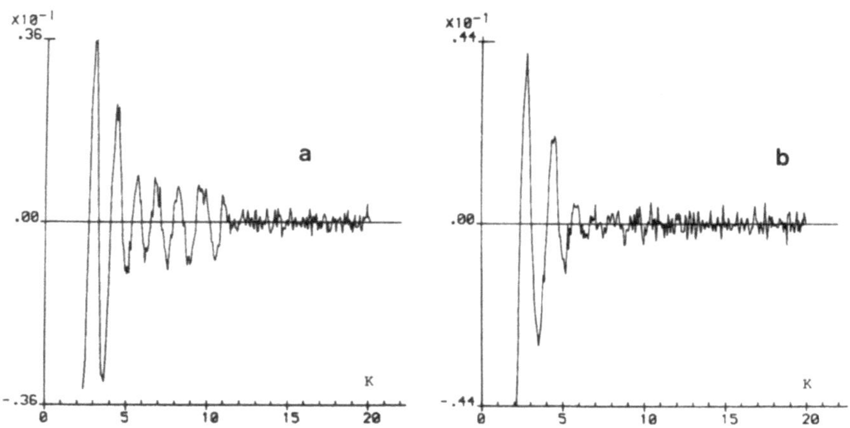

Fig. 1. $Rh(0.5)/Al_2O_3$
a) H_2, 593 K
b) H_2, 593 K + vac. 573 K + CO, 300K

After CO adsorption at room temperature the amplitude of the EXAFS oscillations attributed to a Rh-Rh metal coordination is significantly decreased (see fig. 1b). From this one can conclude that the greater part of the Rhodium-Rhodium metallic bonds are disrupted by the adsorption of CO molecules. This may also explain why the infrared bonds of chemisorbed CO molecules have been interpreted only in terms of isolated Rhodium dicarbonyl species.

In summary, preliminary results of an EXAFS study of the K edge of a highly dispersed Rh/Al_2O_3 catalyst shows that the reduced system is highly but not monatomically dispersed. Adsorption of CO at room temperature resulted in a significant disruption of the Rh crystallites. A more detailed analysis of this system is currently underway.

Acknowledgement

The authors gratefully acknowledge the assistance of the SSRL staff. This work was done at SSRL (Stanford University), which is supported by the NSF through the Division of Materials Research and the NIH (I P41 RR 01209-02). One of us (DES) acknowledges the support of the NSF Division of Material Research under grant DMR 80-11946.

Literature

1. D.J.C. Yates, L.L. Murrell and E.B. Prestridge, J. Catal., 37 (1980) 975.
2. J.T. Yates, T.M. Duncan, S.D. Worley and R.W. Vaughan, J. Chem. Phys., 70 (1979) 1219.
3. R.R. Cavanagh and J.T. Yates, J. Chem. Phys. 74 (1981) 4150.
4. D.C. Koningsberger and J.W. Cook Jr. (see Proceedings of this Conference).

EXAFS Investigation of TiO_2-V_2O_5 Catalysts

R. Kozlowski and R.F. Pettifer
Department of Physics, University of Warwick
Coventry, CV4 7AL, England

J.M. Thomas
Department of Physical Chemistry, University of Cambridge,
Lensfield Road, Cambridge, CB2 1EP, England

1. Introduction

The oxidation of o-xylene to phthalic anhydride is a process for which catalysts containing V_2O_5 supported on the anatase form of TiO_2 have long been employed. The rôle of the support (TiO_2) in this reaction is crucial in producing a material which combines both high selectivity and high activity. V_2O_5 by itself or supported on other substrates produces an indifferent catalyst for this reaction. This system is a classic example of support enhancement of the active phase.

A key concept in the interpretation of this behaviour has been a structural model emphasising a strong active phase-support interaction leading to V_2O_5 layers being epitaxially supported by anatase. The anatase-V_2O_5 epitaxial match is good in comparison with other supports. The resulting homogeneity of the surface thus suggests a reason for the high selectivity of the V_2O_5 anatase combination.

The primary goal of the present EXAFS study was to test this hypothesis and to explain the selectivity of the catalyst on structural grounds.

The reaction of $VOCl_3$ with surface hydroxyl groups leads to a homogeneous single layer of vanadium species on the support. By using ultrafine anatase the concentration of vanadium can be increased so that transmission EXAFS can be performed. Further, the particle size used ($r \sim 30$ Å) enables conclusions to be made concerning the structure of the TiO_2 surface. About 30% of the TiO_2 octahedra lie on the surface for this particular particle size.

As Ti and V are adjacent in the periodic table, the properties of both the support and the monolayer of active phase can be easily examined in a single scan.

2. Experimental

Absorption measurements were performed at Daresbury Laboratory using a channel cut (111) crystal monochromator. By making measurements at different beam energies we could ensure that our measurements are free from harmonic contamination. Further, our samples, consisting of ultrafine particles, could easily be made into homogeneously thin films.

The data was analysed using both Fourier transform and a semiempirical curve fitting approach [1].

3. Results and Discussion

The TiO_2 support surface is not grossly reconstructed but there is evidence for a higher compliance in the inter-octahedra coupling. The radius of coherence of the atomic positions to within 0.1Å includes just 3 octahedra at the surface. It can be thus argued that the surface atoms have a significantly larger disorder associated with either enhanced thermal motion or a static random strain field.

The analysis of the V_2O_5 monolayer on the support proved to be more difficult owing to a poor signal to noise ratio, weak oxygen scattering and high disorder inherent to the sample. The latter could be easily recognised from the rapid decay of the EXAFS envelope with energy.

We attempted, however, to fit the data using Cramer's [1] functions for amplitude and phase. Two independent fitting standards were used, one being VO_4^{-3} anion in alkaline solution the other NH_4VO_3. Both yielded a split first shell for the investigated structure. The radii of shells established using both standards agree closely and well within our conservative error estimate of ± 0.05Å. The EXAFS spectrum and the corresponding fit are presented in Fig.1. Two short and two longer V-O bonds of 1.65 and 1.90Å respectively, were found.

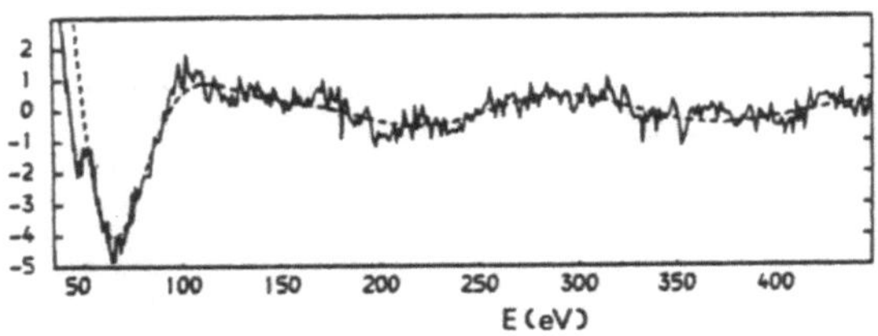

Fig.1. Curve fitting analysis for V-O layer on TiO_2: fit (broken line) of the experimental signal (solid line)

The postulated two-shell structure is further supported by comparing the XANES spectrum with those of vanadium compounds belonging to different co-ordination types illustrated in Fig.2.

The key factor in the classification is the number of the strongly bound terminal oxygens present which may vary from 1 to 4. The postulated model corresponds to type b, and its XANES spectrum shows remarkable similarity to that of $Mn_{1-x}V_{2-2x}Mo_{2x}O_6$;x = 0.36, belonging to the same category [3]- refer to Fig.3. XANES spectra of compounds representing type a (V_2O_5), and d (VO_4^{3-} in solution) differ to a great extent from that of the investigated sample.

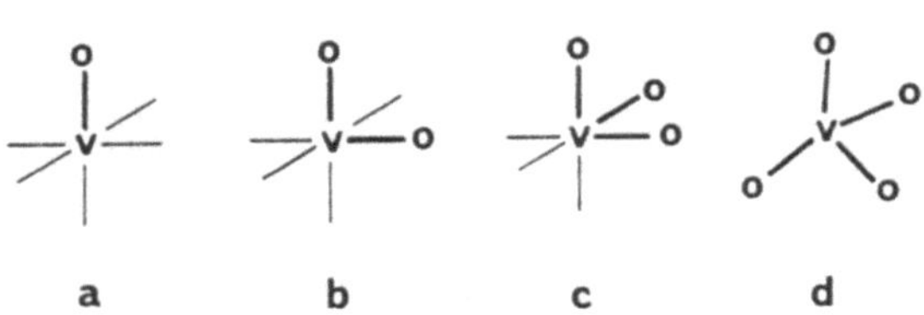

Fig.2 Basic types of V-oxygen co-ordination polyhedra. (Schematic drawing acc. to [2])

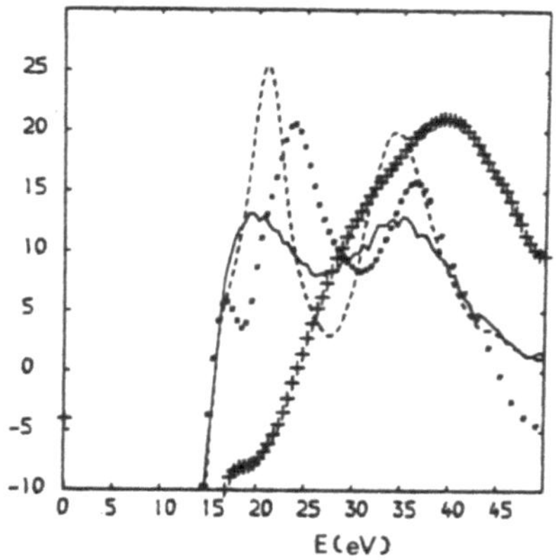

Fig. 3 XANES spectra of V-O layer on TiO_2 (—), $Mn_{1-x}V_{2-2x}Mo_{2x}O_6$; x = 0.36 (--), V_2O_5(.....), VO_4^{3-} in solution (+++).

In conclusion it may be said that the structure observed is not similar to that of V_2O_5 and with the extremely high degree of disorder it clearly eliminated an importance of crystal epitaxy as a consideration when constructing a structural model.

Our tentative model consists of molecular species with two double-bond and two single-bond oxygen ligands bound to the support surface, the surface disorder being coupled to the molecules.

References

1. S.P. Cramer, K.O. Hodgson, E.I. Stiefel, W.E. Newton, J. Am Chem. Soc. 100, 2748, 1978.

2. F.A. Schroder, Acta Cryst. B31, 2294, 1975.

3. R. Kozlowski, K. Stadnicka, J. Solid State Chem. 39, 271, 1981.

Correlation Between Catalytic Activity and EXAFS Studies of "Classical" and "Inorganic-Cluster-Derived" Platinum Catalysts

G. Maire , L. Hilaire , O. Zahraa

Laboratoire de Catalyse et Chimie des Surfaces, Université Louis Pasteur, 4, rue Blaise Pascal, F-67000 Strasbourg, France

M.F. Ravet

Laboratoire de Magnétisme et de Structure Electronique des Solides, Université Louis Pasteur, 4,rue Blaise Pascal, F-67000 Strasbourg, France

1. Introduction

In heterogeneous catalysis, Pt catalysts used for example in the reactions of skeletal rearrangements of alkanes, are usually prepared by impregnating platinum salts on inert carriers like Al_2O_3. A lot of methods have been devised to get a dispersion of Pt atoms as high as possible with a mean particle size of 20 Å or even lower. A new class of alumina-supported catalysts derived from inorganic complexes have been introduced in the past few years. Pt catalysts prepared from $[Pt_3(CO)_6]_n^{2-}$ complexes or from CHINI complexes of the type $[Pt_3(\mu_2CO)_3(CO)_3]^{2-}$ were shown to behave like classical Pt/Al_2O_3 catalysts of very low particle size [1, 2].

However we recently observed specific reactions on catalysts prepared from CHATT complexes. For example a highly dispersed catalyst prepared by deposition of $Pt(P\ Ph_3)_3$ on alumina led to a selective demethylation of methylcyclopentane (up to 60% cyclopentane), which was not observed on other catalysts.

The aim of this study was to understand the structure of the cluster at each step of the preparation of the catalyst and to deduce a mechanism for the observed reactions.

2. Experimental

The EXAFS experiments were performed using the synchrotron radiation provided by DCI at LURE, Orsay, which operates at 200 m A and 1.72 GeV. All spectra were recorded at room temperature and good statistics were obtained after 20 mn accumulation. The EXAFS spectra were recorded near the L_{III} edge of platinum.

3. EXAFS Results

The catalysts were studied by EXAFS first just after deposition on alumina - at this stage they were totally inactive - then after activation under a hydrogen flow at 300°C.

We give in Fig. 1 the EXAFS fits $\chi(K)=f(E)$ for the cluster deposited on alumina before and after activation, respectively. In Table I we have gathered the Pt-P, Pt-O and Pt-Pt distances resulting of the EXAFS analysis for the cluster deposited on alumina, before and after activation by hydrogen. The Pt-P distance given in the literature from X-ray diffraction analysis [3] for the starting material is also indicated.

From these results, two main points may be emphasized :

- The impregnation on alumina and the activation by hydrogen did not dramatically change the structure of the starting material.

- No Pt-Pt distances were detected as first neighbours. Only second neighbours distances were measured after activation.

Table 1. EXAFS determination of the coordination and the distances in $Pt(PPh_3)_3$

	Pt - P		Pt - O		Pt - Pt	
	n	d Å	n	d Å	n	d Å
$Pt(PPh_3)_3$	3	2.26	-	-	-	-
$Pt(PPh_3)_3/Al_2O_3$ before activation	3	2.20	1	2.03	0	-
$Pt(PPh_3)_3/Al_2O_3$ after activation	2	2.20	2	2.05	1st:0	-
					2nd:4	3.45
						3.65

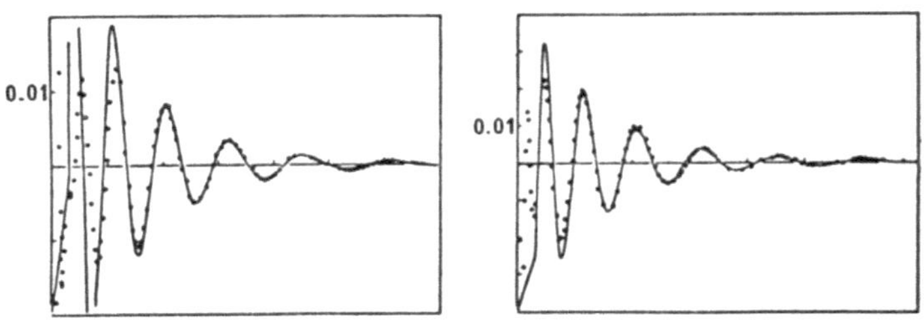

Fig.1. EXAFS fits $\chi(K)=f(E)$, E ranging from 0 to 700 eV
Left : after activation right : before activation

4. Interpretation

These EXAFS determinations enable us to draw tentative structures of the cluster supported on alumina before and after activation.(Fig.2 and 3)

An interesting result of the EXAFS study is the fact that we did not detect any Pt-Pt bond as first neighbours. This is due to the stability of this cluster when deposited on alumina and activated. The presence of Pt-P bonds prevents the Pt atoms to coalesce. We have also detected the presence of P on the surface by ESCA experiments.

Skeletal rearrangements of saturated hydrocarbons are usually explained by mechanisms involving two or several contiguous Pt atoms at the surface of the catalyst [2]. In the present case the selective demethylation, which we observed with methylcyclopentane (and with 2-methylpentane as well)cannot

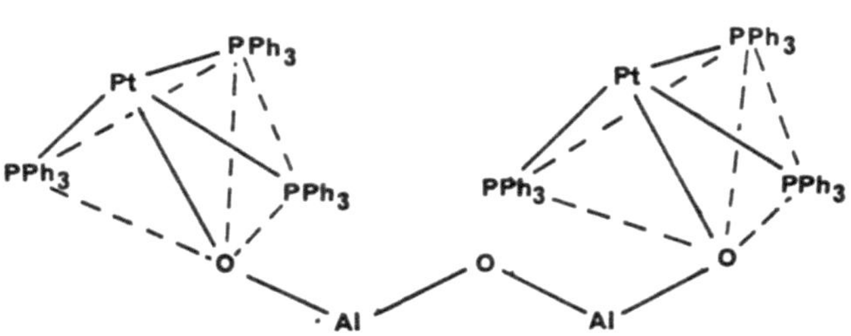

Fig.2. Proposed structure for the cluster on alumina before activation

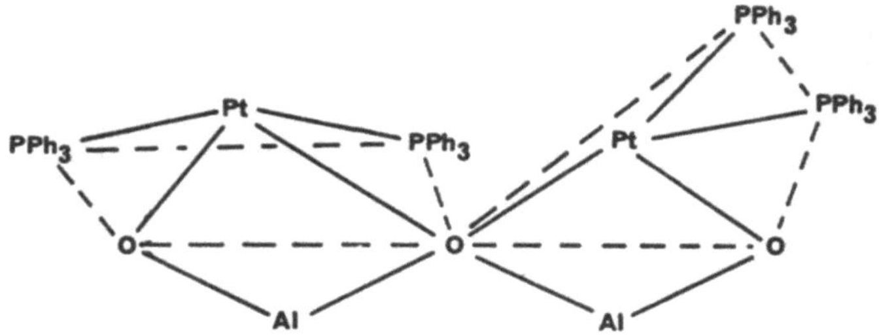

Fig.3. Proposed structure for the cluster on alumina after activation

be explained in this way since no Pt-Pt bonds were observed as first neighbours. It is therefore tempting to invoke another mechanism involving a metallacyclobutane intermediary, which has been proved to take part to several reaction in homogeneous catalysis(the dots refer to ^{13}C atoms) .

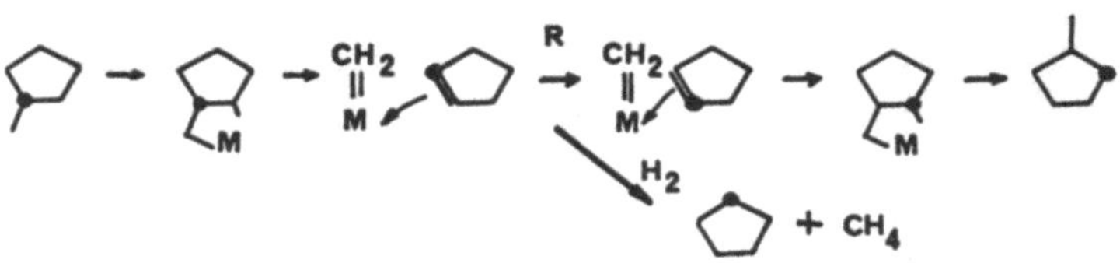

Fig.4. Mechanism for selective demethylation

1. G. Maire, O. Zahraa, F. Garin, P. Légaré, C. Crouzet, S. Aeiyach, P. Braunstein, J. Chim. Phys. 1981, 78, 11
2. O. Zahraa, F. Garin, G. Maire, Faraday discussion 1981 (paper 3) Nottingham sept. 1981
3. V. Albano, P.L. Bellon, V. Scatturin, Chem. Com. 507, 1966

This work was supported by the A.T.P. CNRS "Chimie Fine" n° 0581.

EXAFS Studies of Industrial Pt/Al_2O_3 Catalysts

P. Lagarde, T. Murata, and G. Vlaic
LURE, Bât. 209C, UPS. F-91405 Orsay, France

H. Dexpert, E. Freund, and J.P. Bournonville
IFP, B.P. 311, F-92506 Rueil-Cedex, France

Industrial catalysts for petroleum reforming have been studied by EXAFS at different stages of preparation. At the same time, model compounds have been used as checks of the validity of the electronic parameters. The results give a quite precise model of the structure of the metal in a dryed compound, then after two different calcination processes and finally, they show that this structure is dependent on the support when the sample is reduced on hydrogen.
Samples are 1% Pt on γ or η-Al_2O_3 prepared at the Institut Français du Petrole, and the reduction has been done "in situ" using a specially designed cell [1] . EXAFS data have been taken at LURE-DCI storage ring running at 1.72 GeV, 200 mA with a Si 220 channel-cut monochromator, on the L_{III} edge of Pt, and we have used the standard techniques of data processing, with the help of calculated phase-shifts [2] .

1. Model compounds

We have studied three different models compounds in order to fix some parameters: a solution of hexachloroplatinic acid used for impregnation of the support, a powder of PtO_2 and a 2.5µ metal foil. The fit of the main peak to the crystallographic first distances values allowed us to set the energy shift Eo of the edge, and the mean free path parameter Γ. While the Γ value will be kept constant during the analysis of the catalysts, assuming therefore that it is mainly dependent on the absorbing atom, that of Eo will be fixed to the value of the corresponding model compound.

2. Catalysts studies

A Impregnated and dryed catalyst.
After impregnation and drying of the system at 100°C, the Fourier transform of the data shows a single peak exactly located at the position of the peak in the hexachloroplatinic solution. The fit indicates that the elemental unit Pt/Cl_6 is conserved by drying, with just a slight decrease to 5.5 of the coordination number.

B Calcined samples.
The first step has been a calcination at 530°C for 2 h. In that case, EXAFS shows that the metal is involved in different bonds: while 2.5 chlorine atoms are still linked to the metal, at the correct distance, we find at the same time 5 oxygens at the same distance as in the platinum oxide.

But the most striking feature is the presence of Pt-Pt distances at about 3.1-3.3 Å at the same value as in PtO_2, showing that there is already formation of oxide at this rather low temperature.
By rising the temperature to 750°C, the EXAFS spectrum of the catalyst becomes definitely similar to the metal one, indicating the formation of very large clusters, each atom having a coordination number almost identical to that of the elemental platinum.

C Reduced samples.

We have studied two types of reduced catalysts, just differing by the nature of the support: γ or η-Al_2O_3, and both have been prepared in exactly the same way: reduction "in situ" under H_2 at 480°C. The results show definitely a strong dependence of the nature of the clusters obtained versus the support of the metal. In the case of γ-Al_2O_3, the analysis of the different shells of neighbours indicates that one Pt atom has an average of 6 atoms in the first shell, 2 in the second, 5 in the third and almost no fourth neighbours, at the same distance as in the metal. These results are coherent with a model of a cluster of 13 atoms close packed in the fcc structure.
But in the case of η-Al_2O_3, these numbers become respectively 4, 4, 0 and 0; while the coordination numbers determination is quite imprecise in that case, this is a strong indication for a raft structure, the plane involved being the <100> plane of the elemental metal.
Analyses are currently in process in order to detect any evidence for an interaction between the support and these metal clusters: since the support influences so strongly the shape of the particles, it should be seen as a definite bonding between the alumina and the metal. All the preceding results are listed in Table I.

TABLE I . EXAFS results on 1% Pt/Al_2O_3 catalysts

Sample	Neighbour	R(Å)	N	σ(a.u.)
dryed	Cl	2.31	5.5	0.11
calcined 530°C	O Cl Pt	2.01 2.32 3.1-3.35	5 2.5 12	0.15 0.15
calcined 700°C	Pt	2.76	12	0.13
reduced γAl_2O_3	Pt	2.75 3.9 4.75 5.5	6 2 5 0	0.14
reduced ηAl_2O_3	Pt O	2.72 2.85 2.03	4 1 0.5	0.14 0.15

1 - F.W. Lytle, P.S.P. Wei, R.B. Gregor, G.H. Via and J.H. Sinfelt, J.Chem. Phys. 70, 4849 (1979).
2 - B.K. Teo and P.A. Lee, J.Am.Chem.Soc. 101, 2815 (1979).

EXAFS Study of Short-Range Order in Transition Aluminas

J.F. Larue, B. Moraweck, and A. Renouprez
Institut de Catalyse, 2 avenue A. Einstein, F-69626 Villeurbanne, France

P. Lagarde
LURE, Bât. 209c, Université Paris Sud, F-91405 Orsay, France

INTRODUCTION

Catalytic aluminas as well as calcined aluminas for the production of aluminium metal are mainly obtained from the dehydration of alpha alumina trihydrate (gibbsite), according to the diagram : (1-3)

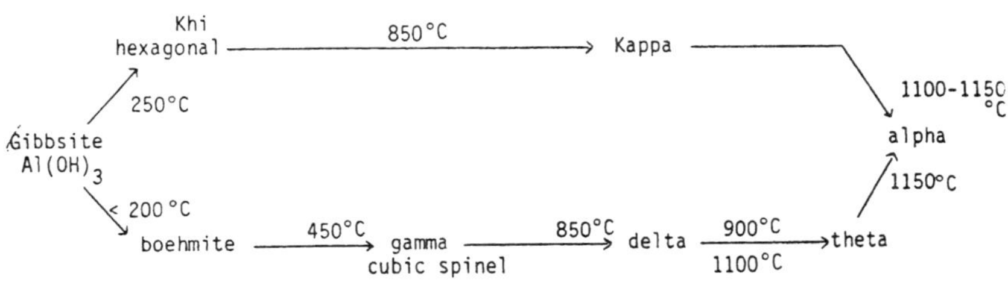

Catalytic aluminas which are issued from the Boehmite branch also contain aluminas from the Khi sequence, far less active, despite a surface area of similar magnitude. The organization of aluminium atoms in the network, i.e. their distribution in the tetrahedral and octahedral oxygen sites, might partly explain the activity of these solids (3, 4). Indeed because of the poor crystallinity of these compounds, a complete structure determination by X-ray diffraction is impossible.

Thus, we focussed our efforts on the determination of the short range order.

EXPERIMENTS

1. X-ray diffractions. Fig. 1 shows the radial distribution functions for two transitions aluminas, gamma and Khi, and for alpha alumina. Obviously the large number of pairs present in the second shell is a serious hindrance for the understanding of short range order.

2. EXAFS experiments. In order to extract the distances and coordination numbers of Al-O and Al-Al we have performed EXAFS experiments on four transitions aluminas, K 700 and K 950 from the khi branch, and BO 700 and BO 950 from boehmite,

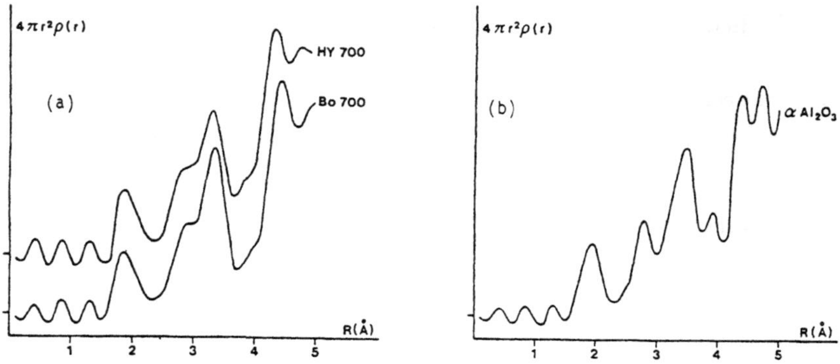

Fig. 1.(a) Radial distribution functions for gamma and khi alumina.
(b) alpha alumina

alpha alumina being taken as a reference for phase shift, back scattering amplitude and edge position. The spectra were recorded using the soft X-ray synchrotron radiation of the ACO storage ring at LURE, on the Al K edge (1559.3 eV). We used a two-crystal KAP monochromator working in high vacuum (10^{-7} torr). Significant EXAFS structure was detected up to 400 eV above the edge.

Because of the high absorption coefficient, the wafer weight should be of the order of 0.3 mg cm^{-2}. Practically, using ultrasonic dispersion of the powder in collodion we were able to reach 3 mg cm^{-2}.

Fig. 2 shows the modulus of the Fourier transform of k^3 $\chi(k)$ for gamma alumina.

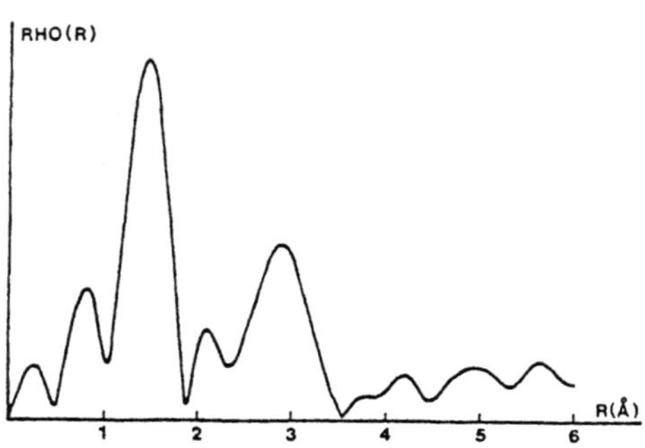

Fig. 2 . Fourier transform of K3xKHI(K) : gamma alumina (Bo700)

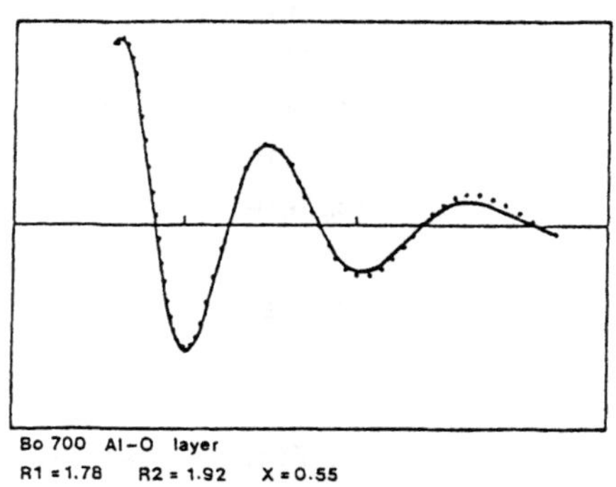

Fig. 3 . First shell fit for gamma alumina

The first peak at 1.50 Å corresponds to oxygen atoms tetrahedrally and octahedrally surrounding aluminium.

The reconstruction of the first shell was performed using the classical formula :

$$\chi(k) = -1/k \sum_i (N_i/R_i^2 \sin(2 k R_i + \Phi(k)) \exp(-2 \sigma^2 k^2)$$
$$\exp(-\gamma R_i/k)). A(k) .$$

RESULTS AND DISCUSSION

After introducing the back scattering amplitude A(k) and the phase shift Φ (k) of alpha alumina, the fitting procedure of the backtransformed EXAFS signal was performed with four parameters : the values R_1 and R_2 of the distances between oxygen and aluminium in tetrahedral and octahedral sites, the percentage of aluminium in each type of site and the edge position (Fig. 2).

Amplitudes cannot be well fitted because of heterogeneities in the wafers. Results are presented in table 1.

Table I. Al-O distances and distribution of Al atoms in tetrahedral and octahedral sites

Sample	T(°C)	$N_{Al\ tetra}/N_{Al\ total}$	R_t(Å)	R_o(Å)	ΔE (ev)
$\alpha\ Al_2O_3$	>1200			1.86 1.97	0
γ	700	0.55	1.78.	1.92	-2 ± 2
δ	950	0.55	1.78	1.92	0 ± 2
χ	700	0.58	1.78	1.92	-6 ± 2
κ	950	0.42	1.78	1.92	0 ± 2

From those results it appears that edge shifts remain small and $\Delta\sigma$ is negligible.

No evolution of the distances and of the distribution of aluminium in the two types of sites is detected in the boehmite branch up to 1000°C, whereas a progressive modification of the short range order is noticed in the khi branch.

CONCLUSION

EXAFS studies of the first shell do not allow one to explain the differences in catalytic activity of the various aluminas. However, they confirm the information yielded by the dark-field images recorded with a scanning transmission electron microscope (STEM) : the faster evolution (sintering) of the khi branch towards alpha alumina is connected with the development of microdomains with lengths up to 1000 Å for kappa alumina whereas this value remains under 200 Å in the case of delta and theta alumina.

REFERENCES

1 - R. TERTIAN and D. PAPEE
J. Chim. Phys. 55, 341 (1958).

2 - G.W. BRINDLEY and J.O. CHOE
Am. Miner. 46, 771 (1961).

3 - H. KNOZINGER and P. RATNASAMY
Catal. Rev. - Sci. Eng. 17, 31 (1978).

4 - S.G. GAGARIN, Y.A. KOLBANOVSKII and Y.V. PLEKHANOV
Kinetics and Catalysis 21 (4), 663 (1980).

Zeolite-Supported Rh Catalysts for Methanol Carbonylation: X-Ray Absorption Measurements

D.R. Denley, R.H. Raymond, and S.C. Tang
Shell Development Company, Houston, TX 77001, USA

1. Introduction

The possible use of Rh catalysts supported by zeolites for the carbonylation of methanol (and related reactions) has attracted considerable attention [1-4]. This reaction takes place in solution with a CH_3I promotor at moderate temperatures. In the case of the homogeneous reaction, the Rh complexes involved form one of the best characterized catalytic systems [5-6]. However, far fewer answers are available to the questions that arise regarding the nature of the heterogeneous system. Infra-red spectroscopy and XPS report Rh valencies of the active catalyst that include Rh^{o}, Rh^{I} and more highly oxidized species [7-11]. However, IR is not a very sensitive probe of cation valence state. At the same time, XPS operates under a disadvantage unique to the zeolite supported catalysts since the exposed surface external to the catalyst grains is generally only a few percent of the combined external surface and internal surface. In this case, the well-known sensitivity of XPS to exposed (external) surfaces may act to obscure the chemical state of the majority of the catalytic metal. With this in mind, we have performed measurements of the extended X-ray absorption fine structure (EXAFS) on a series of zeolite supported Rh catalysts preparations.

The importance of the method of catalyst preparation is well known, but was emphasized in recent work[4] where it was shown that a Rh/NaX catalyst was far more active for the carbonylation of methanol when prepared from an aqueous solution of $[Rh(NH_3)_5Cl]Cl_2$ as compared to the more common $RhCl_3$ starting point. In order to obtain an understanding of the structural reasons that underlie these differences, spectra were taken of systems prepared from solutions of $[Rh(NH_3)_5Cl]Cl_2$ and $RhCl_3$,referred to as "RhA" and "RhC" respectively . Spectra were taken for each system at three stages of preparation. The first stage,*wet* was after impregnation and air drying. Samples were also taken after being *calcined* and finally, after being *activated* by contact with the reactants, CO, CH_3OH, and CH_3I at 150°C.

In preparation, a sufficient amount of each sample was loaded into holders and then sealed in plastic bags while being maintained under an inert atmosphere. The measurements in the region of the Rh K edge (23.2 Kev) were made at the C2 station of the Cornell High Energy Synchrotron Source (CHESS).

2. Summary of Results

The EXAFS spectra do in fact show that the samples underwent important changes at each stage of the preparation and that RhA and RhC are structurally different at each stage of preparation.

Impregnation - Since no Rh-N interaction could be demonstrated for RhA and likewise no Rh-Cl interaction in RhC, it is judged that the initial ligands are not involved directly in the catalyst preparation beyond the initial stage of Rh impregnation into the zeolite lattice. In the impregnated stage, RhA-wet is most likely in the form of a highly dispersed oxide or aquo complex. The lack of secondary neighbor-shells indicates either high static disorder or perhaps cationic mobility. The RhC-wet sample shows very strong similarity to bulk Rh_2O_3-$5H_2O$.

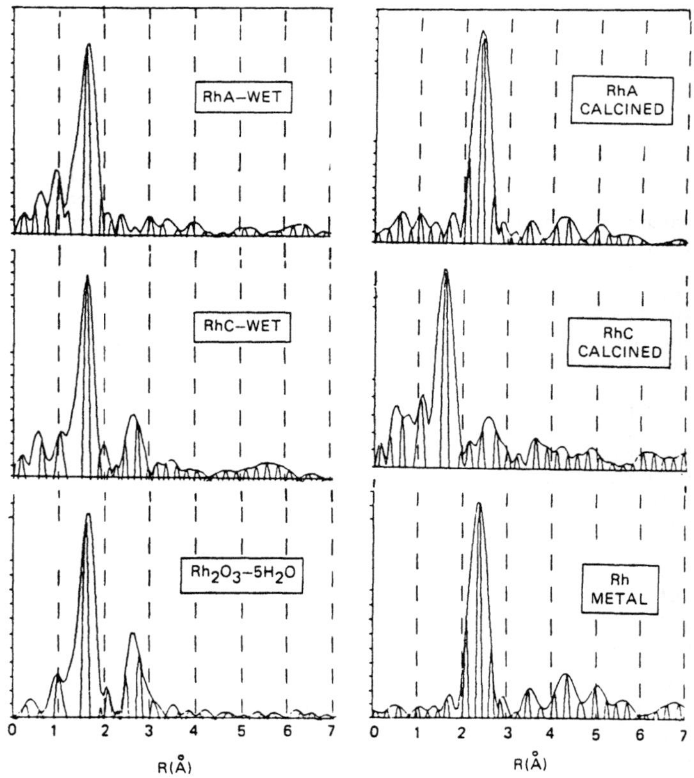

Fig.1. Rdf for RhA and RhC in the wet (left) and calcined (right) forms

Calcination - The RhC-calcined material is not greatly changed aside from some bond contraction that could be due to pressure from the zeolite lattice or perhaps thermal damage to the oxide hydrate. At the same time, RhA-calcined has undergone a dramatic decomposition to Rh metal microcrystallites.

Activation - The Rh metal microcrystallites of RhA do not change substantially upon activation. This suggests that the catalytic activity is not a near-homogeneous catalytic process fixed within the "solid solution" of the zeolite framework. Instead, the zeolite should be viewed as an unusual carrier of high surface area for metal-particle heterogeneous

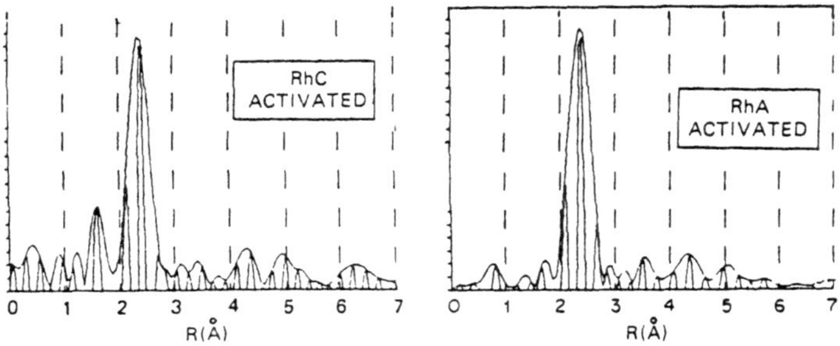

Fig.2. Rdf for RhA and RhC in the activated form

catalysis. The Rh particle sizes in both RhA-calcined and RhA-activated are similar and estimated at an average of ca. 9Å, which is quite compatible with the main cavity size of the faujasites of ca. 12Å. The pattern of a RhC involving a relatively inert Rh species continues into the preparation of RhC-activated, which seems to show only partial conversion from the oxide to active metal.

References:

1. T.Yashima, M.Ebisawa, and N.Hara, Chem.Lett.,6, 473 (1972)
2. T.Yashima, M.Ebisawa, and N.Hara, J. Catal., 36, 320 (1975).
3. B.Christensen, and M.S. Scurrell, J.Chem.Soc., Farraday Trans.I, 74, 2313 (1978).
4. M.S.Scurrell, and R.F.Howe, J. Molecular Catal., 7, 535 (1980).
5. J. Reed, P.Eisenberger, B.K.Teo, and B.M. Kincaid, J.Am.Chem. Soc.99, 5217 (1977).
6. J.F.Roth, J.H.Craddock, A.Hershmann and F.E.Paulik, Chemtech, October 1971, p.600; D.J. Forster, J.Am.Chem.Soc., 98, 846 (1976).
7. J. Yamanis and K-C.Yang, J. Catal.,69, 498 (1981.
8. Y.Okamoto, I.Nobumasa, I.Toshinobu and S.Teranishi, J.Catal,58, (1979)
9. S.M.Kuznicki, E.M.Eyring, J.Catal.,65, 227 (1980).
10. S.L.T.Anderson and M.S.Scurrell, J.Catal., 59, 356 (1979).
11. M.Primet, J.C.Vedrine and C.Naccache, J.Molecular Catal., 4, 411 (1978).

Part VII

Biological Systems

EXAFS of Calcium in Biological Systems

S.S. Hasnain

Science and Engineering Research Council, Daresbury Laboratory, Daresbury, Warrington WA4 4AD, Cheshire, England

1. Introduction

Biologically, calcium is perhaps one of the most important elements. It plays a crucial role in the development of mammals' structure; its importance for the strength of tooth and bone has long been realised. It also plays a very important role in every day essential functions like muscle contraction. In spite of its importance, the physical and chemical structure of calcium in biological and even inorganic systems has remained uncertain. This is partly due to the fact that the calcium ion, Ca^{2+}, having the electronic configuration of a rare gas atom, Ar, is not accessible via other spectroscopic techniques such as electron paramagnetic resonance (EPR); and also that, due to the 'amorphous' nature of the calcium complexes in biological systems, x-ray diffraction technique is of not much help. Over the last four years x-ray absorption spectroscopy (EXAFS and XANES) has been used to probe the structure and chemical nature of calcium in a number of biologically interesting systems. Some of these studies are reviewed here.

2. Bone and its Development

Bone mineral is often referred to as "hydroxyapatite" but its true chemical nature has yet to be established [1-2]. MILLER et al.[4] and EANES et al.[5] have recently used EXAFS to provide new information on the structure of bone mineral. EANES et al.compared the EXAFS spectrum of rat tibia with those of hydroxyapatite, carbonate-apatite, octa calcium phosphate ($Ca_8H_2(PO_4)_6\ 5H_2O$) and calcium phosphate. MILLER et al. compared the EXAFS and Fourier transforms of femurs from three different adult rats with those of a geological apatite, hy-droxyapatite, amorphous and matured amorphous calcium phosphates. MILLER et al.[4] showed that (i) fully matured amorphous calcium phosphate does not have exactly the same structure as crystalline hydroxyapatite, although this has often been assumed to be the case, (ii) the calcium environment in bone mineral is not the same as in apatite structures, i.e. it is not identical to synthetic crystalline hydroxyapatite or to geological apatites and (iii) the structure of bone mineral closely resembles that of matured amorphous calcium phosphate. These results are summarised in Figs.1 and 2.

In a further study, BINSTEAD et al.[6], have recorded the EXAFS spectrum from femurs of freshly killed mice aged three days, one week, one month, two months and seven months. Results were compared, Fig.3, with those obtained from amorphous calcium phosphate and a poorly crystalline hydroxyapatite matured from this amorphous calcium phosphate in the presence of water. BINSTEAD et al. showed that the calcium ion environment changes during the maturation of bone mineral and that the changes are very similar to those which occur when amorphous calcium phosphate matures into poorly crystalline hydroxyapatite.

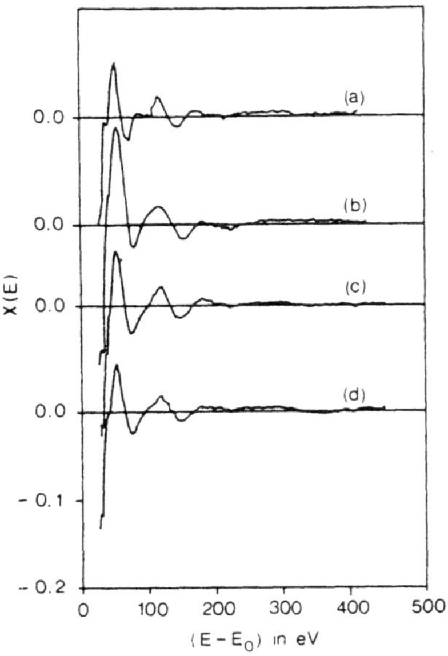

Fig.1. EXAFS spectra, $\chi(E)$, obtained from (a) synthetic crystalline hydroxy-apatite (b) amorphous calcium phosphate (ACP) (c) fully matured ACP and (d) bone mineral [4]

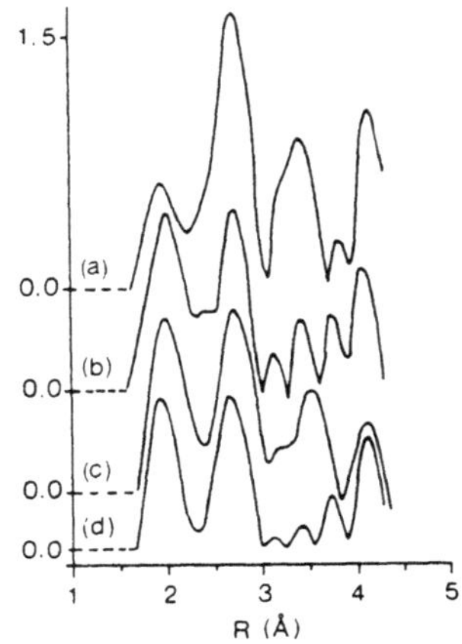

Fig.2. Fourier transform of spectra in Fig.1

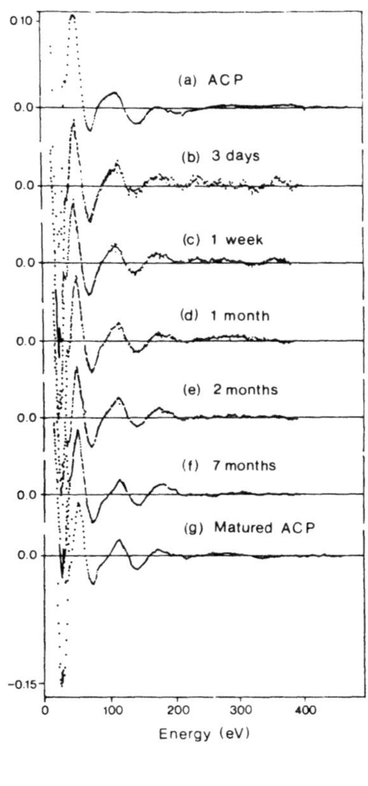

Fig.3. EXAFS spectra of ACP and bone mineral from mice of different ages [6]

Thus it appears that the post natal development changes which occur in the structure of bone mineral closely resemble those accompanying the maturation of amorphous calcium phosphate.

3. Ca Binding to ATP and Proteins

BIANCONI et al. [7] have measured the XANES of Ca^{2+} moles per mole of Troponin C(TnC) for n=2, 4 and 6. TnC is the Ca^{2+}-binding subunit of the troponin complex which is part of the regulatory system of muscle contraction. Four coordination sites are available, low affinity sites I and II are specific for Ca^{2+}, while the high affinity sites III and IV also bind Mg^{2+}. BIANCONI et al. found that the XANES spectra for n=2 (sites III and IV) and n=4 (sites I, II, III and IV) are identical and concluded that the different affinities for Ca^{2+} exhibited by TnC sites are not related to the local chemical bonding of calcium but may depend on several factors such as the total amino acid sequence, fold energy, inter-domain interactions, etc. In a separate study, POWERS et al. [8] obtained the EXAFS and XANES spectra of a number of calcium activated and calcium binding proteins. They also concluded that the high affinity and low affinity sites in the calcium activated proteins are structurally identical but that they are somewhat different from the site in calcium binding proteins.

BIANCONI et al. [9] have also investigated the Ca^{2+} binding to adenosine triphosphate (ATP). They compared the XANES spectrum of CaATP with those of calcium carbonate ($CaCO_3$), monetite ($CaHPO_4$) and calcium acetate. Figure 4 shows the spectra of CaATP and monetite, also included in the figure are the

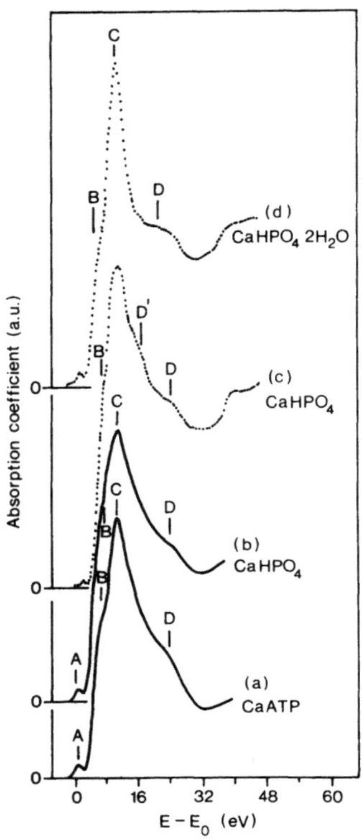

Fig.4. XANES spectra of CaATP, monetite and brushite (a) and (b) are from [9], (c) and (d) are from [10]

spectra of monetite and brushite as recently recorded by HOLT et al.[10]. The two spectra (b) and (c) of monetite are essentially the same except that (c) appears to have better resolution,e.g. peak 'C' is sharper and shoulder 'D'' is clearly visible in the spectrum (c), in fact the higher resolution monetite spectrum looks very similar to that of CaATP. From this close similarity of the XANES spectra, BIANCONI et al.[9] concluded that the Ca^{2+} ion is bound to the O atoms in the $(HPO4)^{2-}$ group of ATP.

4. Calcium in Milk

Milk and its product provides about 70% of dietary calcium in much of the western world. Unlike calcium from vegetable sources, milk calcium is readily absorbed which may be partly due to a specific structure of calcium in milk.

Calcium in milk is mainly in the form of calcium phosphate-phosphoprotein complexes known as casein micelles. These micelles, in contrast to other phosphoprotein complexes in bone and other tissues, can be readily isolated and studied, but conventional techniques have given ambiguous and conflicting evidence on the structure of milk calcium phosphate.

Electron microscopical examination of unstained thin sections of casein micelles reveals electron-dense regions of calcium phosphate, a few nm in size, embedded in the protein matrix [11]. The x-ray diffraction pattern given by native casein micelles shows no evidence of an extensive crystal lattice, which has been interpreted as indicating an amorphous calcium phosphate, analogous to that in bone [12], for which the ratio of calcium atoms to inorganic phosphorus, Ca/P_i, is ~ 1.5. Chemical analyses give Ca/P_i in the range 1.4 - 1.65 suggesting a Ca-deficient apatite [13] $Ca/P_i \leqslant 1.66$) and the heat of solution of milk calcium phosphate from casein micelles resembles the heat of solution of hydroxyapatite gel [14]. In contrast, the ion activity product, which characterizes the solubility behaviour of milk calcium phosphate [13], approximates the form expected of a dicalcium phosphate($Ca/P_i = 1$).

HOLT et al. [10] have recently used the x-ray absorption spectrum of the Ca K-edge to characterise the milk calcium phosphate by comparing its EXAFS and XANES spectra with those of model calcium phosphates such as brushite ($CaHPO_4.2H_2O$), monetite ($CaHPO_4$), β-tricalcium phosphate, hydroxyapatite and amorphous calcium phosphate. Figure 5 shows the EXAFS spectra of freeze-dried casein micelles, a fully hydrated, gelatinous preparation of milk calcium phosphate and brushite. There are no appreciable differences between the spectrum of brushite and those of the two milk calcium phosphate samples with the exception that the feature marked 'a' is slightly smoothed out in the milk samples, possibly due to higher structural disorder. The close similarity is further illustrated by the XANES spectra and the Fourier transforms of the EXAFS spectra. However, HOLT et al.[10] observed clear differences between the spectra of the milk sample and those of other model systems and concluded that casein has a brushite-type structure. This finding is in agreement with the dicalcium phosphate stoichemistry obtained from solubility product measurements [13], but in apparent conflict with other results, notably the Ca/P_i ratio of 1.4 - 1.65. Thus, if the milk calcium phosphate exists in a brushite-type lattice, there is insufficient inorganic phosphorous, P_i, to occupy all the phosphate sites. Holt et al.suggested that these sites are occupied by all, or nearly all, of the phosphate moieties of phosphoserine and phosphothreonine residues from the casein matrix proteins. A domain of calcium phosphate in casein micelles comprises only a few unit cells so there are probably sufficient lattice sites near to a surface that could be occupied by casein PO_4- groups to bring the ratio $Ca/(P_i$ + casein P) to the brushite value of one [15].

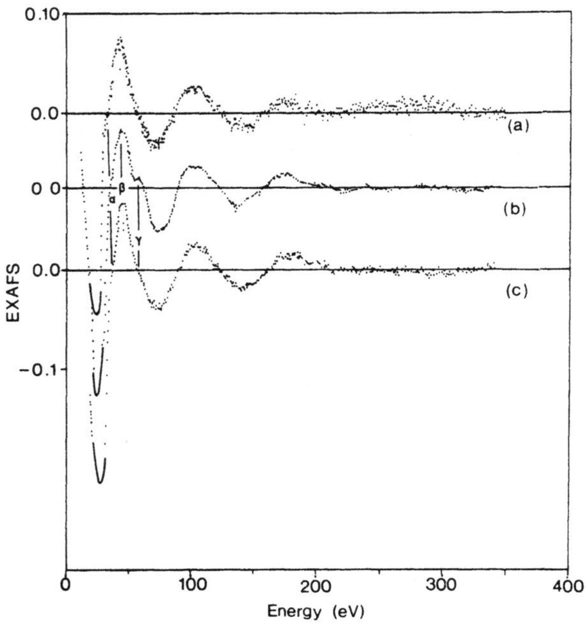

Fig.5. EXAFS spectra of (a) Freeze-dried casein micelles (b) brushite and (c) milk calcium phosphate gel [10]

The presence of protein phosphate groups in the brushite-type lattice adds support to the hypothesis that these act as heterogeneous nucleation sites for calcium phosphate deposition. Phosphoproteins have also been identified in bone, tooth (dentine, enamel and cementum), saliva, calcified tendon, urinary calculi and calcified aortic plaque where it has been suggested that they initiate calcification [14]. Furthermore, it has recently been shown that brushite occurs in the early stages of bone calcification [16], though it is not known whether this phase is associated with phosphoproteins. Although there appear to be some similarities between calcification in milk and that in other tissues and biological fluids, detailed x-ray study of a larger number of biological systems is required to understand the process of calcification.

5. Experimental Pitfalls

The presence of harmonics is a major problem in obtaining correct amplitudes and thus coordination number. This is worse for calcium (K edge = 3.07) because the critical wavelength of most of the storage rings with $E > 1.8$ GeV is < 4 Å and considerable fluxes are present at wavelengths as short as 0.8 Å. Use of an order sorting 2-crystal monochromator has been made to reduce the effect of higher harmonics. The effect of harmonics is clearly evident from Fig.6 where the XANES of CaF_2 has been measured in transmission for different levels of harmonic rejection, harmonic content is reduced by lowering I/I_{max}, which results in an increased absorption. Comparison of 50% and 33% spectra suggests that perhaps $I/I_{max} \sim 50\%$ is optimum for harmonic rejection, after which % harmonic rejection is not much reduced because of the large tails in the rocking curve. However, this is not the case as is evident from Fig.6(e), which is recorded with $I/I_{max} = 33\%$ and a double focusing mirror set at $\lambda_c = 1.30$ Å. Thus, a combination of a mirror and an order sorting monochromator results in a much superior harmonic rejection [17]. It may be worth noting

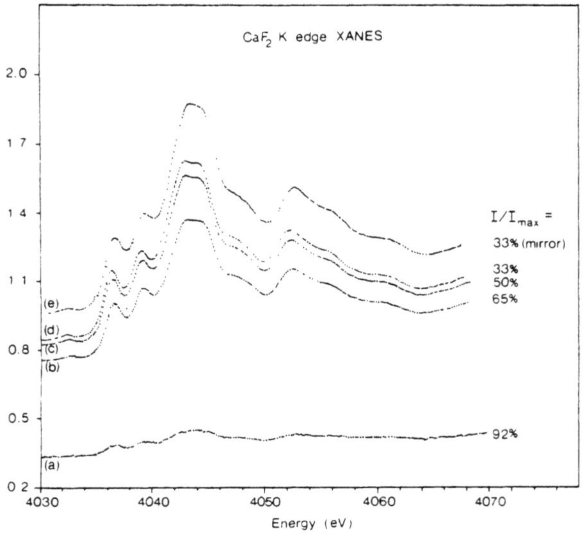

Fig.6. XANES of CaF_2 measured with different amounts of harmonic contamination in the monochromatized beam. I/I_{max} refers to the intensity used, I, compared to the maximum available when the two crystals are parallel, I_{max}. Spectra (e) is recorded when a double focusing mirror is placed between the source and monochromator [17].

from Fig.6(e) that the energy resolution achieved in spite of the use of a double focusing mirror is comparable to that of the unfocused. This has been achieved by using horizontal and vertical slits at the mirror.

It is a great pleasure to thank Drs. Hukins and Holt in whose association most of the work has been done.

References

1. Brown, W.E. and Chow, L.C., Ann. Rev. Mater. Sci. 6, 213-126 (1970).
2. Roufosse, A.H., Landis, W.J., Sabine, W.K. and Glimcher, M.J., J. Ultrastruc. Res. 68, 235-255 (1979).
3. Posner, A.S. and Betts, F., Acc. Chem. Res. 8, 273-281 (1975).
4. Miller, R.M., Hukins, D.W.L., Hasnain, S.S. and Lagarde, P., Biochem. Biophys. Res. Commun. 99, 102-106 (1981).
5. Eanes, E.D., Powers, L. and Costa, J.L., Cell Calcium 2, 251-261 (1981).
6. Binsted, N., Hasnain, S.S. and Hukins, S.S., Biochem. Biophys. Res. Commun. 107, 89-92 (1982).
7. Bianconi, A., Alema, S., Castellani, L., Fasella, P., Giovannelli, A., Mobilio, S. and Oesch, B., J. Mol. Biol. (In Press).
8. Powers, L., Ching, Y., Eisenberger, P.M., Chance, B., D'Ambrosio, C., and Smith, J.C., Stanford Laboratory report SSRL 80/01, pp. VII-47 (1980).
9. Bianconi, A., Doniach, S. and Lublin, D., Chem. Phys. Lett. 59, 121-124 (1978).
10. Holt, C., Hasnain, S.S. and Hukins, D.W.L., Biochem. Biophys. Acta (In Press).
11. Knoop, A.M., Knoop, E., and Wiechen, A.J., J. Dairy Res. 46, 347-350 (1979).

12. Knoop, A.M., Knoop, E., Frede, E., Precht, D. and Wiechen, A., Milchwissenschaft 34, 129-131 (1979).
13. Holt, C., J. Dairy Res. 49, 29-38 (1982).
14. Lyster, R.L.J., Biochem. Soc. Trans. 4, 735-737 (1976).
15. Climcher, M.J., J. Dent. Res. 58B , 790-806 (1979).
16. Roufosse, A.H., Landis, W.J., Sabine, W.K. and Glimcher, M.J., J. Ultrastr. Res. 68, 235-255 (1979).
17. Hasnain, S.S., Diakun, G.P., and Quinn, P.D., Nucl. Inst. & Methods (to be submitted).

EXAFS Studies of Zinc and Copper in Chemical and Biochemical Systems

I. Ross, N. Binstead, N.J. Blackburn[1], I. Bremner[2], G.P. Diakun[3], S.S. Hasnain[3], P.F. Knowles[4], M. Vasak[5], and C.D. Garner

The Chemistry Department, University of Manchester, Manchester M13 9PL, England

1. Introduction

Recent X-ray absorption spectroscopic studies we have undertaken at the Daresbury Laboratory using the dedicated Synchrotron Radiation Source have included investigations of the copper and zinc sites of mammalian liver metallothioneins and bovine erythrocyte superoxide dismutase. Corresponding studies have been accomplished for several copper and zinc compounds, the structure of each having been established by X-ray crystallography and shown to have some relevance to that of the metals in one or other of the above proteins.

2. Metallothioneins

Metallothioneins [1] are a ubiquitous class of proteins which are believed to play important roles in the control of zinc and copper metabolism and in the detoxification of metals, such as cadmium and mercury. Each of these proteins is characterized by its low molecular weight (~6,000), a high cysteinyl content, and the ability to bind substantial amounts of metal ions, such as Cd, Zn, Cu, Ag, Co and Hg. Thus, for the mammalian metallothioneins, cysteine constitutes 20 of the 61 amino acid residues of each protein molecule which is capable of binding up to seven Zn(II) and/or Cd(II) or Co(II) ions. The results obtained from a variety of spectroscopic studies clearly indicate that the metal ions are bound to the protein via the sulphur atoms of the cysteinyl residues. However, a definitive structural characterization of a metallothionein has not yet been achieved. To provide further information concerning the details of metal attachment to metallothionein, we have recorded spectra at the metal K edges for pig liver metallothionein containing copper and zinc, and the K edge of zinc for sheep liver metallothionein 2 containing zinc, and rabbit liver metallothionein containing zinc or zinc and cadmium. Freeze-dried samples were employed and X-ray absorption spectra were recorded for these materials at room temperature and 77K.

[1]Department of Chemistry, University of Manchester, Institute of Science and Technology, Manchester, M60 1QD, England

[2]Rowett Research Institute, Bucksburn, Aberdeen, AB5 9SB, Scotland

[3]The Daresbury Laboratory, Daresbury, Warrington, WA4 4AD, England

[4]Astbury Department of Biophysics, University of Leeds, Leeds, LS3 9JJ, England

[5]Biochemistry Department, University of Zurich, CH-8032 Zurich, Switzerland

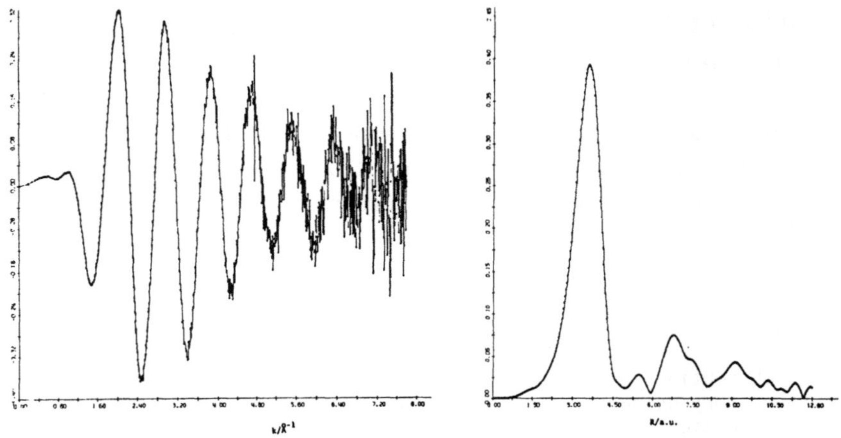

Fig.1. Zinc K edge EXAFS of rabbit liver metallotionein and its Fourier transform

The EXAFS associated with the zinc K edge of rabbit liver zinc-metallothionein at room temperature and its Fourier transform, are shown in Fig.1. Essentially identical data were obtained for all the other zinc edges of the metallothioneins at room temperature. Cooling the samples to liquid nitrogen temperature resulted in a slight (<20%) increase in the amplitude of the EXAFS, with no change in profile or periodicity. Data manipulation and calculation of theoretical spectra were accomplished using the procedures and programs documented previously [2]. Phase shifts were calculated using programs based on those of Pendry and Gurman [3] using wave functions tabulated by Clementi and Roetti [4]; zinc (copper) was assumed to be surrounded by four neutral sulphur atoms and the excited zinc (copper) atom was represented by the wave functions for gallium (zinc) for the $(1s)^1$ configuration.

As reported earlier [5] these EXAFS data are consistent with back-scattering from, essentially, a single shell of four sulphur atoms at ca. 2.28 Å. The chemical interpretation of this information is that zinc is bound by the sulphur atom of four cysteinyl groups, as $Zn(Scys)_4$.

Similar conclusions have been reached for the immediate environment of the copper atoms in pig liver metallothionein containing ca. six copper and two zinc atoms per protein molecule. The Cu - S distance is ca. 2.22 Å and there appear to be three sulphur atoms at this distance. These results are slightly at variance with the interpretation of the copper K edge EXAFS of yeast cu-thionein, for which the best theoretical fit was obtained with four sulphur atoms surrounding the copper, two atoms at 2.22 Å and another two at 2.36 Å [6].

The EXAFS of $Zn(S_2CNMe_2)_2$, $Zn(S_2CNEt_2)_2$, and $Zn_4(SPh)_8(MeOH)$ are virtually identical to one another and, in each case, can be simulated by a single shell of ca. four sulphur atoms at a distance of ca. 2.35 Å from the zinc, in good agreement with the crystallographic data.

The similarity between the X-ray absorption spectra of $Zn(S_2CNMe_2)_2$, $Zn(S_2CNEt_2)_2$, $Zn_4(SPh)_8(MeOH)$ and zinc in metallothionein includes the XANES, and this also extends to ZnS. On this basis, it is suggested that the essentially tetrahedral arrangement of the sulphur atoms about zinc established for ZnS and the other zinc-sulphur compounds also occurs for zinc in metallothionein.

An assessment of the quality of the interpretation of the EXAFS data obtained for zinc in metallothionein and $Zn_4(SPh)_8(MeOH)$ has been obtained by graphically displaying the variation of the function $r_g = 1/\sum_k[k^2(\psi_T(k) - \psi_E(k))]^2$ for different values of the parameters used to interpret the EXAFS data. The reciprocal of the simple least-squares function was chosen since it produces clearer visual displays than the function itself; thus, the optimum interpretation of the experimental data is when r_g is maximised. An iterative procedure was adopted, refining (R and E_0) and (N and α) sequentially, until reasonable self-consistency was obtained. The final value of R (Zn-S) was obtained by plotting a slice through the r_g versus (R, E_0) profile at the optimum value of E_0; the uncertainty in this value being taken as the width of this peak at 70% height, see Figure 2.

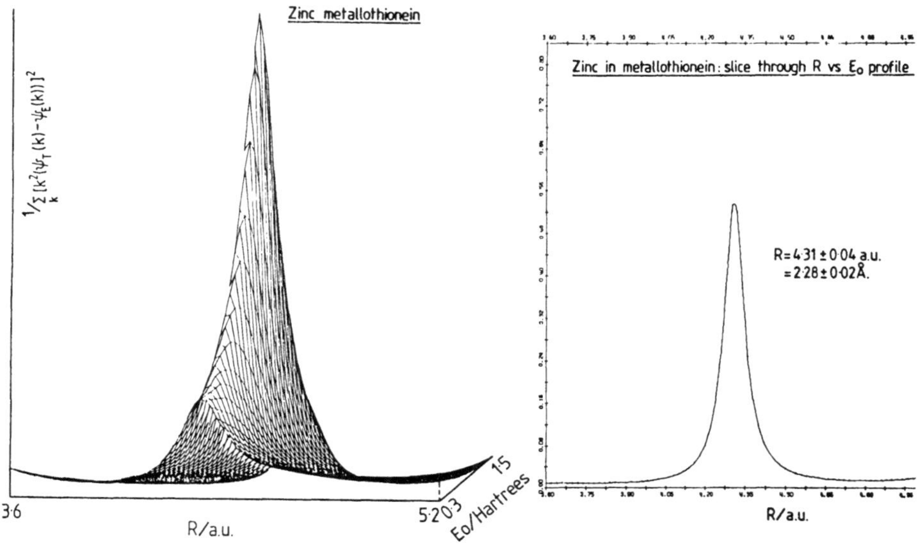

Fig.2. Plot of r_g versus R(Zn-S) and E_0 and a slice through this profile at E_0 = 0.80 Hartrees, for zinc in rabbit liver metallothionein

The analogous procedure was accomplished for E_0, N, and α. For $Zn_4(SPh)_8(MeOH)$, this approach yielded values of R, E_0, N, and α of 2.29±0.01 Å, 0.63±0.07 Hartrees, 3.4±0.6, and 0.047±0.019 (a.u.)^2, respectively; for zinc in rabbit liver metallothionein, the optimum values for the corresponding parameters were 2.28±0.02 Å, 0.80±0.07 Hartrees, 3.8±0.7, and 0.055±0.018 (a.u.)^2, respectively.

3. Bovine erythrocyte superoxide dismutase

Bovine erythrocyte superoxide dismutase has been the subject of a large number of investigations and much of this work has been stimulated by the determination of the crystal structure to 3 Å resolution [7]. The structure shows that each of the two subunits contains one copper and one zinc atom separated by ca. 6 Å and linked through the imidazole ring of a histidine residue. The zinc atom has an approximately tetrahedral co-ordination geometry and is bonded to a nitrogen atom of the imidazole rings of the residues His-61, His-69, and His-78 and a carboxylate oxygen atom of the residue Asp-81. The copper atom is co-ordinated to one nitrogen atom of the imidazole groups of the residues His-44, His-46, His-61, and His-118. This structure alone does not contain the detailed information needed to explain aspects of the enzyme's mechanism as the resolution is too low to detect the subtle structural features which presumably control the activity. Therefore, we have undertaken an X-ray absorption spectroscopic study of the metal centres in this superoxide dismutase, in its native, reduced, and azide bound forms. In each case, the sample was studied as a freeze-dried powder. As an aid to the interpretation of the data from the biological samples, corresponding data were recorded for relevant compounds of a known structure, including $[Zn(imidazole)_4][ClO_4]_2$ and $[Cu(imidazole)_4(NO_3)_2]$. The experimental data was manipulated and interpreted in the manner described above.

The Fourier transform of the EXAFS of $[Zn(imid)_4][ClO_4]_2$ shows clearly that back-scattering occurs from each of the three shells, N_α, C_β and (C_γ,N_γ), of the imidazole group. Although the EXAFS at high (> 4 $Å^{-1}$) k can be interpreted reasonably well with the nitrogen and carbon atoms of the imidazole groups near to their crystallographic values this is not true for the low k region. The interpretation in the latter region can be significantly improved by adjusting C_γ and N_γ to be substantially nearer to the zinc atom than established crystallographically , albeit at the expense of the quality of the agreement in the high k region. This adjustment in the position of C_γ and N_γ is taken to be an artificial way of partially accounting for multiple scattering effects likely to occur between the atoms of the rigid, planar, five-membered imidazole ring. Therefore, this is expected to be a general feature of the EXAFS of metal atoms co-ordinated to this moiety.

The copper K edge EXAFS of $[Cu(imidazole)_4(NO_3)_2]$ is dominated by the back-scattering from the four imidazole groups and, therefore, the overall profile bears a marked similarity to the zinc K edge EXAFS of $[Zn(imidazole)_4][ClO_4]_2$. However, inclusion of back-scattering from two oxygen atoms (of unidentate nitrato-groups) slightly improves the interpretation of the experimental data for $[Cu(imidazole)_4(NO_3)_2]$. The overall agreement between the shell distances required to interpret the EXAFS and the crystallographic values is reasonably good, apart from the short approach of C_γ and N_γ to the copper required to obtain a reasonable interpretation of the data at low k.

The similarities in the profile of the copper K edge EXAFS of native superoxide dismutase and those discussed above for $[Zn(imidazole)_4][ClO_4]_2$ and $[Cu(imidazole)_4(NO_3)_2]$, indicate that in this form of the enzyme, the copper is co-ordinated by four essentially equivalent imidazole groups, with Cu-N_α distances of ca. 2.00 Å. The Cu---C_β and Cu---C_γ(N_γ) distances follow from the Cu-N_α value. The EXAFS data obtained for the zinc site of native superoxide dismutase are reasonably reproduced by an environment comprising one oxygen atom, presumably from the aspartate carboxylate group, at ca. 1.92 Å and one and two imidazole groups with Zn-N distances of ca. 2.06 and 1.97 Å, respectively.

4. References

1 J.H.R. Kägi, and M. Mordberg (Eds.), "Metallothionein" (Birckhaüser Verlag, Basel, 1979) pp. 41-124 and references therein.

2 J. Bordas, R.C. Bray, C.D. Garner, S. Gutteridge, and S.S. Hasnain, Biochem. J., 191, 499 (1980).

3 J.B. Pendry, "Low Energy Electron Diffraction" (Academic Press, New York, 1974); S. Gurman, Daresbury Technical Memorandum, D2/SC1/TM/21T (1980).

4 E. Clementi, and C. Roetti, "Atomic Data and Nuclear Data Tables" (Academic Press, New York, 1974).

5 C.D. Garner, S.S. Hasnain, I. Bremner, and J. Bordas, J. Inorg. Biochem., 16, 253 (1982).

6 J. Bordas, M.H.J. Koch, H.-J. Hartmann, and U. Weser, FEBS Letters, 140, 19 (1982).

7 K.M. Beem, D.C. Richardson, and K.V. Rajagopalan, Biochemistry, 16, 1930 (1977).

A XANES Study of Co(II) in Cobalt Model Compounds and in Some Cobalt-Substituted Enzyme

A. Desideri
Dipartimento di Fisica, Università della Calabria,
I-87036 Arcavata di Rende, Italy

G. Rotilio
Istituto di Chimica Biologica, Università di Roma, I-00185 Roma, Italy

1. Introduction

In this work the K edge of two tetrahedral, three distorted trigonal bipyramidal and one octahedral Co(II) complex and of the cobalt derivatives of superoxide dismutase, carbonic anhydrase, stellacyanin and alcohol dehydrogenase are investigated and the properties of their X-ray absorption spectra are compared. The X-ray absorption data were taken at the synchrotron Radiation Facility PULS of the Frascati National Laboratories. The Co(II) proteins were prepared as described in [1]. Tetrahedral Co(imidazole)$_2$(acetate)$_2$ and Co(aptsz)$_2$I$_2$ were prepared according to [2,3]; trigonal bipyramidal Co(atsz)$_2$Cl$_2$H$_2$O, Co(atsz)$_2$I$_2 \cdot 2H_2O$ and Co(S$_2$PPh$_2$)$_2$C$_9$H$_7$N according to [3]; octahedral Co(bipyridyl)$_3$ perchlorate according to [4].

2. Results and Discussion

Figure 1 shows the XANES spectra of the Co(imidazole)$_2$(acetate)$_2$, Co(aptsz)$_2$I$_2$ and Co(bipyridyl)$_3$perchlorate compounds; Figure 2 shows their first derivative spectra.

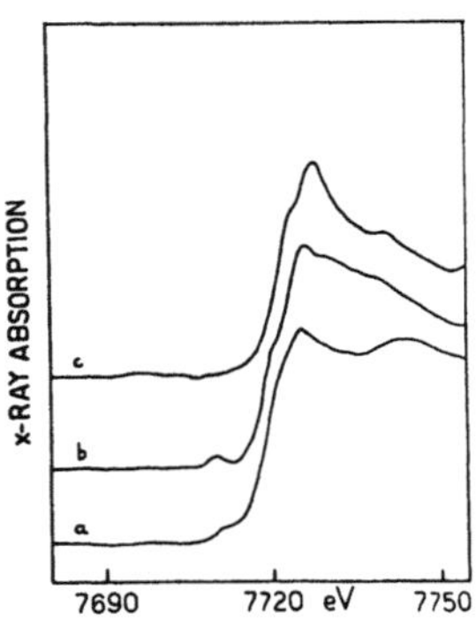

◄ Fig.1. XANES spectra of (a) Co(imidazole)$_2$(acetate)$_2$, (b) Co(aptsz)$_2$I$_2$, (c) Co(bipyridyl)$_3$ perchlorate

Fig.2. First derivative of the spectra shown in Fig.1 ►

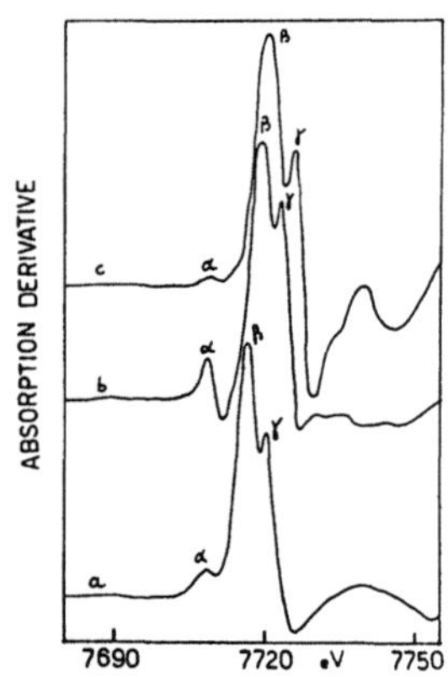

Three main absorptions are observable in the lower energy region of the spectra, which are referred to as α, β and γ (Fig.2). It is immediately apparent that the intensity of the peak labeled α is much lower in the octahedral complex (spectrum c) than in the tetrahedral ones. This may be due to the fact that in a tetrahedral geometry the p-d orbitals can mix with each other, and the forbidden 1s → 3d transition then becomes partly allowed [5]. The spectra of the five-coordinate compounds also display an α peak with an intensity comparable to that of the tetrahedral ones. This peak therefore indicates the presence of a distorted geometry, but it cannot be taken as diagnostic for pseudo-tetrahedral symmetries. Table 1 gives the energy difference between the β and α peaks in Fig.2 for all the compounds studied. It is apparent that the energy difference between these two peaks decreases as the covalency of the metal coordination sphere increases, independently of the geometry of the chromophore. For comparison, Table 2 gives the energy difference between these two peaks for the XANES spectra [1] of the Co(II) derivatives of superoxide dismutase, carbonic anhydrase, stellacyanin and alcohol dehydrogenase.

In carbonic anhydrase and superoxide dismutase this difference is almost the same, and is about 2.0 eV greater than in stellacyanin and alcohol dehydrogenase. These data compare well with the trend observed for the model compounds, since the ligand coordination sphere of superoxide dismutase and

Table 1. Energy difference between the α and β peaks for the Co(II) inorganic complexes

Complex	Coordination sphere	ΔE [eV]
$Co(bipy)_3$	N_6	11.5
$Co(Im)_2(Ch_3CO_2)_2$	N_2O_2	9.8
$Co(atsz)_2I_2\cdot 2H_2O$	N_2S_2O	8.4
$Co(S_2PPh_2)_2C_9H_7N$	S_4N	8.0
$Co(atsz)_2CI_2H_2O$	N_2S_2Cl	7.5
$Co(aptsz)_2I_2$	S_2I_2	5.7

Table 2. Energy difference between the β and α peaks for some Co(II)-substituted proteins

Co(II) protein	ΔE [eV]
Carbonic anhydrase	9.6
Superoxide dismutase	9.2
Stellacyanin	7.8
Alcohol dehydrogenase	7.7

carbonic anhydrase is N_3O [6] and that of alcohol dehydrogenase is NOS_2 [7]. The substantial identity in the shifts observed for stellacyanin and alcohol dehydrogenase supports the idea that sulfur ligands are present in the metal coordination sphere of stellacyanin. Sulfur ligands have been shown by X-ray diffraction analysis to be present for azurin [6] and plastocyanin [6], blue copper proteins whose size and function are comparable to those of stellacyanin. This latter protein is difficult to crystallize. Therefore in this case, the XANES spectra of the cobalt derivative provide a piece of structural information not yet available from X-ray diffraction.

References

1. A. Desideri, F. Comin, L. Morpurgo D, Cocco, L. Calabrese, B. Mondovì, W. Maret, G. Rotilio: Biochim. Biophys. Acta *661*, 312 (1981)
2. A. Dobry-Duclaux, A. May: Bull. Soc. Chim. Biol. *52*, 1467 (1970)
3. C. Bellitto, A.A.G. Tomlinson, C. Furlani, G. De Munno: Inorg. Chim. Acta *28*, 269 (1978)
4. F.H. Burstall, R.S. Nyholm: J. Chem. Soc. 3571 (1952)
5. R.G. Shulman, Y. Yafet, P. Eisenberger, W.E. Blumberg: Proc. Natl. Acad. Sci. *73*, 1384 (1976)
6. G. Rotilio: In *Metalloproteins*, ed. by U. Weser (Thieme, New York 1979) pp. 1-28
7. H. Eklund, C.I. Branden: J. Biol. Chem. *254*, 3458 (1979)

Mn(II) Binding to Nucleotides and Nucleic Acids

M. Belli[1], A. Bianconi[2], E. Burattini[3], S. Mobilio, C.R. Natoli,
L. Palladino[4], A. Reale[5], and A. Scafati[1]

It is well established that divalent metal ions play an important role in the different functions of nucleic acids and nucleotides. Different metal ions may affect the stability of the DNA double helix in different ways; some of them may induce errors in DNA replication. Divalent cations are integral components of the active form of tRNA molecule, where at least four specific sites for Mg ions have been identified.

The adenosine triphosphate (ATP) molecule, which acts as activator of several enzymatic reactions, requires the presence of divalent metal ions, mainly Mg, though it has been shown that Mn can substitute for Mg in many instances.

The solution structure of metal-ATP complexes is therefore of great importance for their intrinsic biological relevance and also for their value as models for metal-DNA (or RNA) binding. The primary reactive sites for metals in ATP complexes are the negatively charged oxygen atoms of the phosphate groups and the nitrogen donor atoms of the adenine. A "backbound" phosphate-metal-ion-adenine ring structure was proposed by Szent-Gyorgyi in 1957, but some important features of these complexes are still unclear after 25 years of research work. Major controversial points are the nature of metal binding to the ring and the occurrence of 1:1 or 1:2 metal-nucleotide complexes.

Most of the studies in solution have been carried out by magnetic resonance technique on Mn-5'ATP complexes [1]. Binding of Mn to all three phosphates was found by NMR of P-31 nuclei. Since interaction of Mn with the adenine ring has also been observed by PMR techniques, a model was proposed

[1]Ist. Superiore di Sanità and INFN, Sezione Sanità, I-00185 Roma, Italy
[2]University of Rome, I-00185 Roma, Italy
[3]I.N.F.N., Frascati and C.N.R., I-00185 Roma, Italy
[4]I.N.F.N., I-00044 Frascati, Italy
[5]I.N.F.N., Frascati and University of L'Aquila, I-00044 Frascati, Italy

in which Mn(II) is octahedrally coordinated to the ATP molecule, where the α, β, γ phosphates and three water molecules are inner sphere and N-7 of the adenine ring is outer sphere, the latter interaction being mediated by the H bond of a water molecule. On the other hand, further studies have indicated that the metal can bind directly to the N-7 nitrogen. The hypothesis of an equilibrium has also been suggested between the 1:1 and the 1:2 (Mn:ATP) complexes. In addition, a transition seems to occur on passing from neutral to acidic conditions, with a change from an α, β, γ chelation to a β, γ one, resulting from protonation of the α phosphate. Thus a quite complicated picture emerges from the experimental results.

A recent investigation of Mn-5'ATP complexes in solution by means of high-resolution X-ray spectroscopy (EXAFS and XANES) carried out with the facility at Frascati supplied further information about this very important complex. The study was performed at several Ph values and Mn:ATP ratios.

From the EXAFS measurements the Mn-O (first coordination shell) and Mn-P distances were measured. The phase shifts in the backscattering amplitude were determined from EXAFS spectra of the model compounds MnO, MnO_2, and MnP, under the assumption of "chemical transferability" from these to the complex under examination. The Fourier transforms of the EXAFS spectra of Mn-ATP (Fig.1) show two significant peaks, and their back-Fourier analysis is consistent with the assumption that the first peak is due to oxygen atoms and the second one to phosphorus atoms. The distances measured by the EXAFS analysis are shown in Table 1, where results for the aqueous hexacoordinated Mn(II) ion in solution are also shown.

Table 1

	1st shell		2nd shell	
	R_I (A^o)	N_I	R_{II} (A^o)	N_{II}
Mn-ATP (pH 9)	2.15 ± 0.05	7±1	3.3 ± 0.05	7±1
Mn-ATP (pH 7)	2.16 ± 0.05	4±1	3.4 ± 0.05	4±1
Mn-ATP (pH 3)	2.15 ± 0.05	5±1	3.4 ± 0.05	2±1
$Mn(ClO_4)$,aq.sol.	2.15 ± 0.05	5±1		

The results indicate that the oxygens of the first shell are at 2.15 Å, the same distance found for the aqueous ion, and that this distance is independent of pH. The Mn-P distance of 3.3-3.4 Å is in agreement with the distance found in LiMnPO crystal [2]. The calculated Mn-O-P bonding angle is about 126°.

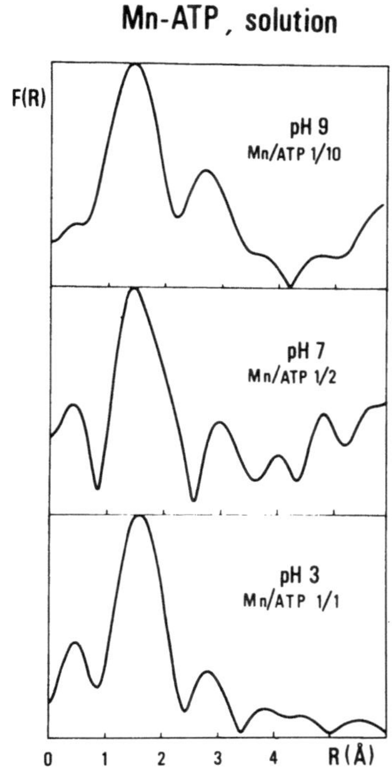

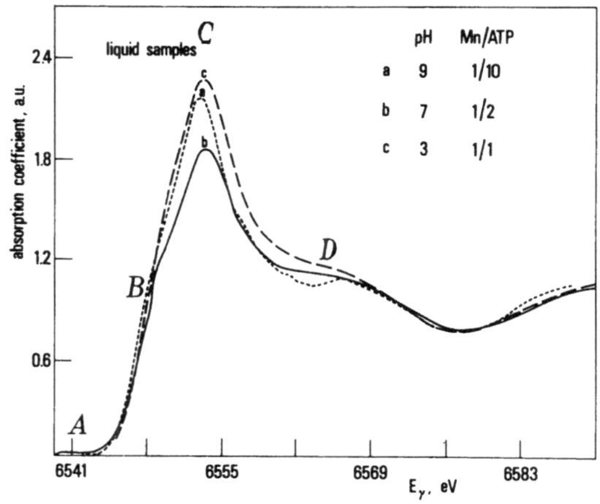

Fig.2. Near-edge spectra

Fig.1. Fourier transforms of EXAFS spectra of Mn-5'ATP complexes

An estimate was made of the coordination numbers for the two peaks. The results, though affected by large errors, indicate that the number of phosphorus atoms increases with pH, which is consistent with the observation that at low pH only the β and γ phosphates are capable of entering the first coordination shell because of the protonation at the α group. This may be related to the formation of a 1:2 complex in which the Mn(II) is coordinated with the phosphate chain of two ATP molecules. Edge analysis supports this hypothesis, as is clear from Fig.2, where the sample at high pH shows a higher peak C and a more evident resolution of the peak D. This feature may be attributed either to a different orientation of the PO_4 groups or to an increase in the number of these groups directly coordinated by Mn(II). EXAFS analysis showed a constant value of the distances Mn-O and Mn-P, so that an increase in the number of PO_4 groups at high pH may be deduced. The formation of 1:2 complexes is also favoured by the low Mn:ATP ratio used at pH 9. In conclusion, the geometry of the Mn binding site and therefore the structure of the complex is dependent on pH and the Mn:ATP ratio in aqueous solutions, and this must be taken into account when comparing experiments conducted under different conditions. The reported results on distances and coordination numbers show that high-resolution X-ray spectroscopy is a very useful tool in the study of Mn-ATP complexes and possibly of other metal-nucleotide and metal nucleic acid complexes.

References

1. R.B. Martin, Y.H. Mariam: Met. Ions Biol. Syst. *8*, 57-125 (1979)
2. S. Geller, J.L. Duran: Acta Crystallogr. *13*, 325 (1960)

On the Investigation by EXAFS Spectroscopy of a Selenocarbonyl Complex (CSe = Fe(II)TPP)

C. Goulon-Ginet[1], J. Goulon[1,2], J.P. Battioni[3], D. Mansuy[3], and J.C. Chottard[3]

1. Introduction

Carbon monoxide is a common ligand of the metalloporphyrin chemistry, whereas thiocarbonyl and selenocarbonyl metal complexes are still rare. We wish to report here the first structural study of a stable and well-identified selenocarbonyl adduct of an iron(II) tetraphenylporphyrin (CSe=Fe(II):TPP) [1]. The EXAFS spectra were carried out at both the selenium (12.7 keV) and iron (7.1 keV) K edges and provide a good illustration of the so-called focusing effects [2] for a colinear arrangement of scattering atoms. The EXAFS spectrum at the iron edge also gives a typical example of destructive interferences causing some strong signals of the porphirinic ring to vanish. The FT spectra $\tilde{\chi}(R)$ reproduced in this paper are all corrected for the amplitudes $F_j(k)$ and phase shifts $\psi_j(k)$ of the shell j (i.e. Se*...C and Fe*...N) according to [3] :

$$\tilde{\chi}_j(R) = \int_0^{\infty} dk\ g(k)\ \chi(k)\ \frac{k\ R_j^2}{N_j\ |F_j|\ A_j}\ \exp\left\{2\ \sigma_j^2 k^2 - 2ikR - i\psi_j(k)\right\} \qquad (1)$$

where g(k) is a Kaiser window function and $A_j(R_j,k)$ is a semi-empirical function accounting for the loss of coherence of the photoelectron and the relaxation phenomena pointed out by STERN et al. [4]. $F_j(k)$ and $\psi_j(k)$ were numerically calculated using the parametrization suggested by TEO et al. [5], E_0 being adjusted according to the LEE-BENI criterion of coincidence of the maxima of $|\tilde{\chi}(R)|$ and $\mathrm{Im}\tilde{\chi}(R)$ [6]. N_j, R_j and σ_j are thus fixed in a self-consistent way or can be extracted from a safe and fast fitting procedure [3]. The additional parameters involved in the definition of $A_j(R_j,k)$ are usually preadjusted by comparison to model compounds (e.g. selenourea, benzylselenocyanate,...).

2. Selenium Edge Data

The FT spectrum $\tilde{\chi}_1(R)$ obtained is reproduced in Fig.1a . The assignment of the first peak to the shell Se*...C and of the second peak to the sequence

[1] Laboratoire de Chimie Théorique, E.R.A. 22 au C.N.R.S., Université de Nancy I, B.P. 239, F-54506 Vandoeuvre-les-Nancy, France

[2] L.U.R.E., Laboratoire Propre du C.N.R.S., Université de Paris-Sud, Bât. 209c, F-91405 Orsay, France

[3] Laboratoire de Chimie de l'E.N.S., 24 rue Lhomont, F-75231 Paris Cedex, France

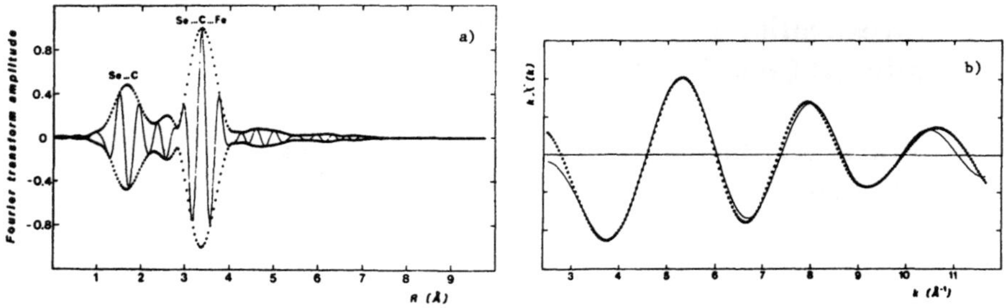

Fig.1. a) Corrected FT spectrum $\tilde{\chi}_1(R)$ of (CSe=Fe:TPP) at the selenium K edge.
b) Least squares fit of the filtered first shell Se*...C

Se*...C...Fe is obvious. On filtering and backtransforming the first peak, we could generate the fit of Fig.1b with $R_1 = 1.64 \pm 0.01$ Å, $N_1 = 1._1$ and $\sigma_1 \simeq 0.035$ Å. Although the distance is quite short, we do not expect for a light scatterer as carbon the plane wave approximation to induce a large error because in the LEE-PENDRY criterion $\delta\Phi \simeq k\,|F_j|^2/R_j < 1$, $|F_j(k)|$ is decaying very rapidly [7]. Our result compares quite reasonably with a few Se...C bond lengths reported recently (R(Se...C)≃1.67 Å) in $RuCl_2(CO)(CSe)(PPh_3)$ [8]. The strong peak at ∿ 3.25 Å is typical of a focusing effect due to the linear sequence Se*...C...Fe. Assuming a similar behaviour for oxygen and carbon, we tried to use the recent theoretical calculations made by TEO [2] for an evaluation of the additional phase shift $\omega_C(k)$. The corrected distance R_2(Se*...C...Fe) ≃ $3.3_7 \pm 0.05$ Å is quite acceptable but the difference $R_2 - R_1 \simeq 1.72 \pm 0.06$ Å is slightly shorter than the Cr...CSe bond length (1.78 Å) reported in reference [9]. The crudeness of the present phase determination is reflected by the need of an additional negative shift of E_0 in order to satisfy the LEE-BENI criterion.

3. Iron Edge Data

If one is trying to compare the FT spectrum shown in Fig.2a to any classical pattern of metalloporphyrin (see for instance reference [10]), it becomes clear that the usually intense signals associated with the carbons α and carbons meso of the porphyrin ring are missing. We understand this result to be caused by a destructive interference of the poorly resolved $C\alpha C\beta$ signals with the Fe*...C...Se signal strongly enhanced again due to the focusing effect.

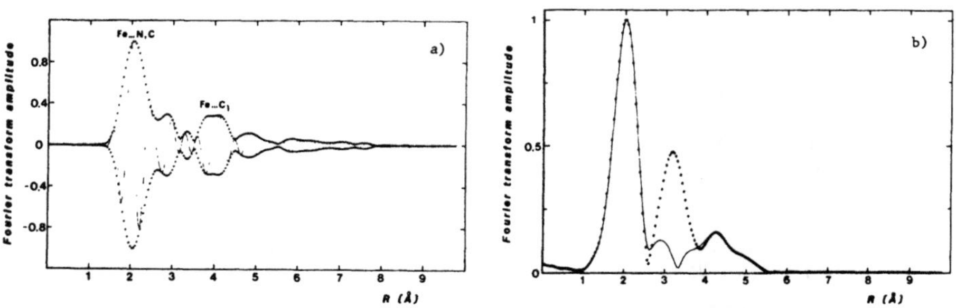

Fig.2. a) Corrected FT spectrum $\tilde{\chi}_1(R)$ of (CSe=Fe:TPP) at the iron K edge.
b) Simulated $|\tilde{\chi}_1(R)|$ spectra. Dots are for the porphyrin pattern without the Fe*...C...Se contribution

Heretofore, this is the first example of such a destructive interference observed in the porphyrin chemistry for the carbons C_α,C_β. Supporting this interpretation is the simulation shown in Fig.1b : starting from a reasonable set of distances Fe...N (2.00_4 Å), Fe...C_α (3.08 Å), Fe...C_β (3.45 Å), Fe...C_γ (4.2 Å), Fe...C_ϕ (4.95 Å), we could reproduce this negative interference simply by choosing Fe*...C...Se = 3.3 Å if the previous additional phase shift $\omega_c(k)$ is taken into account with an enhancement factor $\Omega_c \simeq 7.2$ of the order of the prediction made by TEO, but larger than the experimental value found for the Se*...C...Fe signal. For the sake of comparison, we also reproduced the simulated porphyrin pattern alone. On the other hand, the apparent distance Fe...N deduced from the real spectrum ($2.0_3 \pm 0.0_2$ Å) is a bit larger than our expectation (2.0_0 Å) for a low spin Fe(II) pentacoordinated compound. However, the unresolved distance Fe...C may result in a pseudo-asymmetrical distribution of the first shell with a slight shift $\Delta\bar{R}_1>0$ of the first peak [11].

References

1. J.P. Battioni, D. Mansuy and J.C. Chottard, Inorg. Chem. 19, (1980), 791.
2. B.K. Teo, J. Am. Chem. Soc. 103, (1981), 3990.
3. J. Goulon, C. Goulon-Ginet and M. Chabanel, J. Sol. Chem. 10, (1981), 9.
4. E.A. Stern, S.M. Heald and B. Bunker, Phys. Rev. Lett. 42, (1979), 1372.
5. B.K. Teo, P.A. Lee, A.L. Simmons, P. Eisenberger and B.M. Kincaid, J. Am. Chem. Soc. 99, (1977), 3854.
6. P.A. Lee and G. Beni, Phys. Rev. B15, (1977), 2862.
7. P.A. Lee and J.B. Pendry, Phys. Rev. B11, (1975), 2795.
8. G.R. Clark, K.R. Grundy, R.O. Harris, S.M. James and W.R. Ropper, J. Organometal. Chem. 90, (1975), C37.
9. J.Y. Saillard, D. Grandjean, P. Caillet and A. Lebeuze, J. Organometal. Chem. 190, (1980), 371.
10. J. Goulon, C. Goulon, F. Niedercorn, C. Selve and B. Castro, Tetrahedron 37, (1981), 3707.
11. P. Eisenberger and G.S. Brown, Solid State Commun. 29, (1979), 481.

X-Ray Iron K Edge of Human Hemoglobins

S. Morante[1], M. Cerdonio[1], S. Vitale[1], A. Congiu-Castellano[2], A. Vaciago[2], G.M. Giacometti[3], and L. Incoccia[4]

1. Introduction

The development of the synchrotron radiation sources allows now to extend x-ray absorption spectroscopy to aqueous solutions of biological molecules. Preliminary unpublished results by Yuen et al.[1] reported an energy shift of the K edge of about 5 eV going from the deoxy to ligated derivatives of hemoglobin. Stimulated by this finding we have undertaken a more extensive study of the K-edge absorption spectra for several derivatives of human HbA.

2. Materials and Methods

The hemoglobin derivatives were prepared following the procedure reported in [2]. The required concentration of the protein solution (about 15 mM)was reached by vacuum pressure filtration. The state of the sample was controlled by visible spectral analysis before and after the X-rays measurements.

The X-ray absorption measurements were performed at the PULS radiation facility of the ADONE storage ring in Frascati (Roma) [3]. Data were collected by an on-line computer and the further elaboration was made by using a PDP 11/34 computer. The edge positions of the hemoglobins adduct were evaluated, after the background subtraction and a smoothing procedure, by two different empirical methods; i.e. taking respectively the energy point at half maximum of the absorbance or the energy position of the maximum of the first derivative of the spectrum. All the energies were calibrated relative to a Fe powder standard. The Fe edge position was taken as the first inflection point in the absorbance spectrum, which in our case was found at 7112 eV.

The instrumental resolution in the range covered by our experiment (7000 to 7200 eV) was 0.23 eV.

3. Results and Discussion

The absorption profile of deoxy and carbonmonoxy hemoglobin are compared in fig.1.

[1]Dipartimento di Fisica, Libera Università di Trento, I-38050 Povo, Trento, Ita
[2]Università di Roma, I-00185 Roma, Italy
[3]Istituto di Biologia Animale, I-35100 Padova, Italy
[4]PULS, Laboratori Nazionali di Frascati, I-00044 Frascati, Italy

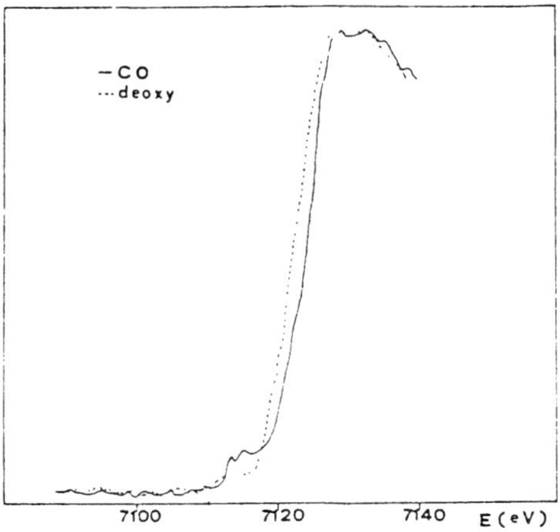

Fig. 1. X-ray edge spectra of the deoxy (dotted line) and carbonmonoxy (full line) derivatives of human hemoglobin. Absorbances are on an arbitrary scale

For all the derivatives explored an edge structure superimposed to the continuum transition is observed, however the amplitudes of this structure are still at level comparable to the noise. Therefore we did not attempt any fitting with theoretical models.

This fact poses serious limitations to the possibilities of extracting unequivocal values for the K edge positions. However within the crude approximation reported above we may reach some significant conclusions. The results obtained for all the derivatives explored are reported in Table 1.

Table 1. K-edge position shift in eV relative to the Fe powder standard; the error is evaluated from the half-dispersion of the data

	deoxy	CO	O_2	CN	H_2O	N_3	OH	F
A	10.2±0.2	12.3±0.5	13.7±0.2	13.6±0.7	13.7±0.2	13.5±0.2	13.7	13.7±0.2
B	10.4±0.5	13.3±0.5	14.0±0.3	14.7±0.5	15.2±0.3	14.7±0.3	15	15 ± 1
C	10.2	14.4	15	16	16			

A from the half maximum of the edge spectrum

B from the derivative maximum

C data taken from Yuen et al. ref.(1)

We may observe that, independently of the method used to evaluate the position, the data for all ligated derivatives group lie within an interval smaller then ± 0.9 eV. Only the edge position of the deoxy form is found outside the interval in the low energy side separated by at least 3 eV. These findings are in qualitative agreement with those of Yuen et al.[1].

Within the ligated forms minor differences are observable both in the edge energy and in the shape of the structure. An interpretation of such differences has to wait for a significant improvement in the quality of the spectra and for test of the current theories on the edge structure in complex systems. In the absence of a general theory for the K edge of complex systems, the simplest possible explanation of the energy gap between deoxy hemoglobin and all the ligated forms either in form of "ferrous" (HbO_2 and HbCO) or ferric derivatives grouped together, demands a positively charged iron nucleus in oxy and carbonmonoxy hemoglobin similar to that present in ferric derivatives. The K-edge X-ray absorption experiments therefore seem to be consistent with a model on the line of that originally proposed by Weiss [4] of a significant charge transfer in ferrous ligated forms of hemoglobin.

References

1. C.Yuen,M.Weissbluth,A.M.Labhardt, SSRP Report n.77/06 (Standfort 1977, unpublished)
2. E.Antonini,M.Brunori, "Hemoglobin and myoglobin in their reactions with ligands" North-Holland Publishing Company Amsterdam-London (1971)
3. A.Balzarotti,F.Comin,L.Incoccia,S.Mobilio,Frascati Internal Reports (1982)
4. J.J.Weiss,Nature 202,83 (1964)

XANES of Calmodulin: Differences and Homologies Between Calcium-Modulated Proteins

A. Bianconi[1], A. Giovannelli[1], I. Ascone[1], S. Alema[2], P. Durham[3], and P. Fasella[4]

Calmodulin is the receptor of Ca^{2+} (acting as a second messenger) in transmission of biological information between cells [1, 2]. It is a calcium binding protein present in every eucariotic cell without any tissual specifity and acts as an enzyme activator modulated by cellular concentration of Ca^{2+}. It is a single peptide chain (P.M.=16700) containing four Ca^{2+} binding sites, whose affinity has been established to be high for sites I, II, and low for sites III, IV, with a difference of a factor 10.

Here we study the structure of Ca^{2+} binding sites of calmodulin by comparing the XANES (X-ray Absorption Near Edge Structure) spectra of calmodulin with those of troponin-C (TnC) and carp parvalbumin (CPa). Troponin-C binds four Ca^{2+} atoms with similar local structure. It has also two Mg^{2+} specific sites whose binding induces conformational changes to the protein [3]. The structure of carp parvalbumin is known from X-ray diffraction [4]. It binds only two atoms of Ca^{2+} in two loops named CD hand (coordination number =6) and EF hand (coordination number =8), and is assumed as a model compound in studying calcium-binding proteins.

X-ray absorption spectroscopy is a powerful technique to study those systems, in fact calcium is silent to many spectroscopies and they escape X-ray diffraction studies because of difficulties to grow and keep crystallized proteins. It has been shown that the XANES (extending over a 40 eV energy range) is the most informative region of X-ray absorption spectra for those systems [3,5]. In fact the EXAFS spectra are not very informative because they give only information about the radial distribution of atoms around the absorber. In these proteins the local structure is very asymmetric. Every oxygen atom is at a different distance from calcium so that only an average distance is obtained by EXAFS.

XANES features [5.6], due to multiple scattering resonances, allow to obtain information relevant for the biological

[1]Istituto di Fisica, Universita' di Roma, I-00185 Roma, Italy
[2]Laboratorio di Biologia Cellulare del CNR, Via Romagnosi, I-00185 Roma, Italy
[3]Daresbury Laboratory, SERC, Warrington WA4 4AD, England
[4]CEE, Bruxelles, Belgium

function: coordination geometry, interatomic distances and orientation of molecular groups bound to calcium [6].

Calmodulin was isolated from beef brain by a method involving DEAE chromatography, gel filtration and pyacrylamide gel electrophoresis. Atomic absorption shows $3Ca^{2+}$ ions bound per protein. Troponin-C was prepared from beef scheletal muscle by the method of Perry and Cole [7]. Carp parvalbumin was prepared from carp white muscle according to the method of Pechere and Kretsinger [8]. 4 and 2 calcium ions per protein were bound to troponin-C and parvalbumin respectively.

The spectra were measured at the Frascati Synchrotron radiation facility PULS.

Fig. 1 shows the comparison between XANES of calmodulin and troponin-C; Fig. 2 between XANES of carp parvalbumin and calmodulin. The energy position of peak E is the same in all spectra, indicating that Ca^{2+} has the same charge in all the proteins. We find that peak A of calmodulin is shifted by 0.8 eV respect to that of troponin-C but is coincident with that of carp parvalbumin.

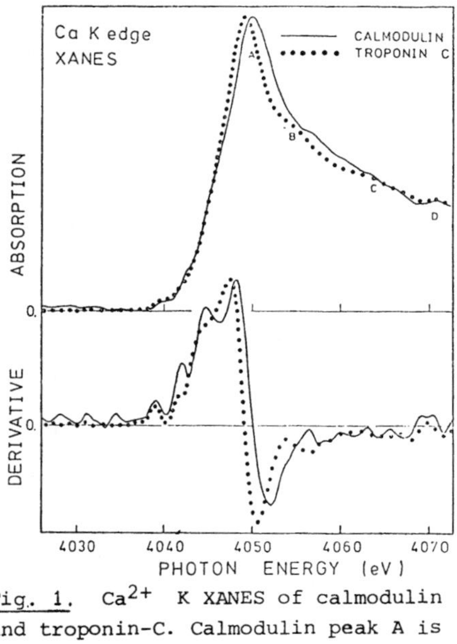

Fig. 1. Ca^{2+} K XANES of calmodulin and troponin-C. Calmodulin peak A is shifted of 0.8 eV

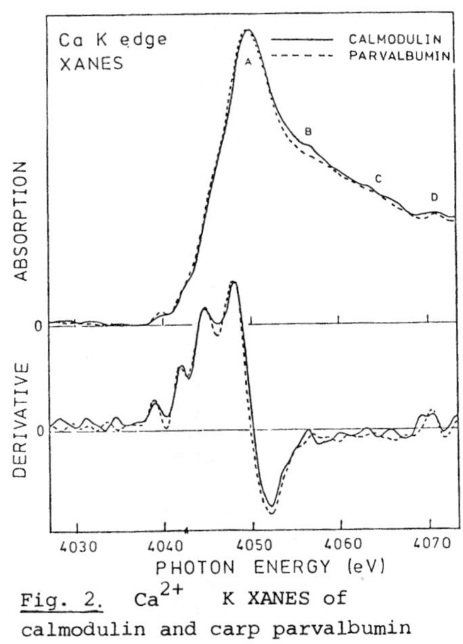

Fig. 2. Ca^{2+} K XANES of calmodulin and carp parvalbumin

The preliminary calculations of XANES of a simple symmetric cluster of $Ca^{2+}(COO^-)$ have been performed using the multiple scattering approach [9], with the aim to establish the contribution of neighbour shells around calcium on the XANES. Fig. 3 shows that the main peak A appears in XANES of a cluster with only the first coordination shell but the full series of resonances A, B and C appear only including the complete

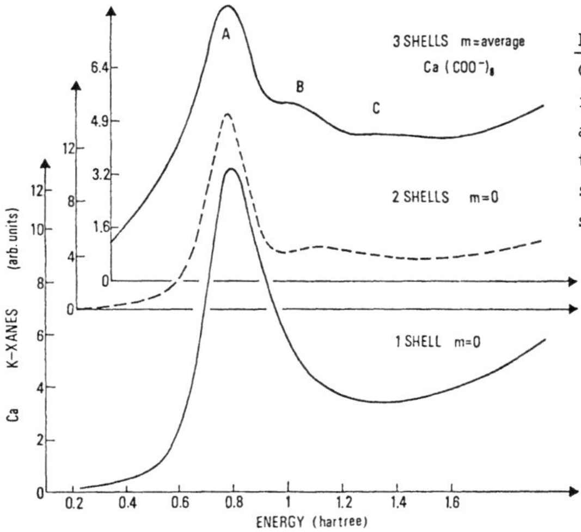

Fig. 3. Calculation of the Ca^{2+} K XANES for a cluster formed by 8 COO^- groups in a cubic configuration showing the effect of including successively each neighbour shell. One Hartree is 27 eV

COO^- molecular groups. Further calculations show that the intensity and the energies of B and C peaks are modified by rotation of COO^- molecular groups.

We assign the difference between the intensities of peaks B and C in the XANES of troponin-C and calmodulin to a different orientation of COO^- groups in calmodulin sites in comparison with calcium sites in troponin-C [3,5]. The shift of peak A toward high energy in calmodulin is related with a shorter $Ca^{2+}-O$ distance of first-neighbour atoms following the $\Delta E/2E = -\Delta R/R$ rule [5,6].

The similarity of XANES of calmodulin with parvalbumin indicates similar calcium sites as detected by XANES. These results are not in agreement with homologies of amminoacid sequence showing a large similarity between troponin-C and calmodulin [2]. Therefore the structures of Ca^{2+} sites are not determined only by the amminoacid sequence.

REFERENCES

1. W.Y. Cheung, Science 207, 19 (1980).
2. R.H. Kretsinger, Crit. Rev. Biochem. 8, 119 (1980).
3. A. Bianconi et al., J. Molecular Biology, (1982) to be published.
4. P.C. Moews and R.H. Kretsinger, J. Mol. Biol. 91, 201 (1975).
5. A. Bianconi in "EXAFS for inorganic systems", C. Garner, S. Hasnain editors (Daresbury Rep. , 1981), p. 13.
6. A. Bianconi, M. Dell'Ariccia, P. Durham and J.B. Pendry Phys. Rev. B 15 November 1982.
7. S.V. Perry and H.A. Cole, Biochem. J. 141, 733 (1974).
8. J.F. Pechere et al., Bioch. Biophys. Acta 236, 391 (1971).
9. P.J. Durham, J.B. Pendry and C.H. Hodges, Comput. Phys. Comm. 25, 193 (1982).

Nickel-Containing Enzyme: Urease

S.S. Hasnain

Science & Engineering Research Council, Daresbury Laboratory, Daresbury, Warrington WA4 4AD, Cheshire, United Kingdom

B. Piggott

Inorganic Chemistry Research Laboratory, Division of Chemistry, School of Natural Sciences, Hatfield Polytechnic, Hatfield, Herts., United Kingdom

Summary

Preliminary results from an x-ray absorption study are presented to show that the local environment of the Ni atoms in jack bean urease resemble closely that of the Ni atom in $Ni(L)_2(L')_1(ClO_4)_1$ where L is 2-α-Hydroxybenzylbenzimidazole and L' the deprotonated form.

Introduction

Urease from jack bean was isolated in 1926 by SUMNER [1]. This was the first enzyme to be crystallised. In 1975, some fifty years later, DIXON et al.[2] showed that jack bean urease (urea amidohydrolase) was in fact a metalloenzyme. This enzyme has a molecular weight of 480,000 daltons and contains 6-8 atoms of nickel per molecule. Prior to this discovery, nickel had been the only metal in the first transition series beyond titanium which had not been shown to have a biological role. Urease catalyses the hydrolysis of urea to carbon dioxide and ammonia with extreme efficiency [3]. It is also highly specific hydroxyurea, dihydroxyurea [3] and aminourea [4]

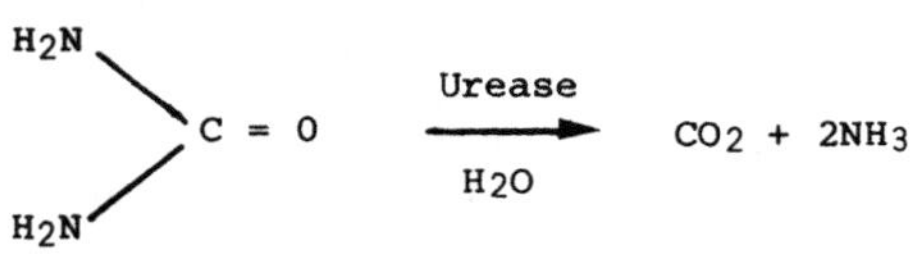

being the only other known substrates. Little is known about the mechanism of this reaction although there has been some speculation as to the direct involvement of the nickel ions [5]. However, without knowing the co-ordination sphere of the nickel ions, suggestions as to their function in the hydrolysis reaction remains speculative. It was with this aim that x-ray absorption (EXAFS and XANES) study was undertaken.

Experimental Details

Urease from jack bean was obtained from Sigma Chemical Company. X-ray absorption spectrum (EXAFS and XANES) was obtained at the Ni K-edge in the standard transmission mode [6] at the Daresbury Synchrotron Radiation Source. The storage ring was operating at 2 GeV and an average beam current of 80 mA. Harmonic contamination (λ/2, λ/3 ...) was minimised by using a double crystal Si 220 monochromator and by offsetting the crystal to give I/I_{max} = 70%, where I_{max} is the intensity when the two crystals are set parallel. For the enzyme, six scans were used to produce the average spectrum shown in Figs.1 and 2 while for the model systems, single scan data is presented.

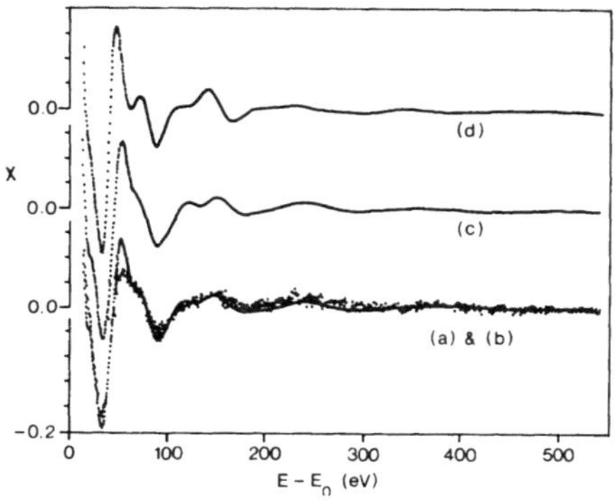

Fig.1. EXAFS spectra of (a) jack bean urease (b) $Ni(L)_2(L')_1(ClO_4)_1$ (c) $Ni(HMB)_3(Br)_2$ where HMB is 2-hydroxymethylbenzimidazole and (d) $Ni(imid)_6(ClO_4)_2$

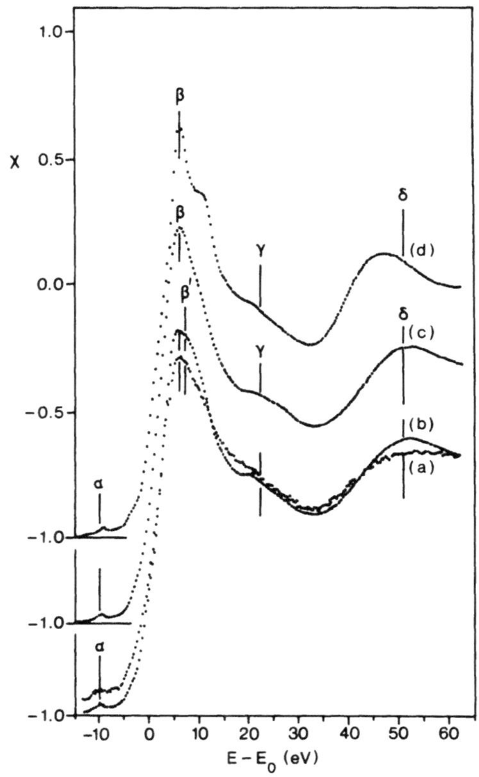

Fig.2. XANES spectra of (a), (b), (c) and (d) as defined in Fig.1

Results and Discussion

The EXAFS spectrum of jack bean urease along with the spectra of three chemical systems is shown in Fig.1. Each of the spectra shows strong beat regions typical for histidine type environment [7]. The spectrum of the enzyme has been superimposed on that of $Ni(L)_2(L')_1(ClO_4)_1$, where L is 2-α-Hydroxybenzylbenzimidazole and L' is the deprotonated form to show the close similarity between the co-ordination sphere of nickel in the two systems. This is further confirmed by a similar comparison of the near edge spectrum (XANES) where contributions from distant shells of atoms are stronger. Thus, from EXAFS and XANES spectra, we suggest that the local environment (up to 4 Å) of the nickel ion in the two cases are very similar.

Acknowledgement

We would like to thank the Science & Engineering Research Council, Daresbury Laboratory, for allocating beam time under exploratory agreement. We would also like to thank Sigma Chemical Company, London, for their generous gift of pure enzyme.

References

[1] Sumner, J.B., J. Biol. Chem. 69, 435 (1926).

[2] Dixon, N.E., Gazzola, C., Blakeley, R.L., and Zerner, B., J. Amer. Chem. Soc. 97, 413 (1975).

[3] Blakeley, R.L., Hinds, J.A., Kuze, H.E., Webb, E.C., and Zerner, B., Biochemistry, 8, 5 (1969).

[4] Gazzola, C., Blakeley, R.L., and Zerner, B., Can. J. Biochem. 51, 1325 (1973).

[5] Dixon, N.E., Gazzola, C., Blakeley, R.L., and Zerner, B., Science, 191, 1144 (1976).

[6] Hasnain, S.S., in EXAFS for Inorganic Systems, Garner, C.D., and Hasnain, S.S., (eds.) pp.23-27, Daresbury Laboratory Report, DL/SCI/R17 (1981).

[7] Perutz, M.F., Hasnain, S.S., Duke, P.J., Sesler, J.L., and Hahn, J.E., Nature, 295, 535 (1982).

Part VIII

Related Techniques – Anomalous Scattering

Anomalous X-Ray Scattering and Its Relationship to EXAFS

J. Kortright[1], W. Warburton[2], and A. Bienenstock[1,2]
Stanford University, Stanford, CA 94305, USA

X-ray anomalous scattering techniques are described which utilize the strong energy dependence of the atomic scattering factor near an atom's absorption edge to separate the total radial distribution function (RDF) into more species-specific distribution functions. Differential anomalous scattering (DAS) techniques and transmission EXAFS have been applied to sputtered amorphous $Mo_{42}Ge_{58}$ films. Although each approach yeilds, in principle, a description of the average coordination distances and numbers of the atomic species whose absorption edge is utilized, there is a significant discrepancy between the EXAFS radial structure functions (RSF) and the DAS differential distribution functions (DDF). The RSF's indicate shorter first neighbor distances and markedly narrower first shells than do the DDF's. The DDF's show more distant coordination shells clearly, whereas the RSF's do not. We attribute these differences to the different k-space ranges accessible to the two experiments and to the significant spreads in interatomic distances. With the combination of experiments, chemical short range order is demonstrated in this material.

1. Introduction

The development of EXAFS analysis has brought us the capability of determining considerable information about the atomic coordinations of individual atomic species in amorphous or crystalline polyatomic condensed matter. These species may be present in from high to very low concentrations and in the bulk or on the surface of a material. The analysis suffers significantly, however, from the absence of data at low k, which leads to a loss of information about more distant than near neighbors in amorphous materials [1] and about near-neighbor coordinations in anharmonic or disordered systems [2].

In this paper, x-ray anomalous scattering techniques are described which also yield the atomic coordinations of individual atomic species. Relative to EXAFS, they provide more information about more distant neighbors as well as anharmonic

---

1 Department of Materials Science and Engineering.

2 Stanford Synchrotron Radiation Laboratory, SLAC, Bin 69, P.O. Box 4349.

or disordered systems, but can only be used when the species to be studied is present at reasonably high concentrations. This limitation arises because in this case the changes in diffracted intensity with photon energy are too small to measure reliably. Reliable estimates of this lower concentration limit have not yet been determined experimentally. In addition, the anomalous scattering techniques are limited to the study of species with atomic number greater than about 25 or 30. For lighter elements the x-ray wavelength is too long to allow for measurements over a sufficiently large range in k-space. The possibility of combining the anomalous scattering techniques with surface diffraction does exist, but no implementation of that approach has been published. To illustrate the methods and their relationship to EXAFS, we will present some results from an extended study of amorphous Mo-Ge alloys. We begin, however, with a brief introduction to anomalous scattering.

Far from an absorption edge, the atomic scattering factor is well represented by its classical value, f_o, which is the Fourier transform of the electron density. When the photon energy is close to that of an absorption edge the scattering factor, f, has real and imaginary contributions added to it so that

$$f = f_o + f' + if'' \qquad (1)$$

where f' and f" are the anomalous scattering factors. f' is small and negative at photon energies well below the absorption edge. As the edge is approached from below it becomes increasingly negative and peaks at the edge. With further increases in photon energy, it becomes less negative and eventually becomes small and positive well above the edge. f" is simply related to the x-ray absorption coefficient μ by the equation

$$f'' = (m\omega c/4\pi Ne^2)\mu \qquad (2)$$

with N the number density for a particular species. Hence, it is small and positive for energies below the edge, rises sharply at the edge and then decreases with increasing photon energy, always remaining positive.

With the advent of synchrotron radiation, which allows scattering experiments to be performed at photon energies close to the edges, this phenomenon has been explored and utilized extensively [3]. f" is measured directly through measurements of the x-ray absorption coefficient. f' can be obtained by the following techniques, as reviewed in [3]: (a) total reflection of x-rays, (b) Kramers-Kronig dispersion relation analysis of f", (c) x-ray interferometry, (d) fitting if Bragg reflection of intensities from a known crystal structure. Nevertheless, the absence of systematic calculations and/or measurements of f' presently impose significant limitations to the general applicability of anomalous scattering structure determination techniques.

The relationship of anomalous scattering to EXAFS techniques may best be understood through analysis of the Debye equation for the intensity of scattering per atom, I(k), from an isotropic array of atoms:

$$I(k) = \sum_A \sum_B \chi_A f_A(k) f_B^*(k) [S_{AB}(k) + \delta_{AB}] \quad . \qquad (3)$$

Here, k is the scattering wave number $(4\pi\sin\theta)/\lambda$, 2θ is the scattering angle, λ the wavelength, and χ_A the atomic fraction of species A. The summations are over atomic species. The partial structure factors, S_{AB}, contain all the structural information provided by diffraction and can be written as

$$S_{AB}(k) = (4\pi/k)\int_0^\infty [\rho_{AB}(r) - \bar{\rho}_B] r\sin(kr)dr \qquad (4)$$

where $4\pi r^2 \rho_{AB}(r)dr$ is the number of B atoms at a distance between r and r + dr from an A atom and $\bar{\rho}_B$ is the average number density of B atoms.

X-ray diffraction radial distribution function studies of polyatomic amorphous materials frequently suffer from ambiguities because they yield a weighted sum of the ρ_{AB}, so that the coordination of each species cannot be separately determined [4]. A virtue of EXAFS is that it provides a weighted sum over the B atoms of the $\rho_{A'B}$ for a specific species, A', whose edge structure is being studied. As a result, some of the ambiguity is often removed.

Several x-ray diffraction methods for decreasing the ambiguity have been discussed in the literature. SHEVCHIK [5] noted that a pair distribution function similar to that obtained from EXAFS could be obtained. This approach, differential anomalous scattering (DAS), is explained here in the modified form of FUOSS et al. [6] who first implemented it. Near the absorption edge of species A', f_A, is varying rapidly with photon energy whereas all other atomic scattering factors are slowly varying. Consequently, one can write

$$\delta I(k)/\delta E = \chi_{A'} \sum_B (\delta/\delta E)[f_{A'} f_B^* + f_{A'}^* f_B] S_{A'B} \quad . \qquad (5)$$

Fourier transformation of this derivative yields a weighted sum of the $\rho_{A'B}$ and, therefore, information about the environment of the specific atomic species, A'. FUOSS et al. [6] obtained considerable information about atomic arrangements in amorphous Ge-Se alloys with this approach. Further utilization of it for the Mo-Ge alloys is described here.

An even more complete removal of the ambiguity for binary alloys by both x-ray and neutron diffraction was proposed by KEATING [7]. X-ray implementation of this approach is discussed extensively by WASEDA [8]. In the case of binaries, there are three independent S_{AB}. Measurement of I(k) at three different photon energies at which there are three independent sets of values of the f_A and f_B yields 3 linear equations for the 3 unknown S's. As discussed previously [3] and illustrated by FUOSS et al. [9], these equations are ill-conditioned, so that it is extremely difficult to obtain reliable results. A variation of these approaches, which appears to be less susceptible to error, has recently been suggested by MUNRO [10] and is now being implemented by us and by FUOSS [11].

The Mo-Ge system is particularly interesting in this context because amorphous alloys can be formed over an extended composition range in which the materials go from semiconducting to metallic to superconducting, and because no structural model has been proposed for the system. Because the crystalline compounds have unusual spreads of near neighbor distances, the capabilities of EXAFS analysis are taxed considerably. As a result, EXAFS and DAS yield different and complementary information. But without the DAS, one might readily be brought to incorrect structural conclusions from the EXAFS.

2. Experimental

A 7 micron thick $Mo_{42}Ge_{58}$ sample was prepared by magnetron co-sputtering onto a kapton substrate floating in temperature. Its composition was determined by electron microprobe analysis. The same material was used for both experiments. Optimal thicknesses for each experiment were obtained by stacking several pinhole-free films.

Because of our inability to obtain "infinitely thick" samples, we chose to perform the x-ray diffraction experiment in a symmetric transmission geometry, which is seldom used in x-ray radial distribution studies. However, with the increasing interest in samples which cannot be made thick, we anticipate that it will become increasingly important. The x-ray scattering experiments were performed at room temperature at the Stanford Synchrotron Radiation Laboratory utilizing a Si (220) incident beam monochromator yeilding an energy resolution of about 2 eV at the Ge edge and 7 eV at the Mo edge. A diffracted beam monochromator was utilized at the Mo edge. It had a band pass of about 150 eV, sufficient to eliminate virtually all of the fluorescent and Compton scattered radiation. Data were collected at several energies below each K edge. Details of data analysis will be described elsewhere [12].

Transmission EXAFS measurements were performed at liquid N_2 temperature using standard SSRL techniques and equipment.

3. Results

3.1 EXAFS

The Mo and Ge K-edge EXAFS, multiplied by k^3, are shown in Fig. 1. Our initial analysis of these data was with a single shell, nonlinear least squares fit of the data to the equation

$$k^3\chi(k) = k^2 \sum_i \frac{N_i}{R_i^2} \exp(-\sigma_i^2 k^2)|f_i(k,\pi)| \sin(2kR_i+\phi_i(k)) \qquad (6)$$

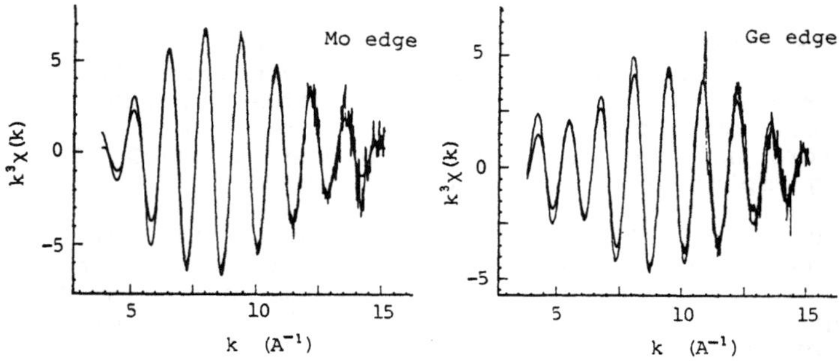

Fig. 1. Mo and Ge edge EXAFS from a-$Mo_{42}Ge_{58}$ multiplied by k^3. Calculated best fits are superimposed. The glitch in the Ge edge EXAFS at 10.7 A^{-1} is not expected to affect results

in which the Debye-Waller factors, the interatomic distances and effective coordination numbers were allowed to vary. The back-scattering amplitudes and phase shifts of TEO and LEE [13] were employed for this purpose. Within the single shell restriction, the only physically meaningful results were obtained under the assumption that Mo is coordinated by Ge and Ge by Mo at a distance of 2.57 $\pm$ 0.02 A, with mean square deviations of 0.015 A^2 and 0.013 A^2 for the Mo and Ge edges, respectively. The calculated EXAFS for the best fits are superimposed on the observed data in Fig. 1.

Two shell fits were also investigated. The EXAFS from neither edge could be fitted with appreciable contributions from a second species. Thus, the EXAFS results appear to imply that there is strong short-range order with each species being surrounded by the other at a distance of 2.57 A.

The Mo and Ge edge radial structure functions (RSF's) are shown in Fig. 2. As is usually the case for amorphous materials, only the first-neighbor peak is definitely observed [1,14] This is because the breadth of a given shell increases, on average, the farther away the shell. This increased breadth strongly damps the k-space information from these more distant

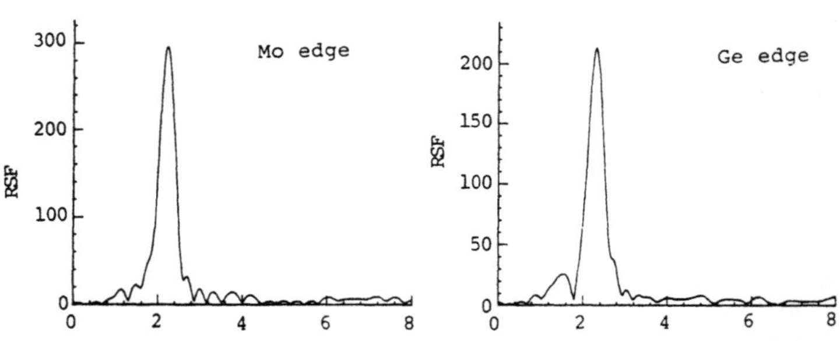

Fig. 2. Mo and Ge edge radial structure functions (RSF's) for a-$Mo_{42}Ge_{58}$ obtained by Fourier transform of the data in Fig. 1

neighbors in the relatively high k range accesible to EXAFS. The RSF's themselves appear to be of high quality, with a minimum of spurious structure at r values too low to be interatomic distances. Note that the first-neighbor peak full widths at half maximum (FWHM's) are about 0.45 A.

3.2 Radial Distribution Function (RDF) Analysis

The RDF calculated from data obtained far from either edge (E = 19.5 keV) is shown in Fig. 3. Note its extremely high apparent quality, as evidenced by the absence of ripples below 2 A, in spite of the thinness of the sample. This shows the effectiveness of the transmission technique and, as well, the virtues of having both an incident and diffracted beam monochromator. The peak maximum occurs at 2.7 $\pm$ 0.05 A, which is about 0.1 A larger than that obtained from the EXAFS. Similar discrepancies between EXAFS and scattering experiments have been noted by other workers [15,16] in their EXAFS studies of metallic glasses. In addition, the peak FWHM is well over 1 A, which is considerably larger than the EXAFS RSF widths. Detailed analysis indicates that this extra RDF width is not due to the k = 18.5 A^{-1} termination of the scattering data. Another possible reason for this extra width is that the RDF is a weighted sum of three, rather than two, partial distribution functions. Hence, the DDF's are of considerable importance to determine, among other things, if this is the case.

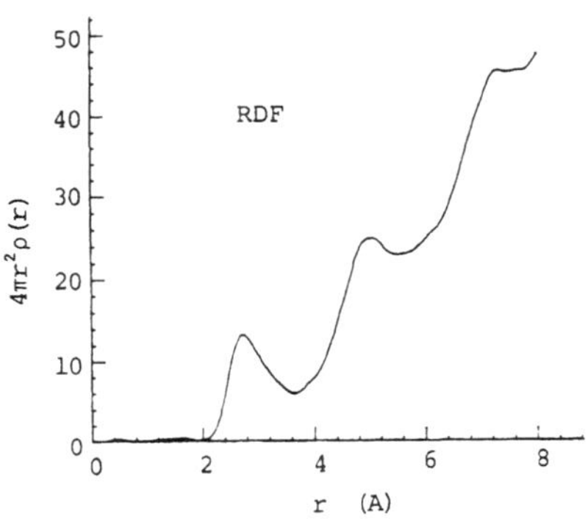

Fig. 3. The radial distribution function (RDF) for a-$Mo_{42}Ge_{58}$ calculated from data taken at 19500 eV

3.3 Differential Distribution Functions (DDF's)

Figure 4 shows the DDF's obtained from the Mo and Ge edge diffraction data, respectively. Again, the low r regions imply that the results are of high quality. The first-neighbor peak of the Ge DDF is well defined, with a maximum at 2.7 A, while the first peak in the Mo DDF has a broad, less well-defined maximum extending to even higher r. These are considerably larger than the corresponding EXAFS peak positions of 2.57 A. No combination of absorber and backscattering phase shifts can make the EXAFS yield the same distances as the DAS.

Note, as well, that the width of the Mo DDF first peak remains over 1 A and that the first peaks in both DDF's are quite

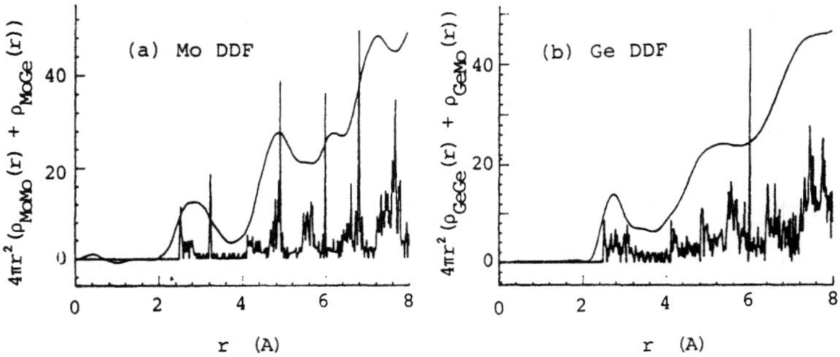

Fig. 4. The differential distribution functions (DDF's) for a-$Mo_{42}Ge_{58}$ obtained from the Mo and Ge edge DAS. The Mo edge DDF is calculated from data taken at 19500 eV and 19985 eV, while the Ge edge DDF is calculated from data at 11003 eV and 11088 eV. Also shown are the corresponding sums of pair distribution functions in the intermetallic compound $Mo_{13}Ge_{23}$ unweighted by the atomic scattering factors

asymmetric, with considerably less fall-off on the large r side. Finally, note the large amount of structure beyond the first-neighbor peak, with strong indications of differences between the average Mo and Ge first- and farther-neighbor coordinations. Such information about the more distant shells is not, in general, obtainable from EXAFS.

4. Comparison of EXAFS and Differential Anomalous Scattering

It is apparent that the EXAFS results and the DDF'S disagree. The DDF's yield longer first-neighbor distances and broader first-neighbor peaks. This leads naturally to the question of what causes this discrepancy.

Since the EXAFS and scattering experiments were performed at liquid N_2 and room temperatures, respectively, it is conceivable that temperature effects are the source of the different widths yielded by the two techniques. One can obtain an approximate upper bound to the thermal mean square displacement at room temperature for the EXAFS case by assuming that: (i) the broadening of the RSF peak, beyond that due to the window function (discussed below), is due entirely to thermal broadening, (ii) the mean square displacement is linear in temperature, (iii) there is no zero point broadening. With these assumptions, the non-window broadening of the 77 K RSF peak yields a thermal mean square displacement of 0.015 A^2. At 300 K, this would be 0.058 A^2, yielding a FWHM of 0.56 A. This constitutes a 25% broadening of the EXAFS RSF peak, which is far from explaining the difference between the widths of the RSF and DDF peaks.

It is instructuve to examine the causes of the widths of the first-neighbor peaks provided by the two approaches. In the

EXAFS analysis, we did not utilize data below $2k = 7\ A^{-1}$ and our data extended to $2k = 30\ A^{-1}$. To avoid large termination ripples we used a Gaussian broadended square window with ends at these values and a Gaussian width of 1 A. Analysis shows that about two-thirds of the width of the RSF's in Fig.2 is due to this finite range of k space used. The other part comes from the actual RMS width of the atomic distribution measured by EXAFS as determined by the fit to the data.

Similarily, the x-ray scattering data are obtained over a range of $0.05\ A^{-1} \leq k \leq 18.5\ A^{-1}$ near the Mo edge and $0.5\ A^{-1} \leq k \leq 10.5\ A^{-1}$ near the Ge edge. In addition, the data are multiplied by a "window" function of the form $\exp(-ak^2)$ with $a = 0.0125\ A^2$ for the Ge edge DDF and the RDF, while $a = 0.0175\ A^2$ for the Mo edge DDF. Such windows in k space would convolute the actual r-space distribution with a Gaussian with a FWHM of about 0.3-0.4 A. Thus, the broadenings introduced by the window functions in both experiments are approximately the same. This might seem counterintuitive because the EXAFS data go out $2k = 30\ A^{-1}$, whereas the scattering data go out only to about $k \approx 15\ A^{-1}$. It can be readily shown, however, that the broadening of a delta function is determined by the total k-space range available in the measurement, independent of where that range is centered. Thus, the effective ranges of the two experiments are not that different. These broadenings of the real space structure due to the finite range of measurement in k space are the major sources of experimental broadening effects for both experiments.

We will now show that the unusually broad DDF peaks may be expected on physical grounds. Plotted in Fig. 4 with the DDF's are the unweighted sums of the corresponding partial distribution functions (e.g., $\rho_{MoMo} + \rho_{MoGe}$ in fig. 4(a)) of crystalline $Mo_{13}Ge_{23}$, the compound of composition closest to that of the amorphous alloy. It is apparent that there are not well-defined first-neighbor shells in this compound. This intermetallic compound has 144 atoms per unit cell and a number of inequivalent Mo and Ge sites. Each site has a distinct coordination with different first-neighbor distances. As a result, the partial distribution functions have the complexity shown. Given the complexity of the crystalline phase, there is no reason to believe that the broad peaks in the observed DDF's for the amorphous material are spurious experimental results. It is not unreasonable to suspect that the spread in coordinations is characteristic of the Mo-Ge system, since it also shows up in crystalline Mo_5Ge_3 and, to a lesser extent, the other intermetallics. In particular such complexity may well hinder nucleation of the crystalline phases and contribute to the stability of the amorphous materials.

These considerations lead us to believe that the differences in widths of the first-neighbor distributions in the two experiments are real consequences of the different techniques and not simply artifacts from the analysis of the data or temper-

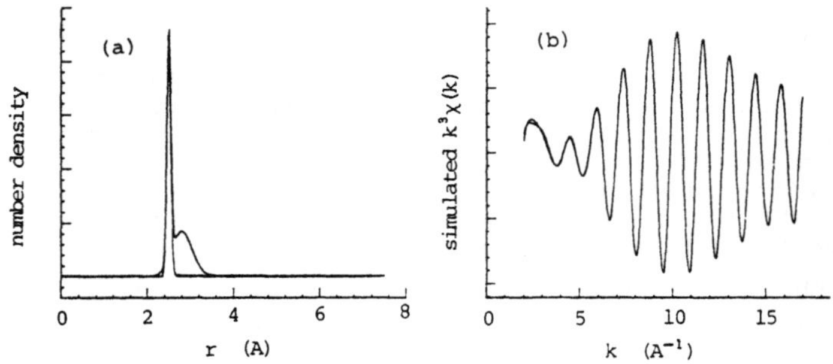

Fig. 5. (a) shows two hypothetical distributions. One consists of a single Gaussian at 2.5 A with FWHM = 0.055 A. The other is a sum of this one plus another Gaussian at 2.8 A with FWHM = 0.28 A. (b) shows the EXAFS spectra calculated from the two distributions using TEO and LEE [13] phases and amplitudes for Ge

ature effects. Indeed, considering the different distances obtained by the two techniques, it appears that the EXAFS experiment is sensitive to only the leading (low r) edge of the first shell of neighbors, a possibility noted by EISENBERGER and BROWN [2], while the DAS experiment shows neighbors over a broader range within the first shell as well as neighbors in farther shells. Ultimately, the problem with EXAFS when used to study disordered systems, as well as the discrepancies between EXAFS and scattering presented here, result from the loss of low k data and from the fact that in k space the interference phenomena resulting from disordered distributions are necessarily concentrated in the low k region.

A simple demonstration of this is shown in Fig. 5. In Fig. 5(a) are shown two hypothetical Gaussian distributions of atoms. One is a single, relatively "sharp" Gaussian with FWHM = 0.055 A, centered at 2.5 A. The second is the sum of two Gaussians, one the same as above while the other is centered at 2.8 A with a FWHM = 0.28 A to represent the situation in the material studied here. Each Gaussian is normalized to have unit area. Fig. 5(b) shows the EXAFS spectra calculated from the two hypothetical distributions using TEO and LEE [13] phase shifts and back scattering amplitudes for Ge. The two spectra are identical and, obviously, are the result of the sharper of the two Gaussians in real space. The broader Gaussian has a very narrow Fourier transform and, hence, contributes primarily at lower k than is accessible in EXAFS. This result follows simply from the breadth of the distribution and does not require assymetric distributions.

Another effective demonstration of the idea that the disordered distribution of atoms shows up preferentially at low k is simply the RDF and DDF's in Figs. 3 and 4. Here, the effective range of the data is confined to low k, yet we see a wealth of information, including significant differences, in the broad, disordered first and higher shells. This is consistent with the classical description of the scattering profile from amorphous materials as showing strong interference peaks, usually one or two, at $k < 6\ A^{-1}$ with the high k data showing essentially a single sinusoid corresponding to the near-neighbor distance. This characteristic, with only one strong interference peak at low k ($k \simeq 2.8\ A^{-1}$), is observed in both the total and differential structure factors for the $Mo_{42}Ge_{58}$ alloy studied here. It is interesting to note that the frequency of the almost pure sinusoid in the high k region of the scattering data is the same as the effective frequency of the EXAFS oscillations if the linear part of the phase is included.

5. Conclusion

In summary, the major difference between the EXAFS and the anomalous scattering techniques, aside from the different physical mechanisms involved in each, is that different regions of k space are available to the two experiments. Scattering experiments are sensitive to low k while the EXAFS experiments are sensitive to a higher effective k range, with little overlap between the two ranges. When studying disordered materials, the implications of this are enormous, with EXAFS being sensitive mainly to sharp, well-defined distances. On the other hand, the DDF's cannot show sharp detail because of the lack of r-space resolution associated with the absence of high k data.

In the case of the highly disordered amorphous $Mo_{42}Ge_{58}$ discussed here, the EXAFS "sees" only the closest atoms in a broad first shell. Even so, the EXAFS data could be analyzed to provide the important chemical identifications of the very nearest neighbors of each species. Indeed, a major virtue of the EXAFS technique in this case is its ability to distinguish unambiguously between the two possible back-scattering species, providing strong evidence for chemical ordering at these short distances. To a degree, this is possible because the data do not contain contributions from the more distant pairs which would make interpretation more difficult.

Anomalous scattering techniques, on the other hand, provide a much more complete picture of the first and farther shells in disordered materials. Had only the EXAFS data been available, an extremely distorted picture of the average near-neighbor environment would have been obtained. The DDF's are strong evidence for chemical short-range order in these materials, with Mo and Ge having very different average environments out to at least the third shell. They do not, however, provide any direct means for the chemical identification of the coordinating species. Our efforts to obtain the individual pair distribution functions, which would provide this information, are continuing.

While the scattering techniques, in some senses, provide much more information than does EXAFS, each technique is limited by a finite k range to yield a somewhat distorted distribution in r space. Thus, the two approaches are complementary at the present time. It seems prudent to employ both in studying amorphous systems when possible. However, it must be remembered that anomalous scattering techniques are limited to the study of species in reasonably high concentrations and with atomic numbers greater than about 25 or 30.

6. Acknowledgements

This work was supported in part by the National Science Foundation through the Metastable Materials Thrust Program of the Center for Materials Research and by the Stanford Synchrotron Radiation Laboratory. The latter is supported by the NSF through the Division of Materials Research and the NIH through the Biotechnology Resources Program in the Division of Research Resources (in cooperation with the Department of Energy).

References

1. S.H. Hunter, A. Bienenstock, T.M. Hayes: in The Structure of Non-Crystalline Materials ed. by P.H. Gaskell (Taylor & Francis LTD, London 1977) pp. 73-76
2. P. Eisenberger, G.S. Brown: Sol. St. Com. 29, 481-484 (1979)
3. P. Fuoss, A. Bienenstock: in Inner-Shell and X-ray Physics of Atoms and Solids, ed. by D.J. Fabian, H. Kleinpopper, L.M. Watson (Plunum Pub. Corp. 1981) pp. 875-884
4. A. Bienenstock: in The Structure of Non-Crystalline Materials ed. by P.H. Gaskell (Taylor & Frances LTD, London 1977) pp. 5-11
5. N.J. Shevchik: Phil. Mag. 35, 805-809 (1979); 35, 1289-1298 (1977)
6. P. Fuoss, P. Eisenberger, W. Warburton, A. Bienenstock: Phys. Rev. Let. 46, 1537-1540 (1981)
7. D.T. Keating: J. Appl. Phys. 34, 923-925 (1963)
8. Y. Waseda: in The Structure of Non-Crystalline Materials, Liquids and Amorphous Solids, (McGraw-Hill Inc., New York 1980)
9. P.H. Fuoss, W.K. Warburton, A. Bienenstock: J. Non-Cryst. Sol., 35 & 36 1233-1238 (1980)
10. R.G. Munro: Phys. Rev. B, 25 5037-5045 (1982)
11. P. Fuoss, private communication
12. J. Kortright, Ph.D Thesis, Stanford University, in preparaton
13. B.-K. Teo, P.A. Lee: J. Am. Chem. Soc. 101, 2815-2832 (1979)
14. M.-L. Theye, A. Gheorghin, H. Launois: J. Phys. C. 13, 6569-6584 (1980)
15. M. De Crescenzi, A. Balzarotti, F. Comin, L. Incoccia, S. Mobillia, G. Motta: Sol. St. Com. 37 921-923 (1981)
16. G.S. Cargill: in Proc. 4th Intl. Conf. on Rapidly Quenched Metals, (Sendai, Japan 1981) pp. 389-392

Simultaneous Measurements of the Real and Imaginary Parts of the X-Ray Anomalous Dispersion Using X-Ray Interferometers

D.P. Siddons and M. Hart

Wheatstone Physics Laboratory, King's College,
London WC2R 2LS, United Kingdom

Introduction

The x-ray interferometer provides a very direct way of measuring x-ray forward scattering factors via the x-ray refractive index [1-3]. The refractive index n is complex, and if we write $n = n_R + in_I$ then

$$n_R = 1 - \frac{e^2\lambda^2}{2\pi mc^2 V}(Z + f')$$

$$n_I = \frac{\lambda\mu}{4\pi}$$

The intensity produced by interference between two beams having amplitudes A_1 and A_2 and a phase difference ϕ is

$$I_0 = A_1^2 + A_2^2 + 2A_1A_2 \cos\phi$$

where A_1 and A_2 are fixed, and ϕ may be varied at will in our realization of the scanning interferometer. If a sample is placed in one of the beams, say that corresponding to A_2, then the intensity becomes

$$I_s = A_1^2 + A_2^2 \exp(-\mu t) + 2A_1A_2 \exp(-\tfrac{1}{2}\mu t) \cos\left(\frac{2\pi t}{\lambda}|1-n_R| + \phi\right)$$

when μ is the conventional linear absorption coefficient. Thus, interference fringes recorded with the interferometer empty and with a sample in one beam can be easily analysed to provide estimates of n_R and n_I, and hence μ and f'.

Results

Figures 1, 2 and 3 show the results of such an analysis in the pre-edge and near-edge regions of three elements, nickel, copper and selenium. Selenium is interesting because it is one of the few elements displaying a strong absorption 'white line' at its K edge in the pure form. Copper and nickel show less pronounced structure, but structure which is more readily visible in f' (compare the split minimum in nickel f' with the barely distinguishable features on the absorption spectrum) than in f''.

Conclusion

For the first time, both real and imaginary parts of the forward scattering factor are made available from a single measurement, using the same sample volume. A straightforward least-squares fitting procedure, using few parameters, is used to analyse the interference pattern. The data for each of

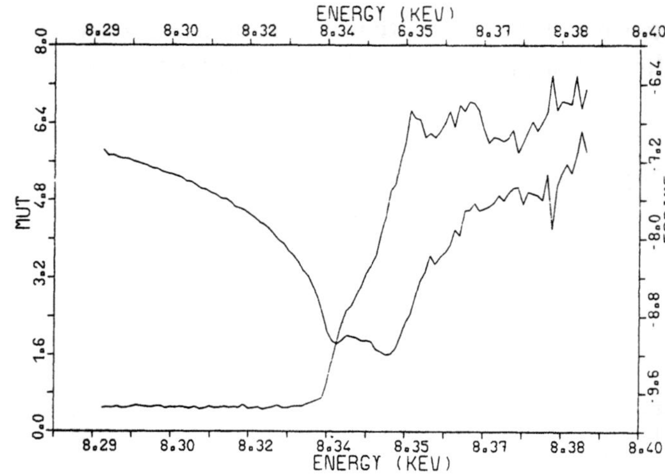

Fig. 1
Caption see opposite page

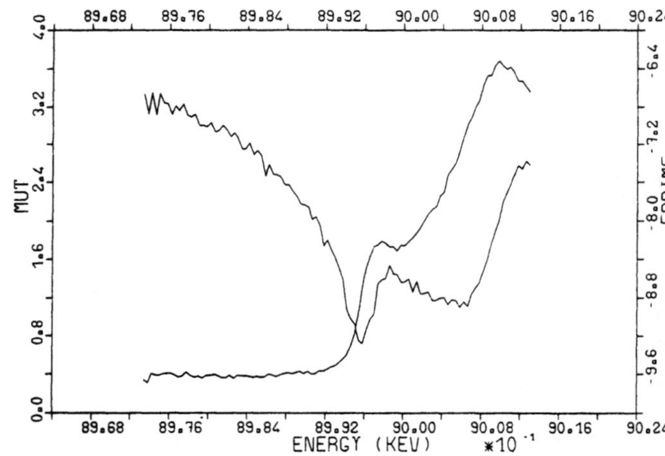

Fig. 2
Caption see opposite page

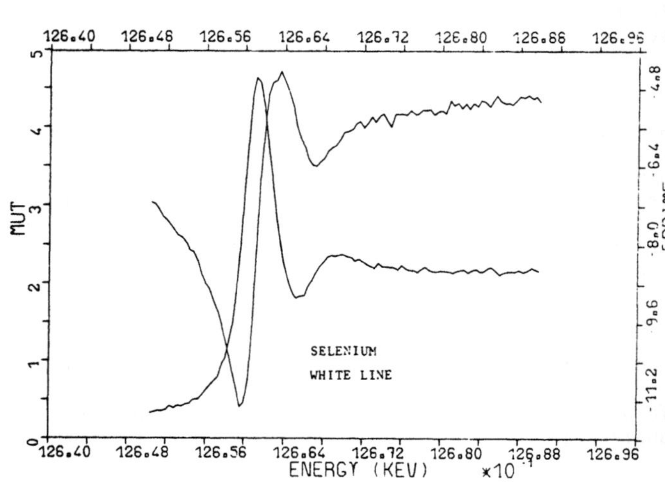

Fig. 3
Caption see opposite page

Figures 1 to 3 was collected in 2 hours using the Daresbury Synchrotron X-ray source running at 2 GeV and with 200 mA circulating current.

Acknowledgement

We are grateful to the Science and Engineering Research Council and to the Director and staff of the Daresbury Laboratory for the provision of Synchrotron Radiation facilities.

References

C. Cusatis and M. Hart, Proc. Roy. Soc. A354, 291 (1977)
D. P. Siddons, Ph D Thesis, London (1979)
M. Hart and D. P. Siddons, Proc. Roy. Soc. A376, 465 (1981)

Fig. 1. Graph of f' and μt near the K edge of nickel

Fig. 2. Graph of f' and μt near the K edge of copper

Fig. 3. Graph of f' and μt near the K edge of selenium, showing the strong absorption peak or 'white line'

Interferometric Measurement of Absorption $\mu(E)$ and Dispersion $f'(E)$ at the K Edge of Copper

U. Bonse, I. Hartmann-Lotsch, and H. Lotsch

Institut für Physik, Universität Dortmund, Postfach 50 05 00, D-4600 Dortmund 50, Fed. Rep. of Germany

The variation with energy E of the X-ray absorption coefficient $\mu(E)$ and that of the anomalous dispersion correction $f'(E)$ at the Cu K edge have been measured under identical experimental conditions using the X-ray interferometer installed in the Hamburger Synchrotronstrahlungslabor (HASYLAB) at the storage ring DORIS at DESY.

The instrument consists of a threefold reflecting groove crystal as a monochromator and a silicon LLL interferometer mounted on a common diffractometer. The energy is scanned by rotating the whole diffractometer with respect to the beam incident from the storage ring [1,2]. In our arrangement the instrumental resolution ΔE is determined by the divergence of the incident beam, which in turn is given by the source size plus height of the entrance slit divided by the source distance. A resolution of about 2 eV has been achieved at E= 9 keV.

Details of the process of measurement were described in [1,3]. f' is given by (1).

$$f'(E) = 2\pi\, p(E) \,/\, (N \lambda r_e t) - Z \quad ; \tag{1}$$

t is the sample's thickness, λ is the wavelength, N is the number of atoms per unit volume, r_e the classical radius of the electron, and Z the atomic number. p is the phase shift of two oscillatory curves gained by rotating a phase shifter plate within the interferometer, one measured with the sample and one measured without the sample in one of the partial beams.

Up to now measurements at the K edges of selenium and copper were carried out. In the present case of measurements at the K edge of copper (E = 8.980 keV) we chose Bragg angles at the silicon interferometer near $\Theta_B = 21^\circ$ giving photon energies of about 9 keV with the 220 reflection and 18 keV with the 440 reflection , respectively . The measurement of f' for E = 18 keV is used to circumvent [4] the necessity of determining directly the actual sample thickness t at the beam position with sufficient accuracy, which in the case of thin foils, may be difficult. The samples were made up of copper foils 8 µm thick, 99.9 % purity, supplied by Goodfellow metals. Two foils behind each other were employed up to E =8.995 keV. At higher energies a single foil had to be used because of rapidly increasing absorption.

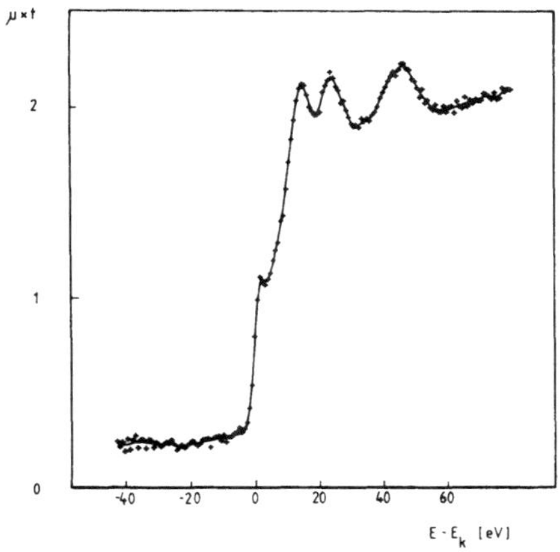

Fig. 1. Absorption μ·t at the copper K edge

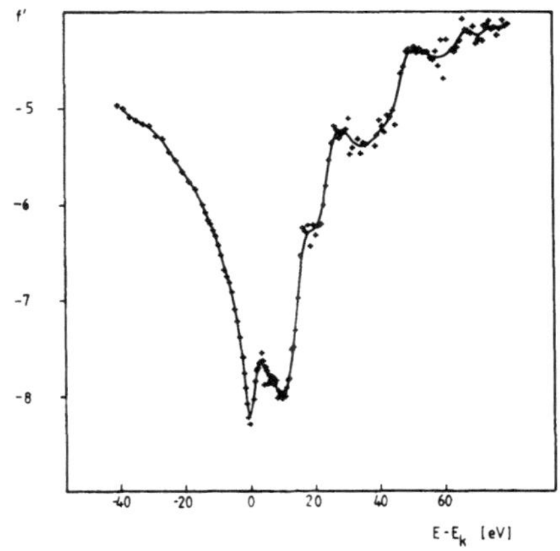

Fig. 2. Anomalous dispersion correction f′ of copper at the copper K edge

The absorption curve (Fig.1) displays the characteristic fine structure at the copper K edge, showing a 2 eV wide peak at the flank of the curve.

The dispersion curve (Fig.2) shows a corresponding fine structure. The two distict minima in f′ correspond to inflecting points in the absorption curve. On the whole, f′(E) seems to behave like the derivative of f"(E) [5].

We thank the Bundesminister für Forschung und Technologie, FKz 05260 Bo P/I and 05114 Bo P/I for financial support.

References

1. U.Bonse, P.Spieker, J.-T.Hein, G.Materlik: Nucl.Inst.Meth. 172,223-226(1980)
2. U.Bonse, G.Materlik: Z.Physik B24,189-191(1976)
3. U.bonse, I.Hartmann-Lotsch, H.Lotsch, K.Olthoff-Münter: Z.Physik B47,in print (1982)
4. C.Cusatis, M.Hart: in Anomalous Scattering, ed.by S.Ramaseshan, S.C. Abrahams (Munksgaard,Copenhagen 1975)pp.57-68
5. T.Kawamura, T.Fukumachi: Jap.J.appl.Phys. 17,Suppl. 17-2, 224-226(1978)

Anomalous Scattering Factors from X-Ray Absorption Data by Kramers-Kronig Analysis

P. Dreier, P. Rabe, W. Malzfeldt, and W. Niemann
Institut für Experimentalphysik, Universität Kiel,
D-2300 Kiel, Fed. Rep. of Germany

The anomalous X-ray diffraction (AXD) provides information about the local structure in amorphous materials comparable to that of EXAFS. The essential parameter for the analysis of AXD is the spectral shape of the scattering factor f close to the absorption edges. In recent years several experiments as specular X-ray reflection (1), X-ray interferometry (2), and absolute integrated intensity measurements (3) have been carried out on the atomic scattering factor f. These techniques provide accurate knowledge of the anomalous dispersion correction f', but they do suffer from high experimental and data analysis efforts. A method comparatively simple is the Kramers-Kronig transform (KKT) of absorption data. In this paper we present calculations of f' around the K edge of copper at 77K.

The total atomic scattering factor f in nonrelativistic theory is written as

$$f = f_o + f' + i f'' , \qquad (1)$$

where f_o is the atomic scattering factor for photon energies much higher than any absorption edge and f' and f" are the real and imaginary parts of the anomalous dispersion correction. f" is related to the photoelectric absorption coeffient at frequency ω by

$$f''(\omega) = \frac{m \cdot c \cdot \omega}{4 \pi \cdot N \cdot e^2} \cdot \mu(\omega) \qquad (2)$$

(m is the electron mass, e the electron charge, c the velocity of light in free space, and N the number of atoms per unit volume.) f' and f" are connected by a Kramers-Kronig relation

$$f'(\omega_o) = \frac{2}{\pi} \int_0^\infty \frac{\omega \cdot f''(\omega)}{\omega_o^2 - \omega^2} \, d\omega . \qquad (3)$$

Equ.(2) and (3) offer a way to calculate the anomalous dispersion corrections from measurements of the absorption coefficient.

The absorption of a 9 μm thick copper foil (Goodfellow metals, 99.97 % purity) has been measured at 77K in the photon-energy range from 8.8 keV to 10.4 keV using a high resolving (2eV) double-crystal monochromator at the storage ring DORIS of the Deutsches Elektronensynchrotron DESY. The resulting f" data are shown in the upper part of fig.1. At the bottom of the edge (8982 eV) a peak is resolved. Due to the low temperature

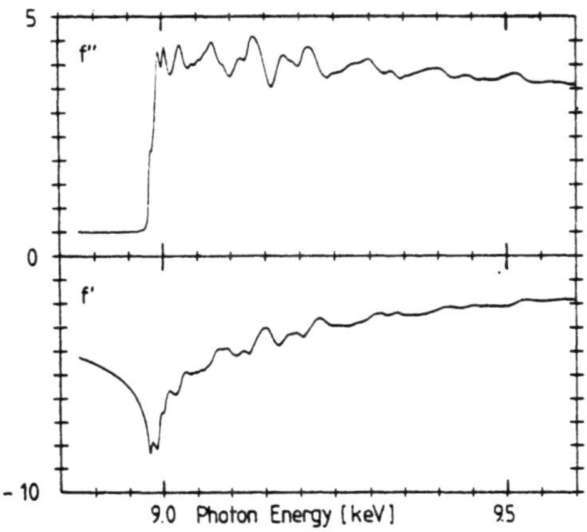

Fig. 1. Anomalous dispersion f' and f" of Cu at the K edge (T = 77 K)

of the specimen the amplitude of the fine structure is very large and small details do come out.

Absorption spectra are available only in a restricted frequency range, so the measured data have to be extrapolated in a physically reasonable way. Equ. (3) shows that the value of $f'(\omega_o)$ is mainly influenced by the data of f" near ω_o. The contribution of f" in regions far away from ω_o will therefore be nearly constant at ω_o. Structures in f' should not be influenced by the extrapolation of the absorption data. Nevertheless the absolute values of the calculated f' will depend strongly on the quality of the extrapolation.

It seems most practicable, simply to append published values of μ on each side of the measured spectrum. We have used the data from (4) in the energy range from 0.1 keV to 1000 keV. From a dataset of 1300 values for μ an ATM 7860 computer calculates a dataset of f' at the same energies in about 45 min.

The calculated spectrum of f' is shown in the lower part of fig. 1. Near the absorption edge in μ there is a minimum in f' at 8.980 keV with f'= -8.33 and another minimum at 8.990 keV with f' = -8.14. These two minima correspond to inflection points of μ. In the energy range below the edge f' is structureless, on the high-energy side oscillations corresponding to the EXAFS in μ show up. To test the quality of our calculations we have backtransformed the f' values by a relation similar to (3). The calculated f" values differ from the initial f" by a constant value of about 0.1. The shape of the two curves is identical.

The anomalous dispersion of copper has also been calculated by FUKAMACHI et al. (5). Their results are confined to an energy range close to the edge (-80 eV $< E - E_K <$ $+40$ eV). The shape of f' in this range is similar to our results. Their values for f" and f' however are 5% larger than ours in the whole range. This difference might be due to an uncertainty in the determination of the absorber film thickness.

In the same way as μ (and f") the spectral shape of f' is determined by the local distribution of atoms around the central atom. Structural investigations by means of AXD therefore necessitate the knowledge of spectral details of f'. A combination of X-ray absorption (to provide f' via KKT) and subsequent X-ray diffraction experiments with the identical sample at various wavelength close to the edge provide the full data set for the evaluation of partial pair distributions.

We would like to thank R.Haensel for his continous interest in this work.

References

1. G.Martens and P.Rabe. J.Phys.C.13,L913 (1980)

2. U.Bonse, G.Materlik, Z.Phys.B.24, 189 (1976)

3. A.Freund, in "Anomalous Scattering" (Ed. S.Ramaseshan and R.C.Abrahams, Munksgaard, Kopenhagen 1975), p.69

4. W. J. Veigele in "Handbook of Spektroskopy",vol.1,Ed. J.W. Robinson (CRC Press, Cleveland, Ohio 1974)

5. T.Fukamachi, S.Hosoya, T.Kawamura, S.Hunter, Y.Nakano Jap.J.appl.Phys. 17-2, 326, (1978)

Part IX

Related Techniques – Electron Energy Loss

Extended Energy Loss Fine Structure (EELFS) Technique: Bulk and Surface Investigation

M. De Crescenzi

Dipartimento di Fisica, Università della Calabria,
I-87036 Arcavata di Rende, Italy

1. Introduction

It is well known that inner shell electrons can be excited not only by the absorption of X-rays but also by the anelastic scattering of electrons. The similarities between these two different probes suggest that structural experiments performed with X-rays (EXAFS spectroscopy)(1) may be carried out with electrons (1).

Recent experiments have demonstrated that Extended X-ray-edge Energy Loss Fine Structure (EXELFS) (2,3) spectra from self-supported thin films can be obtained utilizing fast electrons of 50-200 KeV.

In this paper we shall present a variant use of electron energy loss spectroscopy to give information about local atomic environments.

The Extended Energy Loss Fine Structure (EELFS) technique (4,5) that we shall present is performed in the reflection mode, it may be carried out on massive samples and with low primary beam energy ($E_p \leq 3000$ eV). These improvements with respect to the EXELFS technique are very important in view of the well-known surface sensitivity of energy loss technique and its relatively low level of instrumental sophistication. It will be also pointed out that the counting rates achievable with electrons can be competitive with those from synchrotron radiation experiments expecially for core edges lying below 1 keV and for the lower atomic number elements.

2. Difference between EXAFS and Electron Energy Loss

We shall briefly analyse which physical quantities are involved in the matrix elements both in an electron scattering and in an X-ray absorption process. First of all we must recall that the energy of the EXAFS spectra is determined directly by the incident photon energy while in the EELFS case the energy loss is not varied by the incident electron energy but varies in intensity in a decreasing way with increasing energy loss. Basically, only the electrons for which the momentum transfer is zero carry exactly the fine structure EXAFS information of X-ray absorption.

The cross section per atom for excitation of an inner shell electron from an initial state $|\Psi_i\rangle$ to a final continuum state $|\Psi_f\rangle$ by the absorption of an X-ray of energy $\hbar\omega$ is proportional to:

$$\sigma(\hbar\omega)_{X\text{-ray}} \sim |\langle\Psi_i|\hat{e}\cdot\underline{r}|\Psi_f\rangle|^2 \cdot \delta(E_f - E_i - \hbar\omega) \quad . \tag{1}$$

Here $\hat{e}$ is the polarization unit vector of the electric field. The EXAFS oscillations in the X-ray cross section are due to the interference between the outgoing and the backscattered parts of the photoelectron $|\Psi_f\rangle$ final state. For an electron energy loss process one analyses the count rate of scattered electrons N(E), which have been analysed by the spectrometer and have lost energy E and momentum $\underline{q}$ through the core level excitation:

$$N(E) \sim \frac{d^2}{dE\,dq}\sigma_{el}(E,q) \sim \frac{1}{q^3}\,|\langle\Psi_i\,|e^{iq\cdot r}\,|\,\Psi_f\rangle|^2 \cdot \delta(E_f - E_i - E) \quad . \qquad (2)$$

The differential cross section will be of the same form of Eq.1 in the limit of small angle scattering which requires the momentum transfer $q \ll 1/r_c$, where r_c is the core orbital radius. Thus the electron probe operator in Eq.2 may be expanded in terms:

$$e^{iq\cdot r} \sim 1 + iq(\hat{\varepsilon}_q\cdot r) - q^2(\hat{\varepsilon}_q\cdot r)^2 \quad : \qquad (3)$$

$\hat{\varepsilon}_q$ is a unit vector in the direction of $\underline{q}$ and plays the role of the electric field polarization $\hat{e}$ in the case of X-ray absorption. In this limit the matrix element in Eq.2 will be $|\langle\Psi_i|\hat{\varepsilon}_q\cdot r|\Psi_f\rangle|^2$.
We expect therefore a complete correspondence between energy loss and absorption spectra. Another correspondence must be pointed out between the energy losses of transmitted electrons through a thin film and those that are reflected from the sample surface. Only difference is the mean free path covered by the the electron in the solid before any scattering event. The mean free path is a function of the incoming energy and for the case of the EXELFS technique high primary beam energies ($E_p \geq 50$ keV) and thin films are required. For the reflection this is no longer required and the primary beam can be utilized as a variable probe inside the surface ($\lambda_{el} \sim \sqrt{E_p}$).
In the following we shall present reflection energy loss data with low primary beam energy ($E_p \leq 2000$ eV) that corresponds to a $\lambda_{el} \sim 10$-15 Å. These energies were chosen for mantaining again a surface sensitivity, good instrumental resolution compatible with our electron analyser and for avoiding the superposition of Auger transition.

3. Experiment

EELFS measurements can provide accurate structural information by using a standard set-up like an ultra-high-vacuum chamber (10^{-10} torr range) equipped with an electron gun coaxial to a single pass cylindrical mirror analyser (CMA) with a resolution $\Delta E/E \geq 0.3\%$). Ion argon bombardment cycles for removing the major contaminants on the samples surface followed by heating, Auger and LEED techniques were employed to clean and to check the sample crystallinity. The samples were massive Al, Ni, Cu (4,5), Fe (6) and Pd (7) polycrystalline. Ni(100) single crystal was used for studying the interaction of oxygen(8) at different coverage rates. Graphite single crystal (9) was cleaved in the experimental chamber in order to study anisotropic effects on the carbon K edge due to the momentum transfer. A modulation of 5-10 volts was applied to the outer cylinder of the CMA to record energy loss spectra as the first and second derivative of the electron distribution N(E). The spectra were iden-

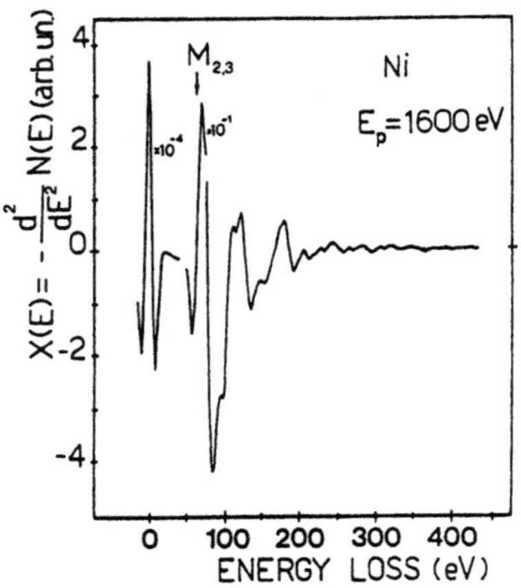

Fig.1. Reflection energy loss spectrum for polycrystalline Ni. The primary beam energy was E_p=1600 eV. The energy loss fine structures extend for 300 eV above the M_{23} core edge

tical for different beam energies (1500-2000 eV) and no broadening effects were observed by increasing the beam energy. The EELFS spectrum for polycrystalline Ni, shown in Fig.1 together with the elastic peak and M_{23} core levels structures, has a good signal to noise ratio up to about 300 eV above the edge.

4. EELFS data analysis

Extended fine structure in the reflection energy loss spectra may be analysed with the same procedure of the EXAFS spectra (1). Structural information around each excited atom can be obtained from the oscillatory part of $X(E)=-\frac{d^2}{dE^2}N(E)$ via the Fourier transform:

$$F(R)=\frac{1}{\sqrt{2\pi}}\int_{k_1}^{k_2} X(k)k^n W(k)\, e^{-2ikR}dk \quad ; \quad (4)$$

k is the photoelectron wave vector ($k(\mathring{A}^{-1})=\sqrt{0.263(E-E_0)eV}$, E_0 is the edge onset), k^n is an appropriate weight factor and W(k) is a window function (usually a Gaussian or an Henning function) to avoid truncation effects on X(K).

5. Bulk EELFS results

In Fig.2a we show the X(k)k EELFS signal for Ni above the M_{23} core level (E_0= 68 eV, $3p_{3/2}-3p_{1/2}$ =1.8 eV), and in Fig.2b the corresponding Fourier transform.

The position of the peaks of F(R), after the appropriate phase shift correction (14), is in good agreement with previous X-ray results carried out above the K edge (4) demonstrating that structural information is obtained also above shallow core levels.

It is interesting to note that due to the second derivative mode utilized in our data collection the Ni F(R) is multiplied by a factor $\sim R^2$.

Fig.3 shows a comparison between the F(R) obtained from the analyses of EELFS oscillations above the Cu and Fe M_{23} edges. The compactness of the Cu fcc lattice compared to that of a bcc appears clearly in the broadening of the outer shells that are still resolved in Fe.

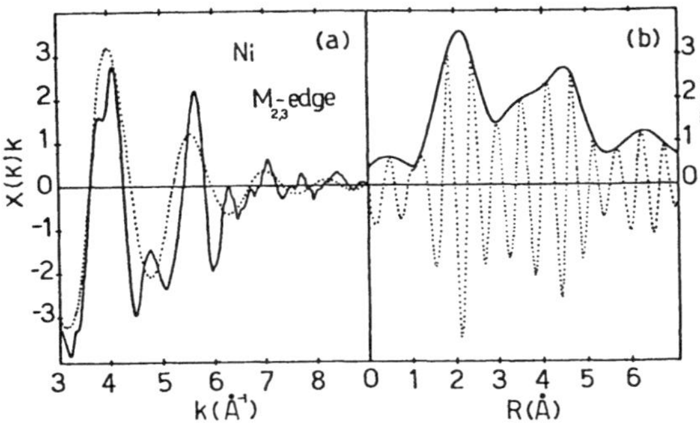

Fig.2.(a)-EELFS for Ni of Fig.1 in k space (E_0=68 eV),dotted line: contribution due to the first-neighbour's shell of F(R).(b)-Fourier integration of the experimental EELFS signal(solid line) and its imaginary part (dotted line)

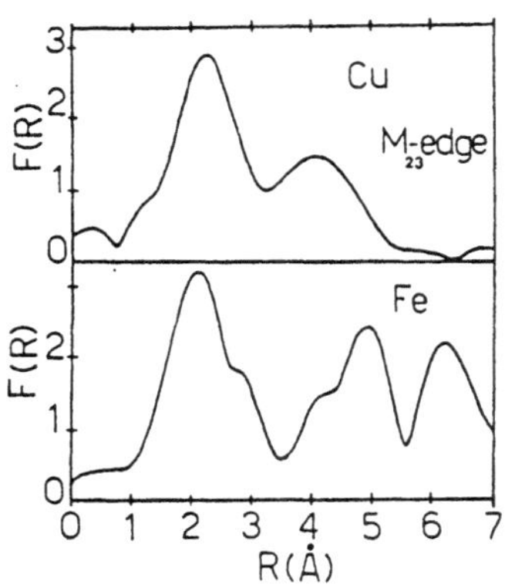

Fig.3. Fourier transform of the EELFS signals above the M_{23} core levels of Cu (upper curve) and Fe (down curve).The integration limits were 3.0 – 9.25 Å^{-1}

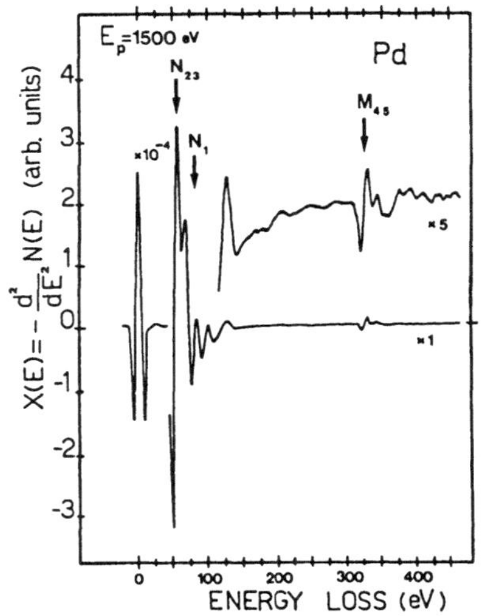

Fig.4. EELFS signal for polycrystalline Pd. The arrows indicate the N_{23}, N_1 and M_{45} edges

The use of low electron primary beam limits the number of core edges that may be excited with our technique.However,as in the case of Pd shown in Fig.4 (7), when different edges are sufficiently separated in energy, the detection and the analysis of the EELFS oscillations become still feasible.
By back-fourier inverting a peak of F(R) one can select the contribution of a single shell in the k space. The envelope function of this contribution gives $\widetilde{A}(k)$ that is related to the backscattering amplitude of each atomic site. The results of our analysis are shown in Fig.5 for different metals.
We pointed out that EELFS can be used not only to determine bond lengths from F(R) but also to identify atomic species around the absorption site.
The dependence of the EELFS structures upon temperature provides their sensitivity to the lattice arrangement. Fig.6 shows the temperature behaviour of

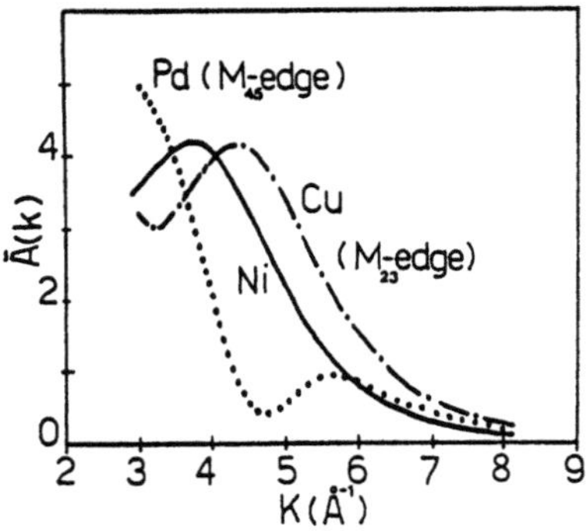

Fig.5. Experimental backscattering amplitude for Ni,Cu and Pd obtained by back-foureir inverting the first peak of F(R)

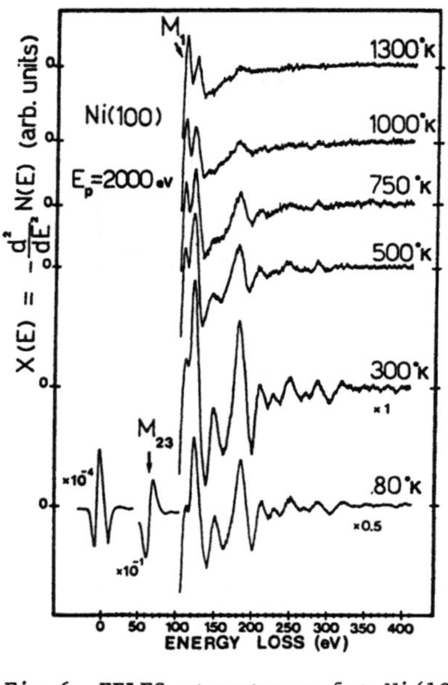

Fig.6. EELFS structures for Ni(100) at different temperatures and with the same experimental settings

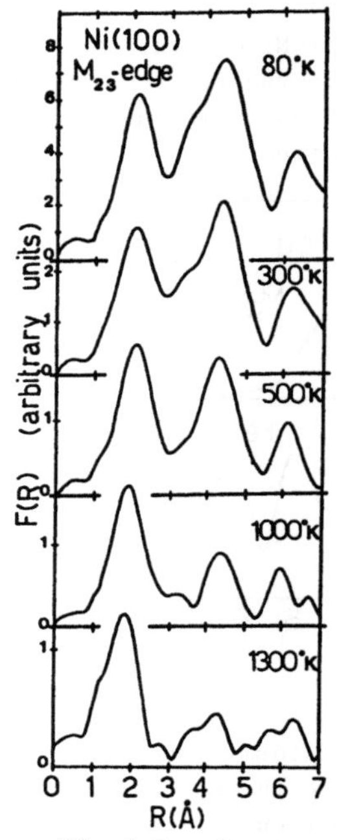

Fig.7. Fourier transforms of the EELFS signals of Fig.6,obtained with the same k-range integration (3.0 -10.0 $Å^{-1}$). Note the apparent bond length contraction of the first peak with the increase of the temperature

the EELFS M_{23} of Ni(100). The F(R) reported in Fig.7 show the progressive loss of the outer shells up to 1300 K,where pratically the first coordination shell is present. A careful analysis of the position of the first shell may give useful information about the large and anisotropic movements of the atomic layers at the surface (10).

6. Surface EELFS

The high degree of surface sensitivity of EELFS technique is shown in Fig.8. The energy loss spectrum of a clean polycrystalline Al sample is reported before and after an exposure of 100 L using identical experimental settings. The oxygen (532 eV) lies in this energy loss range. The aluminium 1s and 2p edges occur at 1560 and 73 eV respectively. Thus the spectrum of the O-Al system above the oxygen K edge is not affected by the interference of the EXAFS modulation of the different edges.

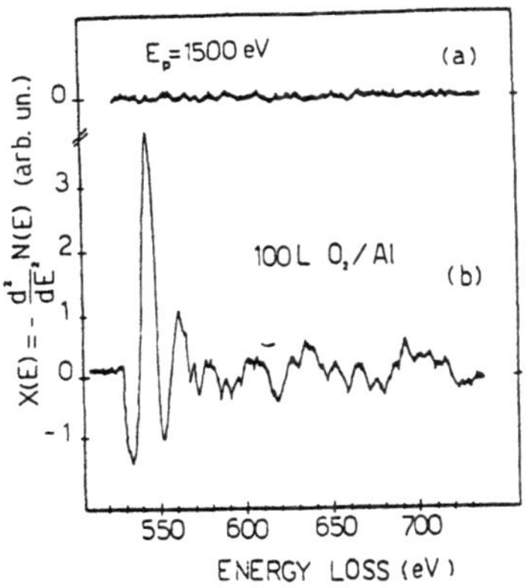

Fig.8 Surface EELFS signal of pure polycrystalline Al obtained (a) before and (b) after an exposure of 100 Langmuir of Oxygen

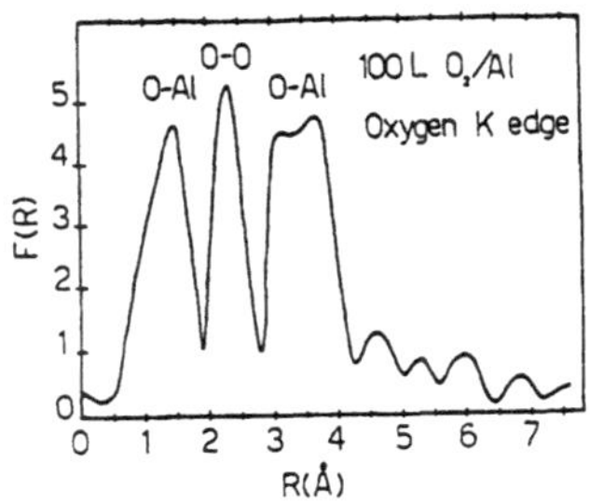

Fig.9 F(R) of the SEELFS spectrum of Fig.8 (b)

The comparison of the two spectra Fig.8 (a) and (b) gives clear evidence that high energy structures (560 - 750 eV) are oxygen induced. No changes were detected in the Al Auger line-shapes or in the SEELFS signal after prolonged beam exposure, showing that the effect of the electron beam current was negligible. The agreement with SEXAFS results for polycrystalline oxidized aluminium (11) is quite good and indicates that the SEELFS technique could be effectively used for structural studies of adsorbates on solid surfaces.

Fig.10 shows the SEELFS measurements on the O_2/Ni(100) system (8). We demonstrate that it is possible to detect the oxygen position at the various stages of oxidation and in particular the transition between the chemisorbed and oxide phase. The Fourier transform of the experimental spectra is shown in Fig.11. The dotted line of Fig.11(a) is the analysis of the clean nickel spectrum within the same integration range and provides the limit of reliability of the observed structures. The solid line in Fig.11(a) shows the F(R) for 0.3 ML (Mono Layer) oxygen coverage. The LEED pattern shows in this stage a well-defined C(2×2) structure. The nearest-neighbour distance (after phase shift (12) correction) of the absorbing oxygen from the nickel atom is 1.95± 0.05 Å which corresponds to an oxygen atom at 0.85 Å above the first Ni plane. This value is in good agreement with that found with SEXAFS experiments ($R_{\perp}$=0.86 Å) (13) and confirms that in the C(2×2) structure the oxygen is still well above the Ni plane without any precursor oxide state (14).

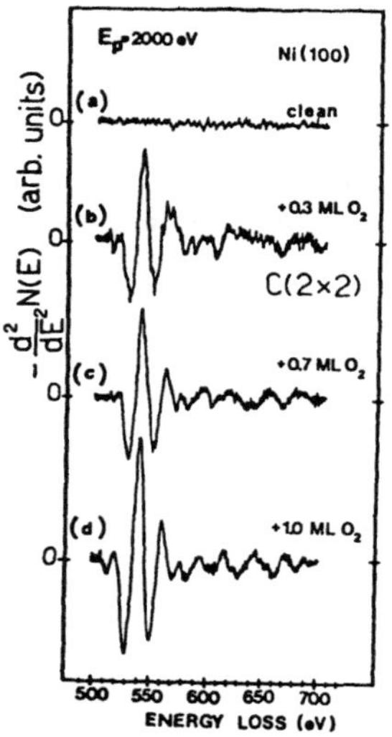

Fig. 10

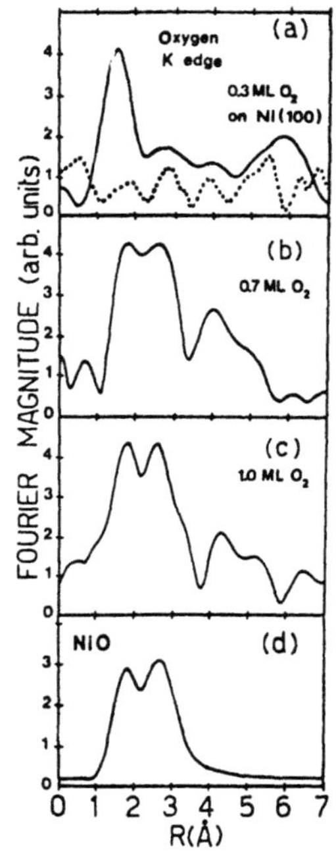

Fig. 11

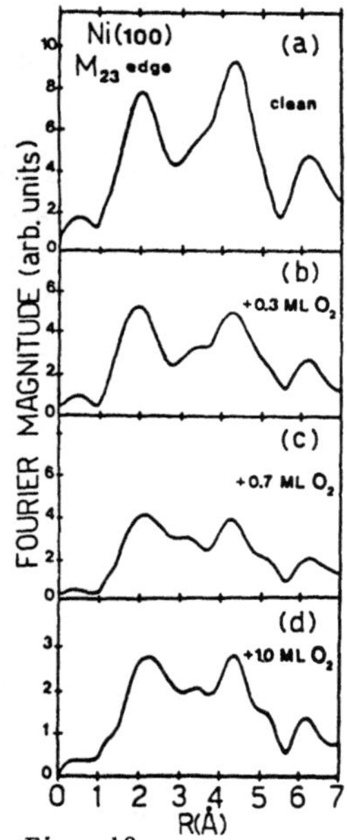

Fig. 12

Fig. 10. Energy loss spectra of Ni(100) around the oxygen K edge for different oxidation rates. The primary beam energy was E_p=2000 eV

Fig. 11. Fourier transform of the SEELFS Spectra of Fig. 10. The dotted line in (a) is the F(R) of the clean nickel spectrum with the same integration limits (2.0-7.0Å^{-1}). (d) EXAFS model calculation (15) for NiO

Fig. 12. Fourier transforms of the SEELFS structures above the Ni M_{23} edge for the same oxygen coverages of Fig. 10. The integration limits were 3.0-10.0Å^{-1}

Fig.11(b) shows that for 0.7 ML oxygen coverage, the F(R) is peaked at 2.10 Å and 3.00 ± 0.05 Å. We interpret these peaks as the nearest-neighbour's distance O-Ni and O-O of nickel oxide, in good agreement with the crystallographic values. Fig.11(c) shows that increasing the oxygen coverage to 1 ML structures in the F(R) function become sharper but do not shift. Fig.11(d) shows an EXAFS model calculation for NiO obtained by using theoretical phase shifts and backscattering amplitudes for the oxygen K edge (15) including the first two (O-Ni and O-O) neighbours shells. The analysis of SEELFS spectra taken above the Ni M_{23} core level for the same oxygen coverage shows that in the C(2×2) chemisorbed phase the adsorbate does not perturb appreciably the Ni substrate. At high oxygen coverage the Ni substrate is strongly affected. This effect is shown in Fig.12 which gives the Fourier transforms. The F(R) shown in Fig.12 (b) is quite similar to the one obtained for the clean Ni sample. The Fourier functions for the oxide phases (0.7 and 1 ML's) show a shift of the first peak towards higher R values and a marked broadening due to the overlap of different Ni-Ni pair distances in the metal (2.50 Å) and in the oxide phase (2.96 Å).

7. Anisotropic effects

In our integrated experimental geometry it is quite difficult to evaluate the amount of $\underline{q}$ transfer as in the case of transmission EXELFS experiments (16,17). However anisotropic effects can be singled out by changing the primary beam energy. In fact, for a K-edge transition, the orientation dependence of X(k) is given by a factor (16):

$$(\hat{\varepsilon}_q \cdot R_j)^2 = \sin^2(\alpha_j + \beta) \tag{5}$$

where $\hat{\varepsilon}_q$ is the unit vector along $\underline{q}$ and R_j is the position of the jth atom respect to the absorbing atom. α_j is the angle between R_j and the normal to the incident beam. The direction of $\underline{q}$ (assumed small so that the dipole approximation applies) depends on the scattering angle ϑ and $\beta = \mathrm{arctg}\,\vartheta/\vartheta_p$ ($\vartheta_p = \Delta E/2E_p$, E_p = incident beam energy) according to the diagram of Fig.14. Fig. 15 gives evidence of anisotropic effects in the F(R) of a graphite (9) single crystal.

The X(E) spectra were recorded above the C K-edge with the c axis oriented parallel to the incident beam direction. For E_p=1000 eV one measures both components $q_\perp$ and $q_\parallel$ easily seen in the near-edge structures ($1s\rightarrow\pi$, $1s\rightarrow\sigma$ have comparable intensity: inset of Fig.14(a)). The Fourier transform of the

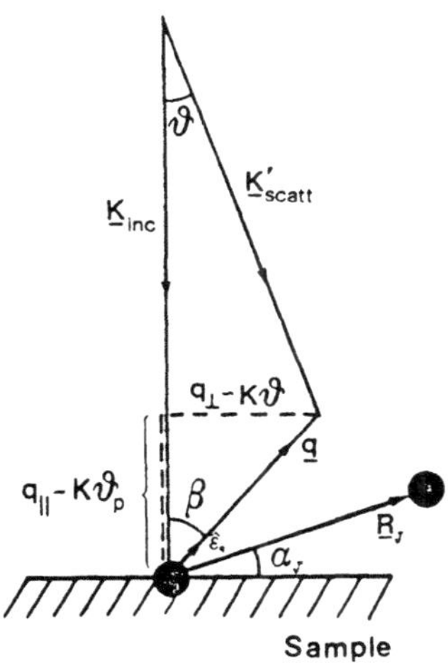

Fig.13. Vector diagram of the orientation dependence of the EELFS spectroscopy

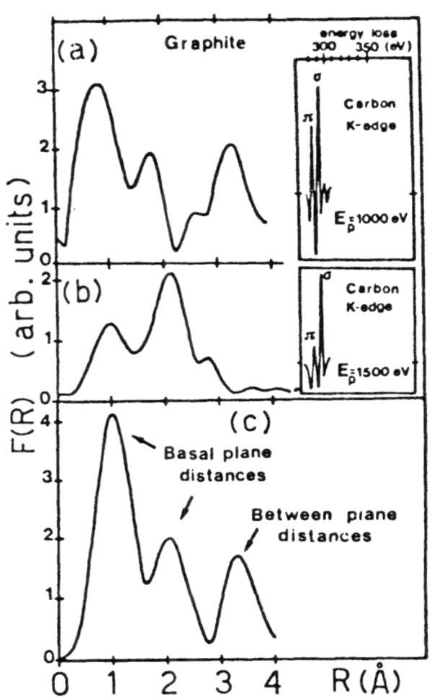

Fig.14. (a) EELFS analysis of the K edge of a graphite crystal. For E_p=1000 eV both basal plane distances and both between plane distances are present. For E_p=1500 eV (b) only distances in the plane are detected. (c) Theoretical EXAFS model for graphite (15)

EELFS region shown in Fig.14 (a) leads to three peaks in the 1-4 Å region related to both distances in the basal plane and distances along the c axis. For E_p=1500 eV the decrease of β ($q_{\parallel} \sim 0$) gives information about the R_j in the basal plane.

8. Conclusion

We have shown that the extended energy loss fine structure technique in the reflection mode is higly sensitive to the bulk local environment.
It can be also used for measuring the geometric surface structure of adsorbated species down the submonolayer coverage. It seems particularly suitable for light elements like oxygen,nitrogen and carbon (18).
The theoretical formalism underlying the technique is quite simple and follows the same steps and procedure used in the EXAFS and SEXAFS analysis.The technique is also extremely attractive and it needs only a CMA and an electron gun. We feel that the combination of the surface EELFS technique together with other surface electron spectroscopies (19) may open an avenue in future to an improved understanding of surface and interfaces.

Acknowledgments

I should like to thank A.Balzarotti for discussions and for a critical reading of this manuscript. The help of F. Antonangeli is greatly and particularly acknowledged.

Thanks are also due to S.Mobilio for the EXAFS analysis programs.

(1)- "EXAFS spectroscopy" edited by B.K.Teo and D.C.Joy,Plenum Press,N.Y. 1981.

(2)- B.M.Kincaid,A.E.Meixner,P.M.Platzman,Phys.Rev.Lett.40,1296(1978).

(3)- R.D.Leapman,L.A.Grunes,P.L.Fejes,Phys.Rev.B26,614(1982).

(4)- M.De Crescenzi,L.Papagno,G.Chiarello,E.Colavita,R.Scarmozzino,R.Rosei, S.Mobilio,Solid State Comm.40,613(1981).

(5)- L.Papagno,M.De Crescenzi,G.Chiarello,E.Colavita,R.Scarmozzino,L.S.Caputi, R.Rosei,Surface Sci.117,525(1982).

(6)- F.Antonangeli,S.Polizzi,G.Chiarello,M.De Crescenzi,to be published.

(7)- M.De Crescenzi,F.Antonangeli,C.Bellini,R.Rosei,to be published.

(8)- G.Chiarello,E.Colavita,M.De Crescenzi,S.Nannarone,to be published.

(9)- L.Papagno,L.S.Caputi,M.De Crescenzi,R.Rosei,Phys.Rev.B26,2320(1982).

(10)-M.De Crescenzi,F.Antonangeli,C.Bellini,R.Rosei,to be published.

(11)-D.Norman,S.Brennan,R.Jaeger,J.Stöhr,Surface Sci.105,L297(1981).

(12)-We have utilized as a shift the linear coefficient of the experimental phase obtained from the oxidized compound. So that:

$$R^{O-Ni}_{real}(1.95\ \text{Å}) = R^{O-Ni}_{EELFS}(1.60\ \text{Å}) + \varphi^{O-Ni}_{shift}(0.35\ \text{Å}).$$

(13)-J.Stöhr,R.Jaeger,T.Kendelewicz,Phys.Rev.Lett.49,142(1982).

(14)-H.Ibach,D.Bruchmann,Phys.Rev.Lett.40,36(1980).

(15)-B.K.Teo,P.A.Lee,Journ.Amer.Chem.Soc.101,2815(1979).

(16)-R.D.Leapman,L.A.Grunes,P.L.Fejes,J.Silcox,in "EXAFS spectroscopy"ed.by B.K.Teo and D.C.Joy,Plenum Press,N.Y.1981.

(17)-M.M.Disko,O.L.Krivanek,P.Rez,Phys.Rev.B25,4252(1982).

(18)-S.Modesti,M.De Crescenzi,F.Sette,C.Quaresima,P.Perfetti,A.Savoia,R.Rosei see these proceedings.

(19)-T.L.Einstein,Applic.of Surface Sci.11/12,42(1982).

Extended Fine Structure in APS *

T.L. Einstein[+], M.J. Mehl[+], J.F. Morar, and R.L. Park

Department of Physics and Astronomy, University of Maryland,
College Park, MD 20742, USA

G.E. Laramore

Department of Radiation Oncology, University of Washington Hospital,
Seattle, WA 98195, USA

With the great success of EXAFS and related X-ray techniques, it is natural to ask why one should bother with a technique that replaces X-rays by incident electrons, sacrificing an intrinsic dipole selection rule and convenient polarization. There are several good reasons for using appearance potential spectroscopy (APS): First, it opens the possibility of absorption fine structure measurement to any modern surface science laboratory. The technique can use single crystal samples, is intrinsically surface sensitive [in most modes of operation], and is ipso facto ultra-high vacuum compatible. The only equipment needed is a low energy electron diffraction system or, better, a cylindrical mirror analyzer (CMA). Second, in a recently developed mode (AMEFS), the Auger decay of a particular element can be selectively monitored. This procedure opens the possibility of studying systems previously inaccessible because fine structure above different core levels overlapped. Third, the excitation matrix element for electron excitation is such that the decay of the fine structure is very slow as a function of energy above threshold. Thus, one obtains excellent data ranges, extending up to 11 or 12 $Å^{-1}$. Accordingly, the first 100 eV can be excluded from the analysis, thereby avoiding the regime where calculated phase shifts are least reliable.

In APS, an incident electron with energy 0.5 - 2.0 keV excites a core electron. The final state is essentially a convolution over energy-conserving combinations of these two "active" electrons in unoccupied states. The excitation probability is monitored as a function of incident energy E_i. Because of the sharp onset of unfilled states at the Fermi energy (or at the bottom of the conduction band in the case of semiconductors), we have shown that the derivative of this probability with respect to E_i is dominated by the case of one electron at the Fermi energy and the other electron carrying the remaining energy [1-4]. Traditionally, APS was operated with the "other" electron restricted to within $\sim$ 10 eV of threshold, in order to study the unfilled density of states [5]. This other electron is analogous to the EXAFS final state electron; thus when it has energy 100-500 eV, it has a small EXAFS-like sinusoidal component due to near-neighbor backscattering [6].

In order to obtain the energy dependence of the excitation probability bearing the extended appearance potential fine structure (EAPFS), one collects the electrons (over some specified energy range) or the photons associated with the core decay. The core hole lives long enough so that this yield provides a passive monitor of their number. Four years ago, at the University of Maryland, ELAM et al. [1,7-9] collected the total electron yield ("Auger electron" or AEAPS) and measured EAPFS above the L_3 edges of polycrystalline Ti, V, and Fe. Using a relatively crude analysis, they obtained nearest-neighbor

[+]Supported by DOE grant AS05-79ER-10427.

spacings consistent with the known values. DENBOER et al. [10] measured the decrease in the elastic yield (disappearance potential spectroscopy or DAPS), thereby directly monitoring the excitation probability. They studied EAPFS above the K edges of O and Al in Al(100) reacted with oxygen. An optical Fourier transform was used, with calculated phase shifts appropriate to the EXAFS-like electron being an s wave. The improved analysis reduced error bars to ± 0.05 Å. Both O and Al K-edge spectra indicated the larger of the two nearest-neighbor spacings of bulk corundum, consistent with other experiments showing that the O bonds between the top and second layers of this face of Al. A thick oxide on Ni was also studied [4]. Since the Ni L_3-edge lies only 323 eV above the O K-edge, the desired data range could not be obtained above the latter; accordingly EAPFS above the Ni L_3 was examined. The importance of good data range is underscored by JACH and DISTEFANO's [11] DAPS study of EAPFS above the L_3 edges in NiO and Cr_2O_3; using data only for $k < 5$ $Å^{-1}$, they could not extract known spacings. When treating ordered samples, the AEAPS and DAPS modes encounter difficulties due to diffraction. Consequently, MORAR et al. [12] developed a sensitive soft X-ray detector to count core holes by their radiative decay (SXAPS). When applied to the O K-edge of oxidized Si, this scheme obtained the O-Si and the O-O nearest spacing, with spectra at least as good as SEXAFS [13]. While the low probability of radiative core decay required moderately high current (100 μA), excellent signal-to-background limited this problem. Moreover, since the photons are not energy analyzed, relatively low current density (1 μA/mm^2) could be used. The SXAPS spectrometer constructed for these experiments relied on a field emission source that is unfortunately rather temperamental and not commercially available.

Very recently MORAR et al. [14] monitored the excitation probability by tuning a CMA to the Auger line of the central atom core-hole decay. Dubbed AMEFS (Auger-Monitored Extended Fine Structure), this new technique offers the impressive prospect of exploring systems with core levels closely spaced in energy, e.g. the O K-edge in the NiO system mentioned above. In EXAFS-like schemes, Auger detection has been thwarted, for Auger lines below 2 keV, by interference from photoemission peaks passing through the acceptance window [13a]. In AMEFS, a modulation technique is used to distinguish, to leading order, features tied solely to the sample energy levels (viz. Auger lines) from features dependent on the incident energy. AMEFS study of the spectrum above the L_3 edge ($\sim$ 450 eV) of a single crystal of Ti yields fine structure very similar to that in EXAFS [15]. Optical transformation of simply filtered data gives a nearest-neighbor spacing of 2.93 or 2.91 Å (vs. a known 2.92 Å) with s- or p-wave phase shifts, respectively, for the central atom. For a Ti (0001) exposed to 50 L of oxygen, data above the O K-edge (530 eV) is analyzed to show an O-Ti spacing of 1.99 or 1.95 Å (vs. 1.98 Å for Ti O_2) with s- or p-wave phase shifts, respectively. The S L_3 edge (165 eV) on this crystal and on Ni (100) is currently being studied. We reemphasize that these experiments were done with a standard commercially available CMA with integral electron gun; the only modification was in the electronics.

A lingering question is the angular momentum ℓ_x of the EXAFS-like final state. Initial calculations of radial integrals using partially orthogonalized wave functions indicated that for deep K levels (e.g. V) the EXAFS-like electron was overwhelmingly $\ell_x = 0$; for shallower K levels, (e.g. O) this monopole behavior still held, though less strongly. For L_3 edges, this approach indicated ℓ_x was below 3, but that the final state could have more than one angular momentum component. We are improving the calculations by taking fully into account orthogonalization, Clebsch-Gordan coefficients, and factors due to antisymmetrization, as well as by performing model calculations to corroborate trends. We are also using wave functions derived from atoms self-consistently embedded in jellium and from SCF-Xα programs. Results depend on the angular momentum of the electron at the Fermi level and are rather insensitive to the energy of the EXAFS-like electron. Detailed exposition requires far more formalism than is appropriate here [16]. A noteworthy result for K edges is that antisymmetrization mutes the $\ell_x = 0$ contribution while angular momentum coefficients enhance the $\ell_x = 1$ part, so that both may play a substantial role. For the low Z elements we have studied, the central atom phase shifts for these two angular momenta have similar slopes over the relevant energy range, with the $\ell_x = 0$ slope a bit greater in magnitude. Thus, analysis with just $\ell_x = 0$, or $\ell_x = 1$ will slightly overestimate or underestimate, respectively, interatomic spacings, bracketing the true value. Such behavior is consistent with our AMEFS results. Note that self-consistent inner potential adjustment tends to diminish the small misestimates, which are a fraction of our error bars.

References

*EAPFS development was supported by ONR grants N00014-81-K-0540, N00014-79-C-0371, N00014-77-C-0485, and N00014-75-C-0292. Computer time and facilities were provided by the University of Maryland Computer Science Center. For more detailed and comprehensive though less recent reviews, see T. L. Einstein, Appl. Surface Sci. 11/12, 42 (1982), and T. L. Einstein and R. L. Park, in Extended X-Ray Absorption Fine Structure, R. W. Joyner, ed. (Plenum, to be published). For figures, see these and our publications below. We are most grateful to those co-authors for fruitful and enjoyable collaboration.

1. P. I. Cohen, T. L. Einstein, W. T. Elam, Y. Fukuda, and R. L. Park: Appl. Surface Sci. 1, 538 (1978).
2. G. E. Laramore: Phys. Rev. B18, 5254 (1978).
3. G. E. Laramore: Surface Sci. 81, 43 (1979).
4. M. L. den Boer: T. L. Einstein, W. T. Elam, R. L. Park, L. D. Roelofs and G. E. Laramore: J. Vacuum Sci. Technol. 17, 59 (1980).
5. R. L. Park: Surface Sci. 48, 80 (1975); R. L. Park, J. E. Houston: J. Vacuum Sci. Technol. 11, 1 (1974).
6. By integration, we see that the [undifferentiated] excitation probability must also have [co-]sinusoidal oscillation on top of a [larger] background; this feature could be derived using an end-point-dominated, stationary-phase argument.
7. W. T. Elam, P. I. Cohen, L. D. Roelofs and R. L. Park: Appl. Surface Sci. 2, 637 (1979).
8. W. T. Elam: Ph.D. Thesis, Univ. of Maryland, College Park (1979), unpublished; W. T. Elam, P. I. Cohen, L. D. Roelofs and R. L. Park: Bull. Am. Phys. Soc. 24, 506 (1979).
9. R. L. Park, P. I. Cohen, T. L. Einstein and W. T. Elam: J. Crystal Growth, 45, 435 (1978).
10. Y. Fukuda, W. T. Elam and R. L. Park: Appl. Surface Sci. 1, 78 (1978).
11. T. Jach and T. H. DiStefano: Phys. Rev. B19, 2831 (1979).
12. T. L. Einstein, M. L. den Boer, J. F. Morar, R. L. Park, and G. E. Laramore: J. Vacuum Sci. Technol. 18, 490 (1981); J. F. Morar: Ph.D. Thesis, Univ. of Maryland, College Park, 1981, unpublished.
13. (a) J. Stöhr, L. I. Johansson, I. Lindau, and P. Pianetta: Phys. Rev. B20, 664 (1979); (b) J. Vacuum Sci. Technol, 16, 1221 (1979).
14. J. F. Morar and R. L. Park: J. Vacuum Sci. Technol. A1 (to be published).
15. D. Denley, R. S. Williams, P. Perfetti, D. A. Shirley, and J. Stohr: Phys. Rev. B19, 1762 (1979).
16. See M. J. Mehl, T. L. Einstein, G. E. Laramore, and G. W. Bryant: to be published.

Geometrical Structure of Graphitic Carbon on Ni(111) Studied by Extended Energy Loss Fine Structure Spectroscopy (EELFS)

S. Modesti[1], M. De Crescenzi[2], P. Perfetti, C. Quaresima, R. Rosei[2], A. Savoia, and F. Sette

PULS-INFN-GNSM, Laboratori Nazionali di Frascati, I-00044 Frascati, Italy

The geometry of adsorbed atoms on ordered surfaces may be determined by different experimental techniques. Among them the most used are ion scattering spectroscopy (ISS) [1,2], surface extended X-ray absorption fine structure (SEXAFS) [3], low energy electron diffraction (LEED) [4] and normal photoelectron diffraction (NPD) [5]. All these techiques show limits in the number of possible systems to be exploited. In particular in the case of adsorbed carbon atoms the ISS suffers of low back-scattering amplitude while the carbon contamination of the optical elements [6] is the most limiting factor in SEXAFS and NPD techniques. LEED analysis is not affected by the above limits, nevertheless it requires careful and time-consuming computer analysis.

We present here a geometrical structure determination of graphitic carbon (C_{graph}) on Ni (111) using extended energy loss fine structure spectroscopy (EELFS) [7]. This technique is quite recent and the information of the absorption coefficient is provided by the analysis of the intensity of a diffracted electron beam (1-2 KeV) impinging on the sample as a function of the energy lost during the scattering. The small sample depth (~10 Å) explored by the reflected electrons makes EELFS extremely powerful in surface analysis. The impact of EELFS in the study of surface-adsorbate interaction is even more relevant if we take into account that it requires standard ultrahigh-vacuum equipments normally used for photoemission or Auger spectroscopy.

In Fig. 1 we report the most meaningful results obtained on a thin (< 1 monolayer, ML) graphite overlayer on Ni (111) [8]. This overlayer was obtained by dissociation of CO on the Ni substrate at 380 °C. Figure 1a shows the experimental EELFS in the second derivative mode in an extended energy region between 50 and 450 eV. The upper curve of Fig. 1a refers to the clean Ni (111) substrate showing the M_1 and $M_{2,3}$ edges and their related extended fine structures. In the region above the C 1s edge (~ 285 eV) the substrate does not show appreciable structures

[1] Istituto di Fisica "G. Marconi", Università di Roma, I-00185 Roma, Italy

[2] Dipartimento di Fisica, Università della Calabria, I-87036 Arcavata di Rende, Italy

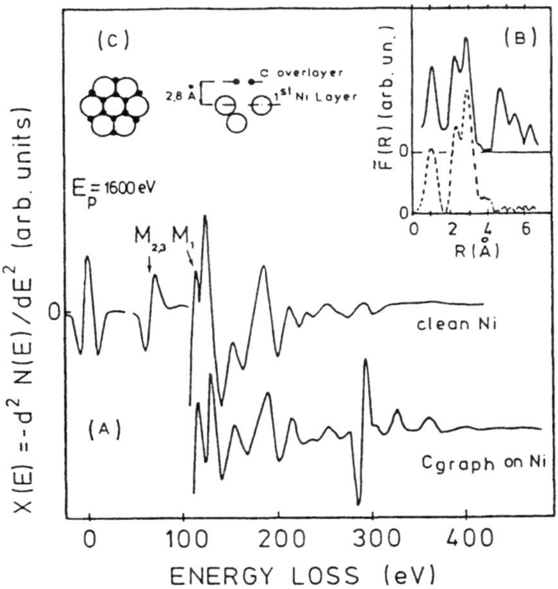

Fig. 1. a) Extended energy loss fine structure of clean Ni (111) (upper curve) and of ~.5 ML of graphitic carbon on Ni (111). b) Fourier transform of the experimental carbon signal (solid line) and theoretical calculation including two C-C and one C-Ni shell (dashed line). c) geometrical model for graphitic carbon

that could interfere with the C 1s fine structure which are observed in the lower curve when C_{graph} is formed.

A Fourier transform of the C_{graph} signal is reported in Fig. 1b (full line) for the momentum k ranging between 2.3 and 8.3 $Å^{-1}$. The theoretical calculation reported in Fig. 1b (dashed curve) has been performed using the standard EXAFS formula [9] and using two C-C shells at 1.42 and 2.46 Å (as in bulk graphite) and a C-Ni shell at 3.1 Å. Taking the C-Ni distance as the fitting parameter the theoretical results reproduce well the experimental one. We conclude that the first two experimental peaks are due to the same C-C distances present in bulk graphite and the third to C-Ni pairs. A possible geometry is that reported in Fig. 1c. The overlayer-substrate distance is measured for the first time in this system within .05 Å.

References

1 H. H. Brongersma and J. B. Theeten, Surface Sci. 54, 519 (1976)
2 D. J. Godfrey and D. P. Woodruff, Surface Sci. 105, 438 (1981)
3 J. Stöhr, L. J. Johansson, S. Brennan, M. Hecht, J. N. Miller, Phys. Rev. B22, 4052 (1980)

4 P. M. Marcus, J. E. Demuth and D. W. Jepsen, Surface Sci. 53 501 (1975)
5 D. H. Rosemblatt, J. G. Tobin, M. G. Mason, R. F. Davis, S. D. Kevan, D. A. Shirley, C. H. Li and S. Y. Tong, Phys. Rev. B23, 3828 (1981)
6 D. Denley, P. Perfetti, R. S. Williams and D. A. Shirley, Phys. Rev. B21, 2267 (1980)
7 M. De Crescenzi, these proceedings; L. Papagno, L. S. Caputi, M. De Crescenzi and R. Rosei, Phys Rev. B26,2320 (1982)
8 A more extensive study will be reported elsewhere.
9 The theoretical curve to be compared to the experimental one has been obtained from the second derivative of the theoretical X(E) after Fourier integration between the same limits used for the experimental curve.

Oxygen on Ni(100): A Surface-Extended Energy Loss Fine Structure (SEELFS) Study

F. Antonangeli, C. Bellini, M. De Crescenzi[1], and R. Rosei[1]

Gruppo PULS, Laboratori Nazionali di Frascati, I-00044 Frascati, Italy

Energy loss fine structures measured above the oxygen K edge and nickel M_{23} core level are reported in function of several oxygen coverages. The experiment is performed in the reflection mode and the primary beam energy was E_p=2000 eV (1). We demonstrate through the high degree of surface sensitivity of the SEELFS technique that it is possible to detect the oxygen position at the various stages of oxidation and in particular the transition between the chemisorbed C(2×2) and oxide phase (2).

The experiment is carried out with a CMA electron analyser in a standard ultra-high-vacuum chamber. The clean and oxygen-covered surfaces were inspected by Auger and LEED. No changes were detected in the Ni Auger line-shapes or in the SEELFS spectra after prolonged beam exposure, showing that the effect of the electron beam was negligible.

Fig. 1 shows SEELFS spectra above the oxygen K edge for O_2/Ni(100) system for 0.3 ML (Mono Layer, 20-30 Langmuirs at 10^{-6} torr), 0.7 and 1.0 ML.

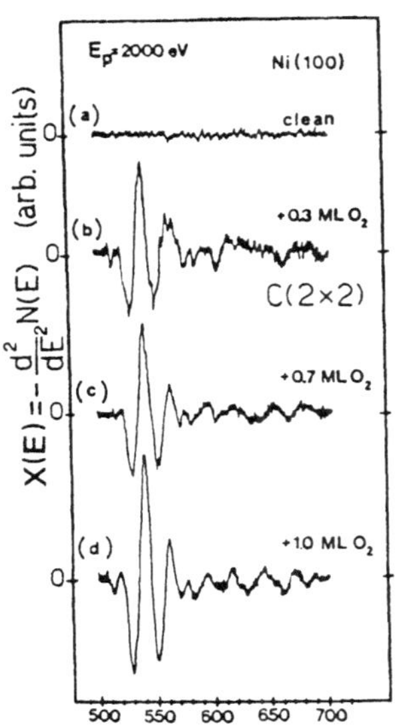

Fig.1. Energy loss fine structures for O_2/Ni(100) system around the oxygen K edge. Clean (a), 0.3 Mono Layer (b), 0.7 ML (c) and 1.0 ML (d) oxygen coverages are shown. The primary beam energy was E_p=2000 eV

[1]Dipartimento di Fisica, Università della Calabria, I-87036 Arcavata di Rende, Italy

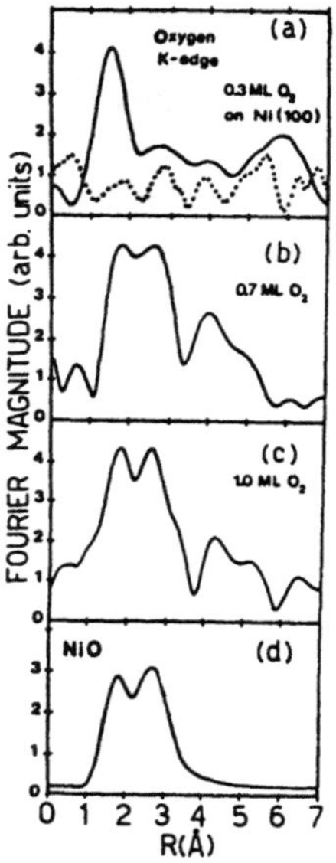

Fig.2. Fourier transforms of the SEELFS spectra of Fig.1. The dotted line in (a) is the F(R) of the clean Ni spectrum with the same integration limits (2.0-7.0 $Å^{-1}$). The curve (d) is an EXAFS model calculation (11) for NiO

Curve (a) shows the loss signal for pure Ni(100) before the oxygen exposure with the same experimental settings used for recording curves (b),(c) and (d), and provides a measurements of the background and noise level.
The data analysis is quite simple and follows the same procedure used for the SEXAFS (3,4) analysis.
Fig.2 shows the direct Fourier transforms of the spectra of Fig.1.
The F(R) gives (after the appropriate phase shift correction (5)) the position in the real space around the absorbing oxygen atom. For 0.3 ML (Fig.2(a) solid line) the distance between oxygen and the nickel first neighbours is 1.95 ± 0.03 Å that corresponds to a 0.85 ± 0.05 Å above the Ni(100) first plane.
This value is in good agreement with that found in SEXAFS (6), LEED (7) and RBS (8) measurements and confirms that in the C(2×2) configuration the oxygen is still well above the first nickel plane without any precursor oxide state (9,10).
The F(R) for 0.7 ML (Fig.2 (b)) oxygen coverage shows peaks at 2.10 and 3.00 Å and for 1.0 ML (Fig.2(c)) these peaks are sharper but their position do not change appreciably. We interpret these peaks as the nearest-neighbours distances O-Ni and O-O of nickel oxide.
Fig.2(d) shows an EXAFS model calculation (11) for NiO using two O-Ni and O-O coordination shells. The F(R) is obtained after second derivative of the theoretical signal and it was integrated in the same k range used for the experimental spectra. The theoretical model agrees in shape with that of Fig.2(c) (1.0 ML) but disagrees in the R position (within 0.05-0.10 Å). We conclude that the theoretical phases (11) are not accurate and reliable for the absorbing light atoms.
The local geometry of O_2/Ni(100) system may be also investigated through the analyses of the Ni M_{23} core edge (E_o = 68 eV) shown in Fig.3.
In the chemisorbed phase (0.3 ML) the SEELFS structures remain unchanged so that the adsorbate does not perturb noticeably the nickel substrate at this

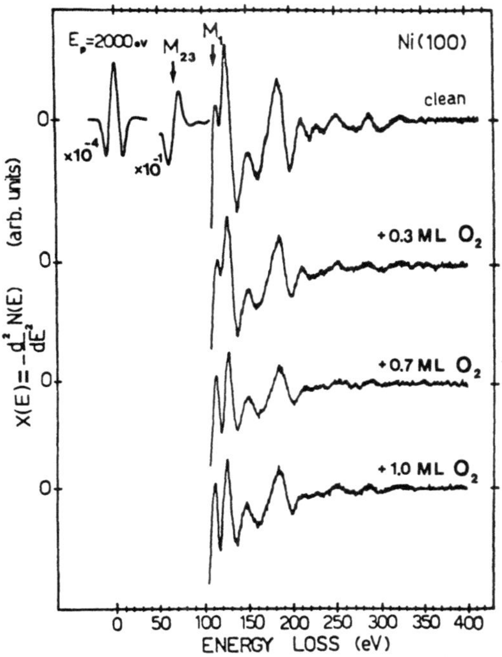

Fig.3. Energy loss fine structures above the Ni(100) M_{23} core level as a function of different oxygen exposures

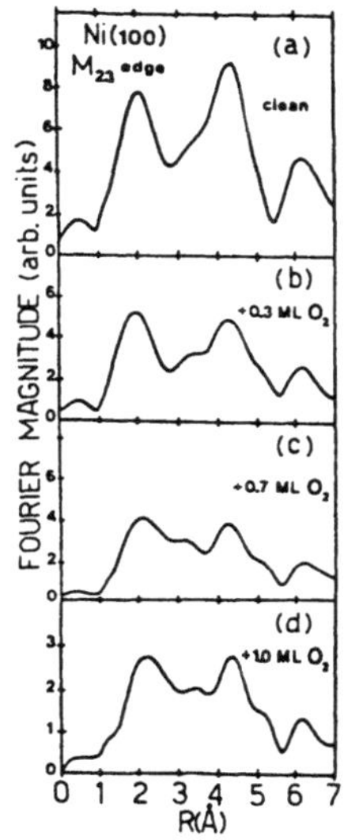

Fig.4. Fourier transforms of the structures of Fig.3 for the same oxygen coverages. The integration limits were 3.0 - 10.0 $Å^{-1}$

coverage. Increasing the oxidation, however, the Ni substrate is strongly affected. The relative Fourier transforms of these spectra are shown in Fig.4. For the 0.3 ML the Ni-Ni pair is exactly at the same position as pure nickel is (Fig.4(a) and(b)). For 0.7 and 1.0 ML the first peak in F(R) is broadened and shifted towards higher R values due to overlap of different Ni-Ni pair distances in pure nickel (R_j=2.50 Å) and in the oxide phase (R_j= 2.96 Å).
In conclusion the SEELFS technique is a highly promising tool for investigating the chemisorbed species on solid surfaces and in particular oxygen, carbon and nitrogen.

(1) - M.De Crescenzi, L.Papagno, G.Chiarello, E.Colavita, R.Scarmozzino, R.Rosei, S.Mobilio, Solid State Commun.40,613(1981).
(2) - M.De Crescenzi, F.Antonangeli, C.Bellini, R.Rosei, to be published.
(3) - P.H.Citrin, P.Eisenberger, R.C.Hewitt, Phys.Rev.Lett.44,309(1978).
(4) - J.Stöhr, L.I.Johansson, S.Brennan, M.Hecht, J.N.Miller, Phys.Rev.B22,4052(1980).
(5) - we have utilized the experimental O-Ni phase shift obtained from the most oxidized sample.
(6) - J.Stöhr, R.Jaeger, T.Kendelewicz, Phys.Rev.Lett.49,142(1982).
(7) - G.Hanke, E.Lang, K.Heinz, K.Müller, Surface Sci.91,551(1980).
(8) - D.J.Godfrey, D.P.Woodruff, Surface Sci.105,438(1981).
(9) - H.Ibach, D.Bruchmann, Phys.Rev.Lett.44,36(1980).
(10)-T.S.Rahman, J.E.Black, D.L.Mills, Phys.Rev.B25,883(1982).
(11)-B.K.Teo, P.A.Lee, Journ.Amer.Chem.Soc.101,2815(1979).

Part X

Instrumentation

Laboratory EXAFS Systems*

G.S. Knapp[1], H.K. Pan[1,2], P. Georgopoulos[3], and T.E. Klippert[1]

Great progress has been made in the experimental technique to make EXAFS measurements both in the laboratory and at synchrotron facilities. While the great majority of work is now being carried out at synchrotrons, quite a number of laboratory facilities are in use or are being constructed. We will briefly describe the facility that we have at Argonne. Very similar facilities exist or are being constructed at other laboratories both in the U.S. and in Europe. The state of the art of laboratory EXAFS systems has been reviewed in 1980 [1], so this paper will discuss the improvements that have been made since that time. Our current spectrometer is much easier to use and has much better resolution.

Figure 1 illustrates the principles of operation of the spectrometer. The point A is placed directly under the anode image. The point C, above which the focusing crystal is located, is moveable by means of a lead screw and a computer controlled stepping motor. The rigid arms RA, RC and RS are all equal to r, the Rowland circle radius of the monochromator crystal (25 cm). The point above which the slit is located is moveable by means of another lead screw and computer controlled

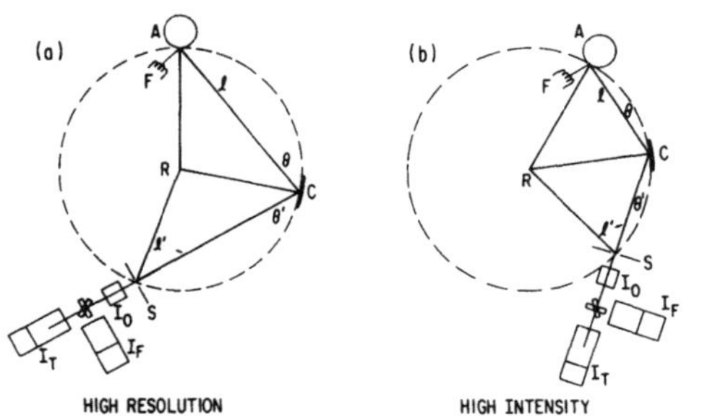

Fig.1. EXAFS spectrometer schematic, illustrating the principles of operation, as well as the sample-detector arrangement: A-anode, F-filament, C-crystal, S-receiving slit, I_0-beam monitor detector, I_F-fluorescence detector, I_T-transmission detector, (a) high resolution, (b) high intensity

*Work supported by the U.S. Department of Energy.

[1] Materials Science Division, Argonne National Laboratory, Argonne, IL 60439, USA

[2] Department of Chemical Engineering, University of Wisconsin-Madison, Madison, WI 53706, USA

[3] Materials Science Department, Northwestern University, Evanston, IL 60201, USA

stepping motor. The $\measuredangle$ ACR = $\measuredangle$ RCS = $\pi/2 - \theta$, when l is made to equal l' via the stepping motors and lead screws.

The resolution of our spectrometer is approximately given [1] by $\Delta E = E\cot\theta\ \Delta\theta$ where

$$\Delta\theta=[8r\sin\theta]^{-1}[(W_a + W_s)^2+(2t_{eff}\cos\theta)^2+(\frac{h^2}{8r\cos\theta})^2]^{1/2} \qquad (1)$$

where W_a is the projected width of the anode image, W_s is the width of the receiving slit, t_{eff} effective depth of penetration of the x-ray into the crystal, which at low energy ($E < 10$ keV) is equal to $\sin\theta \ln 2\ \mu$, where μ is the linear absorption coefficient of the crystal. At high energy ($E > 15$ keV), t_{eff} equals the thickness of the crystal (0.01 cm). Resolution of the various crystals that are used are seen in Fig.2. Note the very high resolution of the order of one volt which possibly near small energy ranges for each crystal. The actual resolutions are ca. 30% worse than predicted by (1). The use of a focusing spectrometer to achieve high resolution, instead of the usual two-crystal monochromators, results in much higher flux. Fluxes of the order of 10^5 are realized in the high resolution mode. As an example, we show in Fig.3 the XANES spectra of $KMnO_4$ and $Na_2MoO_4 \cdot 2H_2O$. EXAFS measurements require resolutions of only 8-10 eV, so we use a crystal with a larger 2d spacing and naturally we achieve much greater fluxes. We achieve between 10^6 and 5×10^6 using a rotating anode generator.

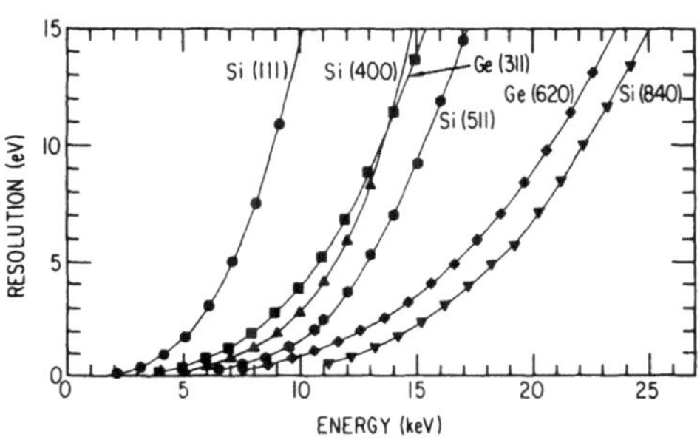

Fig.2. The energy resolution ΔE vs energy for Si(111), Si(400), Ge(311), Si(511), Ge(620), and Si(840)

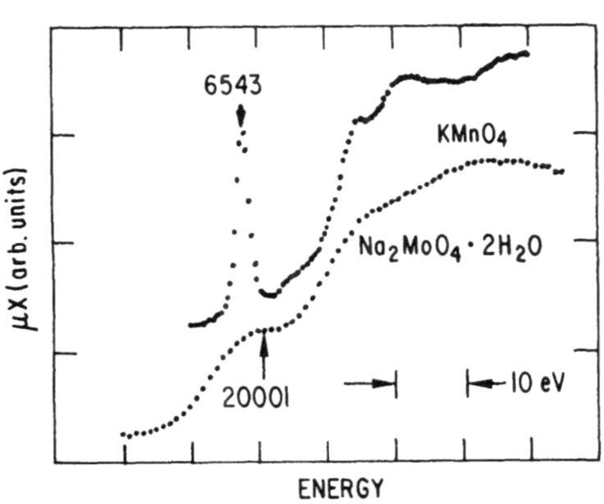

Fig.3. X-ray absorption near edge structure of $KMnO_4$ with ~ 0.8 eV resolution and $Na_2MoO_4 \cdot 2H_2O$ with ~ 7 eV resolution

We employ both ionization and proportional variable pressure detectors on the I_0 channel so the proper % absorption can be achieved. Either ionization or scintillation detectors for I channel are used and both transmission and fluorescence measurements can be done. The crystals we use are diamond shaped and mounted on individual bending fixtures. Crystals can be changed in less than five minutes. The x-ray beam is fully enclosed in the He-filled beam lines for safety and increased flux.

The use of any laboratory access facilities cannot replace synchrotron facilities. However, many experiments now carried out at synchrotrons can easily be done in the laboratory. Any sample with $\mu_K/\mu \geq 0.1$ is quite easy to do, and by using the fluorescence mode, high quality data can be taken on samples with $\mu_K/\mu \sim 0.01$. Even those samples with very small jumps on laboratory facility make the synchrotron work more efficient because the samples can be taken before measurement at the synchrotron facility. The cost of laboratory facilities is quite modest and they are now available commercially.

Reference

1. E. A. Stern, AIP Conference Proceedings, No. 64, 1980.

Time-Resolved EXAFS

A.M. Flank[1], A. Fontaine[2], A. Jucha, M. Lemonnier, and C. Williams[3]
LURE, CNRS, Bât. 209c, F-91405 Orsay-Cedex, France

1. Introduction

When synchrotron radiation became available it has been possible to reduce the data collection time of one EXAFS spectrum from 2 or 3 weeks using the bremsstrahlung of a conventional generator down to a few minutes when using the white beam provided by a magnet of storage ring.

In the last two years at Stanford and at LURE [1],[2] it has been shown that another considerable time reduction can be achieved when EXAFS data are collected in a dispersive mode rather than a point by point scanning method.

2. Principle

The key elements of this new method of collecting EXAFS data are a dispersive optics combined with a position sensitive detector able to work under high flux conditions.

The geometrical arrangement of the dispersive diffractometer is shown in Fig. 1. When the triangle-shaped crystal is over bent, it no longer works as a monochromator. The point source is far out of the Rowland circle whereas the focus image is within the Rowland circle. The glancing angle θ_i of the incident X-ray beam varies continuously across the crystal surface. That yields different X-rays energies from different points on the surface, emerging with different directions. With the relevant parameters indicated

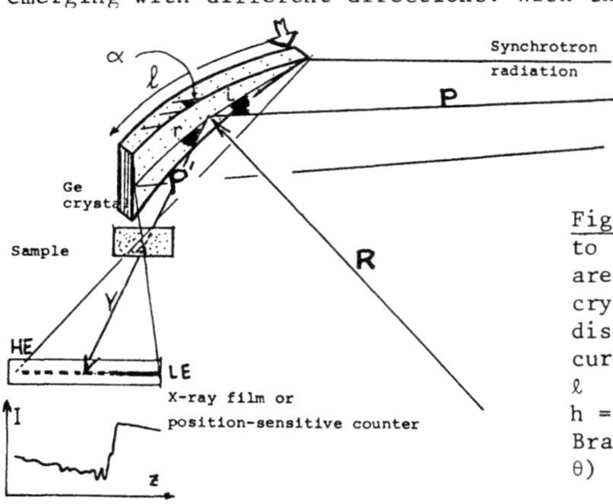

Fig. 1 The relevant parameters to describe the curved crystal are p and p' the source to crystal and crystal to focus distances, R the radius of curvature of the crystal, ℓ the length - $i = \theta+\alpha$ $h = \theta-\alpha$ are the angles of the Bragg reflection (defined by θ) on a blade cut asymmetrically

[1]Lab. Met. Physique ass. au CNRS, Univers. Poitiers, F-86022 Poitiers, France
[2]Lab. Phys. Sol. ass. au CNRS, Univers. Paris-Sud, F-91405 Orsay, France
[3]Lab. Phys. Mat. Cond. ass. au CNRS, Coll. de France, 11, place Marcellin Berthelot, F-75231 Paris Cedex, France

in the caption, following [3], the classical relation between p, p' and R for the cylindrically bent crystal can be written as

$$2/R = (\sin i/p) + (\sin r/p').$$

The covered energy range, without rotating the crystal is given by

$$\Delta E = E \text{ cotg } \theta * \Delta\theta = E \text{ cotg } \theta[(\ell/R) - (\ell \sin i/p)]$$

where the driving term is ℓ/R (~ 0.2) while cotg θ is typically about 4 for the 9 KeV photons.

In our experiment which has been performed on the small angle scattering port the crystal is curved by a motor-driven constraint at the apex. In the Stanford experiment the crystal was made to remain in close contact with a spherical mirror by air pressure.

Furthermore the Stanford group [4] has introduced just behind the sample located at the focus point a flat mirror to cut off high order harmonics ; they have observed a significant improvement of the contrast of the recorded EXAFS spectra.

3. Results

The 550 eV window of the copper spectrum shown on Fig.2 was obtained with one 0.5 mm-thick, 7 cm-long Ge(111) crystal bent down to a radius of about 3 m. With an asymmetric cut α of 6° the focus point was located at about 20 cm from the crystal, while p is 15 meters. Since the finite size of the X-ray source is 6 mm at LURE-DCI the effective size of the focused image is 1.5 mm for the arrangement used in the experiment of Fig.2. This is about 5 times as wide as the value yielded by the ray tracing method assuming a point source.

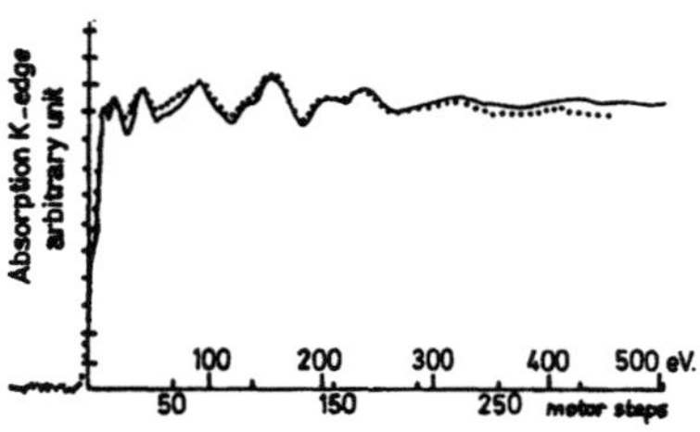

Fig.2. An X-ray absorption of a 10 μ-thick copper foil, recorded on M-Kodak film in dispersive mode, and the EXAFS plot collected by rotating the channel-cut by step of 1/1800 degree

4. Advantages

The advantages of EXAFS in dispersive mode over the conventional point by point data collection are due to two main features.

i. the *rapidity* of recording a complete EXAFS spectrum *simultaneously*. The exposure time on the resolutive M-Kodak film has been about 1 s. We thus may think of time-resolved studies. The data collection is not sensitive to *fluctuations* in the intensity of the *static* X-ray emerging beam.

ii. the *focusing* optics. Small samples can be used which is very suited to high pressure studies. Furthermore, using this optics the new storage ring (ESRF) will provide a *new probe* for local analysis of materials at a size scale of a few ten microns.

5. Calibration, Resolution, Detection

Among these three major points only the last one requires improvements to take the full advantage of this new promising method.

The conversion from spatial position on the detector to energy scale is trivial since the geometry defines unambiguously the energy of the photons via the Bragg angle.

To achieve the best energy resolution requires essentially a good choice of the crystal used to disperse the primary beam. The resolving power is better when using a low Z-number material, a small lattice spacing, no asymmetrical cut, a weak curvature [5]. Besides the finite size of the source and the spatial resolution of the detector two other limitations to the energy resolution are yielded.

The over-all resolution was estimated to be 2 eV when these conditions are fulfilled [1]. The main problem lies with the detector. This difficulty is not unique to energy dispersive EXAFS. Films are currently used as position sensitive detectors for high photon flux(10^8-10^9 cps). But for time-resolved studies to be performed successfully it is necessary to develop the technology of the photodiode arrays which are able to handle large count rates with a good spatial resolution. Promising results have been obtained recently with the RC 256 EC/17 array manufactured by RETICON with a SiO_2-window thickness reduced to one micron. A preliminary XANES experiment was performed using the emerging beam of an almost flat new monochromator positioned across the Ni edge. Even with the highly modulated intensity of the incident beam it has been possible to observe the Ni edge with the well-known peaks separated by 9 eV using a very short exposure time : 1.3 ms to collect the Ni edge and 0.32 ms to record the photon flux without the sample (Fig.3).

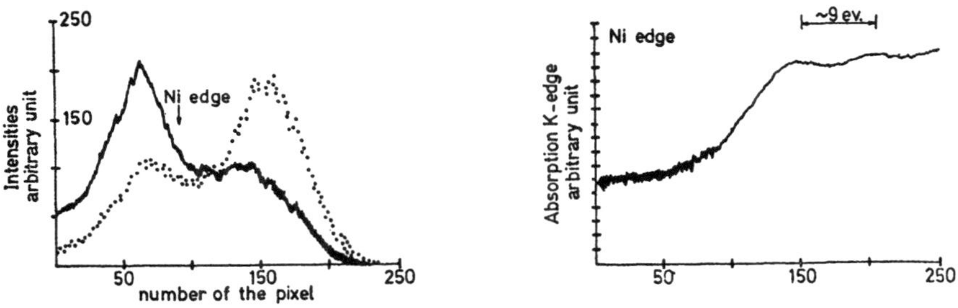

Fig.3. Profiles of the beam with (I) and without (I_0) the Ni sheet and the Log (I_0/I) versus the number of the photodiodes

One draw-back has been the damage observed after four hours of exposure in the beam [2]. Nevertheless the cooled RETICON SFX-array will give an efficient detector. It is currently being developed at SSRL and at LURE.

6. Conclusion

The improved efficiency in data collection opens up the possibility of kinetic studies in material science, chemistry and biophysis. The first data obtained with a self-scanned photodiode array show that XANES study at a few ms time scale can be expected since only 1.3 ms were required to collect the XANES plot of elemental Ni. Let us give the last words to mention the fast EXAFS of aluminium realized with a single pulse of LASER-produced X-rays [6] and the time-resolvedEXAFS set-up proposed by HUANG and LIU [7] who claim that the synchronisation of the pulse clock with sample stimulation may open nanosecond-resolved EXAFS experiment.

References

1 - T. Matsushita, R.P. Phizackerley, Jpn. J. Appl. Phys. 20, 2223.8 (1981).

2 - A.M. Flank, A. Fontaine, A. Jucha, M. Lemonnier, C. Williams, J. Physique Lettres 43, L315, 319 (1982).

3 - M. Lemonnier, R. Fourme, F. Rousseaux, R. Kahn, Nucl. Inst. Meth. 152, 173-177 (1978).

4 - R.P. Phizakerley, Z.V. Rek, G.B. Stephenson, S.D. Conradson, K.O. Hudgson, T. Matsushita, M. Oyanagi, SSRL Act. Rep. 82-01 p. VIII, 82 (1982).
5 - A.M. Flank, A. Fontaine, A. Jucha, M. Lemonnier, D. Raoux, C. Williams, to be published in Proc. Hambourg S.R.I. (1982).
6 - Malozzi, Scherzel, Epstein, Campell, Science V.1206 (oct. 1979).
7 - Huang, Liu, SSRL Act. Rep. 82-01 p. VIII, 137 (1982).

A Versatile Spectrometer for Laboratory EXAFS Studies

W. Thulke, R. Haensel, and P. Rabe

Institut für Experimentalphysik, Universität Kiel,
D-2300 Kiel, Fed. Rep. of Germany

Introduction

Increasing interest in the analysis of the extended X-ray absorption fine structure (EXAFS) as a structural tool has led to increasing demand for EXAFS facilities built up at synchrotron radiation sources. As a result access to such sources is restricted and the use of laboratory facilities for EXAFS studies on non-dilute systems and for training of students is essential. These laboratory facilities should provide high photon flux and good energy resolution and should be flexible in regard to changing experimental requirements.

In this paper we describe a focusing X-ray spectrometer that has been developed in our laboratory. It is adaptable to different types of laboratory X-ray sources and to different types of curved crystals within large ranges of radii of curvature and of Bragg angles. In the following sections we describe the details of the spectrometer design and the spectrometer performance regarding intensity and energy resolution.

Spectrometer Design

The spectrometer is primarily designed for a fixed X-ray source with fixed direction of the primary radiation beam and for Johansson-type monochromator crystals. With a symmetrically cut crystal, optimal focusing and monochromatizing are achieved for the source and the exit slit staying in symmetric positions about the normal of the crystal surface. The focusing circle is defined by the curvature of the crystal surface. The source being fixed, variation of the reflected photon energy in ideal focusing geometry requires two motions of the crystal and two of the exit slit.

In our spectrometer these motions are accomplished by a combination of a small single axis goniometer and three translation stages (Fig.1). The crystal (CR) is turned by the goniometer (GO) which is attached to the sliding carriage of the lower stage (SC1). This lower slide moves the goniometer and the two other translation stages along the primary beam direction. The exit slit (ES) and detectors (IC1, IC2) have to be moved along a lemniscate. This movement is performed by the intermediate translation stage (SC2) lying per-

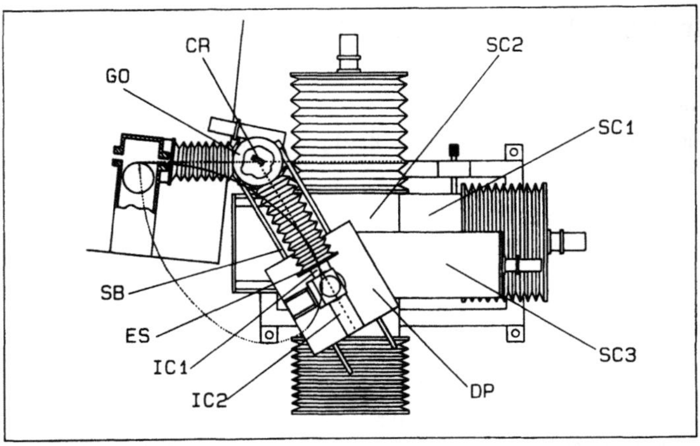

Fig.1. Top view of the spectrometer with rotating anode X-ray generator (explanation in the text)

pendicular to the primary beam direction and by the upper translation stage (SC3) carrying the detector platform (DP). The horizontal axis of this platform is aligned to the reflected beam direction by two slide bars (SB). All motions are driven by stepping motors which are controlled by a PDP-11/23 computer. This computer is also used for the storage and subsequent evaluation of the data.

The minimum angle increments of the goniometer are 0.001 degrees. The total travel lengths of the lower, intermediate, and upper translation stages are 400 mm, 500 mm, and 200 mm respectively. The actual travel lengths are monitored by electronic linear scales with a total accuracy of 0.003 mm. The load capacity of the detector stage amounts to 1 kN providing good feasibility of carrying heavy specimen chambers and detectors. In our spectrometer crystals with focusing circle radii of less than 200 mm up to more than 500 mm can be used. With a crystal radius of 360 mm Bragg angles between 13 and 45 degrees can be achieved without changing any other element. In order to reduce absorption losses the beam path is enclosed in a flexible beam line filled with helium.

Intensity and Energy Resolution

The crystals so far used in our spectrometer are commercially produced 40 mm wide and 10 mm high Si(111), Si(311), and Ge(311) Johansson-type crystals with focusing circle radii of 352 mm ± 5 mm. Using the first-order radiation reflected from these crystals and restricting the energy resolution to values better than 10eV, a photon energy range between 4 and 11 keV can be covered. At energies between 9 and 20 keV a similar energy resolution can be achieved using the same crystals in the third-order reflection mode.

Using a high power rotating anode X-ray generator operated at 24 kV and 200 mA with an Au anode and a Ge(311) crystal, a photon flux of 5×10^6 photons/sec was measured at 8 keV.

The energy resolution in this energy range was evaluated from the energetic width of the Cu $K_{\alpha 1}$ emission line and was found to be 4.6 eV. With the He-filled beam line there is considerable intensity even at lower photon energies. Fig.2a shows a Ti K absorption spectrum measured with a Si(111) crystal in a time of 3 hours. The total number of photons per data point amounts to 10^6 photons above the edge.

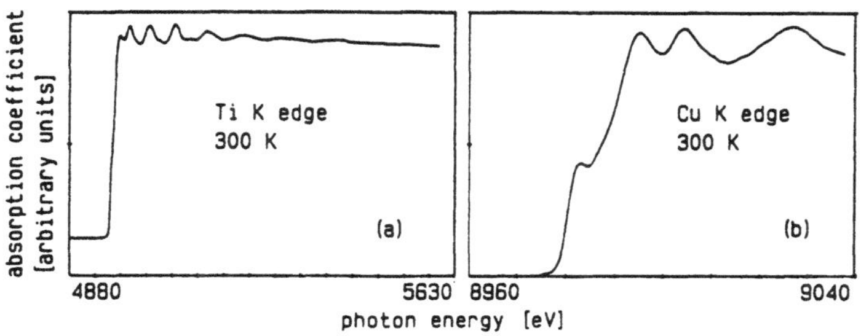

Fig.2. Absorption spectra of a) Ti at the K edge, measured with Si(111) and b) the near-edge region of the Cu K edge, measured with Si(111) in third order

High resolution near-edge spectra can be accomplished by using higher order radiation reflected form low index crystals. Fig.2b shows the near-edge region of the K absorption spectrum of Cu which was measured with a Johansson-type Si(111) crystal in third order. The intensity amounted to 5×10^4 photons/sec. The energy resolution was about 2 eV.

Acknowledgment

We like to thank R. Frahm for his assistance in the design of the experimental control hard and software. This work was partly supported by the Deutsche Forschungsgemeinschaft.

A Sample Cell for in situ EXAFS Measurements

D.C. Koningsberger and J.W. Cook, Jr.[1]

Laboratory for Inorganic Chemistry and Catalysis, Eindhoven University of Technology, NL-5600 MB Eindhoven, The Netherlands

A sample cell has been built in our laboratory to carry out EXAFS experiments on supported metal catalysts. The cell has the capability of heating the sample to 800 K in the presence of specific gases or in vacuo and then performing EXAFS measurements in situ over a temperature range of 77 K to 600 K.

I Introduction

Supported metal catalysts are one class of materials in which it is essential to be able to heat the sample in vacuo or in the presence of specific gases and then perform the X-ray absorption measurements in situ. It is also desirable to make the EXAFS measurements over a wide temperature range to study dynamic effects (i.e. the Debye-Waller factor), the temperature dependence of gas chemisorption or structural changes during catalytic reactions.

This paper describes a sample cell with the capability of heating the sample up to 800 K and then performing X-ray absorption measurements in situ over a temperature range of 80 K to 600 K. Two versions of the cell have been fabricated at the Eindhoven University of Technology and used at the Stanford Synchrotron Radiation Laboratory (SSRL) during dedicated runs of February 1981 and February 1982.

II Description of the cell

A schematic diagram of the EXAFS cell is shown in fig. 1. The cell consists of two compartments (1: sample treatment; 2: measurements). The sample is transferred between the two compartments by a movable inner tube (3), which in its highest position separates compartment 2 from 1 by an O-ring seal (4).

The sample is a self-supporting wafer approximately 0.1 to 0.5 mm thin made by pressing the catalyst powder in a metal die. The sample holder (5) is attached to the base of the movable tube.

The sample can be heated using a heater (6) attached to the base and cooled via nitrogen gas or liquid passing through

---

[1] Department of Physics, North Carolina State University, Raleigh, NC 27650, USA

channels (7). Temperature can be measured or controlled using a thermocouple (8) attached to the base.

Fig. 1

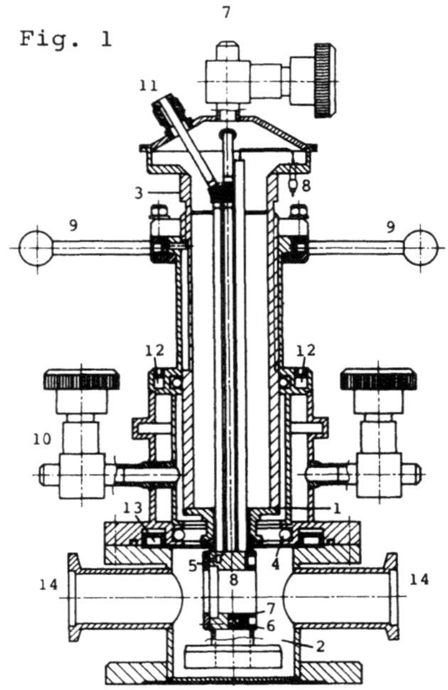

Sample treatment is performed with the movable tube locked in its highest position using the locking ring (9). The sample chamber can be evacuated through the vacuum valve (10) or exposed to specific treatment gases through the inlet (11). During sample heating the walls of the chamber are cooled by flowing water through the cooling channels (12,13). In this way the beryllium windows (14) are not exposed to high temperatures.

X-ray absorption measurements can be carried out with the movable tube in its lowest position. Measurement temperatures between 77 K < T < 300 K are achieved by controlling the flow of liquid nitrogen or cold nitrogen gas through the cooling channel (7) and by using the heating element. It takes about 15 minutes to cool the sample from 300 K to 80 K. It is also possible to heat the sample in this position up to 600 K without damaging the beryllium windows (max. window temperature ∿ 325 K).

Summary

The EXAFS cell described above has the following properties:

I Sample treatment in situ
- reduction, oxidation,
- evacuation (p $\sim 10^{-5}$ Torr),
- adsorption of gases (CO, O_2, H_2, hydrocarbons).

II EXAFS measurements during catalytic reaction.

III Sample temperature variable 77 K < T < 800 K.
(maximum sample temperature during EXAFS experiment 620 K)

IV Rapid and easy sample mounting.

V Sample present as self-supporting wafer prepared by pressing the catalyst powder at high pressure in a metal die.

With this in situ cell EXAFS measurements can be performed on samples under vacuum or exposed to gases. The combination of the variable sample temperature and the possibility of exposure to gases allows us to pretreat the sample at high temperature and to measure in situ the EXAFS spectrum as a function of temperature.

Acknowledgement

The authors are grateful for the technical assistance of P. Brinkgreve, J. Groot and J.J.F.J. Garenfeld. Their technical skill brought the cell to reality.

SSRL Beam Line Wunder

R.Z. Bachrach and I. Lindau

Xerox Palo Alto Research Center, 3333 Coyote Hill Rd., Palo Alto, CA 94304, USA

SSRL, PO Box 4349, Stanford University, Stanford, CA 94305, USA

This report reviews current planning for Beam Line Wunder at the Stanford Synchrotron Radiation Laboratory. The successful operation of a permanent magnet wunder insertion on SPEAR [1] and the results of modeling studies [2,3] show that periodic insertions produce significant flux and brightness advantages in the energy range 10-1000eV relative to a bending magnet. Figure 1 compares the expected beam flux from a bending magnet on SPEAR and various types of periodic insertions. The curves for the wunder represent a tuning range while the wiggler and bending magnets emit a continuum. Relative to the bending magnet source, one gains approximately two orders of magnitude of useful flux. In addition, the horizontal beam divergence is reduced by approximately one order of magnitude so that both the beam transport and optics can be simplified. Based upon these results, a collaborative project between Stanford University and the Xerox Palo Alto Research Center is developing Beam Line Wunder at SSRL with funding provided by the National Science Foundation, Defense Advanced Research Project Agency, and the Xerox Palo Alto Research Center [3]. The significant increase in flux and resolution that will be achieved in the energy range 10-1000 eV should open many new scientific opportunities for soft X-ray science particularly for Surface EXAFS and XANES experiments.

Beam Line Wunder will incorporate multiple branch lines as shown in the functional layout in Figure 2. Three branch lines are proposed following the LOCUST monochromator with one in line and two deflected 8^0 horizontally. Our current schedule aims to have the beam line operational by late 1984. Beam Line Wunder is formulated around the use of a permanent magnet wunder as the light source such as shown in Fig. 3. Fig 3 shows a schematic of the existing SSRL 60 pole wunder that was built as a collaboration between LBL and SSRL [1]. The device is a C-shaped periodic array of permanent magnets which can be easily installed around a straight section vacuum chamber. The attributes of a wunder source which make it attractive are summarized here based on the modeling studies that are presented in more detail in reference 3. Figure 1 presented the first-order power expected from the 60 pole device as well as for possible 30 pole and 20 pole configurations and compares the intensity

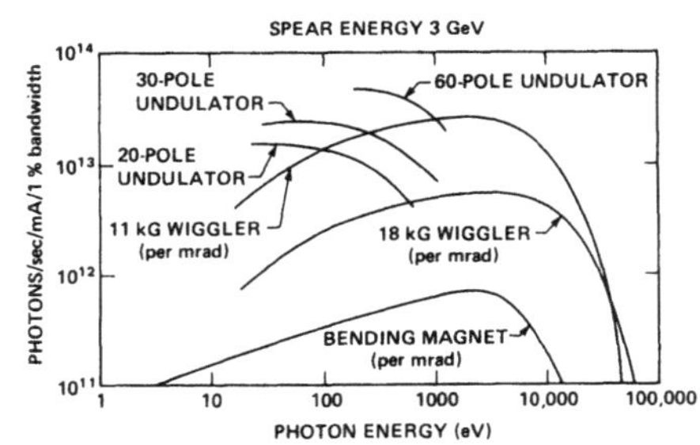

Fig. 1. Estimated first-order flux into the beam line for various sources for SPEAR operating at 3GeV under dedicated conditions. The wiggler and bending magnet curves are a continuum. The undulator curves represent a tuning range with the low energy cut-off determined by the minimum gap between the permanent magnets. This range also shifts with the stored energy of SPEAR beam

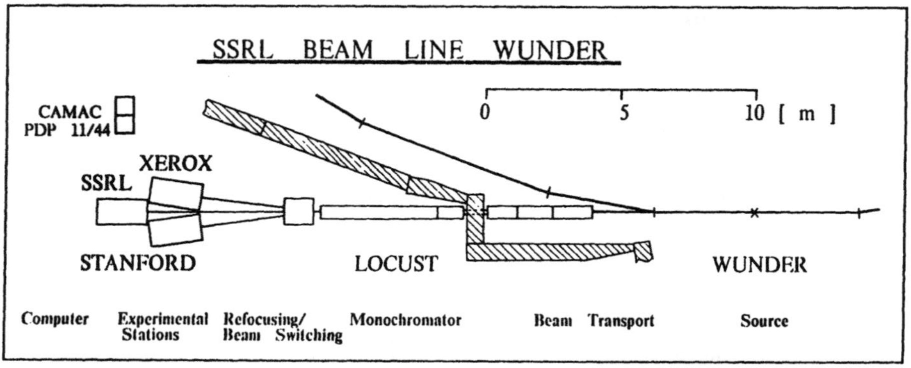

Fig. 2. Functional schematic of the beam line system elements. The wunder insertion is the soft X-ray source and the beam is transported to the LOCUST monochromator. The refocusing system following the monochromator can split the beam into one of three experimental areas. The Locust Monochromator and the experiment will be controlled by a PDP11/44 computer running RSX11M and a CAMAC system utilizing the Xerox Camac Multi-crate Device Driver, XC. A second computer may be installed for on-line data analysis

into a 1% bandpass with a bending magnet and wiggler devices. Because of the high brightness of the wunder source, the beam transport system can be physically small. The major problem is the very high power density present. Issues surrounding this are under investigation.

Beam Line Wunder will incorporate a new design soft X-ray monochromator which will extend the state of the art. This new monochromator will achieve better flux, resolution, and spectral purity than are available with current designs. The monochromator would be a *significant extension* of the Grasshopper optics [4] and would build upon SSRL's previous experience obtained with Grasshopper I [4], Grasshopper II [5], and the Double Crystal Monochromator, Jumbo, [6]. Rather than having a radius arm establishing the Rowland Circle transformation, the Locust would drive the required translation, codling mirror rotation and grating rotation independently under computer control. This aspect of the Locust design is similar to that implemented with the Jumbo Monochromator and with the Grasshopper design extensions developed by F.C. Brown et al. [7].

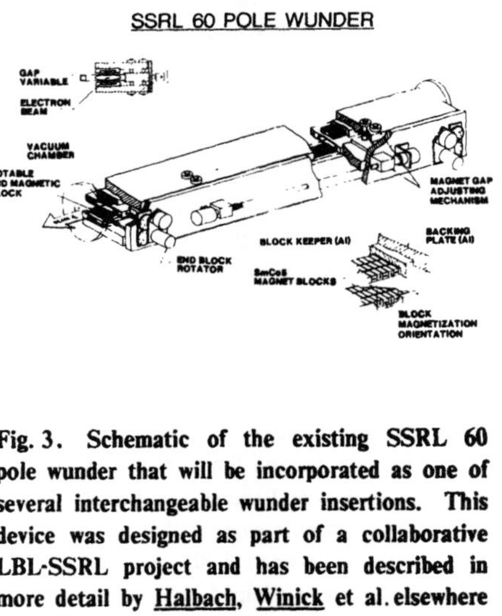

Fig. 3. Schematic of the existing SSRL 60 pole wunder that will be incorporated as one of several interchangeable wunder insertions. This device was designed as part of a collaborative LBL-SSRL project and has been described in more detail by Halbach, Winick et al. elsewhere (ref. 1)

The LOCUST, however, will incorporate multiple gratings on a single fixed length radius arm. The Rowland Circle diameter and primary angle would be varied to optimize the optics while their product would be kept fixed. This geometry with a virtual Rowland Circle center is depicted in Fig. 4. As the primary optical angle is changed, the phase of the Rowland Circle is rotated to keep the exit direction fixed at 2 degrees. The movement of the entrance slit is accomodated by changing the entrance mirror appropriately. Current planning includes four gratings with parameters: (2°,10m;4°,5m;10°,2m;20°,1m).

The optics approach proposed here should allow for the optimization of blaze angles since the optical angle on the grating and the grating blaze can be independently defined. The low energy range will be optimized by using large angles on the grating.

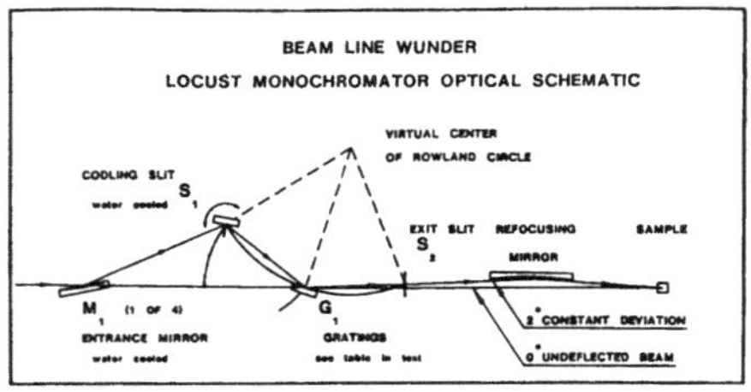

Fig. 4. Proposed variable angle optics for the LOCUST Monochromator utilizing selectable (1 of 4) vertical focusing entrance mirrors and selectable gratings. Note that this design rotates the phase of the Rowland Circle to keep the exit direction fixed

The monochromator and experiment will be controlled as accomplished previously [6,8] with a PDP11 system running software developed under the RSX11M operating system and a CAMAC interface using the XC: Xerox Camac multi-crate device driver [9].

Acknowledgement:

Part of this work is supported by NSF Grant DMR 81-08343, DARPA, and the Xerox Palo Alto Research Center and was performed at the Stanford Synchrotron Radiation Laboratory at SLAC which is supported by NSF Grant No. DMR 77-27489 and the DOE.

REFERENCES:

1. H. Winick, G. Brown, K. Halbach, and J. Harris, Physics Today, 34, 50, May, 1981. and K. Halbach, J.Chin, F. Hoyer, H. Winick, R. Cronin, J. Yang, and Y. Zambre, IEEE Trans. Nucl. Sci. NS28, (1981).
2. R.Z. Bachrach, L.E. Swartz, S.B. Hagstrom, I. Lindau, M.H. Hecht, and W.E. Spicer, Nuclear Instruments and Methods, to be published
3. M.H. Hecht, R.D. Bringans, I. Lindau, and R.Z. Bachrach, Nuclear Instruments and Methods, to be published
4. F.C. Brown, R.Z. Bachrach, and N. Lien, Nuclear Instruments and Methods,152, 73,(1978).
5. R.Z. Bachrach, I. Lindau, V. Rehn, and J. Stohr, SSRL Report 77/14, "Beam Line III - UHV: 4-4000 eV, Report of the Beam Line III Optics Advisory Committee"; and J. Stohr, V. Rehn, I. Lindau, and R.Z. Bachrach, Nucl. Instr and Methods, 152,43,(1978); Leif Johansson and J. Stohr, Nuclear Inst. and Methods,172, (1980).
6. J. Cerino, J. Stohr, N. Hower, and R.Z. Bachrach, Nucl. Instr. and Methods, 172, (1980).
7. F.C. Brown, S.L. Hulbert, and N.C. Lien, Proc Vth International Conference on Vacuum Ultraviolet Radiation Physics, Charlottesville, Va., June (1980).
8. G.V. Hansson, B. Goldberg, and R.Z. Bachrach, Rev Sci Instr., 52, 517, (1981).
9. J. Gallup and R.Z. Bachrach, (unpublished).

Index of Contributors

Electron Spectroscopy for Surface Analysis

Editor: **H. Ibach**

1977. 123 figures, 5 tables. XI, 255 pages
(Topics in Current Physics, Volume 4)
ISBN 3-540-08078-3

Contents: *H. Ibach:* Introduction. – *J. D. Carette, D. Roy:* Design of Electron Spectrometers for Surface Analysis. – *J. Kirschner:* Electron-Excited Core Level Spectroscopies. –*M. Henzler:* Electron Diffraction and Surface Defect Structure. – *B. Feuerbacher, B. Fitton:* Photoemission Spectroscopy. – *H. Froitzheim:* Electron Energy Loss Spectroscopy.

Structural Studies of Surfaces

1982. 120 figures. VII, 180 pages
(Springer Tracts in Modern Physics, Volume 91)
ISBN 3-540-10964-1

Contents: *K. Heinz, K. Müller:* LEED Intensities – Experimental Progress and New Possibilities of Surface Structure Determination. – *T. Engel, K. H. Rieder:* Structural Studies of Surfaces with Atomic and Molecular Beam Diffraction.

O. Madelung

Introduction to Solid-State Theory

Translated from the German by B. C. Taylor

1978. 144 figures. XI, 486 pages
(Springer Series in Solid-State Sciences, Volume 2)
ISBN 3-540-08516-5

Contents: Fundamentals. – The One-Electron Approximation. – Elementary Excitations. – Electron-Phonon Interaction: Transport Phenomena. – Electron-Electron Interaction by Exchange of Virtual Phonons: Superconductivity. – Interaction with Photons: Optics. – Phonon-Phonon Interaction: Thermal Properties. – Local Description of Solid-State Properties. – Localized States. – Disorder. – Appendix: The Occupation Number Representation.

K. Seeger

Semiconductor Physics

An Introduction
2nd corrected and updated edition. 1982. 288 figures. XII, 462 pages
(Springer Series in Solid-State Sciences, Volume 40)
ISBN 3-540-11421-1

Contents: Elementary Properties of Semiconductors. – Energy Band Structure. – Semiconductor Statistics. – Charge and Energy Transport in a Nondegenerate Electron Gas. – Carrier Diffusion Processes. – Scattering Processes in a Spherical One-Valley Model. – Charge Transport and Scattering Processes in the Many-Valley Model. – Carrier Transport in the Warped-Sphere Model. – Quantum Effects in Transport Phenomena. – Impact Ionization and Avalanche Breakdown. – Optical Absorption and Reflection. – Photoconductivity. – Light Generation by Semiconductors. – Properties of the Surface. – Miscellaneous Semiconductors. – Appendix: Physical Constants. – References. – Subject Index.

Springer-Verlag
Berlin
Heidelberg
New York
Tokyo

MIX
Papier aus verantwortungsvollen Quellen
Paper from responsible sources
FSC® C105338

If you have any concerns about our products,
you can contact us on
ProductSafety@springernature.com

In case Publisher is established outside the EU,
the EU authorized representative is:
Springer Nature Customer Service Center GmbH
Europaplatz 3, 69115 Heidelberg, Germany

Printed by Libri Plureos GmbH
in Hamburg, Germany